AF755874

Gold Metallogeny and Exploration

Edited by

R.P. FOSTER
Senior Lecturer in Economic Geology
Department of Geology
University of Southampton

CHAPMAN & HALL
London · Glasgow · New York · Tokyo · Melbourne · Madras

Published by Chapman & Hall, 2-6 Boundary Row, London SE1 8HN

Chapman & Hall, 2-6 Boundary Row, London SE1 8HN, UK

Blackie Academic & Professional, Wester Cleddens Road, Bishopbriggs, Glasgow G64 2NZ, UK

Chapman & Hall, 29 West 35th Street, New York NY10001, USA

Chapman & Hall Japan, Thomson Publishing Japan, Hirakawacho Nemoto Building, 6F, 1-7-11 Hirakawa-cho, Chiyoda-ku, Tokyo 102, Japan

Chapman & Hall Australia, Thomas Nelson Australia, 102 Dodds Street, South Melbourne, Victoria 3205, Australia

Chapman & Hall India, R. Seshadri, 32 Second Main Road, CIT East, Madras 600 035, India

First published in Hardback by Blackie & Son Ltd
First edition 1993

© 1993 Chapman & Hall

Typeset by Pure Tech Corporation, Pondicherry, India
Printed in Great Britain by St Edmundsbury Press Ltd, Bury St Edmunds, Suffolk

ISBN 0 412 56960 4

Apart from any fair dealing for the purposes of research or private study, or criticism or review, as permitted under the UK Copyright Designs and Patents Act, 1988, this publication may not be reproduced, stored, or transmitted, in any form or by any means, without the prior permission in writing of the publishers, or in the case of reprographic reproduction only in accordance with the terms of the licences issued by the Copyright Licensing Agency in the UK, or in accordance with the terms of licences issued by the appropriate Reproduction Rights Organization outside the UK. Enquiries concerning reproduction outside the terms stated here should be sent to the publishers at the London address printed on this page.

The publisher makes no representation, express or implied, with regard to the accuracy of the information contained in this book and cannot accept any legal responsibility or liability for any errors or omissions that may be made.

A catalogue record for this book is available from the British Library
Library of Congress Cataloging-in-Publication Data available

Preface

Within the last decade, the high and continuing demand for gold has prompted a global gold rush on a scale never before seen, not even in the heady days of Ballarat, California and the Yukon. Gold is being sought on every continent and, with very few exceptions, in every country around the world. Such interest and fierce competition has demanded considerable innovation and improvement in exploration techniques paralleled by a rapid expansion of the geological database and consequent genetic modelling for the many different types of gold deposits now recognized.

This proliferation of data has swamped the literature and left explorationist and academic alike unable to sift more than a small proportion of the accumulating information. This new book represents an attempt to address this major problem by providing succinct syntheses of all major aspects of gold metallogeny and exploration, ranging from the chemical distribution of gold in the Earth's crust, and the hydrothermal chemistry of gold, to Archaean and Phanerozoic lode deposits, epithermal environments, chemical sediments, and placer deposits, and culminates in chapters devoted to geochemical and geophysical exploration, and the economics of gold deposits. Each chapter is written by geoscientists who are acknowledged internationally in their respective fields, thus guaranteeing a broad yet up-to-date coverage. In addition, each chapter is accompanied by reference lists which provide readers with access to the most pertinent and useful publications.

Quite clearly this book is not intended to be the ultimate authoritative statement on gold metallogeny and exploration, nor can it hope to embrace all aspects of gold mineralization. For access to a wealth of new information the reader is encouraged to turn to the relevant journals and to the excellent publications arising from many major symposia held within the last decade. The more recent of these include the proceedings of the Gold '86 meeting in Toronto, the Pacific Rim Congress '87 in Gold Coast, Australia, and the Gold '88 symposium held in Melbourne.

The stimulus provided by the demand for gold in the 1980s has opened up exciting new avenues of research and generated a wealth of hard facts, innovative ideas, conceptual models, and advances in instrumental techniques. The resultant databases and the success arising from this surge in exploration are a timely reminder to the world at large of the importance of applied research and of exploration underpinned by sound science. I hope that this volume provides both a useful state-of-the-art review and an insight into the scientific and industrial excitement inspired by the 1980s gold rush.

Finally, I should like to thank the following for providing the all-important refereeing of contributions which is essential to ensure the quality of such a publication: B.R. Berger, H.F. Bonham, Jr., C.A. Boulter, P.R.L. Browne, A.H.C. Carter, H. Colley, D.S. Cronan, M.J. de Wit, P.I. Eimon, W.R. Fitches, A.O. Fuller, A.P. Gize, D.D. Hawkes, J.W. Hedenquist, J.J. Hemley, R.W. Henley, P. Lhotka, A. Panteleyev, D.A. Pretorius, R.H. Sillitoe, M. Smith, D.A.V. Stow, J.F.H. Thompson.

R.P.F.

Contributors

Dr W.C. Bagby *US Geological Survey, Menlo Park, California, USA*

William Bagby received a Ph.D. in Earth Science from the University of California, Santa Cruz, on petrogenetic research of Tertiary volcanic rocks in the Sierra Madre Occidental, Mexico. His mining industry experience includes porphyry copper exploration, geologic evaluation of volcanic-hosted uranium deposits, epithermal silver vein deposits, and hot spring gold deposits. Research done at the USGS has been focused on gold deposits in Nevada and a mineral-resource assessment of Costa Rica. Dr. Bagby has served as a programme coordinator for the USGS Strategic and Critical Minerals Program and presently is serving as the Chief of the Branch of Western Mineral Resources.

Dr B.R. Berger *US Geological Survey, Spokane, Washington, USA*

Byron Berger has been involved in exploration for and research on epithermal and Carlin-type gold deposits since 1971, after completing an MS in geology at the University of California, Los Angeles, and a BA in economics at Occidental College, Los Angeles. Initially with Conoco Inc. in exploration and minerals exploration research, he joined the USGS, Office of Mineral Resources in 1977. His industrial experience focused on Carlin- type deposits in northern Nevada, and during his tenure with the USGS he has conducted research on volcanic-rock-related epithermal deposits in the Western Great Basin and Pacific Northwestern United States. Administratively, he has served as an Adjunct Professor, University of Colorado, and Chief, Branch of Geochemistry in the USGS, and is currently Deputy Chief, Office of Mineral Resources and on the Editorial Board of Economic Geology.

Dr W.B. Coker *Geological Survey of Canada, Ottawa, Canada*

Bill Coker graduated from Carleton University in 1971 with an Honours B.Sc. in Geology. Specializing in geochemistry, he received his Ph.D. from Queen's University, Ontario in 1974. He has over 20 years experience with the mineral exploration industry, most recently with Kidd Creek Mines Ltd., and in government research. Dr Coker is currently a Research Geochemist with the Geological Survey of Canada conducting research on geochemical methodologies, primarily overburden geochemistry, in glaciated terrains.

Dr J.H. Crocket *Department of Geology, Mc Master University, Hamilton, Ontario, Canada*

James Crocket is a graduate of the University of New Brunswick, Oxford University, and the Massachusetts Institute of Technology where he obtained a Ph.D. in 1961. He joined the Department of Geology at McMaster University in Hamilton, Ontario, in the same year and is currently involved in teaching economic geology and geochemistry at McMaster. Much of his research is concerned with Archaean gold deposits and platinum group metal mineralization, with modest diversions into the Cretaceous/Tertiary boundary problem. Dr. Crocket is a Fellow of the Geological Association of Canada.

Dr R.P. Foster *Department of Geology, University of Southampton, UK*

Bob Foster graduated from the Universities of Durham in 1970 (B.Sc.) and Manchester in 1973 (Ph.D.) and was then employed as a mine and exploration geologist by Falcon Mines Ltd in Rhodesia (now Zimbabwe). In 1977 he moved to the Institute of Mining Research at the University of Rhodesia (Zimbabwe) where he was involved in problem solving and applied research in collaboration with the local mining industry. In 1984 he moved to the University of Southampton, where he is now Senior Lecturer in Economic Geology and also acts as a consultant on gold mining and exploration in Europe, Africa, and the Middle East.

Professor D.I. Groves *Key Centre for Teaching and Research in Strategic Mineral Deposits, University of Western Australia, Australia*

David Groves is currently Professor (personal chair) in Geology and Director of the Key Centre for Strategic Mineral Deposits, Geology Department, University of Western Australia. He is a B.Sc. Honours (First Class) and Ph.D. graduate of the University of Tasmania. Formerly a geologist with the Geological Survey of Tasmania, where he had research interests in specialized granitoid and tin deposits, Professor Groves joined the University in 1971. Since that time, his primary interest has been Archaean metallogeny, with particular emphasis on komatiite-associated Ni-Cu deposits and mesothermal gold deposits. Currently he heads a team of ten staff and postdoctoral fellows and about 20 postgraduate students who are mostly researching Precambrian ore deposits.

Dr P.G. Hallof *Consultant Geophysicist, Toronto, Canada*

Philip Hallof attended university and graduate school at Massachusetts Institute of Technology, Cambridge, Massachusetts and graduated with a Ph.D. degree in 1956. He then joined McPhar Geophysics Ltd., Toronto, Canada, as a geophysicist and later became president of the company. He subsequently served as president of Phoenix Geophysics Ltd. and now works in Toronto as a consulting geophysicist. Throughout his professional career he has specialized in the use of electrical geophysical methods in the search for base metal deposits and precious metal deposits, and has planned, supervised and interpreted geophysical surveys in mining areas throughout the world.

Dr M.D. Hannington *Geological Survey of Canada, Ottawa, Canada*

Mark Hannington is a graduate of Queen's University at Kingston, Ontario, and received his Ph.D. in Geology from the University of Toronto in 1989. He is currently a Visiting Fellow in the Mineral Resources Division of the Geological Survey of Canada, where he is studying active hydrothermal systems on the modern seafloor and their relationship to ancient sulphide deposits on land. Dr Hannington's research has focused on the geochemistry of gold in volcanogenic massive sulphides, and processes of secondary enrichment of gold in sulphide gossans.

Dr R.W. Henley *Consultant Economic Geologist, Canberra, Australia*

Dick Henley has had close involvement with the development of understanding of the controls on the distribution and origin of a wide range of hydrothermal ore deposits for over 20 years. Since obtaining his Ph.D. from the University of Manchester in 1971 he has held positions at Otago University, New Zealand, and Memorial University, Newfoundland and from 1980 was Head of the Geothermal Chemistry Group within DSIR in New Zealand. From 1986 to 1988, he was Chief Scientist of the Division of Petrology and Geochemistry at the Bureau of Mineral Resources in Canberra, and has since consulted privately to the minerals industry on exploration.

Dr P.M. Herzig *Institute of Mineralogy and Economic Geology, Aachen University of Technology, West Germany*

Peter Herzig received his Ph.D. in 1986 from the Aachen University of Technology, where he is now a Research Associate in the Institute of Mineralogy and Economic Geology. He has worked extensively on sulphide deposits in the Troodos Ophiolite and in particular on sub-seafloor stockworks associated with fossil hydrothermal systems. His current research includes exploration for hydrothermal vents in the Indian Ocean and studies of the geology and mineralogy of sulphide deposits in modern back-arc basins of the Southwest Pacific.

Dr C. Leduc *Applied Geochemistry Branch, Bureau de Recherche Geologie et Minerales, Orleans, France*

Claudy Leduc received a Masters degree from the University of Paris. He spent several years working for exploration companies in Canada and subsequently went to Tunisia where he was mostly involved in geochemical exploration. He joined the BRGM in 1981 where he is now in charge of various geochemical exploration programmes, in France and North Africa, where he has gained considerable experience in the application of geochemical techniques to exploration for gold and base metal deposits.

Professor B.W. Mackenzie *Centre for Resource Studies, Department of Geological Sciences, Queen's University, Kingston, Canada*

Brian Mackenzie is a Canadian mining engineer interested in the economics of mineral supply. He received his formal education at McGill University, the University of the Witwatersrand, and Columbia University. In his early years, he worked for Sheep Creek Mines, Union Corporation, the South African Department of Mines, and Energy, Mines and Resources Canada. From 1968 to 1977 he was responsible for teaching and research in mineral economics and was Director of the Mining Program at McGill University. Since 1978 he has been a Senior Research Associate of the Centre for Resource Studies and a Professor in the Department of Geological Sciences at Queen's University, Kingston.

Professor W.E.L. Minter *Department of Geology, University of Cape Town, South Africa*

Lawrie Minter is Professor of Economic Geology at the University of Cape Town. He is a South African and a graduate of the University of Cape Town and the University of the Witwatersrand. Dr Minter has spent most of his career working for mineral exploration and mining companies, and for the past 25 years has been actively engaged in gold and diamond palaeoplacer research, both in southern Africa and Brazil. He is best known for his publications on sedimentological aspects of Witwatersrand gold placers.

Professor B.E. Nesbitt *Department of Geology, University of Alberta, Canada*

Bruce Nesbitt is a Professor of Economic Geology in the Department of Geology, University of Alberta, Canada. After receiving his BA in 1973 from Carleton College, he undertook graduate studies in economic geology and metamorphic petrology at the University of Michigan, receiving his MS degree in 1976 and Ph.D. in 1979. From 1979 to 1980, he held an NSF Postdoctoral Fellowship at Pennsylvania State University. From 1980 to the present, he has been on the staff of the University of Alberta. Dr Nesbitt's early research interests centred on the study of chemical effects of metamorphism on ore deposits. More recently, his research has focused on the genesis of mesothermal Au±Ag deposits and the role of crustal fluid-flow in the generation of such deposits.

Dr N.R. Paterson *Paterson, Grant and Watson Ltd., Toronto, Canada*

Norman Paterson received his Ph.D. from the University of Toronto in 1955. After short periods with major mining and petroleum companies, he was with the Canadian Hunting Airborne Survey Group and later president of Huntec Ltd. He obtained world-wide experience in all types of airborne, ground and marine geophysics, as well as instrument development and manufacture. In 1969 he started a consulting business, reorganized in 1973 into Paterson, Grant and Watson Ltd. As a consultant he has provided advice, technical management, and interpretation services to mining companies, governments, and aid agencies. He has managed major programmes in the developing countries for CIDA, World Bank and the UN. In 1977 he was made a Fellow of the Royal Society of Canada.

Professor S.D. Scott *Marine Geology Research Group, Department of Geology, University of Toronto, Canada*

Steve Scott received his Ph.D. from Pennsylvania State University, where he studied sulphide mineralogy and chemistry. He joined the University of Toronto in 1969, where he is now a Professor in the Department of Geology, and with his students has published extensively on the geology and geochemistry of volcanogenic massive sulphides in Canada and Japan. Since 1983, he has been actively involved in the study of modern seafloor hydrothermal systems. He now directs the Marine Geology Research Group at the University of Toronto and has been involved in the Ocean Drilling Program.

Professor T.M. Seward *Chemistry Division, Department of Scientific and Industrial Research, Lower Hutt, New Zealand*

Terry Seward obtained his B.Sc. from McMaster University in Canada and his Ph.D. from the University of Manchester, UK. He was formerly section leader and senior research scientist with the Chemistry Division of the New Zealand DSIR. From 1976 to 1978, he was an Alexander von Humboldt Fellow at the University of Karlsruhe (BRD). Dr Seward was visiting professor at Pennsylvania State University in 1982 and guest professor at the Swiss Federal Institute of Technology (ETH) in Zürich in 1988. He is at present professor of geochemistry at the ETH. His main research interests are in the chemistry and geochemistry of aqueous solutions at high temperatures and pressures.

Dr W.W. Shilts *Geological Survey of Canada, Ottawa, Canada*

William Shilts received his BA degree from DePauw University (1963), his MS from Miami University, Ohio (1965) and his Ph.D. from Syracuse University (1970). He has worked for the Geological Survey of Canada since 1970 and is currently Senior Quaternary Scientist in the Terrain Sciences Division. Dr Shilts has been particularly interested in practical applications of research to the chemistry, mineralogy and physical properties of glacial drift, especially till. Although aimed primarily at improving or developing techniques of drift prospecting, much of his recent research has been applied to problems in environmental geochemistry, radioactive waste disposal, and an improved understanding of the configuration of the North American continental ice sheet.

Dr R.H. Sillitoe *Consultant Geologist, Highgate Village, London, UK*

Richard Sillitoe received his B.Sc. and Ph.D. degrees from the University of London, UK and went on to work as an economic geologist with the Geological Survey of Chile. During the last 19 years he has worked as an independent geological consultant to international mining companies, foreign governments and international agencies in some 60 countries world-wide. Since 1979 his activities have been almost exclusively devoted to gold exploration, especially in Cordilleran and island-arc environments.

Dr H. Zeegers *Applied Geochemistry Branch, Bureau de Recherche Geologie et Minerals, Orleans, France*

Hubert Zeegers is now Head of the Exploration Department of the BRGM, after having been in charge of the Applied Geochemistry Branch. He received a Masters degree from the Louvain University, Belgium, and was first employed in Burkina Faso, West Africa, as exploration geochemist in a United Nations Development Program, Mineral Exploration Project. He joined the BRGM in 1974, and participated in numerous exploration and research projects involving exploration geochemistry. Most of his experience was gained in tropically weathered terrains, and he has been deeply involved in developing exploration techniques suitable to such environments.

Contents

1 Distribution of gold in the Earth's crust

J.H. CROCKET

1.1 Introduction

The distribution of gold in the lithosphere bears critically on models for gold ore deposits. One group of workers advocates derivation of gold from large masses of rock through fluid–rock reactions in the deep crust (Kerrich, 1983; Groves and Phillips, 1987; Colvine *et al.*, 1988) while others stress the importance of relatively restricted petrogenetic groups of rocks termed source rocks that concentrate gold above the norm (Viljoen *et al.*, 1970; Keays, 1984). It is therefore important to know the average gold contents for rock types of the lithosphere, and what concentrations are potentially indicative of mineralization. Tilling *et al.* (1973) noted that unaltered igneous rocks generally have less than 5 parts per billion (ppb) gold and that 10 ppb is exceeded only rarely. A survey of more recent data presented here confirms that statement. As ore deposits may concentrate gold 10^4 times the rock background, gold contents of only a few tens of ppb may indicate ore-forming processes.

This chapter reviews gold in igneous, sedimentary and metamorphic rocks and rock-forming minerals. Based mainly on post-1972 data, it also incorporates selected major data bases published earlier. The data are compiled in Section 1.7 at the end of this chapter and cover minerals, igneous, sedimentary and metamorphic rocks respectively.

The igneous rock data are grouped according to regional tectonic setting except for Precambrian rocks which are treated as a specific group for reasons noted in subsequent text. Previous reviews emphasizing abundance data include Korobeynikov (1986, 1982, 1981, 1980, 1976), Barnes *et al.* (1985), Boyle (1979), Crocket (1978) and Gottfried *et al.* (1972).

The data in Section 1.7 represent extensive averaging of original literature. Averages for large groups of analyses given in original literature are used directly, but generally the appendices represent data from several papers and are averages calculated by the author. The data estimate abundances of gold applicable to non- mineralized average crustal rocks. Omissions represent deletions of mineralized and altered samples. Author judgement (and bias) is unavoidable in some cases of data selection.

1.2 Gold in rock-forming minerals

The mineral data in Section 1.7.2 draw heavily on papers by Korobeynikov (1976, 1980, 1981, 1982) which present original work and reviews of the Russian literature. Averages refer to the number of localities or geological units (rock types, plutons, intrusive complexes) represented. For example, the first entry under Quartz gives an average of 0.23 ppb gold for five granitic plutons from the western US. Averages for each pluton were first calculated from the original literature (Gottfried *et al.*, 1972) and then a group average was computed giving equal weight to each pluton. The entire group consists of 13 samples from 5 plutons. The mineral data, except for skarns, are illustrated in Figure 1.1.

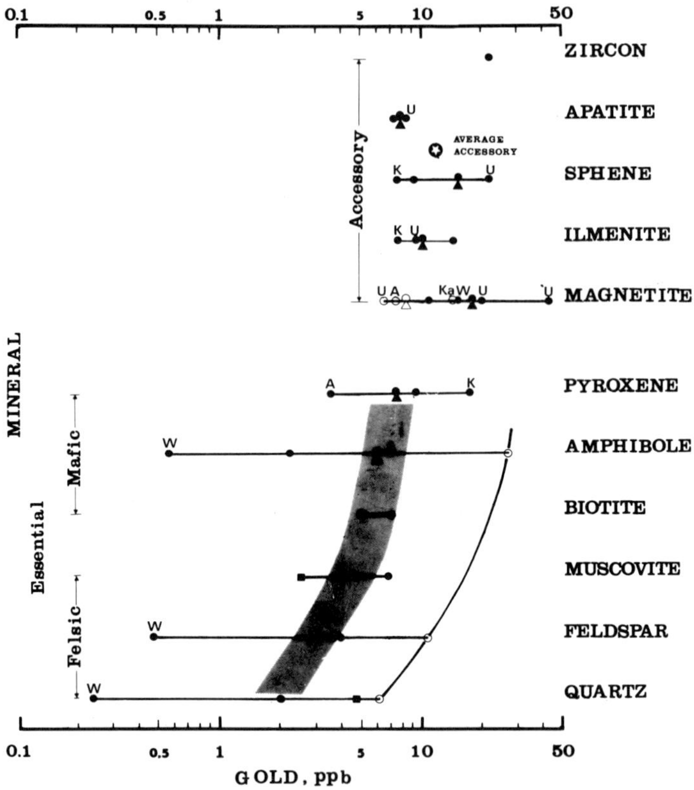

Figure 1.1 Range and average gold content of rock-forming minerals (ppb). Symbols represent geological occurrence or locality averages (see text): granitoids ●(an overall average for USSR granitoids, which represent much of the data base, is indicated by a linked triangle/circle symbol) mafic rocks ❍, sedimentary rocks ■; trends, granitoids shaded, mafic, solid line; granitoid localities, A-Altai-Sayan, K- Kazakhstan, Ka-Kamchatkha, W-Western US. Data are from Section 1.7.2, but skarns are omitted.

The main trends and characteristics of gold distribution in rock-forming minerals are as follows:

(i) The gold content of the same mineral, especially accessory and ferromagnesian minerals, may vary from 2 to 10 times, and reflects the whole-rock gold level. In general, if the whole-rock gold content is above average for the rock type, the same trend is shown by all the essential minerals.

(ii) There is a general trend of decreasing gold content from accessory (magnetite, sphene, zircon), through ferromagnesian (biotite, hornblende), to felsic (quartz, feldspar) minerals in a given geological host (see granitoid trend, Figure 1.1).
(iii) The concentrator minerals, those with higher gold than the whole rock, are the accessory and ferromagnesian minerals, mainly magnetite and biotite. The carrier minerals, with gold contents close to or below the whole rock, are quartz and feldspar. In silicic and intermediate rocks, the carrier minerals are of high modal proportion and account for 50 to 80% of whole-rock gold.
(iv) Accessory minerals often carry 1.5–10 times the gold content of the essential minerals, and in granitic rocks a large variation in gold content of accessory minerals is typical of productive plutons (Korobeynikov, 1980).

A well-established trend in igneous rocks is a decrease in gold from mafic silicates and iron oxides, usually early crystallizing minerals, to late crystallizing quartz and feldspar (Shcherbakov and Perezhogin, 1964; Tilling *et al.*, 1973). A less well documented trend is a decrease in the gold content of specific minerals such as feldspar (Korobeynikov, 1981) and biotite (Shilin and Osipov, 1978) from mafic to felsic host rocks. Such trends led Tilling *et al.* (1973) to conclude that gold does not concentrate in residual magmatic fluids during fractional crystallization.

Accessory minerals such as sphene, apatite, zircon and magnetite are often high in gold relative to their host rocks (Korobeynikov, 1980). As a variety of different crystal structures and chemical compositions are involved, and as the gold contents of such minerals are highly variable but generally correlated with whole-rock gold abundance, accessory minerals may be gold concentrators due simply to early crystallization.

1.3 Gold in igneous rocks

1.3.1 *Introduction*

Igneous rock data are divided into orogenic, non-orogenic and Precambrian greenstone categories. Orogenic environments are those associated with plate convergence and compressive stress regimes. They include island arcs and active continental margins, and encompass the obductive emplacement of ophiolites. Non-orogenic environments include stable craton settings and divergent plate margins where extensional stress regimes predominate. Rifting and the early stages of continental rupture are also included in non-orogenic regimes. The Precambrian greenstone subdivision is without tectonic connotation as correlation between Archaean and Phanerozoic regimes is tenuous.

1.3.2 *General characteristics of gold distribution in the lithosphere*

The data base for igneous rocks is summarized in Table 1.1 and Figure 1.2. A restricted range of gold is typical of fresh igneous rocks. Gottfried *et al.* (1972) found that a 10- fold variation from 0.5 to 5 ppb is typical of igneous rock averages. The data in Figure 1.2 tend to confirm this generalization. Averages for all Phanerozoic volcanic rocks range from 1.2 ppb (MORB) to 3.5 ppb (flood basalt), decreasing to 0.5 ppb (spinel lherzolites) when plutonic rocks are considered. Ranges applicable to individual groups are larger with variations from 0.1 to 10 ppb typical of volcanic rocks, although MORB has a greater range (0.04–15 ppb). Precambrian volcanic rocks have higher group averages (4.2–12.4 ppb) than Phanerozoic volcanics and, on a locality basis, somewhat greater ranges (0.5–37 ppb).

Table 1.1 Average gold content of igneous rock groups (ppb)

Rock group	Comment	$\overline{X}$	*R*	*N*	Ir/Au
Rocks from non-orogenic environments					
MORB	MAR, 98; EPR, 112	1.2	0.04–15	210	0.03
Intraplate basalt	Mainly Hawaii	2.0	0.2–6.6	69	0.38
Flood basalt	3 Provinces	3.5	0.5–11	49	0.02
Initial magma; layered gabbroic complex	Anakita, Bushveld, Jimberlana, Skaergaard	4.6	2.8–8.0	4	0.08[1]
Kimberlites and mantle xenoliths					
Kimberlites	Siberia; 50 pipes	3.1	0.8–9.1	55	
Garnet peridotite	Siberia	2.7	0.6–8.1	47	
Garnet peridotite[2]	Lesotho	0.85	0.08–2.7	10	7.3
Eclogite	Siberia	3.4	0.8–9.1	25	
Spinel lherzolites	Alkali basalt host	0.5	0.1–1.1	27	7.0
Alkaline plutons		2.8	0.64–4.5	291/21	
Rocks from orogenic environment					
Ophiolite harzburgite		2.8	0.3–6.4	138/14	3.2[3]
Mafic volcanics, convergent margins		2.2	0.5–5.6	315/10	
Felsic volcanics, convergent margins		1.55[4]	0.56–4.2	305/7	
Granitic plutons		2.6	0.5–6.9	969/66	
Igneous rocks of Precambrian greenstone belts					
Perioditic komatiite		4.2[5]	0.49–13.5	156/10	1.5
Komatiitic basalt		12.4	1.0–36	44/6	0.03[6]
Tholeiitic basalt		5.7	1.3–37	323/13	
Granitic plutons		1.5	1.1–2.3	232/45	

1. The Ir/Au ratio for the Bushveld initial magma. 2. Ir/Au for Lesotho suites from Mitchell and Keays (1981) and Morgan *et al.* (1981). 3. Omits Ir/Au ratio of 14 for the Ronda complex. 4. Omits data of Yudin *et al.* (1972) for rhyolites and ignimbrites (9.6 ppb), and data of Mints (1975) for leucogranites (20 ppb). 5. The average of 4.2 for komatiitic peridotites uses a value of 13.5 ppb (Saagar *et al.*, 1982) for Belingwe greenstones. The average drops to 3.0 ppb if a value of 1.4 (Keays, 1984) is used. 6. The Ir/Au ratio is for Kambalda rocks from Keays *et al.* (1981).

There is a weak trend of decreasing gold from mafic to felsic rocks (Shcherbakov and Perezhogin, 1964; Tilling *et al.*, 1973; Shilin, 1980; Korobeynikov, 1981) consistent with higher gold in early crystallizing mafic silicates and Fe–Ti-oxides compared with late crystallizing quartz and feldspar. Korobeynikov (1981) found a decrease in the gold content of feldspars of intrusive rocks from basic to felsic varieties which supports this general trend. However, several studies (Voskresenskaya and Zvereva, 1968; Anoshin and Kepezinskas, 1972) found little fractionation of gold between rocks of varying composition. In addition, layered mafic–ultramafic complexes such as the Bushveld show no systematic variation of gold with igneous stratigraphy (Lee and Tredoux, 1986).

Two factors suggest that the mafic-to-felsic trend is not a direct consequence of fractional crystallization. Concentrator minerals such as magnetite and biotite are often not sufficiently abundant to account for much of the whole-rock gold (Davletov and Dzhakshibayev, 1970; Zvereva and Gavrilenko, 1971; Korobeynikov, 1981;

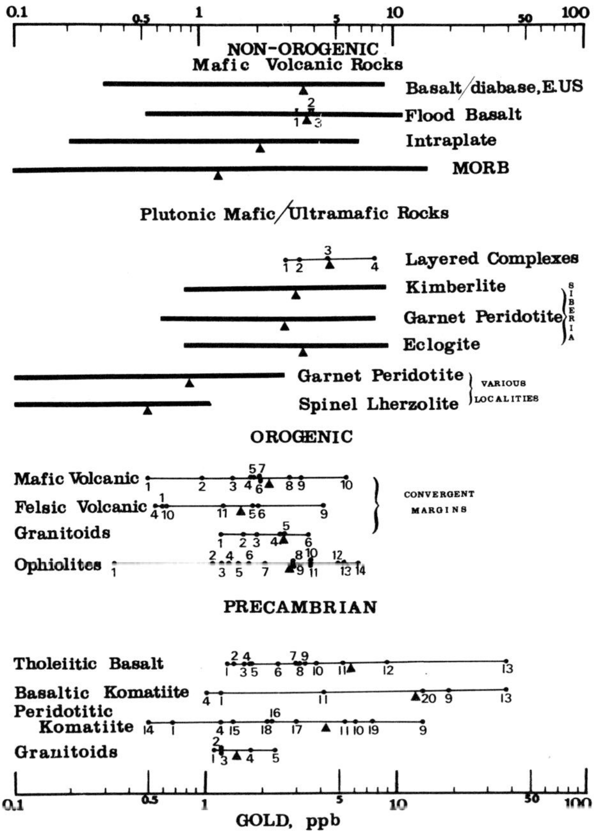

Figure 1.2 Range and average gold contents of igneous rocks (ppb). Thick lines are ranges for individual rock-types, and thin lines represent ranges for geological occurrences (stocks, complexes) or localities (see text). Large triangles indicate group averages. Numbered circles represent averages for geological occurrences or localities, and the small triangles indicate averages for individual flood-basalt provinces. The number of samples represented by any range or occurrence/locality average can be obtained from Section 1.7.3.

Number key to occurrence/locality averages:

Non-orogenic *Layered complexes:* 1.Bushveld; 2.Jimberlana; 3.Skaergaard; 4.Anakita.

Orogenic *Mafic and felsic volcanics:* 1.Snake River Gp., Idaho; 2.Victoria greenstones, Australia; 3.Cape Vogel, PNG; 4.Western US; 5.Kurile/Kamchatka; 6.Okhotsk/Chukotka (Ul'insk); 7.Mariana-Bonin arc; 8.King Island, Tasmania; 9.Okhotsk; 10.Cascades, US; 11.Okhotsk/Chukotka (Tottinsk). *Granitoids*: 1.Western US; 2.S.Kazakhstan; 3.E.Kazakhstan; 4.Urals; 5.Marysville, US; 6.S.Verkhoyan, USSR. *Ophiolites:* 1.Ronda; 2.Mt.Albert; 3.Josephine; 4.Troodos; 5.Thetford; 6.Yenisey Ranges, USSR; 7.Idzhim, W.Sayan; 8.Arabian Shield; 9.Urals; 10.Fomkinsk, USSR; 11.Timetrine, Mali; 12.New Caledonia; 13.Bou Azzer, Morocco; 14.Borus, W.Sayan.

Precambrian *Volcanics*: 1.Barberton; 2.Burkina Faso; 3.Theo's Flow, Abitibi; 4.Pietersburg; 5.Wabigoon (Kakagi); 6.Timmins, Abitibi; 7.Boufoyo, CAR; 8.Tassendjanet, Algeria; 9.Belingwe (Saager *et al.*, 1982); 10.Lunnon, Kambalda; 11.Red Lake, Uchi; 12.Shoal Lake/Lake-of-the- Woods; 13.Borgoin, CAR; 14.Mt.Clifford, Yilgarn; 15.Belingwe (Keays, 1984); 16.Alexo, Abitibi; 17.Pyke Hill, Abitibi; 18.Fred's Flow, Abitibi; 19.Lunnon/Long Shoots, Kambalda; 20.Kambalda. *Granitoids*: 1.Wabigoon, Kakagi; 2.Kaapvaal Craton; 3.E.Aldan Shield, Archaean; 4.E.Aldan Shield, Proterozoic; 5.Matachewan, Abitibi.

Grabezhev *et al.*, 1986). Also comparisons may be based on rocks not necessarily related through fractional crystallization. Thus, Mints (1975) argued that gold variations in Cretaceous volcanic rocks of the Okhotsk-Chukotka belts, USSR, were probably more related to source region and magma generation than to magma/mineral fractionation.

The Ir/Au ratio is included in data tables as an index of magma differentiation with respect to mantle composition. In general, neither fractional crystallization nor partial melting strongly fractionate gold, whereas fractional crystallization of silicate magma removes iridium with early crystallizing phases (e.g. Lac de l'Est cumulate sequence in the Thetford Mines ophiolites; Oshin and Crocket, 1982). During partial melting of mantle peridotite, iridium behaves as a compatible element in that magmas such as komatiite produced by high-percentage partial melting usually have higher iridium contents than those generated by lower degrees of partial melting such as tholeiite. The ratio of unfractionated material is taken as the C1 carbonaceous chondrite ratio of Ir/Au = 481/140 = 3.4 (Anders and Grevesse, 1989).

1.3.3 *Gold abundance in mafic and ultramafic rocks*

Estimates of gold in the mantle are obtained by analysis of mantle rocks or derivative mafic melts. Those analysed for gold include ultramafic nodules from kimberlite and alkali basalt, and non-cumulus ultramafic and mafic rocks of ophiolites.

Ophiolites. Ophiolites are the most massive mantle rock available for study. Two non-cumulus peridotites, harzburgite and lherzolite, may occur in ophiolites but the latter is uncommon. Most petrological opinion considers harzburgite to be a residue of partial melting of a more primitive peridotite, probably lherzolite, from which a basaltic component has been extracted.

Average gold content of 14 ophiolitic harzburgites is 2.8 ppb. The average Ir/Au ratio for three occurrences – Josephine, Mt. Albert and Thetford Mines – is 3.2, or essentially chondritic, but it is unlikely that ophiolitic harzburgite represents unfractionated mantle. This Ir/Au average omits Ronda with Ir/Au=14, the high ratio being due to an unusually low gold content rather than high iridium. The Mt. Albert, Thetford Mines, Josephine and Ronda occurrences average 2.4, 3.2, 4.6, and 7.3 ppb iridium respectively, so Ronda is not exceptional. However, average gold for Ronda is 0.33 ppb, the lowest of all 14 occurrences. If ophiolitic harzburgite and lherzolite are related as restite and source rock, then basalt partial melts may, on occasion, inherit a significantly different Ir/Au ratio than the source rock due to preferential partition of gold into the melt.

Peridotite nodules. Garnet and spinel-bearing peridotite nodules in kimberlite and alkali basalts are additional sources of mantle rock. Two averages are reported in Table 1.2: 2.7 ppb for nodules from Siberian kimberlites and 0.85 ppb gold for nodules from Lesotho pipes (Mitchell and Keays, 1981; Morgan *et al.*, 1981). These differences imply that garnet peridotite mantle has a variable gold content on a regional scale.

Table 1.2 Estimates of gold abundance in the mantle

Source	Estimate (ppb)	Model
Ringwood and Kesson (1977)	4.2	Pyrolite
Mitchell and Keays (1981)	0.9	Spinel lherzolite/pyrolite
Sun (1982)	0.5–1.5	Peridotitic komatiite
Chou *et al.* (1983)	1.3	Pyrolite

That the mantle is a heterogeneous source of noble metals is supported by Mitchell and Keays (1981), who found that only 20 to 40% of whole-rock noble metal content (Pd, Ir and Au) is accounted for by the essential rock-forming minerals, and that sulphides and platinum group minerals are present in these nodules. They found considerable dispersion of noble metals in garnet lherzolites and concluded that sulphides or platinum group minerals were major hosts. Supporting opinion by Ivanov *et al.* (1977), based on Siberian nodules, contends that gold, silver and copper are zoned vertically and horizontally in the mantle.

Spinel lherzolite nodules are more homogenous in gold than garnet lherzolites. Samples from the western US, Australia and European localities (Jagoutz *et al.*, 1979; Mitchell and Keays, 1981; Morgan *et al.*, 1981) average 0.49 ± 0.26 ppb gold for 27 whole-rock analyses. Samples from Tien Shan are apparently higher, averaging 1.2 ± 0.3 ppb. The consensus of petrological opinion places the mainly alkali basalt-hosted spinel lherzolites at lesser mantle depths than kimberlite-hosted garnet lherzolites. Both spinel and garnet lherzolites have an Ir/Au ≈ 7 (Morgan *et al.*, 1981; Mitchel and Keays, 1981), roughly twice the chondritic value, suggesting selective removal of gold from unfractionated peridotite sources.

1.3.4 *Gold in Precambrian rocks*

Volcanic rocks. Gold contents of tholeiitic and komatiitic volcanic rocks of Precambrian greenstone belts are summarized in Table 1.1. Observations and inferences from these data are:

(i) Average gold contents of peridotitic komatiites and tholeiitic basalts are similar, 4.2 and 5.7 ppb respectively. The apparent differences in percentage partial melting of mantle lead to little difference in gold contents of melts.

(ii) Although the data base is small, komatiitic basalts may represent rocks of higher than usual gold content. Of the six locality averages, three – the high-magnesium series footwall basalts from Kambalda (Keays *et al.*, 1981), the Reliance Formation, Belingwe, greenstone belt (Saager *et al.*, 1982), and komatiitic basalts of the Borgoin belt (Dostal *et al.*, 1985) – exceed 10 ppb. Other komatiitic basalt averages are < 4.2 ppb.

(iii) Two greenstone belts, the Belingwe and Borgoin, are anomalous. The Borgoin rocks average 37 ppb gold for both tholeiitic and komatiitic basalts. The Belingwe data of Saager *et al.* (1982) average 13.5 ppb, although an average of only 1.4 ppb was obtained by Keays (1984) for Belingwe greenstones.

Whether the gold contents of greenstones represent parental magmas is problematic. Kaapvaal greenstones are low in gold (0.67–1.7 ppb) but host economic gold deposits, whereas the Belingwe greenstones apparently have high background gold levels but no gold deposits. A suggested explanation is that strongly altered and metamorphosed Kaapvaal greenstones have yielded gold to fluids during alteration and/or metamorphic events, thus forming small deposits through dispersal and concentration in structural traps (Saager *et al.*, 1982; Keays, 1984; Viljoen, 1984).

The Ir/Au ratio may reflect such redistribution with gold experiencing a more severe depletion than iridium. The two localities with lowest average gold, Barberton and Mt. Clifford, have higher Ir/Au ratios (4.85 and 4.8) than any other peridotitic komatiite localities.

In the two greenstone belts with the high background gold (Belingwe and Borgoin), there is no preference for a specific rock type. Both peridotitic komatiites and komatiitic basalts of the Belingwe belt are high in gold, and in the Borgoin belt both komatiitic and tholeiitic basalts average 36 and 37 ppb gold (Dostal *et al.*, 1985). As these high contents seem locality – but not petrologically – specific, they may depend more on the gold contents of mantle source rocks than on magma-generation parameters such as degree of partial melting.

Granitoid rocks. Most Precambrian granitic plutons average 1–2 ppb gold. The most extensive study is that of Saager and Meyer (1982) on granitic rocks of the Johannesburg Dome and the Eastern Transvaal which yielded arithmetic and geometric means of 1.2 and 1.0 ppb gold. The cumulative frequency distribution identified a 6% excess value population consisting of samples with 4.1 ppb gold. Correlations between gold and major elements or differentiation trends of Transvaal granites were absent.

1.3.5 *Estimates of mantle gold abundance*

Estimates of gold abundance in the upper mantle are based on analyses of both mantle nodules and ultramafic flows. The large variation in gold content of spinel and garnet lherzolites, the presence of noble metal-rich minerals, and the need for accurate estimates of percentage partial melting of mantle peridotites introduce a significant level of uncertainty. Nevertheless, recent estimates converge around 1 ppb gold (Table 1.2).

1.3.6 *Mafic volcanic rocks in non-orogenic environments*

Mid-ocean ridge basalts (MORB). The average gold content of MORB is 1.2 ppb, the lowest of any mafic rock group. Tholeiitic and alkalic MORB have similar averages, 0.8 and 0.6 ppb gold (Gottfried *et al.*, 1972), suggesting that gold in MORB does not depend critically on the partial melting or fractional crystallization history of the magma.

The rock average, 1.2 ppb, may be a lower limit for gold in MORB magma. Keays and Scott (1976) showed that the average gold content of glass rims from MOR pillow basalt is three times that of crystalline interiors. They argued that the glass rim values, 1.35 ppb gold for five samples, represent the gold content of the original magma and that the crystalline interiors have lost gold.

General consensus prevails that MORB is sulphur-saturated at eruption (Mathez, 1976). Sulphur saturation would exercise a dominant control on gold in MORB magma in that sulphide globules in MORB glasses carry up to 13 ppm or 10^4 times the average gold content of whole-rock MORB (Peach and Mathez, 1988).

Intraplate basalts. Intraplate basalts, mainly Hawaiian tholeiite, average 2.0 ppb gold, about twice that of MORB. The few analyses of alkalic intraplate basalts suggest a gold content about half that of intraplate tholeiites. Alkaline magma generation may result from deeper melting than for tholeiitic magma generation (Basaltic Volcanism Study Project, 1981).

The higher gold contents of intraplate basalts compared with MORB may result from different gold contents of the respective source regions, differences in partial melting regimes, or differences in associated hydrothermal regimes. Variation in the gold content of mantle source rock may be significant if MORB source regions have sustained previous episodes of partial melting (Basaltic Volcanism Study Project, 1981). The fact that intraplate mantle source regions are more pristine and fundamentally different from MORB source regions is suggested by the high incompatible element contents of intraplate basalts and an average Ir/Au ratio of 0.38 compared with 0.03 for MORB. A further important difference between MOR and intraplate environments is the prevalence of hydrothermal activity at mid-oceanic ridges which may lead to large gold losses from MORB. However, the average gold content of oceanic basalts ranges from only 1 to 2 ppb. Thus all variations due to magma generation, fractional crystallization, and alteration result in only small average differences in the gold contents of seafloor basalt.

Continental flood basalts. Tholeiites of three flood basalt provinces, the Columbia River (USA), the Deccan (India), and the Parana Basin (Brazil), average 3.5 ppb gold with little variation in any one province (3.7, 3.8 and 3.1 ppb respectively). In comparison with oceanic basalts, flood basalts are higher in gold. The Ir/Au ratio for the Deccan traps is 0.023, similar to MORB. Average Deccan iridium is 0.092 ppb (Crocket, 1981), a value between MORB and intraplate basalt. Nd/Sm and Rb/Sr studies led Carlson *et al.* (1981) to model the Columbia River basalts by binary mixing of a primary basalt melt and Precambrian sialic crust.

Flood basalts and MORB are generated in extensional tectonic environments, but flood basalts are erupted through sialic crust. They have similar Ir/Au ratios to MORB and MORB-hosted magmatic sulphide (0.033 and 0.034 respectively), suggesting that gold in flood basalts may be carried mainly in MORB-like immiscible sulphide. However, unlike the MORB setting, pervasive hydrothermal activity is lacking so that sulphide is more readily preserved in the flood basalt setting. Thus a MORB Ir/Au ratio prevails, but the concentration of both metals is higher.

1.3.7 *Mafic plutonic rocks in non-orogenic settings*

Studies of mafic plutons from stable cratons have dealt with gold in initial magmas, the effect of fractional crystallization, and the role of immiscible sulphides. Estimates of initial magma composition are difficult. Marginal rocks may be subject to unusual thermal and dynamic regimes (Keays and Campbell, 1981), and may reflect more than one magma composition. Bushveld initial magmas are best represented by a magnesian series basalt with roughly 3 ppb gold and a tholeiitic melt with about 2.5 ppb (Davies and Tredoux, 1985). Blending of such magmas during the crystallization history of the Bushveld complex is therefore unlikely to have generated major changes in bulk-magma gold content.

Fractionation of gold in the Bushveld and other layered gabbros seems minimal. In the Bushveld Lower and Critical Zones, Lee and Tredoux (1986) found little correlation of gold with stratigraphy or silicate mineralogy, and argued that distribution of the gold was controlled by small quantities of immiscible sulphides precipitated throughout the fractional crystallization process.

1.3.8 *Orogenic environments*

Volcanic rocks. Volcanic rocks typical of plate convergence include the dominantly calc-alkaline rocks of the basalt–andesite–rhyolite association, as well as island arc boninites and continental, bimodal, basalt–rhyolite assemblages. Mafic and felsic types yield average gold contents of 2.2 and 1.55 ppb respectively. The reasons for the lower gold contents of the felsic rocks are ambiguous. Two studies of Soviet Far East rocks (Anoshin and Kepezhinskas, 1972; Mints, 1975) found no significant differences in the gold contents of mafic and felsic units. The lowest gold contents in felsic volcanics are from western US localities (Gottfried *et al.*, 1972), where averages of <1 ppb apply. No averages of <1 ppb are evident from the Soviet Far East suites. This difference may reflect regional variation or analytical bias at very low gold contents.

Plutonic rocks. Very limited variation in gold contents is shown by granitic plutons of convergent regimes. The group average is 2.6 ppb and the range for single plutonic units is 0.5–6.9 ppb. The most extensive data base is for the Hercynian granitoids of the Urals where 28 plutons average 2.5 ppb (range 1.4–4.4 ppb).

The coastal batholiths of the western US are apparently lower in gold with a group average of 1.2 ppb. The origin of batholithic granitoids in continental margin convergent environments is often attributed to partial melting of lower crust linked in various ways to the subduction process. The uniformly low gold contents of this group suggest that magma generation in the deep crust does not usually result in a transfer of major quantities of gold to higher crustal levels.

1.4 Gold in sediments and sedimentary rocks

1.4.1 *Introduction*

Precambrian clastic sedimentary rocks have produced much of the world's gold. The premier example, the Witwatersrand Basin, South Africa, averages some 8 ppm gold. Additional deposits include the Jacobina Series, Brazil, and the Tarkwa Goldfields, Ghana, with average grades of 8–15 ppm and 1.5–9 ppm respectively (Boyle, 1979). The data base for the sedimentary domain is Section 1.7.4 with a summary of averages in Table 1.3. Broad generalizations from these data are as follows:

(i) Of the major sedimentary rock groups (conglomerate/sandstones, shales and limestones), the highest average gold contents are in coarse-to medium-grained detrital rocks.

(ii) Shales have an average gold content comparable to intermediate igneous rocks, and probably reflect the gold contents of upper crustal source rocks better than other major classes of sedimentary rocks. However, the nature of the depositional environment may strongly influence the gold content of shales as seen in the higher gold contents of carbonaceous shale.

(iii) Carbonates have the lowest gold of any sedimentary rock, although the difference with respect to shales is only 17%.

(iv) Precambrian iron formation and ferruginous sedimentary rocks are highly variable in gold. Some sulphide-rich types exceed 100 ppb. Oxide-rich rocks have much lower gold contents and appear to reflect differing depositional environments.

1.4.2 *Gold in sediments*

Fine-grained clay-rich sediment of deep sea and terrigenous origin averages about 3 ppb gold compared with 1.5 ppb for deep-sea biochemical sediment (Table 1.3). Generally, biochemical sediment is lower in gold by roughly a factor of 2 relative to clay-rich sediment originating by weathering of crustal rocks.

Table 1.3 Average gold in sediments and sedimentary rocks (ppb)

Group	$\overline{X}$	N samples	$\overline{X}$	N areas
Sediments				
Clay-bearing deep-sea sediments	3.0	50	3.6	7
Globigerina and siliceous oozes	1.5	33	1.4	3
Terrigenous sediments, <500m water depth	3.2	75	3.5	6
Sedimentary rocks				
Conglomerate, sandstone, siltstone	8.1	1212	6.5	13
Shale	2.3	288	2.4	9
Carbonaceous shale	6.7	553	7.4	9
Carbonate rocks, associated evaporites	1.9	251		
Precambrian iron formation			38	21
Archean, Canadian Shield				
Proterozoic, Canadian Shield			19	7

1.4.3 *Gold in clastic rocks*

The data base for coarse- to medium-grained detrital rocks (Section 1.7.4.2) is dominated by the study of some 2000 Mesozoic continental sedimentary rocks of the Undino-Dainsk Basin in Eastern Transbaykalia, USSR, by Polikarpochkin and Korotayeva (1976). From Table 2 of their paper, an average gold content of 8.75 ppb is obtained for 1074 analyses representing eight localities in the Undino-Dainsk Basin. Other detrital rocks in this conglomerate-sandstone-siltstone size range do not exceed 4.6 ppb gold on average. Conglomerates have the lowest gold contents and fine-grained sandstones and siltstones the highest. High gold in fine-grained rocks correlates with organic matter and iron oxides. If coarse-grained conglomerates are high in gold, red cementing oxides are often present and their gold contents may exceed whole-rock gold by a factor of 3. Polikarpochkin and Korotayeva (1976) contended that mechanical transport of rock fragments and particulate gold from external source regions was the main source of gold in the coarse-grained rocks.

Fine-grained clay-rich sedimentary rocks have nearly identical locality and total population averages of 2.4 and 2.3 ppb gold respectively (Table 1.3). These averages are similar to mafic volcanic rocks (mainly andesite) at active continental margins with 2.2 ppb gold. The shale average is comparable to a Precambrian crustal average of 1.8 ppb obtained by Shaw *et al.* (1976) from the Canadian Shield.

A cursory evaluation by Crocket *et al.* (1973) found little difference in the gold contents of various soils and their source rocks. Tentatively, shales seem characterized by average gold contents similar to other estimates for the upper continental crust. However, the influence of depositional environment is important. Carbonaceous shales have three times the gold content of shales with normal organic carbon levels (Korobeynikov, 1986), indicating that organic matter or the attendant reducing conditions lead to enhanced gold concentration in the marine environment.

1.4.4 *Gold in chemical sedimentary rocks*

Limestones. Carbonates and associated evaporites are slightly lower in gold than other sedimentary rocks. *Pure* carbonate rocks are probably lower than indicated by the data base average as clay mineral components may be significant gold contributors.

Evaporites associated with reef and shelf carbonates (Popov, 1975) concentrate gold slightly relative to carbonate. Minor carnallite-bearing evaporites average 3.1 ppb gold, but pure halite and mixed sylvinite rocks do not exceed 1.7 ppb on average.

Precambrian iron formation. Much variation in the gold contents of iron-rich sedimentary rock is apparent in Section 1.7.4.2. Gold exceeds 100 ppb in sulphidic interflow sediments in volcanic sequences at Kambalda (Bavinton and Keays, 1978). Oxide-dominated iron formation occasionally hosts replacement gold deposits in association with discordant quartz or felsic porphyry intrusions (Fyon *et al.*, 1983; Groves *et al.*, 1987). In unmineralized oxide facies such as the Sherman and Adams Mines iron formations of the Abitibi belt, iron oxide mesobands average < 10 ppb gold in contrast to stratiform pyrite carrying 180 ppb (Crocket *et al.*, 1984). Depositional environment apparently exercises a control on gold content in that Algoman or volcanic-hosted oxide facies iron formation has approximately twice the gold content of Superior-type or shelf sediment-hosted iron formation (Gross, 1988).

The gold contents of metal-rich chemical sediments and volcanogenic sulphides are discussed in detail by Hannington *et al.* in Chapter 8 and so are not considered here.

1.5 Gold in metamorphic rocks

1.5.1 *Introduction*

Much interest in metamorphic rocks is focused on gold mobility in response to prograde devolatilization reactions accompanying regional metamorphism. Averages for greenschists, amphibolites and granulites are compiled in Table 1.4. The greenschist and amphibolite averages represent rocks derived from pelitic protoliths. Protoliths for granulite facies rocks cannot be evaluated with much confidence.

Table 1.4 Average gold content of regionally metamorphosed pelites and granulites (ppb)

Facies	Average gold weighted by:		Refs[2]
	Number of samples ( )	Number of localities ( )[1]	
Greenschist metapelites	7.1 (284)	6.3 (6) / 1.8–9.6	1, 4, 6, 8, 13, 15
Amphibolites	5.9 (276)	6.4 (9) / 1.9–10.4	1, 6, 8, 13
Granulites	2.2 (531)	3.2 (6) / 1.5–7.4	2, 6, 8, 11, 14

1. The numerator is the average for all localities and the denominator is the range of individual locality averages.
2 References: 1. Anoshin *et al.* (1982); 2. Brooks *et al.* (1982); 3. Buryak *et al.* (1972); 4. Crocket *et al.* (1983); 5. Crocket *et al.* (1986); 6. Gavrilenko *et al.* (1974); 7. Golovnya (1974); 8. Li and Shokhina (1974); 9. Moiseenko *et al.* (1971); 10 Moravek and Pouba (1984); 11. Mushkin *et al.* (1974); 12. Pchelintseva and Fel'dman (1973); 13; Petrov *et al.* (1972); 14. Sighinolfi and Santos (1976); 15. Thorpe and Thomas (1976).
Superscripts:
1. Omissions from averages: 1a, two slates–37 and 161 ppb Au, the latter with arsenopyrite; 1b, two greywackes–57 and 80 ppb; 1c, two slates–57 and 102 ppb.

Greenschists and amphibolites show virtually no difference in average gold contents on a locality basis and only a 17% difference on a total sample basis. This small difference presumably represents the integrated effects of metamorphism and variation in gold contents of the protoliths, if not additional factors. There is no indication that greenschist and amphibolite grade metamorphism exercise a dominant control on the gold content of pelites. In fact, the range in locality averages of 2–10 ppb suggests that sedimentary signatures are not erased at these levels of metamorphism. In contrast, granulite facies rocks contain less gold than the lower grade rocks. Five of six locality averages for granulites range from 1.5 to 3.7 ppb gold. Thus, granulites are probably depleted in gold relative to lower grade rocks by the metamorphic processes that generate the granulites.

1.5.2 *Regional metamorphism*

Several studies have concluded that gold is mobilized in response to regional metamorphism (Moiseenko *et al.*, 1971; Buryak *et al.*, 1971; Petrov *et al.*, 1972; Mushkin *et al.*, 1974). Buryak *et al.* (1971) concluded that granitized rocks in the Mama-Oron complex, Vitim-Patom highlands, USSR, with gold contents of 2.2–2.8 ppb gold, contained significantly less gold than the Patom series schist/gneiss protolith which averaged 3.69 ppb.

Studies involving the progressive metamorphism of pelitic rocks from greenschist through amphibolite facies have also concluded that redistribution of gold occurs in response to regional metamorphism (Petrov *et al.*, 1972). There is, however, no unanimity of opinion that gold is mobile in response to regional metamorphism (Pchelintseva and Fel'dman, 1973; Gavrilenko *et al.*, 1974; Li and Shokhina, 1974; Sighinolfi and Santos, 1976; Anoshin *et al.*, 1982). Often the distribution of gold in lower grade, younger rocks of metamorphic complexes is compared with that in higher grade older rocks from lower structural levels (Li and Shokhina, 1974) and many investigators ascribe differences in gold contents to differences in the gold contents of the unmetamorphosed protoliths.

A study of Archaean granulites from Brazil (Sighinolfi and Santos, 1976) found a low average gold content but a relatively high dispersion ($\overline{X} = 1.51$ ppb; $\sigma = 2.9$; range, < 0.4 to 18 ppb). There is a correlation of lower gold with 'acid type' granulites (low CaO + MgO) and higher gold with 'intermediate type' rocks (higher CaO + MgO), as is the case with igneous rocks (Tilling *et al.*, 1973). The authors interpreted this correlation as preservation of a pre-metamorphic trend which would not be expected if granulite facies metamorphism caused loss of gold from protoliths.

1.5.3 *Granitoid intrusions in metamorphic terranes*

The effect of regional scale igneous intrusion on redistribution of gold in metamorphic rocks is exemplified by the Bohemian Massif, Czechoslovakia (Moravek and Pouba, 1984, 1987). The Bohemian Massif is a mainly Precambrian block of juxtaposed low (Barrandian) and high-grade (Moldanubian) metamorphic terranes which have been subjected to Variscan (and earlier) tectonism and granitoid intrusion. Gold in the low-grade rocks is a factor of 4 higher than in high-grade crystalline rocks (4.1 vs. 1.0 ppb, geometric means). Variscan granitic intrusions invade both Barrandian and Moldanubian terranes, especially at the boundary between them where many gold deposits occur. Moravek and Pouba (1987) reviewed previous models which maintain that differentiation of granitoid magma is the cause of gold mineralization and argued that deposits had developed preferentially in low-grade rocks of the Upper Proterozoic Barrandian Block where background gold levels were highest.

1.6 Concluding summary

There is little reason to dispute the view of Tilling *et al.* (1973) that gold varies by roughly one order of magnitude in the crust and upper mantle, and that mafic rocks are usually higher in gold than felsic rocks. This difference may ultimately reflect fractional crystallization but the most direct cause is probably derivation of magmas from different reservoirs. In most instances, parental magmas to mafic rocks are of mantle origin, whereas felsic rocks arise from a spectrum of partial melting processes, mainly in crustal regimes. Thus the contents of gold and the availability of fluids particular to the site of magma generation are probably the more critical controls on gold abundances in mafic and felsic rocks. This chapter concludes with generalizations emphasizing recent contributions.

(i) A few rock-forming minerals, notably magnetite, may carry gold contents higher than their host rock. The role of gold concentrators in fractionating gold during magma crystallization may be minor unless modal proportions are high. In some cases, such as MORB, sulphides may be of critical importance.

(ii) Recent estimates of gold in mantle rocks converge around 1 ppb. Differing gold distributions in spinel and garnet lherozlites, apparent geographic variability in garnet lherzolite data, and the presence of sulphide and metallic concentrator phases in mantle rocks suggest that the upper mantle is a heterogeneous reservoir with respect to gold.

(iii) The gold contents of Phanerozoic basic volcanic rocks erupted in MOR, intraplate, convergent margin, and continental flood basalt tectonic settings vary from 1.2 to 3.5 ppb gold. These differences depend partly on magma-generation processes as suggested by differing Ir/Au ratios. However, the eruptive setting and the extent of hydrothermal activity may significantly determine whether or not rocks record the gold contents of their parental magmas.

(iv) If the growth of Phanerozoic continental crust can be modelled by andesite-dominated magmatism at continental margins, the gold contents of new crustal additions are probably approximated by the average for convergent plate mafic volcanics (2.2 ppb).

(v) The average gold content of large granitic complexes from plate convergent zones may be < 1 ppb (Gottfried *et al.*, 1972). If so, the partial melting of large masses of continental crust is unlikely to enrich gold in the accessible upper crust. However, specific instances in which such magmatism generates fluid-rich melts may lead to mineralization.

(vi) Precambrian mafic–ultramafic volcanism generated rocks with somewhat higher average gold contents than Phanerozoic mafic rocks. The average for Archaean komatiitic basalts is well above that of Phanerozoic basalts; however, the data base is small (n=42) and the average gold contents of individual occurrences are highly variable.

(vii) Greenstones with relatively high gold contents appear to lack mineralization, and the high gold contents are not lithologically specific. The implications for gold-poor greenstones are that deformation and metamorphism cause subsequent dispersal of gold.

(viii) Clastic-dominated continental basin settings may locally concentrate gold through both mechanical and chemical processes. Gold is often concentrated in carbonaceous shales, with gold contents commonly three times those of normal shales.

(ix) Unmineralized, Precambrian, oxide facies iron formations contain only 1 to 5 ppb gold in oxide minerals but pervasive pyrite may contain 100 ppb. In the Canadian Shield, volcanic-hosted Archaean iron formation contains approximately twice as much gold as shelf sediment-hosted Proterozoic iron formation, suggesting that depositional environment may exert some control on gold distribution.

(x) Regional metamorphism leads to gold loss if granulites are generated. Whether lower grade metamorphism causes significant change in gold contents is uncertain. Average gold contents of greenschist and amphibolite facies rocks are similar but some studies do document correlations of gold with metamorphic isograds. Low gold in granulites supports the concept that loss of gold to fluids during granulite formation can generate ore-forming fluids.

1.7 Mineral data

1.7.1 *Explanatory notes*

Column headings

$\overline{X}$ Arithmetic mean; or, if superscripted 'g', geometric mean
All data in parts per billion, ppb

σ Standard deviation

R Range

N *x*/*y*, *x* = number of samples and *y* = localities, occurrences or geological units such as plutons, complexes, formations; single entry, number of samples; if both *x* and *y* are given, the average refers to *y* (number of localities/occurrences)

Ref. last column, reference number.

Abbreviations

Geographic: MAR, EPR – Mid-Atlantic Ridge, East Pacific Rise

Petrological: kom, komatiite (ic); thol, tholeiite (ic); c.a., calc-alkaline; alk, alkaline; MORB, mid-ocean ridge basalt

Lithological: cong, conglomerate; ss, sandstone

Mineralogical: alb, albite; bi, biotite; chl, chlorite; cord, cordierite; gar, garnet; ky, kyanite; mus, muscovite; ortho, orthoclase; plag, plagioclase; ser, sericite; str, staurolite

Analytical methods

About 70–80% of the data are from neutron activation, either radiochemical or instrumental. Fire assay, before or occasionally after irradiation, is employed in some cases. Spectrochemical analysis, usually a wet chemical procedure with optical emission spectrographic finish, accounts for most remaining data. Atomic absorption and polarographic methods are used in a few cases.

1.7.2 *The gold content of rock-forming minerals*

Host Rock Description	$\overline{X}$	σ	*R*	*N*	Ref
Quartz					
Granitic plutons, western US	0.23	0.11	0.1–0.6	13/5	6
Granitoid-hosted[1]	2.0			71/19	1,2,3,18
Granitoids and skarns	4.7	2.9		118	8
Gabbro-hosted	5.85		3.4–86	12/4	7,15,18
Sedimentary rocks, USSR[2]	4.7		1.0–63	236	13
Feldspar					
Plagioclase					
Granitoids associations	3.8		1.2–8.6	526/105	10
Mafic intrusions, Siberia	10.5		1.1–21	22	10
Potassium Feldspar					
Granitic plutons, western US	0.46	0.46	0.02–1.4	8/4	6
Granitoids associations	3.0		1.1–5.6	316/76	10

Host Rock Description	$\bar{X}$	σ	R	N	Ref
Feldspar					
Sedimentary rocks	3.6		1.6–41	37	13
Biotite					
Granitoid-hosted	7.0		1.8–11	128/84	*
Granitoids and skarns	6.8	5.5		146	8
Sedimentary rocks[3]	5.1		2.4–6.9	42	13
Muscovite					
Granitoid-hosted	6.7	2.7	2.8–14	16	5
Sedimentary rocks[4]	2.5		1.1–6.4	58	13
Hornblende, Amphibole					
Granodiorite, Urals	23	9.7	9.7–35	6/6	5
Granitoids, USSR	2.2		1.6–2.7	43/42	2,3,7,18
Granitoids, Altai Sayan	6.8	5.1		176	8
Granitic plutons, western US	0.55	0.03	0.2–1.3	9/4	6
Gabbro, granodiorite, Kamchatka	26	21	8.9–46	6/6	15
Pyroxene					
Granitoids, Kazakhstan	17	18	3.0–43	4/4	7
Granitoids, Sayan-Altai	3.4	1.9	1.0–12	38/10	11
Skarns, various areas, USSR	8.6		0.9 31	123/12	11
Gabbro-dolerite, Siberia	9.0		2.2–34	38/4	4
Magnetite					
Granitoids, Urals	42	31	3.0–112	16	5
Granitoids, various areas, USSR	10.4		5.3–20	542/53	9
Granite association, Urals	19		18–35	26/5	4
Granitic plutons, western US	15	23	0.4–49	8/4	6
Gabbro association, Urals	6.4		3–20	21/5	4
Gabbro, dolerite, Siberia	7.4	2.1	0.3–28	45/10	9
Gabbro, granodiorite, Kamchatka	14.5	14	4.3–46	7	15
Ilmenite					
Granitoids, Kazakhstan	7.5	2.1	5.2–9.2	3/3	7
Granitoids, Urals	9.2	6.5	2.3–22	14	5
Granitoids, various areas, USSR	14		2.2–40	29/4	9
Sphene					
Granitoids, Kazakhstan	10	7.2	3.6–19	6/6	7
Granitoids, Urals	21	11	3.7–37	12	5
Granitoids, various areas, USSR	8.8		6.9–9.9	48/7	9
Apatite					
Granitoids, Urals	8.0	2.5	4.1–13	16	5
Granitoids, various areas, USSR	7.3			10/2	9
Zircon					
Granitoids, various areas, USSR	21			14/3	9

References: 1.Anoshin and Potap'yev, (1966); 2. Bushlyakov, (1971); 3. Davletov and Dzhakshibayev, (1970); 4. Fominykh and Znamenskiy (1974); 5. Grabazhev *et al.* (1986); 6. Gottfried *et al.* (1972); 7. Khitrunov and Mel'tser, (1979); 8. Korobeynikov (1976); 9. Korobeynikov (1980); 10. Korobeynikov (1981); 11. Korobeynikov (1982); 12. Moiseenko *et al.* (1971); 13. Nikitin and Yasyrev (1974); 14. Shcherbakov (1967); 15. Shilin (1980); 16. Shilin and Osipov (1978); 17. Zlobin *et al.* (1981); 18. Zvereva and Gavrilenko (1971).

Superscripts: 1. All localities are USSR unless otherwise designated.
2. All sedimentary samples are from the Central Russian Platform.
3. Biotite and glauconite.
4. Muscovite and sericite.
* References 2, 3, 5, 6, 7, 12, 14, 18

1.7.3 *The gold content of igneous rocks*

1.7.3.1 *Igneous rocks from non-orogenic environments*

Mafic volcanic rocks from tensional environments

Description	$\overline{X}$	σ	*R*	*N*	Ir/Au	Ref.
Mid-ocean ridge basalt						
Basalts; MAR (98), EPR (112)	1.2	1.8	0.04–15	210	0.03[1]	*
Sulphide globules, MORB glass	12.1[2]	2.2	8.9–16	5	0.034	13
Intraplate basalt						
Tholeiite, Hawaii	2.2	1.2	0.5–6.6	56	0.21[3]	**
Alk. basalt, Hawaii, Tahiti	1.0	1.0	0.2–3.9	13	0.28[4]	5,7,15
Continental flood basalt						
Columbia River Group, US	3.7	2.8	0.5–11	11		7
Deccan traps, India	3.8	2.0	0.9–7.5	18	0.023	2
Parana Basin, Brazil	3.1	1.9	0.7–6.7	20		2
Rift-related volcanic rocks, intrusions						
Picrite basalt, Disko Island	4.2	2.1	2.0–8.6	10		10
Triassic basalt/diabase, US	3.4	1.8	0.3–9.0	87		7
Great Lakes dolerite, Tasmania	4.8	3.3	1.0–12	18		7

References: 1. Chou *et al.* (1983); 2. Crocket (1981); 3. Crocket and Kabir (1988); 4. Crocket and Teruta, (1977); 5. Crocket *et al.* (1973); 6. Ehmann *et al.* (1970); 7. Gottfried *et al.* (1972); 8. Hertogen *et al.* (1980); 9. Keays (1984); 10. Keays and Scott (1976); 11. Laul *et al.* (1972); 12. Nesbitt *et al.* (1987); 13. Peach and Mathez (1988); 14. Vincent and Crocket (1960b); 15. Wasson and Baedecker (1970); 16. Zentilli *et al.* (1985).
Superscripts: 1. Keays and Scott (1976) and Hertogen *et al.* (1980).
2. Sulphide globule values are in ppm.
3. Crocket and Kabir (1988) and Gottfried and Greenland (1972).
4. Crocket *et al.* (1973) and Wasson and Baedecker (1970).
* References 1, 4, 6, 7, 8, 10, 11, 12, 16
** References 3, 6, 7, 14, 15

Mafic-ultramafic layered complexes

Bushveld Complex, South Africa						
Marginal rocks and syn-Bushveld sills						
Ultramafic sills, East lobe	1.75	0.06	1.7–1.8	3	1.2	1
Magnesian basalt sills	3.1	0.89	1.8–4.4	9	0.11	1
Thol.basalt, marginal rock	2.4	0.55	1.6–3.2	5	0.04	1
Gabbros and ultramafic rocks, layered sequence, Eastern Bushveld						
Gabbros, Middle Zone	6.9	7.0	1–22	9		4
Norite, Upper Critical Zone	5.5	3.6	2–14	10		4
Bronzite, dunite, L Critical Zone	2.7	–	<1–13	39	1.2	4,3
Bronzite, Lower Zone	4.4	5.2	<1–21	28		6
Bronzite, Lower Zone	0.8	0.8	0.01–3	17		3

Description	$\overline{X}$	σ	*R*	*N*	Ir/Au	Ref.
Jimberlana Intrusion, Southwest Australia[1]						
Lower layered series; centre of microrhythmic units						
Gabbro cumulates	2.6	3.4	0.5–8.4	5	0.12	2
Ultramafic cumulates	3.6	7.2	0.3–29	15	0.49	2
Lower layered series; marginal rocks						
Ultramafic cumulates	3.2	3.0	0.8–8.3	5	0.26	2
Skaergaard Intrusion, East Greenland						
Chilled marginal gabbro	4.6			1		6
Layered series gabbros	4.8	2.3	2.4–9.0	7		6
Anakita Massif, Siberian Platform, USSR						
Chilled border gabbro	8.0			2		5
Layered series gabbros	10		4.0–17	8		5
Zlatogorsk Intrusion, USSR						
Layered gabbros and ultramafic	1.5		1.0–49	12		7

References: 1. Davies and Tredoux (1985); 2. Keays and Campbell (1981); 3. Lee and Tredoux (1986); 4. McCarthy *et al.* (1984); 5. Shcherbakov and Perezhogin (1964); 6. Vincent and Crocket (1960a); 7. Voskresenskaya *et al.* (1970).
Superscripts: 1. Averages are based on 20 of 56 samples from Keays and Campbell (1981), and exclude samples with Cu > 105 ppm and/or S > 1 wt.%.

Alkaline rocks

Alkaline complexes in continental settings					
Mesozoic plutons; syenite, syeno-diorite; East Aldan Shield, USSR	3.8			157/3	5
Mesozoic alkalic massifs, potassium-rich syenite, Aldan Shield, USSR	6.8		4–14	10/4	12
Alkalic syenite and gabbro; potassic type; various areas, USSR	4.2	2.2	2.5–7.9	5/3	12
Palaeozoic alkalic massifs; sodic type; Baykal, Kuznetsk Altai, USSR	3.7	1.6	2.0–5.7	8/3	12
Khibiny complex (Hercynian); nepheline syenite; Kola Peninsula, USSR	1.7		0.5–4.4	31/1	1
White Mountain plutonic series, New Hampshire; alkalic plutons, North Carolina, US. Mainly granite, quartz monzonite	0.64		0.2–2.3	41/5	2
Shonkin Sag and Boxelder laccoliths; mafic syenite; Montana, US	2.85	1.9	0.4–6.7	24/2	2
Trans-Pecos province; syenogabbro, syenite, granite; Texas, US	0.74	0.91	0.2–4.3	19/1	2

Kimberlite, alkali basalt and cognate mantle xenoliths

Kimberlite and alkali basalt host rocks					
Kimberlites					
Siberian Platform, USSR	3.1	2.3	0.5–8.6	55	3,9
Alkali basalt					
Siberian Platform, USSR	6.1	3.6	1.5–9.3	5	8
Tien Shan, Hissar, Nuratau, USSR	2.0	1.1	0.9–7.5	41	8

Description	$\overline{X}$	σ	*R*	*N*	Ref.
Cognate mantle xenoliths					
Garnet peridotite and pyroxenite					
Siberian Platform, USSR	2.7	1.85	0.6–8.1	47	*
Matsoku, Thaba Putsoa; Lesotho	0.85	0.95	0.08–2.7	10	6, 7
Eclogite					
Siberian Platform, USSR	3.4	2.3	0.8–9.1	25	**
Spinel lherzolite					
Mainly Kilbourne Hole (US) and Mt. Porndon (Australia)	0.49	0.26	0.08–1.1	27	4, 6, 7
Spinel peridotites and pyroxenites					
Siberian kimberlite host	4.5	2.5	1.1–7.1	9	3
Tien Shan, alkali basalt host	1.2	0.3	0.4–2.3	18	8

References: 1. Gavrilenko *et al.* (1982); 2. Gottfried *et al.* (1972); 3. Ivanov *et al.* (1977); 4. Jagoutz *et al.* (1979); 5. Karelin *et al.* (1974); 6. Mitchell and Keays (1981); 7. Morgan *et al.* (1981); 8. Mushkin and Yaroslavskiy (1974); 9. Rozhkov *et al.* (1973); 10. Sveshnikova *et al.* (1978); 11. Ukhanov and Pchelintseva (1972).
* References 3, 9, 11; ** References 3, 10, 11

1.7.3.2 *Active orogenic environments*

Plate convergence (island arcs, active continental margin)

Volcanic and subvolcanic rocks					
Cenozoic volcanic rocks; Kuril-Kamchatka, USSR					
Rhyolite, dacite	1.79	2.25		48	1
Andesite	1.88	1.35		82	1
Basalt	1.73	1.70		38	1
Cretaceous volcano-plutonic complexes: Okhotsk-Chukotka. USSR					
Magey complex					
Tottinsk subcomplex					
Rhyolite, and ash flows	1.23	1.56		99	7
Granite (subvolcanic)	0.98	0.60		41	1
Ul'insk subcomplex					
Andesite, basalt	1.98	2.3		34	7
Mafic intrusive	1.97	1.92		15	7
Dacite	1.95	1.99		33	7
Rhyolite	1.47	0.33		6	7
Dzhugdzhur intrusive complex					
Gabbro and diorite	3.89	5.78		18	7
Diorite porphyry	5.1	5.3		23	7
Leucogranite	20	32		8	7
Cretaceous to Palaeocene volcanics; central Okhotsk (Ola-Seymkan)					
Basalt, andesite	3.2	2.2		46	9
Rhyolite dacite; Narauli and Khol'chanskaya suites	4.2	2.9		30	9
Rhyolite, ignimbrite	9.6	8.0	1.5–22.5	53	9

Description	$\bar{X}$	σ	R	N	Ref.
Tertiary boninites (magnesian low-titanium lavas)[1]					
Mariana-Bonin arc (DSDP 458)	1.95	0.85	0.7–2.9	12	4
Cape Vogel, Papua New Guinea	1.4[2]	0.31	1.1–1.9	5	4
Cambrian mafic and ultramafic boninites, Australia					
Victoria greenstone belts	0.97	1.6	0.15–5.4	11	4
Cambrian low-Ti picrites,					
King Island, Tasmania[1]	2.8	1.9	0.48–7.0	13	4
Miocene, Pleistocene c.a. volcanic rocks; Cascades, W. United States					
Basalt, andesite	5.6		0.1–19	35	2
Dacite, rhyolite, rhyodacite	0.63		0.1–1.7	19	2
Volcanic rocks mainly bimodal suites: W. United States					
Mainly basalt	1.8		0.2–6.1	20	2
Rhyolite, qtz latite[3]	0.56		0.1–1.9	64	2
Snake River Group, Idaho, US					
Basalt	0.5		0.1–1.3	19	2
Rhyolite	0.6		0.2–1.2	6	2
Plutonic rocks					
Mesozoic and Tertiary c.a. plutons; W. United States					
Granodiorite, qtz monzonite	1.2	0.23	0.5–1.5	115/7	2
Marysville granodiorite stock	2.6	1.8	1–7	47/1	6
Orogenic granitoids (Hercynian), Eastern Urals, USSR					
Mainly granite, granodiorite	2.5	0.84	1.4–4.4	427/28	3
Altered granitoids, greisen	2.6	0.7	1–7	69/7	3
Mesozoic granitoids; S. Verkhoyan, In'yali-Deba Syncline, USSR					
Granite, granodiorite, diorite	3.5		1.0–6.9	229/22	5
Granitoids; Transili Alatau, Dzhungar Alatau, S. Kazakhstan, USSR					
Kunushsk and Kalbinsk granitoid complexes					
(Hercynian); E. Kazakhstan	1.6	0.74	0.9–2.5	40/4	5
Mainly biotite granite	1.9	0.9	0.9–3.1	111/4	8

References: 1. Anoshin and Kepezhinskas (1972); 2. Gottfried *et al.* (1972); 3. Grabezhev *et al.* (1986); 4. Hamlyn *et al.* (1985); 5. Khitrunov and Mel'tser (1979); 6. Mantei *et al.* (1970); 7. Mints (1975); 8. Uvarov *et al.* (1972); 9. Yudin *et al.* (1972).

Superscripts: 1. Ir/Au:, Mariana suite; 0.07, Cape Vogel; 0.16, Australian greenstone belts and 0.12, King Island picrites.

2. One value of 10 ppb omitted.

3. Samples noted as altered omitted.

Ophiolites, alpine peridotite, related rocks

Description	X	σ	R	N	Ir/Au	Ref.
Ophiolites with associated cumulate[1] sequences						
Troodos, Cyprus						
Harzburgite	1.35	0.2	1.2–1.55	3		1
Cumulates	1.6[2a]	1.2	0.5–3.7	6		1
Lower Pillow Lavas	0.08			1		1
Upper Pillow Lavas	2.3[2b]	0.8	1.3–3.2	6	<0.04	1, 7
New Caledonia, South Pacific						
Lherzolite	10.5	7		4		6

Description	$\bar{X}$	σ	*R*	*N*	Ir/Au	Ref.
Harzburgite	5	4		4		6
Dunite	1	0.9		3		6
Cumulates	4		1–10	11		6
Dolerite dykes	3	2		5		6
Basalt flows	1	0.9		3		6
Basalt, low-Ti lava	0.67			1	0.09	7
Thetford Mines complex, Quebec, Canada						
Harzburgite	1.5	1.35	0.2–3.4	7	2.1	10
Cumulates	1.35	1.6	0.2–6.4	31	0.83	10
Non-cumulus gabbro	7.0	12	0.4–11	7	0.01	10
Basalts	2.9	3.0	0.4–9.5	28	0.04	11
Bon Azzer, Morocco						
Serpentinite	5.4	5.7	0.5–29	50		4
Arabian shield ophiolites						
Serpentinite	2.8	2.6	1–10	14		4
Timetrine Massif, Mali						
Serpentinite	3.6	4.8	1–16	11		4
Altai-Sayan region, USSR						
Idzhin intrusion, W. Sayan						
Dunite and peridotite	2.1		1.3–2.1	4		8
Borus intrusion, W. Sayan						
Harzburgites	6.4		4.8–8.1	5		8
Serpentinites	1.6		1.3–2.1	4		8
Severnaya-Zelenaya hill intrusions, Kuznetsk Alatau						
Dunite	2.7		1.0–6.9	12		8
Serpentinites	15		3.0–24	4		8
Urals, USSR						
Harzburgite, dunites	2.9	0.55	2.0–3.7	15		8
Ophiolites of the alpine peridotite association						
Josephine peridotite, Oregon, US						
Harzburgite and dunite	1.2	0.78	0.12–2.2	9	5.3	15
Ronda complex, Spain						
Lherzolite	1.5	0.48	0.95–2.0	4	3.7	15
Harzburgite	0.33	0.20	0.20–0.63	4	14	15
Mt. Albert pluton, Quebec, Canada						
Harzburgite and dunite	1.1	0.4	0.6–1.6	5	2.2	5
Serpentinites	2.4[2c]	0.7	0.55–5.9	7	0.88	5
Yenisey range, USSR						
Fominsk Massif						
Dunite	4.85		4.8–4.9	2		9
Dunite and harzburgite	3.6	1.3	2.5–5.1	3		9
Berezovyy, Osinonskiy, Porozhnisk and Severnyy Massifs						
Serpentinite	1.7	0.7	0.7–2.5	4		9
Ultramafic\mafic complexes of possible ophiolitic origin						
Finero complex, N.Italy/Switzerland						
Peridotite, serpentinite	0.80	0.75	0.2–3.0	27		13
Amphibolite	0.50	0.33	0.1–1.3	15		13

Description	$\overline{X}$	σ	R	N	Ir/Au	Ref.
Hochgrossen and Kraubath ultramafic complexes, Austria						
Serpentinized rocks	5.7	7.1	0.4–25	14		13
Ivrea-Verbano mafic-ultramafic complex, W. Italian Alps						
Lherzolite	1.3	0.84	<0.6–2.1	4		14
Harzburgite	<0.6			1		14
Layered series (cumulates)	3.9	3.2	0.6–11	14		14
Gabbro (non-layered)	1.5	0.85	0.6–2.5	5		14

References: 1. Agioritis and Becker (1979); 2. Buisson and Leblanc (1987); 3. Crocket and Chyi (1972); 4. Dupuy *et al.* (1981); 5. Hamlyn *et al.* (1985); 6. Korobeynikov and Goncharenko (1986); 7. Li and Kornev (1972); 8. Oshin and Crocket (1982); 9. Oshin and Crocket (1986); 10. Saager *et al.* (1982); 11. Sighinolfi and Gorgoni (1977); 12. Stockman (1982).

Superscripts: 1. Cumulates include dunite, pyroxenite and gabbro. An average is calculated by assigning equal weight to each rock type.

2. Data omitted: 2a, one pyroxenite – 106 ppb; 2b, one analysis – 43 ppb; 2c, one serpentinite – 165 ppb.

Alaska-type complexes

Tulameen ultramafic complex, British Columbia, Canada						
Dunite, peridotite	0.29	0.05	0.1–0.9	19	1.7	3
Serpentinized dunite	4.1	3.4	0.1–4.6	20	0.51	3
Clinopyroxenite	0.4	0.2	0.1–1.9	8	6.8	3
Hbld. clinopyroxenite	3.4	2.9	0.1–12	4	0.03	3
Chromite-rich (> 10%) rock	8.2	5.4	1.2–42	14	12	3

Mafic to ultramafic intrusive rocks

Tholeiitic subtypes						
Gabbro-peridotite complex (Pechenga Syncline), Baltic Shield, USSR	2.9[g]		0.9–6.9	35		1
Anorthositic subtypes						
Fiskenaesset complex, W. Greenland						
Anorthosite	0.067	0.043	0.02–0.12	5	0.29	2
Gabbro anorthosite	0.58	0.50	0.07–1.1	3	0.42	2
Gabbro-anorthosite complex (Kolmozer-Voron'ya zone), Baltic Shield	1.68[g]		0.5–7.4	36		1

References: 1. Gavrilenko *et al.* (1982); 2. Morgan *et al.* (1976); 3. St. Louis *et al.* (1986).

1.7.3.3 *Igneous rocks of Precambrian greenstone belts*

Ultramafic and mafic volcanic rocks

Kaapvaal and Zimbabwe Cratons, southern Africa						
Barberton, Komatii Formation						
Peridotitic komatiite	0.67	0.50	0.07–1.8	30	4.9	1, 12, 18
Basaltic komatiite	1.2	0.46	0.5–1.9	8		1, 18
Tholeiitic basalt	1.3	0.57	0.9–2.1	4		18

Description	$\overline{X}$	σ	R	N	Ir/Au	Ref.
Pietersburg, Eersteling area						
Peridotitic komatiite	1.2	1.4	0.3–4.1	7		18
Basaltic komatiite	1.0	0.75	0.4–1.0	6[1a]		18
Tholeiitic basalt	1.7	1.0	0.4–3.5	13[1b]		18
Belingwe, Upper Greenstones, Bulawayan Group						
Peridotitic komatiite	13.5	19.6	0.4–67	15[1c]		18
Peridotitic komatiite	1.4	0.47	0.65–2.2	15		12
Basaltic komatiite	18.3	21	0.1–64	13		18
Tholeiitic basalt	3.3	6.7	0.1–7.4	20		18
Yilgarn craton, Western Australia						
Mt. Clifford						
Peridotitic komatiite	0.49[1d]	23	0.2–0.98	21	4.8	12
Kambalda						
Peridotitic komatiite flows overlying Long and Lunnon Shoots	7.3	6.5	0.63–25	22	0.40	12
Thick ultramafic komatiite flows, Lunnon Shoot	6.0			18	0.42	13
High-Mg basalt, hanging wall	3.8			20	0.10	13
High-Mg series basalt, footwall	13.6			5	0.03	13
Dunitic komatiite intrusives, W. Australia	2.2	2.85		50	3.3	14
Superior Province craton, Canada						
Abitibi subprovince, Ontario localities						
Peridotitic komatiite, Alexo flow	2.15	0.50	1.4–3.5	8	1.1	2
Peridotitic komatiite, Pyke Hill, Munro Township	2.97	1.56	1.4–7.3	11	0.39	3
Thick (125m) komatiitic flow, Fred's flow, Munro Township	2.1	1.4	0.44–5.9	19	0.23	3
Thick (115m) tholeiitic flow, Theo's flow, Munro Township	1.6	1.4	0.44–4.2	13	0.28	3
Ultramafic and mafic komatiite flows, Timmins area	1.9	1.9	0.5–6.7	19		7
Magnesium tholeiitic basalts, Timmins area	2.4	1.25	0.2–5.3	35		7
Wabigoon and Uchi subprovinces						
Shoal Lake-Lake of the Woods (Wabigoon) and Birch-Uchi areas[2]						
Komatiitic volcanics	16	30	0.4–69	5		8
Basalt, thol	8.9	116	0.2–225	116		8
Andesite, c.a.	6.3	9.8	0.2–58	95		8
Dacite, c.a.	6.3	16	0.2–113	52		8
Rhyolite, c.a.	5.1	6.5	0.2–24.5	24		8
Kakagi Lake, western Wabigoon						
Basalt; thol and c.a.	1.75	1.7	0.3–8.1	27		15
Mainly c.a. dacite	1.5	1.55	0.09–6.8	35		15

Description	$\overline{X}$	σ	*R*	*N*	Ir/Au	Ref.
Red Lake area (Uchi subprovince)						
Ultramafic komatiites	5.3	4.8	0.1–13	5		4
Basaltic komatiites	4.1	5.5	0.4–16	7		4
Iron tholeiitic basalts	15	15	4.3–44	7		4
Thol. basalt, andesite	5.2	3.4	1.3–10	5		4
Central and northern African cratons						
Central African Republic, Archaean greenstone belts						
Borgoin greenstone belt						
Thol., komatiitic basalt	37	11	18–53	10		6
Boufoyo greenstone belt						
Thol., komatiitic basalt	3.0	2.3	1–7	6		6
Burkina Faso, Proterozoic; Hounde and Boromo greenstone belts						
Basalt and andesite	1.4		0.1–4.4	52		10
Algeria, Tassendjanet area; Proterozoic						
Unaltered c.a. andesite	3.1	1.3	2.1–4.4	8		5
Altered c.a. andesite	1.55	0.85	0.8–5.9	31		5
Precambrian felsic plutonic rocks						
Kaapvaal Craton, South Africa; E. Transvaal and Johannesburg Dome						
Tonalite, granodiorite	1.2	1.1	0.3–7.8	190/19		17
Superior Province craton, Canada						
W.Wabigoon subprovince; Kakagi Lake area						
Granodiorite, diorite	1.1	1.2	0.1–4.9	18/2		15
Abitibi subprovince; Matachewan area						
Syenite	2.3	2.0	<1–6	6		19
East Aldan Shield, USSR						
Archaean granites	1.2			8		11
Proterozoic granite	1.7			10		11
South Yenisey Range, Taimyr Peninsula, USSR						
Proterozoic granite	9.9	11	1–40	28/3		16

References: 1. Anhaeusser *et al.* (1975); 2. Brugmann *et al.* (1987); 3. Crocket and MacRae (1986); 4. Crocket *et al.* (1980); 5. Dostal and Dupuy (1987); 6. Dostal *et al.* (1985); 7. Fyon and Crocket (1982); 8. Goodwin (1984); 9. Gottfried *et al.* (1972); 10. Huot *et al.* (1987); 11. Karelin *et al.* (1974); 12. Keays (1984); 13. Keays *et al.* (1981); 14. Keays *et al.* (1982); 15. Kwong and Crocket (1978); 16. Li and Datsenko (1973); 17. Meyer and Saager (1985); 18. Saager *et al.* (1982); 19. Sinclair (1982).

Superscripts: 1. Data omitted from averages: 1a, one komatiite, 20.2 ppb; 1b, pyrite-rich basalt, 16.8 ppb; 1c, one peridotitic komatiite, 372 ppb; 1d, two samples, 2.96 and 6.83 ppb.

2. Median values for these groups are 3.6, 4.1, 3.3, 2.1, 2.8, and 3.2 ppb from komatiite to rhyolite respectively.

1.7.4 *The gold content of sediments and sedimentary rocks*

1.7.4.1 *Sediments*

Deep-sea sediments

Description	$\overline{X}$	σ	*R*	*N*	Ref.
Calcareous, clay-bearing sediment;					
Pacific 12° South	3.1	1.9	1.3–7.6	14	8

Description	$\overline{X}$	σ	R	N	Ref.
Calcareous sediment, some (16%) clay; Pacific 39° South	1.3	065	0.4–2.8	19	18
Lutites (Pacific, Antarctic, Indian Oceans)	2.9	2.4	0.6–9	10	6
Mainly deep-sea clays, Black Sea	6.2		3.5–9	7	1
Calcareous and globigerina ooze; Caribbean and Antarctic	1.8	1.0	0.9–3.5	24	6
Siliceous ooze; Pacific Antarctic	0.85	0.3	0.5–1.5	9	6
Terrigenous sediments					
Silt, clay, sand, 30% carb; Atlantic, Baltic, Black, Mediterranean	3.2		0.1–24	61	1
Sandy and clay mud; Arctic Ocean (Canadian Arctic Islands)	2.8	0.8	1.6–4.2	9	8
Silt with mafic pyroclastic detritus, Atlantic near Iceland	4.2		0.5–15	5	1
Biogenic carbonate-rich (>30%) sediments					
Mainly clay and silty clay-sized sediment; Mediterranean	3.1		1.1–6.5	46	1
Clay, silt and sand-sized sediment; Atlantic	1.5		0.7–3.4	9	1

1.7.4.2 *Sedimentary rocks*

Description	$\overline{X}$	σ	R	N	Ref.
Mainly clastic sedimentary rocks					
Conglomerate, sandstone, siltstone					
Ss, siltstone; Tuva; Taymyr Peninsula; Yenisey Range, USSR	2.7		0.5–7.0	59/3	14
Cong, ss, siltstone; Undino-Dainsk Basin; Mesozoic; Tranbaykalia, USSR	8.75, 4.7[g]		0.5–19	1074/8	19
Siltstone and greywacke; Bohemian Massif, Czechoslovakia	4.6		1–19	55	16
Tuffaceous siltstone; Superior Province (Archaean), Canada	1.1	1.95	0.1–8.8	24	15
Shales, argillite					
Non-carbonaceous shale; Kuznetsk, Sayan, Tuva, Taymyr, Yenisey, USSR	2.2		0.1–8.3	235/7	14
Carbonaceous shale; same regions of USSR as above	6.7		0.1–29	548/8	14
Shale; Witwatersrand Basin; S. Africa; away from mineralization	4.1	5.3	0.3–23	19	15
Shale; Turonian to Palaeocene; Gubbio, Italian Appenines	1.85	1.6	0.4–5.8	16	5

Description	$\overline{X}$	σ	R	N	Ref.
Mainly chemical sedimentary rocks					
Carbonates and evaporites					
Limestones, dolomites, marbles; Kuznetsk Alatau, Sayan, Tuva, USSR	2.5		0.2–5.5	124/6	14
Pelagic limestone; Turonian to Palaeocene, Gubbio, Italian Appenines	1.1	1.4	0.3–4.6	10	5
Basinal facies limestones; Late Jurassic, Soviet Central Asia	0.71		0.6–1.0	14	20
Halite, sylvite, anhydrite-bearing evaporites; Jurassic, Central Asia	1.4		0.4–7.0	103	20
Precambrian iron formation and iron-rich sedimentary rocks					
Mainly Superior province, Canadian Shield					
Algoma-type BIF; Ontario, Northwest Territories	57	129	<5–430	47	3
Oxide facies BIF; Superior Province, Quebec–Labrador geosyncline					
Algoma-type, mainly Archaean	38		1–130	52/21	12
Superior type, Proterozoic	19		1–74	17/7	12
Algoma-type BIF; Abitibi subprovince; Sherman and Adams Mines, Ont.					
Magnetite–chert mesobands	4.4	6.7	0.5–34	42	9
Jasper–chert mesobands	8.5[1a]	4.5	2.6–20	28	9
Pyrite mineral separates	180	240	12–880	19	9
Algoma-type BIF; Abitibi subprovince; Carshaw-Malga iron fm, Timmins					
Magnetite mesobands	29[g]	1	0.9–5900	12	10
Carbonate mesobands	131[g]	0.7	1.6–660	10	10
Chert mesobands	9.8[g]	0.8	0.7–1400	19	10
Algoma-type BIF, Uchi subprovince, Red Lake area, Ontario					
Oxide facies BIF	40	55	0.7–196	15	7
Chert from BIF	1.7	7	0.4–3.5	7	7
Mixed oxide and silicate BIF	8.2	11	0.8–43	19	7
Siderite–pyrite Algoma-type BIF; Helen iron fm, Wawa subprovince					
Banded chert members					
Magnetic chert, oxide facies	1.1[1b]	0.6	0.2–20	8	11
Carbonaceous, sulphidic chert	9.4	17	0.2–49	8	11
Siderite–pyrite members	1.2[1c]		0.3–1.9	9	11
South Africa					
Ferruginous sedimentary rocks, oxide BIF; Kaapvaal Craton.					
Algoma-type; Barberton, Pietersberg, Sutherland greenstone belts	172	175	1.0–667	23	21
Superior-type; Pongola Group, Witwatersrand Basin	3.7[1d]	2.9	0.5–8.0	8	21
Western Australia					
Archaean BIF and sulphidic sediments, Yilgarn Block, W. Australia					
Interflow sulphidic sediments					
Kambalda	146			44	2
15 localities	31			97	3

Description	$\bar{X}$	σ	R	N	Ref.
Archaean BIF; regional background, Murchison, Southern Cross	2.2[1e]			13	13
Archaean BIF; mineralized areas, Murchison, Eastern Goldfields	12	21	2–68	11	13

References: 1. Anoshin *et al.* (1969); 2. Bavinton and Keays (1978); 3. Boyle (1979); 4. Chapman and Groves (1979); 5. Crocket (unpublished); 6. Crocket and Kuo (1978); 7. Crocket *et al.* (1980); 8. Crocket *et al.* (1973); 9. Crocket *et al.* (1984); 10. Fyon *et al.* (1983); 11. Goodwin *et al.* (1985); 12. Gross (1988); 13. Groves *et al.* (1987); 14. Korobeynikov (1986); 15. Kwong and Crocket (1978); 16. Moravek and Pouba (1984); 17. Phillips (1987); 18. Piper and Graef (1974); 19. Polikarpochkin and Korotayeva (1976); 20. Popov (1975); 21. Saager *et al.* (1982).
Superscripts: 1. Omissions of data: 1a, one analysis – 70 ppb Au; 1b, one BIF sample in contact with mafic volcanics – 45 ppb Au; 1c, one sample – 27 ppb Au; 1d, one sample – 162 ppb; 1e, one sample – 15 ppb Au.

1.7.5 *The gold content of metamorphic rocks*

1.7.5.1 *Regional metamorphic rocks*

Description	$\bar{X}$	σ	R	N	Ref.
High-grade rocks; gneiss, granulite, migmatite					
Mama-Oron complex; L.Proterozoic, Vitim-Patom highlands, USSR					
Primary gneiss and schist, Patom series	3.69			262	3
Mama-Oron complex; granitized equivalents, primary rocks above	2.80			66	3
Mama-Oron complex; metamorphic granite and granite gneiss	2.24			315	3
Mama-Oron complex; metamorphic plagioclase granite pegmatite	2.33			38	3
Mama-Oron complex; microcline and plagioclase granite pegmatite	2.82			117	3
Granulites, charnockite affinity; Archaean, Bahia state, Brazil	1.51	2.87	<0.4–18	105	14
Granuite facies gneiss xenoliths in basic dykes; Nuratau, Tien Shan					
Two-pyroxene gneisses	7.4	0.56		6	11
Eclogized (garnet-bearing) two-pyroxene gneisses	2.5	1.08		9	11
Greenschist to mainly amphibolite grade sequences					
Boydaba synclinorium; Proterozoic, Patom uplands, USSR					
Ugakham area, metapelites and metasiltstones					
Greenschist; chl, ser, bi	6.7			19	13
Epidote amphibolite; bi, gar	14.7			16	13
Amphibolite; gar, str subfacies	9.1			15	13
Amphibolite; gar, ky subfacies	3.3			14	13
Ugakham area, metamorphosed carbonate rocks					
Greenschist	2.6			13	13
Epidote amphibolite	7.1			11	13
Amphibolite; gar, str subfacies	7.6			14	13

Description	$\overline{X}$	σ	*R*	*N*	Ref.
Amphibolite; gar, ky subfacies	13.8			7	13
Zhuy area, metapelites and metasandstones					
Greenschist; chl, ser, bi	8.5			46	13
Epidote amphibolite	9.85			33	13
Epidote amphibolite, high temperature subfacies	13.2			7	13
Metamorphosed arenaceous, argillaceous rocks; Yenisey Range, USSR					
Greenschist; late Proterozoic					
Chl, ser subfacies	9.75	7.8	0.5–44	88	8
Chl, bi subfacies	9.31	7.4	0.7–27	31	8
Epidote amphibolite; late and middle Proterozoic	9.43	7.6	0.7–22	24	8
Amphibolite; early, middle and late Proterozoic	7.48	4.9	0.5–18	30	8
Granulite facies; Archaean					
Angara-Kansk bloc	3.74	1.75	0.6–5.8	16	8
Kokchetav Uplift, central Kazakhstan, USSR					
Yefimovo series, late Proterozoic; mainly greenschist					
Marbles	0.9[g]		<0.5–5.6	12	12
Quartzites	0.9[g]		<0.5–3.5	81	12
Mainly phyllites and schists after clay-rich protoliths	1.7[g]		<0.5–6.5	51	12
Metamorphosed basic igneous rocks	1.0g		<0.5–8.7	4[1]	12
Lower Zerenda series, pre–late Proterozoic; amphibolite facies					
Quartzite	0.8[g]		<0.5–6.9	24	12
Schists (bi, cord rock; sill, ky, gar rock; amphibolite; eclogite)	2.0[g]		<0.5–28	4[1]	12
Amphibolite, after basic rock	2.6[g]		0.8–17	20	12
Eclogite, after basic rock	0.9[g]		<0.9–3.5	31	12
Kolmozero-Voron'ya complex; mainly amphibolite grade; Proterozoic, Kola Peninsula, USSR					
High-alumina schist, carbonaceous pelite protolith	7.3		0.7–50	24	6
Two-mica gneisses, bi, amphibole, plag schist; Voron'yetundra Fm	2.1		0.4–24	30	6
Mainly amphibolites; Polmostundra Fm	2.2		0.3–12	55	6
Amphibole, garnet rocks; Lyavozero Fm	4.8		1.2–34	17	6
Archaean basement granite gneiss	1.6		0.2–7.2	14	6
Tongulak, S.Chuya, Sangilen metamorphic complexes; Altai-Sayan, USSR					
Metapelites					
Greenschist facies; chl-mus	3.6			17	1
Epidote amphibolite facies	2.9			92	1
Amphibolite facies	1.9			35	1
Metamorphosed mafic rocks					
Greenschist facies	3.1			9	1
Epidote amphibolite facies	3.5			25	1

Description	$\overline{X}$	σ	R	N	Ref.
Amphibolite facies	2.4			18	1
Meguma Group, Palaeozoic, Canada; greenschist metagreywacke, slate					
Unmineralized areas (background)					
Metagreywacke	0.95	1.1	<0.2–3.6	74	2,4,15
Slate	1.5[1a]	2.2	<0.2–14	60	4,15
Mineralized district (Harrigan Cove)					
Metagreywacke	8.7[1b]	9.4	1.7–41	30	5
Slate	8.2[1b]	10	2.4–50	26	5
Greenschists, granitoids; Bohemian Massif, Czechoslovakia					
Very low grade greenschist metasediments and volcanics; Barrandian Block, upper Proterozoic	3.9[g]		<1–19	156	10
Greenschist metasediments and volcanics; Central Mobile Zone	3.1[g]		<1–20	54	10
Granulites, gneisses; Moldanubian Block, pre-upper Proterozoic	1.2[g]		<1–280	233	10
Variscan granitoids (tonalite, granodiorite, K-granite, syenite)					
Intruding Barrandian Block	4.1[g]		1–12	19	10
Intruding Moldanubian Block	1.0[g]		<1–100	92	10

References: 1. Anoshin *et al.* (1982); 2. Brooks *et al.* (1982); 3. Buryak *et al.* (1972); 4. Crocket *et al.* (1983); 5. Crocket *et al.* (1986); 6. Gavrilenko *et al.* (1974); 7. Golovnya (1974); 8. Li and Shokhina (1974); 9. Moiseenko *et al.* (1971); 10. Moravek and Pouba (1984); 11. Mushkin *et al.* (1974); 12. Pchelintseva and Fel'dman (1973); 13. Petrov *et al.* (1972); 14. Sighinolfi and Santos (1976); 15. Thorpe and Thomas (1976).
Superscripts: 1. Omissions from averages: 1a, two slates – 37 and 161 ppb Au, the latter with arsenopyrite; 1b, two greywackes – 57 and 80 ppb; 1c, two slates – 57 and 102 ppb.

References

Agiorgitis, G. and Becker, R.(1979) The geochemical distribution of gold in some rocks and minerals of the Troodos complex, Cyprus. *Neues Jahrb. Mineral. Monatsh.* **7**, 316–320.

Anders, E. and Grevesse, N. (1989) Abundance of the elements: meteoritic and solar. *Geochim. Cosmochim. Acta* **53**, 197–214.

Anhaeusser, C.R., Fritze, K., Fyfe, W.S. and Gill, R.C.O. (1975) Gold in 'primitive' Archaean Volcanics. *Chem. Geol.* **16**, 129–135.

Anoshin, G.N. and Kepezhinskas, V.V. (1972) Petrochemical features related to gold distribution for the Cenozoic volcanic rocks of the Kuril-Kamchatka province. *Geochem. Internat.* **9**, 618–629.

Anoshin, G.N. and Potap'yev, V.V. (1966) Gold in granites of the Kolyvan' (Altai) and Khangilay-Shilinskiy (Transbaikalia) massifs (according to radioactivation analysis data). *Geochem. Internat.* **3**, 850–854.

Anoshin, G.N., Zolotarev, B.P. and Tsimbalist, V.G. (1971). Gold in rocks of tholeiite-basalt formations. *Akad. Nauk. SSSR Dokl.* **201**, 700–703.

Anoshin, G.N., Yemel'yanov, Ye.M. and Perezhogin, G.A. (1969) Gold in the modern sediments of the northern part of the Atlantic Ocean basin. *Geochem. Internat.* **6**, 879–905.

Anoshin, G.N., Lepezin, G.G., Mel'gunov, S.V., Miriyevskaya, O.S. and Tsimbalist, V.G. (1982) Behavior of gold and silver during progressive metamorphism, as illustrated by metamorphic complexes of the Altai-Sayan fold system. *Dokl. Akad. Sci. USSR, Earth Science Section* **262**, 210-214.

Barnes, S.J., Naldrett, A.J. and Gorton, M.P. (1985) The origin of the fractionation of platinum group elements in terrestrial magmas. *Chem. Geol.* 53, 303-323.

Basaltic Volcanism Study Project (1981) *Basaltic Volcanism on the Terrestrial Planets*, Chapter 1. Pergamon Press, New York. 1286pp.

Bavinton, O.A. and Keays, R.R. (1978) Precious metal values from interflow sedimentary rocks from the komatiite sequence at Kambalda, Western Australia. *Geochim. Cosmochim. Acta* **42**, 1151-1164.

Boyle, R.W. (1979) The geochemistry of gold and its deposits. Geol. Surv. of Can., Bull. 280. 584pp.

Brooks, R.R., Chaterjee, A.K., Smith, P.K., Ryan, D.E. and Zhang, H.F. (1982) The distribution of gold in rocks and minerals of the Meguma Group in Nova Scotia. Can. *Chem. Geol.* **35**, 87-95.

Brugmann, E., Arndt, N.T., Hofmann, A.W. and Tobschall, H.J. (1987) Noble metal abundances in komatiite suites from Alexo, Ontario, and Gorgona Island, Columbia. *Geochim. Cosmochim. Acta* **51**, 2159-2170.

Buisson, G. and Leblanc, M. (1987) Gold in mantle peridotites from Upper Paleozoic ophiolites in Arabia, Mali, and Morocco. *Econ. Geol.* **82**, 2091-2097.

Buryak, V.A., Sizykh, Yu.I., Chernenko, A.L., Rusina, L.D., Klyachina, R.M., Volynets, T.P. and Yelizar'yeva, T.I. (1972) Behavior of gold during granitization and pegmatization in the Vitim-Patom highlands. *Dokl. Acad. Sci. USSR, Earth Science Section* **207**, 230-233.

Carlson, R.W., Lugmair, G.W. and Macdougall, J.D. (1981) Columbia River volcanism: the question of mantle heterogeneity or crustal contamination. *Geochim. Cosmochim. Acta* **45**, 2483- 2500.

Chapman, D.G. and Groves, D.I. (1979) A preliminary study of the distribution of gold in Archean interflow sulphidic metasediments, Yilgarn Block, Western Australia. In *Gold Mineralization*, eds. Glover, J.E. and Groves, D.I. Publication No. 3, Geol. Dept. Univ. Extension, Univ. of W. Australia, 76-88.

Chou, C.L., Shaw, D.M. and Crocket, J.H. (1983) Siderophile trace elements in the earth's oceanic crust and upper mantle. *J. Geophys. Res.* **88**, A507-A518.

Colvine, A.C., Fyon, J.A., Heather, K.B., Marmont, S., Smith, P.M. and Troop, D.G. (1988) Archean lode gold deposits in Ontario. Ontario Geol. Surv. Misc. Paper 139. 136pp.

Crocket, J.H. (1981) Geochemistry of platinum group elements. In L.J. Cabri (ed.), *Platinum-Group Elements: Mineralogy, Geology, Recovery*. Can. Inst. of Mining and Metallurgy, Special Vol. 23, 47-64.

Crocket, J.H. (1978) Gold. In *Handbook of Geochemistry*, ed. Wedepohl, K.H. Vol. II/5, Chapter 79, Sections B–O.

Crocket, J.H. and Chyi, L.L. (1972) Abundances of Pd, Ir, Os and Au in an alpine ultramafic pluton. In: *Proc. of 24th Internat. Geological Congress*, Section 10, 202-209.

Crocket, J.H. and Kabir, A. (1988) PGE in Hawaiian basalt: implications of hydrothermal alteration on PGE mobility in volcanic fluids. In *Geo-Platinum 87*, eds. Prichard, H.M., Potts, P.J., Bowles, J.F.W., and Cribb, S.J. Elsevier, Barking, 259.

Crocket, J.H. and Kuo, H.Y. (1979) Sources of gold, palladium and iridium in deep-sea sediments. *Geochim. Cosmochim. Acta* **43**, 831-842.

Crocket, J.H. and MacRae, W.E. (1986) Platinum-group element distribution in komatiitic and tholeiitic volcanic rock from Munro Township, Ontario. *Econ. Geol.* **81**, 1242-1251.

Crocket, J.H. and Teruta, Y. (1977) Palladium, iridium and gold contents of mafic and ultramafic rocks drilled from the Mid- Atlantic Ridge, Leg 37, Deep Sea Drilling Project. *Can. J. Earth Sci.* **14**, 777-784.

Crocket, J.H., Cowan, P. and Kusmirski, R.T.M. (1980) Gold content of volcanic-hosted interflow sedimentary rocks in the Red Lake area: implications on ore genesis at the Dickenson Mine. In *Genesis of Archean, Volcanic-Hosted Gold Deposits*, ed. Roberts, R.G., Ontario Geol. Surv. Open-File Report 5293, 66-91.

Crocket, J.H., Macdougall, J.D. and Harriss, R.C. (1973) Gold, palladium and iridium in marine sediments. *Geochim. Cosmochim. Acta* **37**, 2547-2556.

Crocket, J.H., Fueten, F., Kabir, A. and Clifford, P.M. (1986) Distribution and localization of gold in Meguma Group rocks, Nova Scotia: implications of metal distribution patterns in quartz veins and host rocks on mineralization processes at Harrigan Cove, Halifax County. *Maritime Sed. and Atlantic Geol.* **22**, 15-33.

Crocket, J.H., Clifford, P.M., Fueten, F. and Kabir, A. (1983) Distribution and localization of gold in Meguma Group rocks, Nova Scotia; Part 2: Implications of background geochemistry and cleavage development – a preliminary report. In *Current Research, Part B*. Geol. Surv. of Can., Paper 83-1B, 285-290.

Crocket, J.H., Blum, N., Hurley, T.D., Bowins, R.J., McRoberts, G.D., Fyon, J.A., McNutt, R.H. and Schwarcz, H.P. (1984) Geological and geochemical studies of the Boston and Temagami iron formations and their contiguous volcanosedimentary piles. In ed. Milne, V.G. Ontario Geol. Surv., Misc. Paper 121, 72-83.

Davies, G. and Tredoux, M. (1985) The platinum-group elements and gold contents of the marginal rocks and sills of the Bushveld Complex. *Econ. Geol.* **80**, 838-848.

Daveltov, I.K. and Dzhakshibayev, Sh. (1970) The mineral balance and the behaviour of gold during the emplacement of an intrusive body. *Geochem. Internat.* **7**, 997-1005.

Dostal, J. and Dupuy, C. (1987) Gold in Late Proterozoic andesites from northwestern Africa. *Econ. Geol.* **82**, 762-766.

Dostal, J., Dupuy, C. and Poidevin, J.L. (1985) Geochemistry of Precambrian basaltic rocks from the Central African Republic (Equatorial Africa). *Can. J. Earth Sci.* **22**, 653-662

Dupuy, C., Dostal, J. and Leblanc, M. (1981) Distribution of copper and gold in ophiolites from New Caledonia. *Can. Mineral.* **19**, 225-232.

Ehmann, W.D., Baedecker, P.A. and McKowan, D.M. (1970) Gold and iridium in meteorites and some selected rocks. *Geochim. Cosmochim. Acta* **34**, 493-507.

Fominykh, V.G. and Znamenskiy, N.D. (1974) Distribution of platinum-group metals and gold in magnetite from genetically different Ural granitoids. *Dokl. Acad. Sci. USSR, Earth Science Section* **217**, 216-217.

Fyon, J.A. and Crocket, J.H. (1982) Gold exploration in the Timmins district using field and lithogeochemical characteristics of carbonate alteration zones. In *Geology of Canadian Gold Deposits*, eds. R.W. Hodder and W. Petruk. Can. Inst. Mining and Metallurgy, Special Vol. 24, 113-129.

Fyon, J.A., Crocket, J.H. and Schwarcz, H.P. (1984) The Carshaw and Malga iron-formation-hosted gold deposits of the Timmins area. In *The Geology of Gold in Ontario*, ed. Colvine, A.C. Ontario Geol. Surv., Misc. Paper 110, 98-110.

Gavrilenko, B.V., Bolotov, V.I. and Zhangurov, A.A. (1982) The distribution of gold in magmatic differentiation products. *Geochem. Internat.* **19**, No.1, 23-34.

Gavrilenko, B.V., Belolipetskiy, A.P., Bolotov, V.I. and Vorobets, A.A. (1974) Geochemistry of gold in the Kolmozero- Voron'ya metamorphic complex (Kola Peninsula). *Geochem. Internat.* **11**, 111-117.

Goodwin, A.M., Thode, H.G., Chou, C.-L. and Karkhansis, S.N. (1985) Chemostratigraphy and origin of the late Archean siderite- pyrite rich Helen Iron Formation, Michipicoten belt, Canada. *Can. J. Earth Sci.* **22**, 72-84.

Goodwin, A.M. (1984) Archaean greenstone belts and gold mineralization, Superior Province, Canada. In *Gold'82*, ed. Foster, R.P. Geol. Soc. of Zimbabwe Special Pub. No.1. Balkema, Rotterdam, 71-98.

Gottfried, D. and Greenland, L.P. (1972) Variation of iridium and gold in oceanic and continental basalts. *Proc. 24th Internat. l. Geol. Congress*, Section 10, 135-144.

Gottfried, D., Rowe, J.J. and Tilling, R.I. (1972) Distribution of gold in igneous rocks. US Geol. Surv., Prof. Paper 727. 42pp.

Grabezhev, A.I., Levitan, G.M., Vigorova, V.G., Chashchukhina, V.A., Neykur, T.A. and Yershova, N.A. (1986) Gold in orogenic granitoids of the Ural eugeosyncline. *Internat. Geol. Rev.* **28**, 449-460.

Gross, G.A. (1988) Gold content and geochemistry of iron- formation in Canada. Geol. Surv. of Can., Paper, 86-19. 54pp.

Groves, D.I. and Phillips, G.N. (1987) The genesis and tectonic control on Archean gold deposits of the Western Australian shield – a metamorphic replacement model. *Ore Geol. Rev.* **2**, 287-322.

Groves, D.I. and Keays, R.R. (1979) Mobilization of ore-forming elements during alteration of dunites Mt. Keith – Betheno, Western Australia. *Can. Mineral.* **17**, 373-390.

Groves, D.I., Phillips, G.N., Falconer, L.J., Houstoun, S.M. Ho., S.E., Browning, P., Dahl, N. and McNaughton, J. (1987) Evidence for an epigenetic origin for BIF-hosted gold deposits in greenstone belts of the Yilgarn Block, Western Australia. In *Recent Advances in Understanding Precambrian Gold Deposits*, eds. Ho S.E. and Groves D.I. Publication no.11, Geology Dept. Univ. Extension, Univ. of W. Australia, 167-180.

Hamlyn, P.R., Keays, R.R. Cameron, W.E. Crawford, A.J. and Waldron, H.M. (1985) Precious metals in magnesian low-Ti lavas: implications for metallogenesis and sulfur saturation in primary magmas. *Geochim. Cosmochim. Acta* **49**, 1797-1812.

Hertogen, J., Janssens, M.-J. and Palme, H. (1980) Trace elements in ocean ridge basalt glasses: implications for fractionations during mantle evolution and petrogenesis. *Geochim. Cosmochim. Acta.* **44**, 2125-2144.

Huot, D., Sattran, V. and Zida, P. (1987) Gold in Birrimian greenstone belts of Burkina Faso, West Africa. *Econ. Geol.* **82**, 2033-2044.

Ivanov, V.V., Ilupin, I.P., Starozhitskaya, M.I. and Gross. Yu.I. (1977) Distribution of Cu, Ag, and Au in Yakutia plutonic xenoliths and kimberlites. *Geochem. Internat.* **14**, No. 4, 48-60.

Ivensen, Yu P., Amuzinskiy, V.A., Korobitsyn, A.V., Kukhtinskiy, G.G. and Trunilina, V.A. (1974) Gold distribution in igneous rocks. *Dokl. Acad. Sci. USSR, Earth Science Section* **216**, 242-244.

Jagoutz, E., Palme, H., Baddenhausen, H., Blum, K., Cendales, M., Dreibus, G., Spettel, B., Lorenz, V. and Wanke, H. (1979) The abundances of major, minor and trace elements in the earth's mantle as derived from primitive ultramafic nodules. *Proc. of 10th Lunar and Planetary Science Conf.*, 2031-2050.

Karelin, V.V., Maksimov, Ye.P., Nikitin, V.N. and Uyutov, V.I. (1974) Gold in Mesozoic igneous rocks in the eastern part of the Aldan Shield. *Geochem. Internat.* **11**, 399-405.

Keays, R.R. (1984) Archaean gold deposits and their source rocks: The upper mantle connection. In *Gold '82*, ed. Foster, R.P. Geol. Soc. of Zimbabwe, Special Pub. No.1. Balkema, Rotterdam, 17-52.

Keays, R.R. and Campbell, I.H. (1981) Precious metals in the Jimberlana Intrusion, Western Australia: implications for the genesis of platiniferous ores in layered intrusions. *Econ. Geol.* **76**, 1118-1141.

Keays, R.R. and Davison, R.M. (1976) Palladium, iridium and gold in the ores and host rocks of nickel sulfide deposits in Western Australia. *Econ. Geol.* **71**, 1214-1228.

Keays, R.R. and Scott, R.B. (1976) Precious metals in ocean ridge basalts: implications for basalts as source rocks for gold mineralization. *Econ. Geol.* **71**, 705-720.

Keays, R.R., Ross, J.R. and Woolrich, P. (1981) Precious metals in volcanic peridotite-associated nickel sulfide deposits in Western Australia. II: Distribution within the ores and host rocks in Kambalda. *Econ. Geol.* **76**, 1645-1674.

Keays, R.R., Nickel, E.H., Groves, D. McGoldrick, P.J. (1982) Iridium and palladium as discriminants of volcanic exhalative, hydrothermal, and magmatic nickel sulphide mineralization. *Econ. Geol.* **77**, 1535-1547.

Kerrich, R. and Fyfe, W.S. (1981) The gold-carbonate association: source of CO_2, and CO_2 fixation reactions in Archaean lode deposits. *Chem. Geol.* **33**, 265-294.

Kerrich, R. (1983) *Geochemistry of Gold Deposits in the Abitibi Greenstone Belt.* Can. Inst. Mining and Metallurgy, Special Vol. 27. 75pp.

Khitrunov, A.T. and Mel'tser, M.L. (1979) Geochemistry of gold in the granitoid process. *Geochem. Internat.* **16**, No.5, 37-42.

Korobeynikov, A.F. (1986) Gold distributions in black shale associations. *Geochem. Internat.* **23**, 114-124.

Korobeynikov, A.F. (1986) The gold distribution in differentiation products of basic and acid magmas of various ages. *Geochem. Internat.* **23**, No.7, 101-110.

Korobeynikov, A.F. (1982) Gold in pyroxenes in intrusive and contact-metasomatic rocks. *Geochem. Internat.* **19**, No.2, 13-24.

Korobeynikov, A.F. (1981) Gold in feldspars of intrusive and metasomatic rocks. *Geochem. Internat.* **18**, 130-152.

Korobeynikov, A.F. (1981) Gold fractionation in crystallizing magma. *Dokl. Acad. Sci. USSR, Earth Science Section* **258**, 193-196.

Korobeynikov, A.F. (1980) Gold in accessory minerals of intrusive rocks. *Geochem. Internat.* **17**, No.4, 112-125.

Korobeynikov, A.F. (1976) Geochemical behavior of gold in contact-metasomatic processes in granitoid intrusions. *Dokl. Acad. Sci. USSR, Earth Science Section* **227**, 213- 216.

Korobeynikov, A.F. and Goncharenko, A.I. (1986) Gold in ophiolite complexes in the Altai Sayan folded region. *Geochem. Internat.* **23**, No.6, 43-54.

Kwong, Y.T.J. and Crocket, J.H. (1978) Background and anomalous gold in rocks of an Archean greenstone assemblage, Kakagi Lake area, northwestern Ontario. *Econ. Geol.* **73**, 50-63.

Laul, J.C., Keays, R.R., Ganapathy, R. Anders, E. and Morgan, J.W. (1972) Chemical fractionations in meteorites – V. Volatile and siderophile elements in achondrites and ocean ridge basalts. *Geochim. Cosmochim. Acta* **36**, 329-346.

Lee, C.A. (1983) Trace and platinum-group element geochemistry and the development of the Merensky Unit of the Western Bushveld Complex. *Mineral. Deposita* **18**, 173-190.

Lee, C.A. and Fesq, H.W. (1986) Au, Ir, Ni and Co in some chromitites of the eastern Bushveld Complex South Africa. *Chem. Geol.* **62**, 227-237.

Lee, C.A. and Parry, S.J. (1988) Platinum-group element geochemistry of the Lower and Middle Group chromitites of the eastern Bushveld Complex. *Econ. Geol.* **83**, 1127-1139.

Lee, C.A. and Tredoux, M. (1986) Platinum-group element abundances in the Lower and the Lower Critical Zones of the Eastern Bushveld Complex. *Econ. Geol.* **81**, 1087-1095.

Li, L.V. and Datsenko, V.M. (1973) The gold distribution in the granitoids of Proterozoic intrusions in the south Yenisey ridge. *Geochem. Internat.* **10**, 649-655.

Li, L.V. and Kornev, T.Ya. (1972) Gold content in the ultramafic rocks of the Yenisey Range. *Geochem. Internat.* **9**, 949-951.

Li, L.V. and Shokhina, O.I. (1974) Behavior of gold in progressive regional metamorphism of Precambrian rocks on the Yenisey range. *Geochem. Internat.* **11**, 314-321.

Mantei, E., Bolter, E. and Al Shaieb, Z. (1970) Distribution of gold, silver, copper, lead and zinc in the productive Marysville stock, Montana. *Mineral. Deposita* **5**, 184-190.

Mathez, E.A. (1976) Sulfur solubility and magmatic sulfides in submarine basalt glass. *J. Geophys. Res.* **81**, 4249- 4275.

McCarthy, T.S., Lee, C.A., Fesq, H.W., Kable, E.J.D. and Erasmus, C.S. (1984) Sulphur saturation in the lower and critical zones of the Eastern Bushveld Complex. *Geochim. Cosmochim. Acta* **48**, 1005-1019.

Meyer, M. and Saager, R. (1985) The gold content of some Archaean rocks and their possible relationship to epigenetic gold-quartz vein deposits. *Mineral. Deposita* **20**, 284-289.

Mints, M.V. (1975) Gold in igneous rocks of southwestern part of Ul'insk superposed downwarp (Okhotsk-Chukotka volcanic belt). *Internat. Geol. Rev.* **17**, No.5, 604.

Mitchell, R.H. and Keays, R.R. (1981) Abundance and distribution of gold, palladium and iridium in some spinel and garnet lherzolites: implications for the nature and origin of precious metal-rich intergranular components in the upper mantle. *Geochim. Cosmochim. Acta* **45**, 2425-2442.

Moiseenko, V.G., Shcheka, S.A., Fat'yanov, I.I. and Ivanov, V.S. (1971) *The Geochemical Peculiarities of the Distribution of Gold in the Rocks of the Pacific Belt.* Izd. 'Nauka' Publishing House, Moscow. 207pp.

Moravek, P. and Pouba, Z. (1984) Gold mineralization and granitoids in the Bohemian Massif, Czechoslovakia. In *Gold '82*, ed. Foster, R.P. Geol. Soc. of Zimbabwe Special Pub. No.1. Balkema, Rotterdam, 713-729.

Moravek, P. and Pouba, Z. (1987) Precambrian and Phanerozoic history of gold mineralization in the Bohemian Massif. *Econ. Geol.* **82**, 2098-2114.

Morgan, J.W., Ganapathy, R., Higuchi, H. and Krahenbuhl, U. (1976) Volatile and siderophile trace elements in anorthositic rocks from Fiskenaesset, West Greenland: comparison with lunar and meteoritic analogues. *Geochim. Cosmochim. Acta* **40**, 861-888.

Morgan, J.W., Wandless, G.A., Petrie, R.K. and Irving, A.J. (1981) Composition of the earth's upper mantle – I. Siderophile trace elements in ultramafic nodules. *Tectonophys.* **75**, 47-67.

Mushkin, I.V. and Yaroslavskiy, R.I. (1974) Gold in the alkali basaltoids of southern Tien Shan and in their xenoliths of mantle material. *Geochem. Internat.* **11**, 718-721.

Mushkin, I.V., Yaroslavskiy, R.I. and Kotsyuba, A.S. (1974) Behavior of gold in eclogitization of gneisses of the granulite facies as seen in xenoliths in alkali basaltoid dikes in the southern Tien Shan. *Geochem. Internat.* **11**, 813-819.

Nesbitt, B.E., St. Louis, R.M. and Muehlenbachs, K. (1987) Distribution of gold in altered basalts of DSDP hole 504B. *Can. J. Earth Sci.* **24**, 201-209.

Nikitin, N.M. and Yasyrev, A.P. (1974) Mean gold contents of sedimentary rocks. *Geochem. Internat.* **11**, 855-857.

Oshin, I.O. and Crocket, J.H. (1986) Noble metals in Thetford Mines ophiolites, Quebec, Canada. Part II: Distribution of gold, silver, iridium, platinum and palladium in the Lac de l'Est volcano-sedimentary section. *Econ. Geol.* **81**, 931- 945.

Oshin, I.O. and Crocket, J.H. (1982) Noble metals in Thetford Mines ophiolites, Quebece, Canada. Part I: Distribution of gold, iridium, platinum, and palladium in the ultra-mafic and gabbroic rocks. *Econ. Geol.* **77**, 1556-1570.

Pchelintseva, N.F. and Fel'dman, V.I. (1973) Gold in metamorphic rocks in the Kokchetav uplift. *Geochem. Internat.* **10**, 1357-1365.

Peach, C.L. and Mathez, E.A. (1988) Gold and iridium in sulphides from submarine basalt glasses. In *Geo-Platinum 87*, eds. Prichard, H.M., Potts, P.J., Bowles, J.F.W. and Cribb, S.J., Elsevier, Barking, 409-410.

Petrov, B.V., Krendelev, F.P., Bobrov, V.A., Tsimbalist, V.G. (1972) Behavior of radioelements and gold during metamorphism of sedimentary rocks in the Patom Uplands. *Geochem. Internat.* 647-655.

Phillips, G.N. (1987) Anomalous gold in Witwatersrand shales. *Econ. Geol.* **82**, 2179-2186.

Piper, D.Z. and Graef, P.A. (1974) Gold and rare-earth elements in sediments from the East Pacific Rise. *Marine Geol.* **17**, 287-297.

Polikarpochkin, V.V. and Korotayeva, I.Ya. (1976) Distribution of gold in the sedimentary rocks of the Undino-Dainsk basin (eastern Transbaykaliya). *Geochem. Internat.* **13**, No. 2, 57- 62.

Popov, V.S. (1975) Geochemistry of gold in the Upper Jurassic evaporite formation on the southern part of Soviet Central Asia. *Dokl. Acad. Sci. USSR, Earth Science Section* **224**, 204-207.

Ringwood, A.E. and Kesson, S.E. (1977) Basaltic magmatism and the bulk composition of the moon II. Siderophile and volatile elements in Moon, Earth and Chondrites: Implications for Lunar origin. *The Moon* **16**, 425-464.

Ross, J.R. and Keays, R.R. (1979) Precious metals in volcanic- type nickel sulfide deposits in Western Australia.I. Relationship with the composition of the ores and their host rocks. *Can. Mineral.* **17**, 417-436.

Rozhkov, I.S., Kaminskiy, F.V. and Frantsesson, Ye. V. (1973) Gold in kimberlites and ultrabasic inclusions. *Geochem. Internat.* **10**, 1385-1389.

Saager, R. and Meyer, M. (1984) Gold distribution in Archaean granitoids and supracrustal rocks from southern Africa: a comparison. In *Gold '82*, ed. Foster, R.P. Geol. Soc. of Zimbabwe, Special Pub. No.1. Balkema, Rotterdam, 53-70.

Saager, R., Meyer, M. and Muff, R. (1982) Gold distribution in supracrustal rocks from Archean greenstone belts of southern Africa and from Paleozoic ultramafic complexes of the European Alps: metallogenic and geochemical implications. *Econ. Geol.* **77**, 1-24.

Shaw, D.M., Dostal, J. and Keays, R.R. (1976) Additional estimates of continental surface Precambrian shield composition in Canada. *Geochim. Cosmochim. Acta* **40**, 73-84.

Shcherbakov, Yu.G. (1967) *Distribution and Concentration Conditions of Gold in Ore Provinces.* Nauka Press, Moscow. 268pp.

Shcherbakov, Yu.G. and Perezhogin, G.A. (1964) Geochemistry of gold. *Geochem. Internat.* **6**, 489-496, 1964.

Shilin, N.L. (1980) Mineral balance of gold during crystallization of an intrusion under hypabyssal conditions (Central Kamchatka). *Internat. Geol. Rev.* **22**, 145-153.

Shilin, N.L. (1968) Gold in rocks of Central Kamchatka. *Izv. Akad. Nauk SSSR Ser. Geol.* **11**, 69-76.

Shilin, N.L. and Osipov, B.V. (1978) Gold in biotite from central Kamchatka granitoids. *Geochem. Internat.* **15**, 144-155.

Sighinolfi, G.P. and Santos, A.M. (1976) Geochemistry of gold in Archean granulite facies terrains. *Chem. Geol.* **17**, 113-123.

Sighinolfi, G.P. and Gorgoni, C. (1977) Gold distribution in the Ivrea-Verbano gabbroic complex. *Chem. Geol.* **20**, 99-107.

Sinclair, W.D. (1982) Gold deposits of the Matachewan area of Ontario. In *Geology of Canadian Gold Deposits*, eds. Hodder, R.W. and Petruk, W. Can. Inst. of Mining and Metallurgy, Special Vol. 24, 83-93.

St. Louis, R.M. Nesbitt, B.E. and Morton, R.D. (1986) Geochemistry of platinum-group elements in the Tulameen ultramafic complex, southern British Columbia. *Econ. Geol.* **81**, 961-973.

Stockman, H.W. (1982) Noble metals in the Ronda and Josephine peridotites. Ph.D. thesis, Massachusetts Institute of Technology, Cambridge. 236pp.

Sun, S.-S. (1982) Chemical composition and origin of the earth's primitive mantle. *Geochim. Cosmochim. Acta* **46**, 179-192.

Sveshnikova, Ye.V., Andreyeva, Ye.D., Yeremeyev, N.V. and Latysh, I.K. (1978) Clarke and anomalous gold and silver contents of alkaline rocks. *Internat. Geol. Rev.* **20**, No.5, 587- 592.

Thorpe, R.J. and Thomas, G.M. (1976) Gold content of greywacke and slate of the Goldenville Formation, Nova Scotia, as determined by neutron activation analysis. In *Reports of Activities, Part A*. Geological Survey of Canada Paper 76-1A, 314-328.

Tilling, R.I., Gottfried, D. and Rowe, J. (1973) Gold abundance in igneous rocks: bearing on gold mineralization. *Econ. Geol.* **68**, 168-186.

Ukhanov, A.V. and Pchelintseva, N.F. (1972) Gold contents of peridotite and eclogite xenoliths from a kimberlite pipe. *Geochem. Internat.* **19**, 157.

Uvarov, V.V., D'yachkov, B.A., Burmistrov, V.R., Stepanov, V.M. and Madiyanov, T. (1972) Geochemistry of gold in the magmatic complexes of the Kalba-Narymsk region (eastern Kazakstan). *Geochem. Internat.* **9**, 774-776.

Viljoen, R.P., Saager, R. and Viljoen, M.J. (1970) Some thoughts on the origin and processes responsible for the concentration of gold in the Early Precambrian of southern Africa. *Mineral. Deposita* **5**, 164-180.

Vincent, E.A. and Crocket, J.H. (1960a) Studies in the geochemistry of gold – I. The distribution of gold in rocks and minerals of the Skaergaard Intrusion, East Greenland. *Geochim. Cosmochim. Acta* **18**, 130-142.

Vincent, E.A. and Crocket, J.H. (1960b) Studies in the geochemistry of gold – II. The gold content of some basic and ultrabasic rocks and stone meteorites. *Geochim. Cosmochim. Acta* **18**, 143-148.

Voskresenskaya, N.T. and Zvereva, N. (1968) Geochemistry of gold and its distribution in the magmatic complexes of northern Kazakhstan. *Geochem. Internat.* **5**, 373-381.

Voskresenskaya, N.T., Zvereva, N.F., Konkina, O.M. and Fel'dman, V.I. (1970) Behaviour of gold in the process of basic magma differentiation. *Geokhimiya* **12**, 1438-1445.

Wasson, J.T. and Baedecker, P.A. (1970) Ga, Ge, In, Ir and Au in lunar, terrestrial and meteoritic basalts. *Proc. of Apollo 11 Lunar Science Conf.* **2**, 1741-1750.

Yudin, S.S., Yudina, V.N. and Shilin, N.L. (1972) Gold concentrations in volcanic series in the central part of the Okhotsk volcanic belt. *Dokl. Acad. Sci. USSR, Earth Science Section* **207**, 39-441.

Zentilli, M., Brooks, R.R., Helgason, J., Ryan, D.E. and Zhang, H, (1985) The distribution of gold in volcanic rocks of eastern Iceland. *Chem. Geol.* **48**, 17-28.

Zlobin, V.A., Ponomareva, A.P., Ponomarchuk, V.A. and Mel'nikova, R.D. (1981) Geochemical properties of biotite from gold- and rare metal-bearing granitoid rocks of the Kalba range as indicators of metallogenic specialization of igneous complexes. *Dokl. Akad. Sci. USSR, Earth Science Section* **259**, 180-182.

Zvereva, N.F. and Gavrilenko, B.V. (1971) Gold in the rock- forming minerals of the Rykkuduk intrusive complex (Northern Kazakhstan). *Geochem. Internat.* **8**, 76-80.

2 The hydrothermal geochemistry of gold

T.M. SEWARD

2.1 Introduction

Hydrothermal ore-forming fluids are multicomponent aqueous electrolyte solutions which transport gold to an ore-depositing environment somewhere in the Earth's crust. Deposition of elemental gold and gold-containing minerals occurs in response to changes in the chemical and physical environment through which the hydrothermal ore fluid is migrating. For example, a buoyantly ascending hydrothermal fluid may encounter a permeable zone which induces volatile phase separation and/or boiling, or a deep fluid may encounter and mix with a cooler, steam-heated water of lower pH. Whether or not gold precipitates in these situations has to do with the changing stability of the gold-containing complex ions in response to these new conditions. Hydrothermal gold deposition may, therefore, take place over a wide range of temperature, pressure and fluid composition comprising diverse environments extending from amphibolite facies metamorphism to lower temperature epithermal mineralization and sea floor massive sulphides. It is thus the purpose of this chapter to consider our current state of knowledge concerning the high-temperature, high-pressure aqueous chemistry of gold and its relevance to gold ore formation.

2.2 The inorganic chemistry of gold

2.2.1 *Oxidation states*

Gold has the electronic configuration, Xe $4f^{14}5d^{10}6s^{1}$, and is unique in the group Ib elements (i.e. Cu, Ag, Au) in that it forms a number of high oxidation states (Table 2.1). Its aqueous chemistry is in fact dominated by the Au(III) oxidation state and to a lesser extent by Au(I). However, other oxidation states have been reported (Puddephatt, 1978). Organo Au(II) complexes have been studied in which the electron donors are sulphur-donor ligands (e.g. dithiocarbamate) or in which organic ligands coordinate Au(II) in a complex having a dimeric gold–gold bond.

The higher oxidation states are rather more rare. The only known Au(IV) complex is the poorly characterized dinitrosylgold (IV) hexafluoride, $(NO)_2AuF_6$ (Sunder *et al.*, 1979). A few Au(V) complexes have been reported with the first synthesis of an Au(V) compound, $CsAuF_6$, having been reported by Leary and Bartlett (1972). Complexes containing gold in the +6 oxidation state have not been reported but recently Timakov *et al.* (1986) claimed to have synthesized the compound AuF_7, in which gold is present as Au(VII).

Table 2.1 The oxidation states of gold

Oxidation state	Examples of compounds
+1	$Au(CN)_2^-$; $Au(HS)_2^-$
+2	*Au $[R_2(NCS_2)]_2$
+3	$AuCl_4^-$
+4	$(NO)_2AuF_6$
+5	AuF_5 ; $CsAuF_6$
+7	AuF_7

*R = alkyl group such as C_2H_5 or iso–C_3H_7.

2.2.2 *Relativistic effects*

Effects which are a direct consequence of relativity play a significant role in the chemistry of heavy elements such as gold (see, for example, Pyykkö, 1988). Because electrons in s orbitals and, to a lesser extent, p orbitals spend much of their time in the vicinity of the nucleus (i.e. the largest radial densities are near the nucleus), they have average velocities approaching that of light. Hence they have substantially increased mass and are even more attracted to the nucleus. This leads to a contraction of s (and p) shells up to the valence shell as a result of more strongly bound, relativistically stabilised, s (and p) electrons. The electrons in the contracted s atomic orbitals screen the nucleus more effectively and hence the d and f orbitals expand radially. Their electrons are thus more loosely bound and energetically destabilized.

These relativistic effects, together with lanthanide contraction (itself, relativistically assisted), are at a maximum for gold and fundamentally affect its chemistry. For example, the yellow colour of gold arises from relativistically modified 5d to Fermi level electronic transitions and, furthermore, relativistic bond contraction causes an increase in the density of metallic gold by 18% (Schwerdtfeger *et al.*, 1989). Non-relativistic gold would have the metallic white appearance of silver.

The aqueous chemistry of gold is thus strongly influenced by relativistic effects. The 'nobility' of gold (i.e. it is difficult to oxidize) arises to a considerable extent from the relativistically enhanced ionization potential due to the contraction of the 6s atomic orbital. Some insight into the extent to which these effects influence gold chemistry may be gained by inspection of bond lengths and dissociation energies for a number of gold species (Table 2.2, Figure 2.1).

Firstly, it is interesting to note that the bond lengths of all the gold(I) halide complexes given in Table 2.1 are relativistically contracted. Furthermore, the dissociation energies are decreased which indicates a reduction in the stability of the gold(I)–halide bond. The relativistic destabilization of the AuF bond is especially large. The electronegativities of the halide ligands are all greater than that of gold. However, in the case of AuH, the ligand electronegativity is less than that of gold and the dissociation energy is, therefore, increased by relativistic effects.

If one now considers the Au(III) halide complexes, it is observed (Table 2.2) that the relativistic bond contractions are smaller than for Au(I) complexes. This arises from the increased contribution of relativistically destabilized 5d electrons to the Au(III)–halide bond. In addition, the dissociation energies for the Au(III) halide complexes are appreciably increased by relativistically assisted d-orbital shielding which accounts for their unusually high thermodynamic stability. The existence of the other higher oxidation states of gold is also a direct consequence of relativistic shielding.

Table 2.2 Calculated relativistic (R) and non-relativistic (NR) internuclear (bond) distances, r (in 10^{-2} pm), an dissociation energies, D_e (in kJ/mol), for some Au(I) and Au(III) molecules

Molecule	r (10^{-2} pm)			D_e (kJ/mol)		
	R	NR	Δr	R	NR	ΔD_e
AuF	2.010	2.174	0.164	88.6	163.1	+ 74.5
AuCl	2.333	2.541	0.208	160.7	217.9	+ 57.2
AuBr	2.448	2.647	0.299	157.2	202.7	+ 45.5
AuI	2.616	2.824	0.208	152.4	191.4	+ 39.0
AuH	1.578	1.831	0.253	152.7	92.3	– 60.4
$AuCl_2^-$	2.398	2.630	0.232	416.1	428.0	+ 11.9
$AuCl_4^-$	2.344	2.424	0.080	496.4	327.9	–168.5
$AuBr_4^-$	2.492	2.549	0.057	396.0	292.9	–104.6
AuI_4^-	2.716	2.830	0.114	308.2	220.3	– 87.9

Note: 10^{-2} pm = 1 Å

Data from: Schwerdtfeger *et al.* (1989) and Schwerdtfeger (1989).

It is interesting that the non-relativistic (NR) dissociation energies indicate that $AuCl_2^-$ is more stable than $AuCl_4^-$. However, there is a large relativistic increase in D_e for $AuCl_4^-$ which leads to the stability sequence:

$$AuCl^0 << AuCl_2^- < AuCl_4$$

A similar stability sequence will exist for the hydrosulphide complexes of gold, i.e.

$$AuHS^0 < Au(HS)_2^- < Au(HS)_4^-$$

Data from Renders and Seward (1989a) (see Table 2.3) confirm that $AuHS^0$ is less stable than the dihydrosulphidogold(I) complex, $Au(HS)_2^-$. However, there are, at present, no experimentally determined thermodynamic data or theoretical calculations (e.g. relativistic Hartree-Fock) pertaining to the stability of $Au(HS)_4^-$.

2.2.3 *Coordination chemistry of Au(I)*

In hydrothermal systems in the Earth's crust, mineral solution equilibria give rise to ore-forming fluids which are characterized by low oxidation potentials (Seward, 1984; Henley *et al.*, 1984) such that the dominant oxidation state of dissolved gold is +1. Thus, it is the coordination chemistry of Au(I) with which we are concerned when considering the transport and deposition of gold in hydrothermal ore-forming systems.

Most Au(I) complexes are linear, triatomic molecules (e.g. $AuCl_2^-$) but higher coordination number compounds with ligands such as tertiary phosphines and arsines are known (Parish *et al.*, 1981). ^{197}Au Mössbauer spectroscopy has been particularly useful in the elucidation of three- and four-coordinate Au(I) complexes (Parish, 1984).

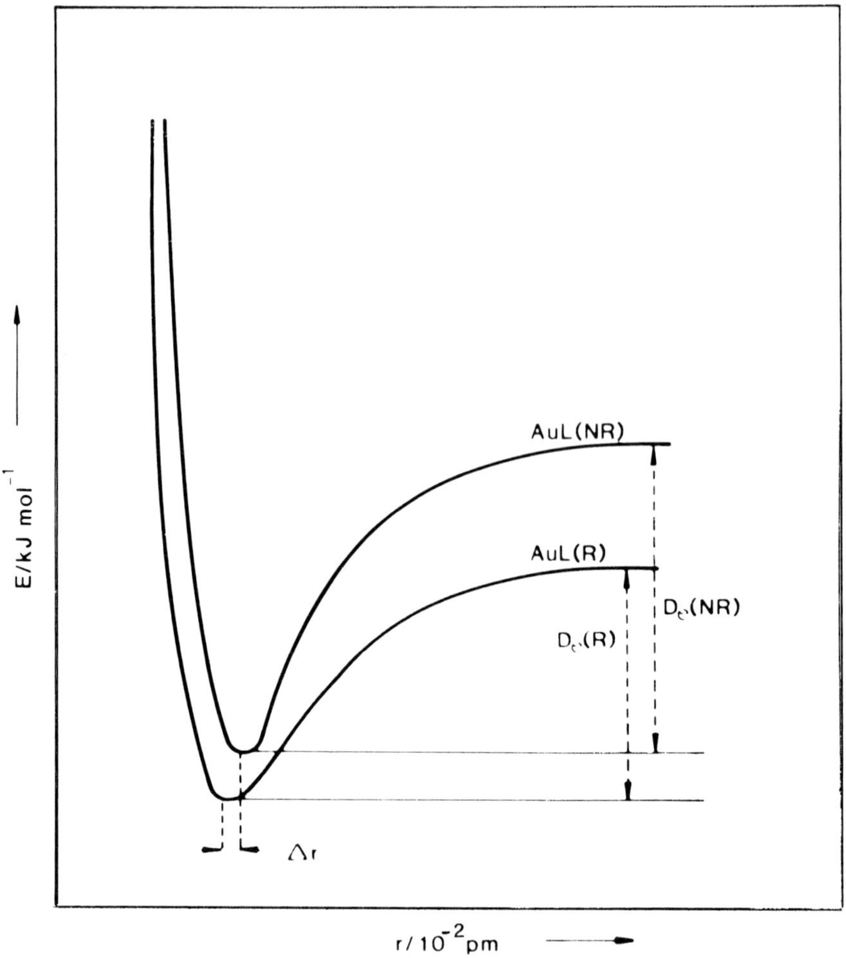

Figure 2.1 A schematic plot of energy (E) versus internuclear distance, r, for a simple diatomic gold(I) molecule such as AuCl or AuHS; D_e(NR) and D_e(R) are non-relativistic and relativistic dissociation energies as shown in Table 2.2; Δr and ΔD are the changes in internuclear distance and dissociation energies respectively, arising from relativistic effects.

The Au^+ ion with a filled d^{10} shell is the archetype class 'b' metal ion or 'soft' Lewis acid. It therefore prefers coordination with more polarizable, 'soft' Lewis bases or class 'b' electron donors. This is well illustrated by consideration of the stability of the simple triatomic halogen complexes (Table 2.3), the stabilities of which increase in the sequence,

$$F^- \ll Cl^- < Br^- < I^-$$

The interaction of Au^+ with the 'soft', polarizable iodide ion gives rise to the covalently bonded di-iodogold(I) complex, AuI_2^- which is about ten orders of magnitude more stable than the dichloridogold(I) species, $AuCl_2^-$. Interaction with the 'hard' fluoride ion to give the weak AuF_2^- is not favoured and the fluoride complex is unknown.

Table 2.3 Thermodynamic equilibrium formation constants for a selection of Au(I) complexes at 25°C

	log β		log β
AuI_2^-	19.0	$Au_2S_2^{2-}$	41.1
$AuBr_2^-$	12.4	$Au(HS)_2^-$	30.1
$AuCl_2^-$	9.2	$AuHS^0$	24.5
AuF_2^-	–	$Au(S_2O_3)_2^{3-}$	26.0
$Au(CN)_2^-$	38.7	$Au(H_2NCSNH_2)_2^+$	22.2
$Au(SCN)_2^-$	16.8	$Au(SO_4)_2^{3-}$	–

Data from: Jorgensen and Pouradier (1970), Peshchevitsky *et al.* (1970), Renders and Seward (1989a); the equilibrium constants are consistent with standard reduction potential given by Johnson *et al.* (1978) (i.e. E^0 – 1.695 V).

Au(I) complexes are generally not very stable in aqueous solutions and disproportionate to elemental gold and Au(III). For this reason, the inorganic co-ordination chemistry of Au^I complexes in aqueous media has not been extensively studied. Even the value of the standard reduction potential, E^0, has been under debate for many years. The first and only direct experimental determination of $E^0 = -1.695$ volt for

$$Au = Au^+ + e^{-1} \qquad (2.1)$$

was only reported relatively recently by Johnson *et al.* (1978).

The dichloridogold(I) complex, $AuCl_2^-$, is potentially of geochemical interest in so far as $AuCl_2^-$ has been considered to be of some importance in the transport of gold in hydrothermal ore-depositing systems. However, $AuCl_2^-$ decomposes in water at 25°C according to the reaction

$$3AuCl_2^- = 2Au + AuCl_4^- + 2Cl^- \qquad (2.2)$$

In the presence of excess chloride, $AuCl_2^-$ should become more stable but its existence in aqueous solutions at 25°C has been the subject of some debate. Braunstein and Clark (1973) synthesized crystalline tetraethylammonium and tetra-n-butylammonium dihalogenidogold(I) compounds which are soluble in methanol and behave as 1 : 1 electrolytes in this solvent. That is, they disproportionate to give methanolic solutions containing the simple dihalogenido complexes, e.g.

$$[Et_4N]\,[AuCl_2] = Et_4N^+ + AuCl_2^- \qquad (2.3)$$

where Et_4N^+ is the tetraethylammonium ion.

Thus, $AuCl_2^-$ exists as a stable species in methanol. This is an important observation because methanol is a protic, water-like solvent having a dielectric constant ($\varepsilon = 33.1$ at 25°C) which is similar to that of high temperature water at 210°C. ($\varepsilon = 33.11$; Uematsu and Franck, 1980). This suggests that $AuCl_2^-$ must occur as a stable, identifiable complex in high-temperature aqueous solutions, especially those containing excess chloride ligands. As will be discussed later, there are at present few if any experimental data which unambiguously verify the presence of $AuCl_2^-$ in hydrothermal solutions.

The preference of Au(I) for 'soft', class 'b' electron donors is further emphasized by the equilibrium formation constants for some of the simple Au(I) complexes with sulphur-containing ligands (Table 2.3). The simple sulphido and hydrosulphido complexes are very stable with the equilibrium formation constant for $Au(HS)_2^-$ being about twenty-one orders of magnitude greater than that of $AuCl_2^-$. Other sulphur-containing ligands in which the electron donor is the sulphur (e.g. thiosulphate and thiourea) also form quite stable complexes with Au(I). However, the 'hard' class 'a' sulphate ion in which the electron donors are oxygen atoms does not form stable complexes with Au(I) and there are no reported thermodynamic data pertaining to $Au(SO_4)_2^{3-}$ stability.

2.3 Gold complexing in hydrothermal solutions

2.3.1 *Which complexes are important?*

Hydrothermal ore-forming fluids are multicomponent electrolyte solutions which contain a number of species which may form stable complexes with gold(I). Table 2.4 gives some of the possible ligands, which, under differing conditions, may play an important role in hydrothermal gold transport.

Table 2.4 Ligands of potential importance in the complexing of gold(I) in natural hydrothermal fluids

Cl^-	Br^-	I^-		
HS^-	S^{2-}	S_n^{2-}	$S_2O_3^{2-}$	$S_nO_6^{2-}$
$A_3S_6^{3-}$,	$Sb_3S_6^{3-}$	Te_2^{2-}		
NH_3	OH^-			
CN^-	SCN^-			

In deciding which complex (or complexes) is most important in determining the mechanism of transport and deposition of gold, there is a 'trade-off' in terms of complex stability and ligand availability. For example, the gold(I) iodide complex, AuI_2^- is much more stable than $AuCl_2^-$ but would generally be unimportant in gold transport because of the low concentrations of iodide found in hydrothermal ore solutions.

Much consideration by numerous workers has been given to the role of chloride and reduced sulphur (e.g. HS^-) in hydrothermal gold chemistry because both chloride and hydrosulphide ions are known to form complexes with gold(I) and both are present in appreciable concentrations in hydrothermal ore fluids. For these reasons, almost all of the experimental work conducted at elevated temperatures and pressures has focused on gold solubility and complexing in chloride and reduced sulphur-containing solutions.

2.3.2 *Halide complexes of gold(I)*

Much of the early research on gold solubility in chloride solutions at elevated temperatures and pressures is summarized by Ogryzlo (1935) who demonstrated that gold was soluble in acid chloride solutions up to 300°C. The redox state of these early experiments was not controlled and, hence, these data are of only qualitative interest. More recently, Anderson and Burnham (1967) and Glyuk and Khlebnikova (1982) demonstrated that gold was soluble in chloride solutions (HCl, NaCl, KCl) over a wide range of temperatures and pressures up to 800°C and 3 kbar but lack of knowledge about the redox state (e.g. f_{H_2}) meant that no conclusions about complex stoichiometry or stability could be drawn. In addition, Vilor (1973) has measured the solubility of gold in chloride solutions over a wide range of temperatures (100 – 550°C). Henley (1973) has also determined the solubility of gold in KCl solutions from 300 to 550°C and up to 2000 bar. Rytuba and Dickson (1977) have measured the solubility of gold in 1M NaCl at 1000 bar and at 450°C (Au = 0.1 mg/kg) and 500°C (Au = 1.5 mg/kg) in the presence of a pyrite–pyrrhotite assemblage. The role of reduced sulphur arising from pyrite–pyrrhotite dissolution in complexing gold under these conditions is unknown but could contribute significantly to the above analytical gold concentrations. Wood *et al.* (1987) measured the solubility of gold in NaCl–CO_2 solutions from 200 to 350°C in the presence of a complex mixture of sulphide and oxide minerals in which the hydrogen fugacity was controlled by the pyrite–pyrrhotite–magnetite buffer assemblage. Recently, Zotov and Baranova (1989) have obtained thermodynamic data for the solubility of gold (as $AuCl_2^-$) from 350 to 500°C and 500 to 1500 bar.

The most important of the above mentioned data have been recalculated in terms of the equilibrium reaction

$$Au_c + 2Cl^- + H^+ = AuCl_2^- + \tfrac{1}{2}H_{2(g)} \qquad (2.4)$$

and are presented in Figure 2.2. In addition, the low temperature values (25–80°C) computed from Nikolaeva *et al.*'s (1972) emf data together with unpublished data by Henley (1985) and Bloom and Seward (1989) are also shown. The calculated curve derived from Helgeson (1969) is given for comparison.

Figure 2.2 illustrates the discrepancies in the available published data and indicates that it is not possible to reconcile the low-temperature data of Nikolaeva *et al.* (1972) with the data of Henley (1973) and Wood *et al.* (1987). It is instructive to briefly consider the reliability of Nikolaeva *et al.*'s (1972) data. The following values of log K for reaction (2.4) at 25°C have been calculated from three completely different sets of experimental data.

(a) $\log K = -19.5$ (Helgeson (1969) and based on data from Latimer (1952), Bjerrum and Kirschner (1918), and Bjerrum (1948));
(b) $\log K = -19.5$ (Nikolaeva *et al.* (1972));
(c) $\log K = -19.47$ (calculated from data of Pouradier *et al.* (1965) and Johnson *et al.* (1978)).

They are in perfect agreement and it may be concluded that the 25°C datum point for log K of reaction (2.4) is known with some confidence. Nikolaeva *et al.*'s (1972) data to 80°C are thus quite consistent with the 25°C data. It should be further noted that the data given above, together with those of Nikolaeva *et al.* (1972) to 80°C, Helgeson

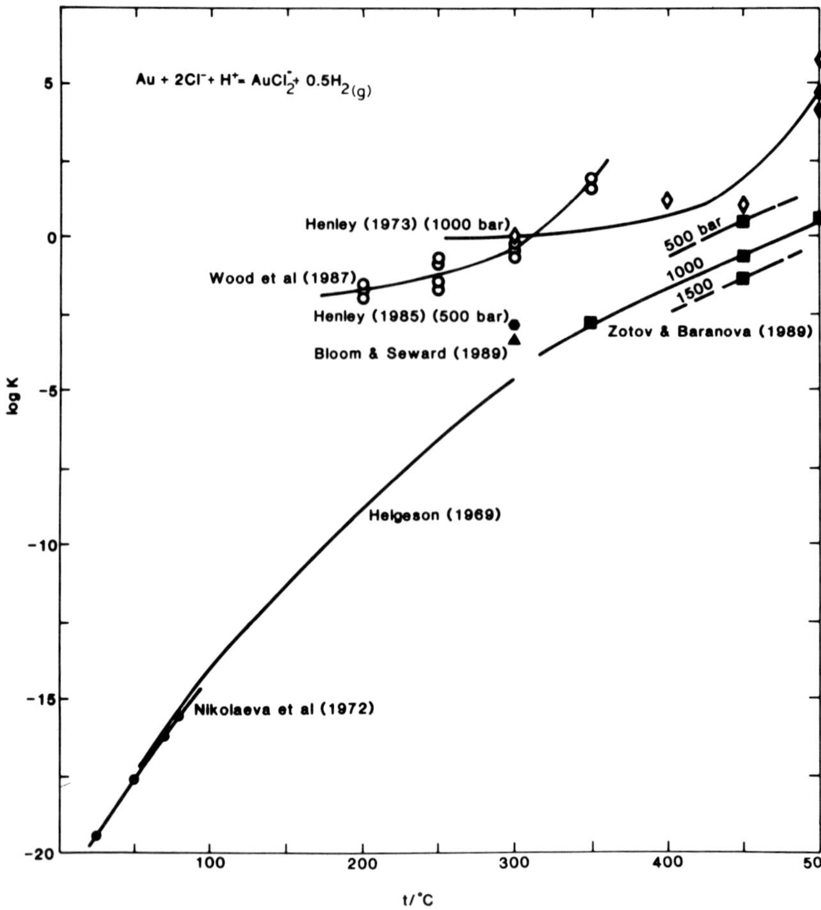

Figure 2.2 The variation of log K for reaction (2.4) with temperature up to 500°C; the data refer to the saturated vapour pressure except where otherwise indicated.

(1969), Bloom and Seward (1989), Henley (1985) and Zotov and Baronova (1989) comprise an approximately self-consistent data set which gives the best indication of the true values of log K at the saturated vapour pressure from 25 to 374°. There is a desperate need for additional experimental data from 150 to 500°C over a range of pressures.

As far as the other halide complexes are concerned, the dibromido- and di-iodogold (I) species are more stable than $AuCl_2^-$ at least at 25°C. However, the bromide and iodide complexes are not normally expected to play a significant role in the transport of gold because of the low concentrations of bromide and iodide in most natural hydrothermal fluids. For example, fluids of the Broadlands (Ohaaki) geothermal system (Weissberg *et al.*, 1979) are typical epithermal ore-forming fluids and contain ~ 4 mg/kg Br^- and ~ 0.5 mg/kg I^-.

Bromide and iodide may, however, play some role in gold complexing in the Cheleken brines, USSR ($t \sim 100°C$; $Cl^- \sim$ 4.4 M). These brines have high concentrations of bromide and iodide up to 672 mg/kg and 31.7 mg/kg respectively (Lebedev, 1972). Paramonova *et al.* (1988) have recently analysed the Cheleken brines for gold, obtaining an average value of 0.62 µg/kg. They suggested that the gold might be present as $AuCl_2^-$ as well as mixed complexes such as $AuClBr^-$ and $AuCl(I)^-$. However, the presence of reduced sulphur ($H_2S \sim$ 7 mg/kg) indicates that bisulphide complexes cannot be ignored.

It should be noted also that appreciable concentrations of bromide (700 mg/kg) and iodide (18 mg/kg) have been reported for the brines of the Salton Sea geothermal system (Skinner *et al.*, 1967; Ellis, 1967).

2.3.3 *Hydrolysis*

It should be mentioned that Au^+ and its complexes may hydrolyse to form simple and mixed hydroxo complexes. Gadet and Pouradier (1972) have studied the hydrolysis of $AuCl_2^-$ at 25°C and have determined the equilibrium constants for the successive reactions

$$AuCl_2^- + OH^- = AuCl(OH)^- + Cl^- \tag{2.5}$$

and

$$AuCl(OH)^- + OH^- = Au(OH)_2^- + Cl^- \tag{2.6}$$

for which log K (2.5) = –6.66 and log K (2.6) = 6.0. Unfortunately, there are no high-temperature data for these equilibria. Baranova *et al.* (1977) have measured the solubility of gold in alkaline solutions up to pH = 14 and 250°C in the presence of a magnetite–hematite redox buffer assemblage. They concluded that gold was present as $AuOH^0$, $Au(OH)_2^-$ and Au^0. Further work by Baranova *et al.* (1984), Zotov *et al.* (1985) and Ryabchikov *et al.* (1985) has indicated that $AuOH^0$ is an important species in alkaline solutions and/or in the absence of other ligands up 750°C and 1500 bar. However, $AuOH^0$ is not considered to play a significant role in gold transport if other ligands are present in the ore-transporting hydrothermal fluid.

2.3.4 *Hydrosulphido and sulphido complexes*

Sulphur is an ubiquitous component of hydrothermal ore solutions and, because of the generally prevailing low oxidation potential, is predominantly in the – 2 oxidation state as H_2S and HS^-. The second ionization constant of hydrogen sulphide is very small and the presence of S^{2-} in solution is negligible (see, for example, Giggenbach, 1971). As previously discussed (see also Table 2.3), reduced sulphur ligands form very stable complexes with Au^+ and such species are, therefore, of considerable importance in hydrothermal gold transport.

Earlier workers (Lehner, 1912; Ogryzlo, 1935; Lindner and Gruner, 1939; Zviagincev and Paulsen, 1940; Smith, 1944; Krauskopf, 1951) recognized the potential importance of sulphide complexes of Au(I) and carried out experiments and calculations of gold solubility, primarily in alkaline sodium sulphide solution. Weissberg (1970) measured the solubility of gold in alkaline sulphide solutions over the range from 150 to 290°C and at 1000 bar. His data further emphasized the increased solubility in high-temperature alkaline sulphide solutions but there was still no indication as to the nature of the gold complexes responsible for the measured solubilities. In addition, most thinking on the role of sulphide in gold transport had focused on alkaline solutions and there were some doubts as to the importance of sulphide complexing in the weakly acid to near-neutral region of pH typical of most ore-forming fluids.

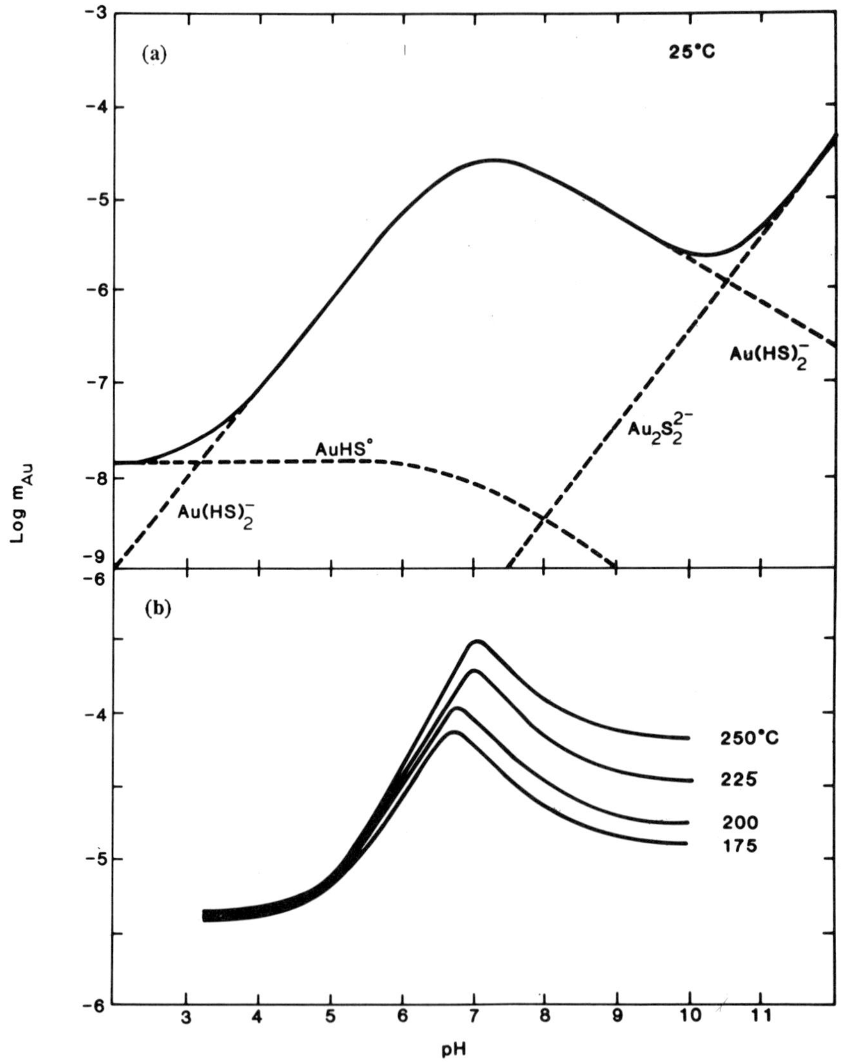

Figure 2.3 The concentration of gold in equilibrium (a) with Au_2S at 25°C, saturated vapour pressure and $m_{S_t} = 0.01$ (Renders and Seward, 1989) and (b) with elemental gold at higher temperatures p=1000 bar, $m_{S_t} = 0.50$ (Seward, 1973); the dashed lines in (a) refer to the concentration of the various complexes which contribute to the overall solubility curve.

Seward (1973) was the first to demonstrate that sulphide complexes of gold (I) were of major importance in the near-neutral region of pH and concluded that hydrosulphido complexes played a fundamental role in the transport and deposition of gold by hydrothermal fluids in the Earth's crust. His experiments extended from 160 to 300°C and up to 1500 bar and showed that three complexes, $AuHS^0$ (acid pH),

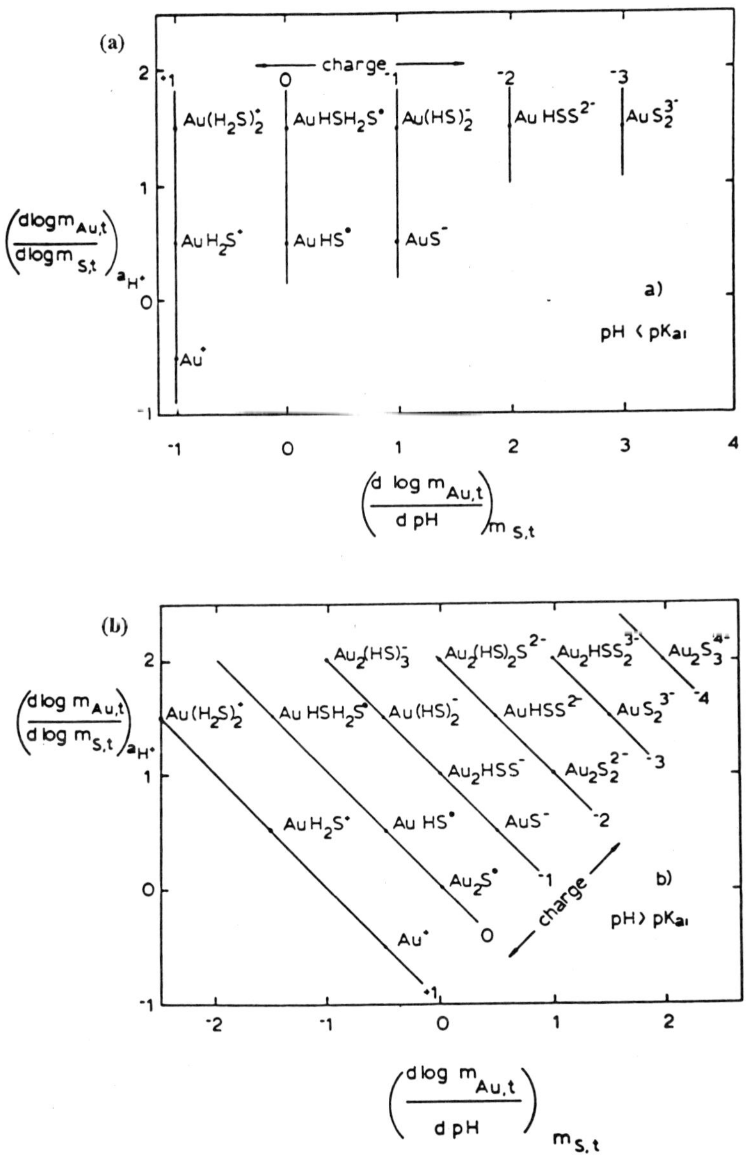

Figure 2.4 Some of the possible stoichiometrics of Au(I) complexes which could occur in aqueous sulphide solutions (from Renders and Seward, 1989a); (a) refers to acid solutions where pH < $pK_1(H_2S)$ and (b) refers to alkaline solutions where pH > $pK_1(H_2S)$.

$Au(HS)_2^-$ (near-neutral to weakly alkaline pH), and $Au_2(HS)S^-$ (alkaline pH) (see also Renders and Seward, 1989a), were present in high-temperature, high-pressure aqueous sulphide solutions (Figure 2.3). Belevantsev *et al.* (1981) experimentally demonstrated the presence of $Au(HS)_2^-$ in aqueous sulphide solutions at 25°C and, more recently, Renders and Seward (1989a) have further confirmed the stoichiometry of $AuHS^0$ and $Au(HS)_2^-$ at 25°C by measuring gold(I) sulphide solubility in aqueous solutions over a wide range of pH (Figure 2.3). In the more alkaline pH region, the species $Au_2S_2^{2-}$ was identified. The number of possible stoichiometries of aqueous sulphide complexes is shown in Figure 2.4 but in both the acid and near-neutral pH region, the only two species identified over a wide temperature range are $AuHS^0$ and $Au(HS)_2^-$. Shenberger (1985) and Shenberger and Barnes (1989) have measured the solubility of aqueous sulphide solutions from 150 to 350°C and at the equilibrium saturated vapour pressure of the system. Values of log K for the reaction

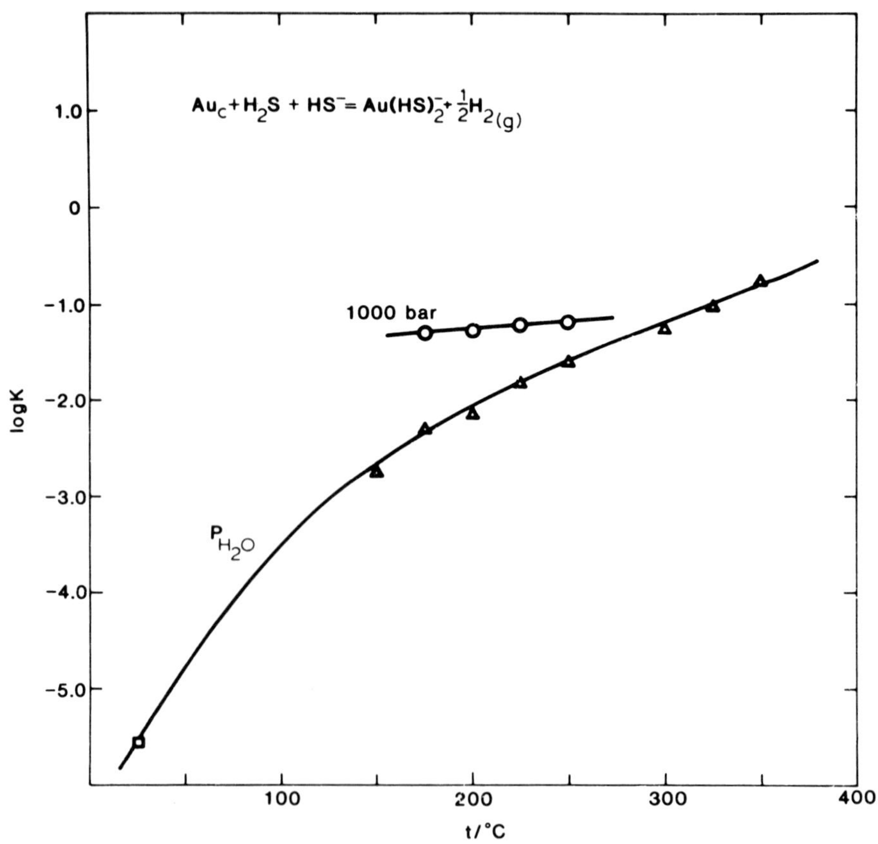

Figure 2.5 The variation of log K for reaction (2.7) with temperature; P_{H_2O} refers to the equilibrium saturated vapour pressure. Key: ❍ Seward (1973); Δ Shenberger (1985); ❑ Ronders and Seward (1989).

$$Au_c + H_2S_{(aq)} + HS^- = Au(HS)_2^- + \tfrac{1}{2}H_{2(g)} \quad (2.7)$$

are summarized in Figure 2.5.

The solubility maximum in the near-neutral region of pH is due to $Au(HS)_2^-$ and many studies of gold transport in specific ore-depositing environments have assumed that this is the dominant hydrosulphido complex over a wide range of temperature. This becomes a spurious assumption at temperatures >300°C. Figure 2.6 illustrates how the stability field for $Au(HS)_2^-$ and, hence, the solubility maximum migrates to higher pH with increasing temperature. The solubility maximum occurs where pH = pK_1 of H_2S at any given temperature. For example, as temperature changes from 25 to 360°C, the solubility maximum shifts from pH = $pK_1(H_2S)$ = 7.0 to pH = pK_1 = 9.2. In other words, the region of stability of $Au(HS)_2^-$ shifts to more alkaline conditions. The pH of hydrothermal ore fluids is typically in the range from 5 to 6.5 because of

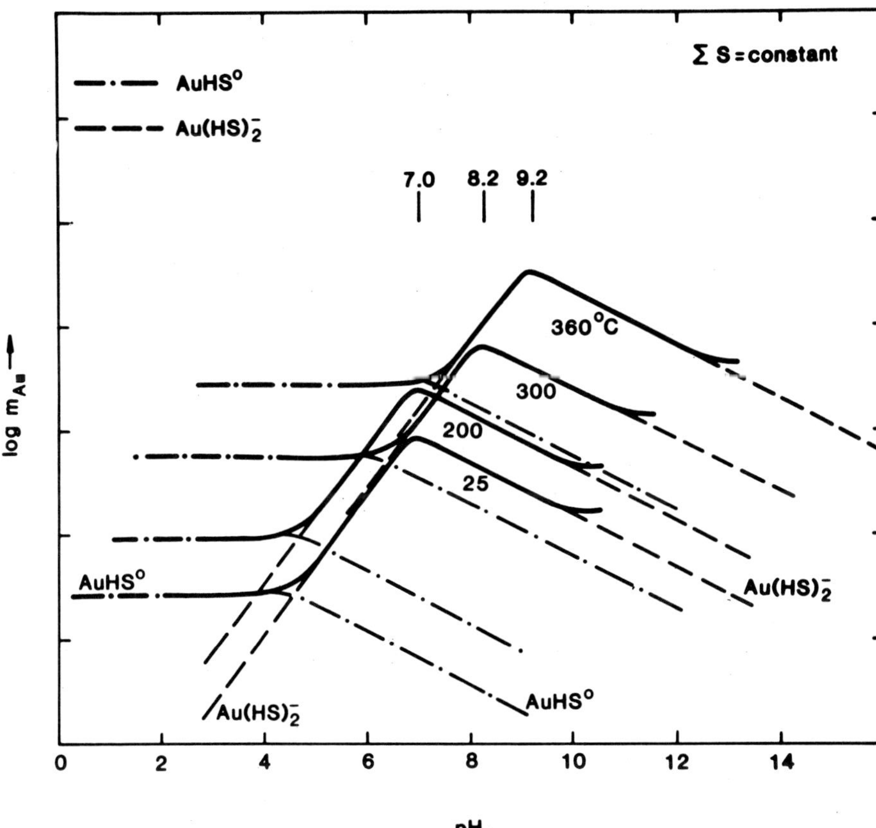

Figure 2.6 The variation of gold solubility in equilibrium with elemental gold at constant total sulphur (ΣS) concentration from 25 to 360°C as a function of pH; the numbers indicated by the vertical bars (e.g. 7.0, 8.2 and 9.2) are the values of pH = $pK_1(H_2S)$ at 25, 200, 300 and 360°C.

buffering by mineral equilibria and, hence, at high temperatures ($t > 300$°C), $Au(HS)_2^-$ will play an increasingly diminished role in gold transport. However, that is not to say that sulphide complexes become less important. Inspection of Figure 2.6 indicates that for an ore fluid at, say, 360°C, and at pH~ 5, the concentration of $Au(HS)_2^-$ will be at least two orders of magnitude less than that of $AuHS^0$. The point to be made here is that in modelling the transport and deposition of gold in an ore-depositing hydrothermal system (fossil or active), one must include the dominant gold species in one's model. Note, that in the case of the very stable $AuHS^0$ complex, there are no experimentally based thermodynamic data pertaining to its stability at high temperatures, although Renders and Seward (1989a) have reported thermodynamic data at 25°C.

2.3.5 *Additional sulphur-containing ligands*

Gold(I) complexes in which the ligand or the electron donor part of the ligand is a sulphur atom are quite stable. Hydrosulphidogold(I) complexes (e.g. $Au(HS)_2^-$) are very stable and, because ore-forming fluids contain appreciable concentrations of reduced sulphur (i.e. $H_2S + HS^-$), these complexes are very important in effecting gold transport in many hydrothermal systems. However, sulphur species in which the sulphur is in an intermediate oxidation state, between $-2(H_2S)$ and $+6(SO_4^{2-})$, are also present in cooler hydrothermal fluids and may play some role in gold transport and deposition in the upper portions of geothermal systems.

Webster (1987) has reported thiosulphate concentrations of up to 80 mg/kg from hot spring waters in New Zealand as well as polysulphide concentrations of ~ 3 mg/kg (J.G. Webster, pers. comm.). Subzhiyeva and Volkov (1982) have also reported thiosulphate and sulphite in thermal waters. In addition, Takano (1987) and Takano and Watanuki (1988) have reported large concentrations of polythionate ions ($S_nO_6^{2-}$; $4 \leq n \leq 9$) (up to 2800 mg/kg) in hot spring and volcanic crater-lake waters in Japan. Complexes of gold(I) with polysulphide and the oxyanions of sulphur (e.g. $S_2O_3^{2-}$ and polythionates such as $S_4O_6^{2-}$) will not be important in the high-temperature transport of gold (i.e. $t > 200$°C) by deep geothermal fluids because these anionic sulphur species disproportionate, although they may persist as metastable transitory intermediates (e.g. Pryor, 1960; Giggenbach, 1974). However, these sulphur species could play a significant role in gold complexing in the upper levels of hydrothermal systems where deep, flashed fluids mix with oxygenated, steam-heated waters. Gold transport from the near surface environment to thermal springs and pools such as at Waiotapu and Rotokawa in New Zealand (Hedenquist and Henley, 1985; Krupp and Seward, 1987) may be in part as complexes with sulphur oxyanions (e.g. $S_2O_3^{2-}$) and polysulphides (e.g. S_3S^{2-}). Kakovskiy and Tyurin (1962) and Grigor'yeva and Sukneva (1981) have reported enhanced gold solubilities in the presence of thiosulphate and polysulphides at temperatures up to 200°C. Furthermore, in some volcanic-hosted gold deposits in which magmatic SO_2 has migrated to the upper levels of a hydrothermal ore-depositing system to interact with groundwaters or in which hot, crater-lake waters permeate to depth, sulphur-species of intermediate oxidation state could be important in the transport and deposition of gold.

Finally, we should consider the role of thioarsenite and thioantimonite anionic species (see Table 2.4) in the complexing and transport of gold. A number of workers

(Grigor'yeva and Sukneva, 1981; Nekrasov *et al.*, 1982; Nekrasov and Konyshok, 1982) have measured the solubility of gold in aqueous sulphide solutions in the presence of stibnite (Sb_2S_3) and orpiment (As_2S_3) in the temperature range from 200 to 300°C in order to determine the effect of thioantimonite and thioarsenite species on gold solubility. However, uncertainties in the hydrogen fugacities prevailing in these experiments together with ambiguities in the stoichiometry of the thioantimonite and thioarsenite complexes make their results inconclusive.

Akhmedzhakova *et al.* (1988) have studied the solubility of gold in aqueous sulphide solutions in the presence of orpiment and the Fe/Fe_2O_3 redox buffer assemblage at 200 and 300°C. Their results indicate that the solubility of gold increases with increasing arsenic concentration. However, the mechanism for this solubility enhancement is unclear. Their experiments were conducted in the presence of 0.1 *N* HCl (i.e low pH) which suggests that the predominant arsenic species in solution should be the undissociated arsenious acid (H_3AsO_3) (see Webster, 1989). Furthermore, the possible reaction of the iron–magnetite assemblage with arsenic-containing sulphide solutions to form iron sulphides and/or arsenides as well as perhaps arsine (AsH_3) makes the interpretation of their data difficult. Thus, we are still unable to say whether gold(I) thioarsenite or thioantimonite complexes are of any importance in ore transport.

2.3.6 *Other ligands*

There are a number of other ligands which could play an important role in hydrothermal gold transport but unfortunately there are few if any high-temperature experimental data which provide much insight into their importance in hydrothermal systems. A few potentially important ligands are those involving tellurium (e.g. Te_2^{2-}, TeS^{2-} and related species), NH_3 and CN^-. In the case of tellurium, there are no available experimentally based thermodynamic data pertaining to aqueous tellurium chemistry at elevated temperatures, although D'yachkova and Khodakovskiy (1968) have calculated the stability of H_2Te, HTe^- and Te_2^{2-} as a function of pH and temperature up to 300°C. The reliability of these calculations is unknown. There are some data on the aqueous chemistry of reduced tellurium species at 25°C (e.g. Panson, 1963, 1964), but the stability of gold(I) complexes such as $Au(Te_2)_2^{2-}$, $Au(TeS)_2^{3-}$, $Au(Te_2)HS^{2-}$ or $AuTe_2^{3-}$ is unknown at any temperature. However, Smith's (1944) experiments indicate that gold is soluble in high-temperature aqueous solutions containing tellurium. He dissolved tellurium and gold in an aqueous sulphide/polysulphide solution at 250°C and precipitated crystals of $AuTe_2$ (calaverite).

There is virtually no information on the concentration of tellurium in geothermal fluids with the exception of the recent data by Goguel (1988). He found (using ICP–MS) that the concentration of tellurium in fluids of the Broadlands (Ohaaki) and Wairakei geothermal systems in New Zealand were 50 ng/kg (BR45) and 10 ng/kg (W116) respectively. These are very low concentrations, suggesting that, at least in an epithermal ore-depositing fluid such as at Broadlands, tellurium is of little importance. However, some hydrothermal gold deposits are characterized by an abundance of gold telluride minerals, implying that higher tellurium concentrations may occur in some ore-forming fluids.

Another complex which, in some circumstances, could be of considerable importance in hydrothermal gold transport is the diamminogold(I), $Au(NH_3)_2^+$. It is quite stable at 25°C (i.e. log $\beta_2 = 26.5$; Skibsted and Bjerrum, 1974) but there are no thermodynamic data available for this species at elevated temperatures. Many hydrothermal fluids contain appreciable ammonia concentrations. For example, the deep fluids of the Ngawha and Broadlands (Ohaaki) geothermal systems in New Zealand contain 0.013 M (178 mg/kg) (NG9 well) and 0.808×10^{-3} M (11.3 mg/kg) (BR22 well), of NH_3 respectively (Barnes and Seward, 1989; Seward, 1989). In order to calculate the solubility of gold in deep geothermal fluids where gold is assumed to be present as the simple diammine complex, high temperature thermodynamic data for the reaction

$$Au_c + 2NH_3 + H^+ = Au(NH_3)_2^+ + \tfrac{1}{2} H_{2(g)} \tag{2.8}$$

are required. However, equilibrium data for the above reaction are premised on some knowledge of the stability of $Au(NH_3)_2^+$. It may only be suggested at this point that in some ore-forming fluids, the $Au(NH_3)_2^+$ complex could account for appreciable gold concentrations. There is a need for experimental data.

Finally, the possible presence of cyanide complexes should be considered. The dicyanogold(I) complex, $Au(CN)_2^-$, is very stable at 25°C (Table 2.3) but there are no high-temperature thermodynamic data pertaining to its stability. In addition, there have been no data published on the concentration of HCN in hydrothermal fluids. Recently, however, the writer and colleagues have determined the concentration of HCN in the Broadlands (Ohaaki) geothermal system and found ~ 8 μg/kg in the deep reservoir fluid (290°C) of BR35 well. Bearing in mind that $Au(CN)_2^-$ is approximately nine orders of magnitude more stable than $Au(HS)_2^-$ at 25°C (see Table 2.3), this raises the possibility of the dicyanogold (I) complex playing some role (as yet unknown) in hydrothermal gold transport. Thiocyanate (SCN^-) was not detected in BR35 discharge.

2.4 Gold deposition

It is not the purpose of the chapter to discuss in detail the mechanisms and conditions of gold deposition from hydrothermal fluids in ore-depositing environments. Nevertheless, it is instructive to consider several mechanisms of gold deposition in order to emphasize the importance of gold(I) complex equilibria in determining how and where gold precipitation occurs. To facilitate such a discussion, let us consider the chemistry of gold in the high-temperature fluids of the active ore-depositing geothermal system at Rotokawa, New Zealand (Krupp and Seward, 1987).

Table 2.5 gives the calculated concentrations of two potentially important gold(I) complexes in the deep fluid of well RK4. A detailed analysis of the major components of this fluid is given by Krupp and Seward (1987). The concentration of the dihydrosulphidogold(I) complex, $Au(HS)_2^-$, is approximately six orders of magnitude larger than that of $AuCl_2^-$, and therefore $AuCl_2^-$ is not considered to play any significant role in the ore solution chemistry of gold in this hydrothermal fluid. This result is similar to that obtained by Seward (1989) for the Broadlands geothermal system. Note that even if the sodium chloride concentration of the Rotokawa fluid was

~ 3.0 M (i.e. $Cl^- = 0.75$ M) and assuming the same pH, f_{H_2} and reduced sulphur concentration, the calculated concentration of $AuCl_2^-$ (= ~ 1×10^{-3} μg/kg) is still small in comparison to $Au(HS)_2^-$.

Table 2.5 Concentrations of gold(I) complexes in Rotokawa well 4 (RK4) deep fluid at 311°C; pH = 5.8, $Cl^- = 2.3 \times 10^{-2}$ M, $H_2S_{(aq)} = 6.3 \times 10^{-3}$ M, $HS^- = 4.2 \times 10^{-5}$ M, $P_{H_2} = 0.47$ bar

$AuCl_2^-$ (μg/kg)	$Au(HS)_2^-$ (μg/kg)
1.6×10^{-6} (~ 1×10^{-3})*	3.7

*For Cl^-(free chloride) = 0.75 M or $\sum$ Cl(as Nacl) ≈ 3.0 M.

2.4.1 *Boiling*

There are a number of fundamental processes (e.g. fluid-rock interaction, boiling, mixing) which operate in hydrothermal systems and which are important in the

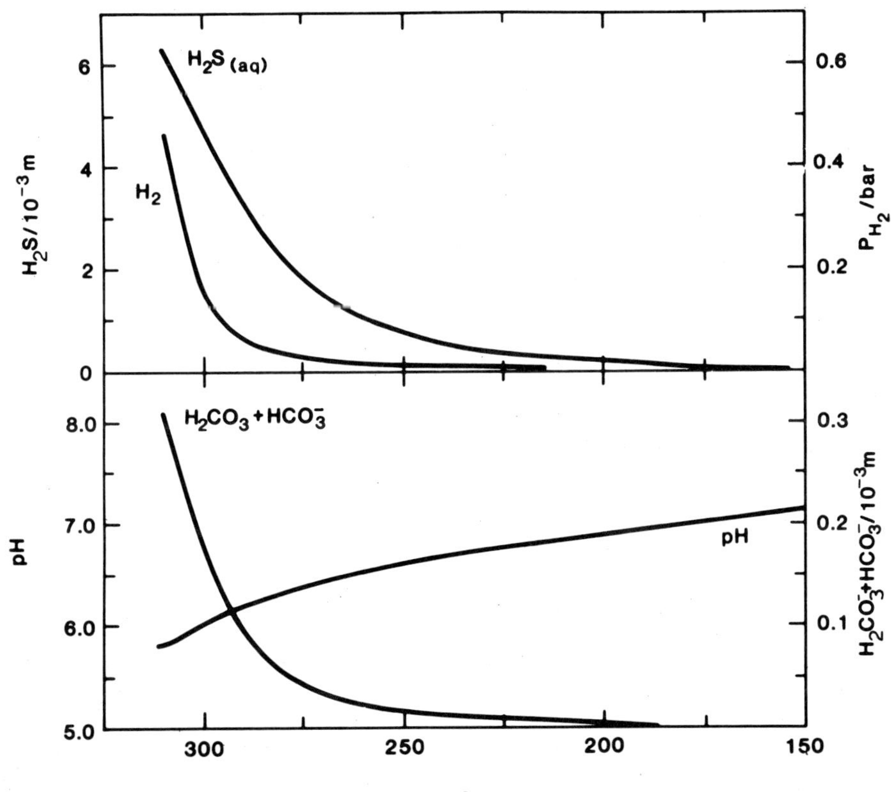

Figure 2.7 Changes in $H_2S_{(aq)}$, H_2, ΣCO_2 (as $H_2CO_3 + HCO_3^-$), and pH (of the residual liquid phase) when Rotokawa reservoir fluid in RK4 well undergoes adiabatic, closed-system boiling.

chemical evolution of fluids rising buoyantly through the Earth's crust. In this discussion, we will confine our attention to boiling and its effect, not only on the bulk fluid composition but also, in particular, on the response of the gold(I) complexes present in the hydrothermal fluid.

An ascending hydrothermal fluid will reach a depth at which boiling and/or phase separation occurs. The depth at which this takes place depends upon the gas content of the fluid and commences at the point where the confining hydrostatic pressure becomes equal to the equilibrium saturated vapour pressure. With further rise to the surface, progressive boiling will occur and two phase conditions will prevail. The loss of volatiles during boiling has a major effect on the chemistry of the residual liquid phase. This is illustrated by Figure 2.7 which shows a few of the important changes which occur when the deep reservoir fluid of the Rotokawa geothermal system undergoes boiling. Details of the calculation procedure are given in Giggenbach (1980), Henley *et al.* (1984) and Seward (1989). The changes shown in Figure 2.7 as a consequence of gas partitioning are in response to adiabatic (isoenthalpic), closed-system boiling. The total CO_2 (shown as H_2CO_3 + HCO_3^-) in the residual liquid falls dramatically and this causes a significant increase in pH. In addition, the concentration of reduced sulphur decreases and the oxidation potential of the residual liquid phase is diminished significantly by hydrogen loss. These changes have a major effect on the stability of various gold complexes and, hence, on gold solubility. We have demonstrated (Table 2.5) that the dominant gold complex in the Rotokawa (RK4) reservoir fluid is $Au(HS)_2^-$. How does the solubility of gold change in response to the changes depicted in Figure 2.7?

The initial gold concentration at 311°C is 3.7 μg/kg. With the onset of adiabatic, closed-system boiling, the gold solubility will increase (Figure 2.8, curve a), primarily in response to increasing pH and associated increase in the stability of $Au(HS)_2^-$ (refer to Figure 2.3). The loss of H_2 (reaction 2.7) will also enhance the gold solubility. Note that the actual gold concentration of the residual liquid will not actually increase along curve a because there would be a lack of available gold in the immediate environment to maintain equilibrium. The residual liquid would be merely undersaturated and could not precipitate gold. The 'real' gold concentration would, of course, increase along curve c as a result of steam loss. Further adiabatic, single-step boiling eventually leads to a decrease in gold solubility as the loss of reduced sulphur (ligand) overrides the increasing pH as the dominant effect on gold solubility. With continued boiling, gold precipitation would commence at 250°C where curves a and c intersect. Boiling over the temperature interval from 250 to 180°C would precipitate much of the gold. Curve b represents the decrease in gold solubility as a result of conductive cooling. Conductive cooling is not considered to play any significant role in ore formation but is shown here for comparison. The same amount of gold would be deposited by conductive cooling from 311 to 180°C, but it would be precipitated over the permeability path through which the fluid migrated. This might give rise to a geochemical anomaly but not an ore deposit.

It is also important to emphasize at this point that reference to reaction (2.7) alone in order to try to understand the mechanisms of hydrothermal gold deposition is insufficient. One must consider the chemistry of the bulk, multicomponent fluid in order to gain a detailed insight. For example, the change in pH on boiling of Rotokawa fluid (and most other hydrothermal ore fluids) arises predominantly from CO_2 loss. Of

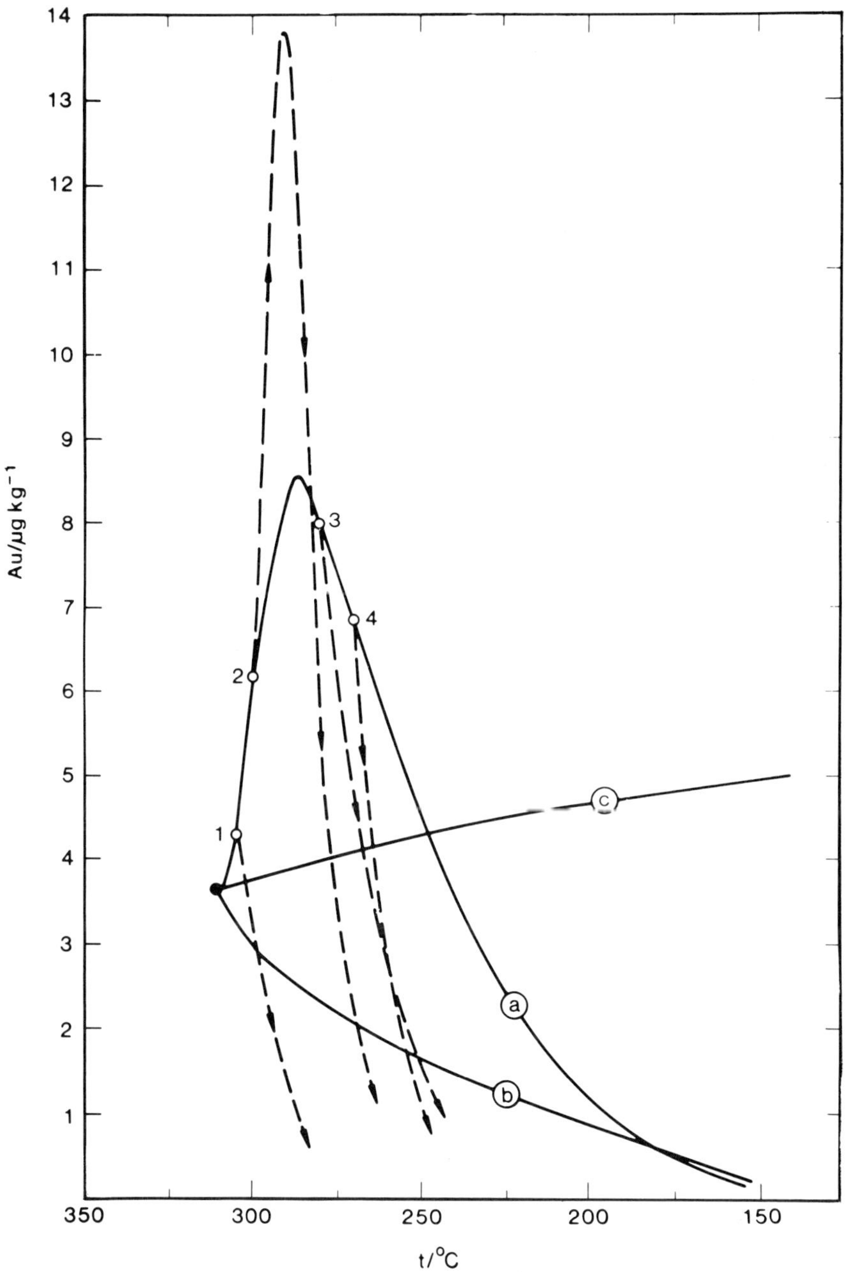

Figure 2.8 The effect of boiling on the solubility of gold (as $Au(HS)_2^-$) in Rotokawa deep fluid; curve a refers to adiabatic closed-system boiling; curve b refers to conductive cooling; curve c represents the increase in initial gold concentration due to steam loss. The dashed lines departing from points 1,2,3 and 4 (305, 300, 280 and 270°C) represent open-system boiling following adiabatic, closed-system boiling to that point.

course, the loss of H_2S will make a small contribution to the increase in pH but this is a minor effect in comparison to the actual decrease in concentration of the complexing ligand which eventually leads to gold precipitation.

Thus far, we have considered only closed-system boiling. However, we know that in active hydrothermal systems, open-system boiling occurs intermittently. A gas-rich steam phase may rise separately to the surface to be 'trapped' by near-surface, oxygenated water. This gives rise to steam-heated, low chloride, bicarbonate/sulphate waters of low pH which may cause extensive advanced argillic alteration, a common phenomenon in most hydrothermal systems.

Figure 2.8 also illustrates the effect of multistep, open-system boiling on the solubility of gold. If the deep fluid is boiled adiabatically (closed system) to 305°C (point 1) and then the volatile phase (gases + steam) is continually removed as further boiling proceeds, most of the gold will be precipitated over a small temperature interval of ~ 20°C. Other examples (points 2, 3 and 4) are also shown and in all cases the effect is to quite dramatically precipitate gold over a relatively small temperature interval, thus providing a potent mechanism for the localized deposition of gold (i.e. ore deposition) which is even more effective than simple closed-system, adiabatic boiling (curve a).

Even though the $AuCl_2^-$ complex is a minor contributor to the total gold concentration of hydrothermal fluid at Rotokawa, it is nevertheless instructive to consider the effect of boiling on this species. Closed-system, adiabatic boiling leads to immediate deposition of gold which is present as $AuCl_2^-$ (Figure 2.9).

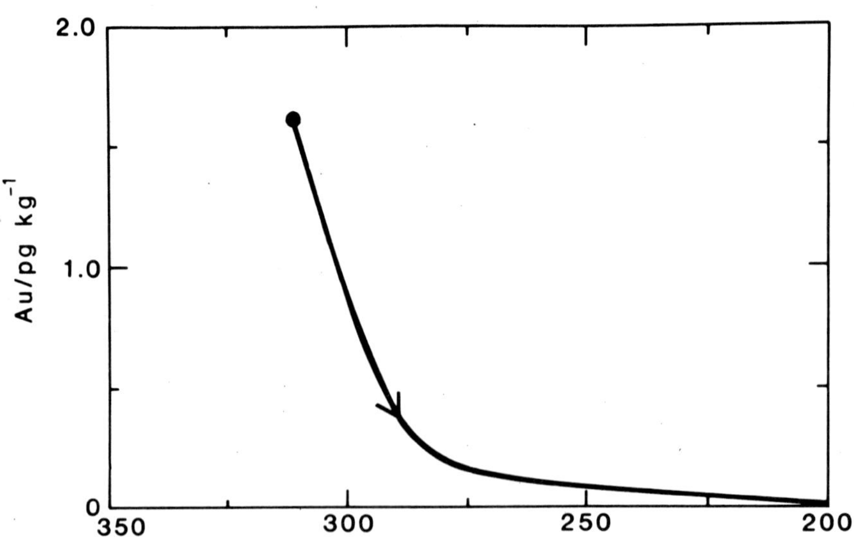

Figure 2.9 The effect of closed-system, adiabatic boiling on gold solubility when gold is present as $AuCl_2^-$ in Rotokawa fluid; 1 pg = 10^{-12} g.

These examples demonstrate the importance of gold complex equilibria in determining not only the nature of transport but also the importance of knowing which complex (or complexes) has been dominant in a particular hydrothermal ore fluid. The extent and location of gold deposition will be dependent upon the interrelation between the hydrothermal system hydrology and permeability and relevant gold complex equilibria. In the case of $Au(HS)_2^-$ (or $AuHS^0$), any processes which decrease the activity of reduced sulphur in the hydrothermal ore solution will ultimately lead to gold precipitation. These include sulphide mineral deposition, boiling, mixing, and oxidation but only boiling has been considered in this discussion. Boiling may also lead to precipitation of base metal sulphides, which enhances the removal of reduced sulphur from the ore fluid. In addition, mixing of deep fluids with oxygenated meteoric waters in the upper parts of hydrothermal systems may cause oxidation of H_2S and HS^- to sulphur and sulphate, thus diminishing the activity of reduced sulphur. It is worth remembering, however, that the formation of intermediate oxidation state species (e.g. S_nS^{2-} and $S_2O_3^{2-}$), which can form stable complexes with gold, could facilitate gold transport to the surface environment such as at Rotokawa (Krupp and Seward, 1987) despite the disappearance of hydrosulphido complexes.

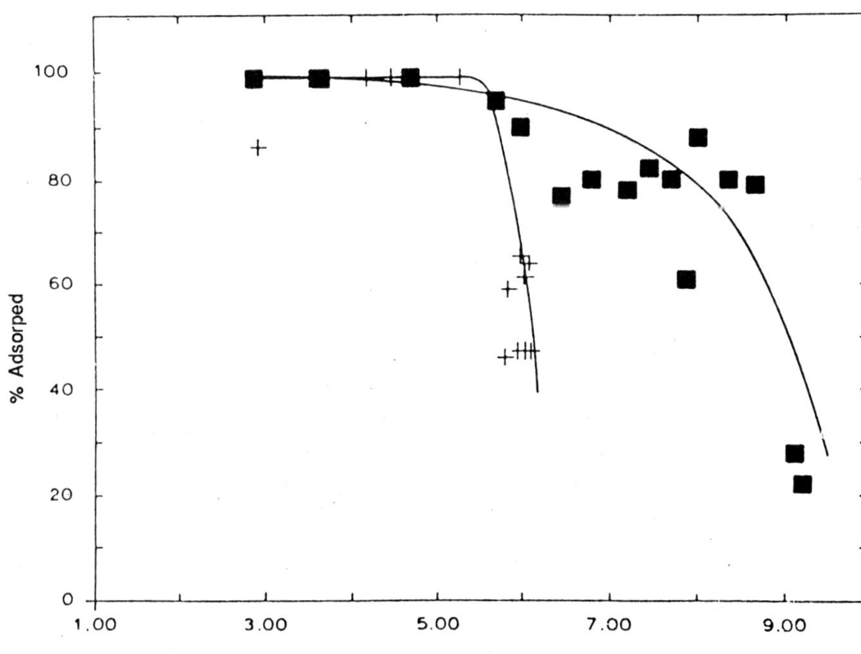

Figure 2.10 The percent adsorption of hydrosulphido-complexed gold(I) at 90°C by colloidal arsenic (+) and antimony sulphide ■ (from Renders and Seward, 1989b).

2.4.2 *Precipitation on colloid and mineral surfaces*

One aspect of gold deposition which has received little attention is the role of surface chemistry in the co-precipitation of gold with sulphide minerals in ore-depositing environments. Bancroft and co-workers (Bancroft and Jean, 1982; Jean and Bancroft, 1985; Hyland and Bancroft, 1989) have studied the precipitation of gold on sulphide mineral surfaces (pyrite, galena, sphalerite) using X-ray photoelectron spectroscopy (XPS) and Starling *et al.* (1989) have emphasized the potential importance of gold precipitation on pyrite surfaces under hydrothermal conditions. In addition, Renders and Seward (1989b) have recently studied the adsorption of gold on to colloidal As_2S_3 and Sb_2S_3 up to 90°C (Figure 2.10). They demonstrated that gold (as $AuHS^0$) can be quantitatively removed from solution by arsenic and antimony sulphide colloids within a specific range of pH. Colloidal arsenic and antimony sulphides may form at higher temperatures (e.g. 200°C). If a sulphide colloid forms in a hydrothermal ore-depositing environment, it may eventually crystallize to give an ordered compound. However, the important point is that if the initial sulphide precursor is a colloid (despite eventual crystallization), then scavenging on to the charged surface of a colloid could play an important role in the deposition of gold from solution. Precipitation on to mineral surfaces such as pyrite at elevated temperatures could also be important. There is a great need for experimental studies in this field.

2.5 Summary

We have some insight into the coordination chemistry of gold(I) under hydrothermal conditions but the quality and extent of thermodynamic data pertaining to Au(I) complexes at elevated temperatures and pressures is still very inadequate. A major barrier to our further understanding of the formation of hydrothermal gold deposits is, in fact, the current lack of experimentally based thermodynamic data.

Gold(I) hydrosulphido complexes play a fundamental role in the formation of the majority of hydrothermal gold deposits. However, depending upon the composition, temperature and pressure of the ore-forming fluid, complexes with other ligands such as chloride may also contribute to, or in some cases comprise, the total gold in solution. Unfortunately, the necessary data required to model the various different transport and deposition mechanisms are still unavailable. Experimental data on the stability of gold(I) complexes in low dielectric fluids in the system CO_2–CH_4–H_2O at high temperatures and pressure are also required. In addition, further knowledge of the chemistry of transport and deposition of other elements frequently associated with hydrothermal gold (e.g. As, Sb, Ag, Hg, Tl, Te) would appreciably enhance our cognizance of the processes involved.

Although it has seldom been considered an important effect in hydrothermal ore-depositing environments, surface chemistry may also be a significant factor in the deposition of gold. Sulphide colloids may act as extremely efficient scavengers of gold and provide an additional potent mechanism for concentrating gold. Even if a hydrothermally precipitated sulphide colloid undergoes recrystallization shortly after having formed, if the initial sulphide precursor was a colloid, then gold adsorption on to the charged surface could be important. In this situation, the localization of gold is not only a function of the sulphide colloid surface chemistry but also of the conditions under which a particular metal sulphide colloid (or colloids) precipitates. The adsorption and subsequent reduction of gold complexes on mineral surfaces such as pyrite may also be important. Such adsorption effects in general may be of major importance in the deposition of gold in seafloor polymetallic sulphides such as those described by Hannington *et al.* (1986) as well as in volcanogenic massive sulphides.

References

Akhmedzhakova, G.M., Nekrasov I., Ya. Tikhomirova, V.I. and Konyshok, A.A. (1988) Solubility of gold in sulphide–arsenic solutions at 200–300°C. *Dokl. Akad. Nauk SSSR* **300**, 1453-1455.

Anderson, G.M. and Burnham, C.W. (1967) Reactions of quartz and corundum with aqueous chloride and hydroxide solutions at high temperatures and pressures. *Amer. J. Sci.* **265**, 17- 27.

Bancroft, G.M. and Jean, G. (1982) Gold deposition at low temperature on sulphide minerals. *Nature* **298**, 730- 731.

Baranova, N.N., Barsukov, V.L., Dar'yina, T.G. and Bannykh, L.N. (1977) *Geochem. Internat.* **14**, 126-132.

Baranova, N.N., Zotov, A.V., Bannykh, L.N., Dar'yina, T.G. and Savel'yev, B.V. (1984) The effects of redox conditions on the solubility of gold in water at 450°C and 500 atm. *Geochem. Internat.* **21**, 117-122.

Barnes, H.L. and Seward, T.M. (1990) Mercury deposits and related geothermal systems. In *Hydrothermal Processes–Applications to Ore Genesis*, eds. Barnes, H.L. and Ohmoto, H. Reidel, Dordrecht (in press).

Belevantsev, V.I., Peshchevitskiy, B.I. and Shamovskaya, G.I. (1981) Gold(I) sulphide complexes in aqueous solution. *Isvest. Sib. Otd. Akad. Nauk SSSR, Ser. Khim.*, **1**, 81-87.

Bjerrum, N. and Kirschner, A. (1918) Die Rhodanide des Goldes und das freie Rhodan mit einem Anhang über das Goldschlorid. *Kgl. Danske Vidensk. Selsk. Skr., Naturvidensk. math. Afdel* **5**, 1-77.

Bjerrum N. (1948) La stabilitè des chlorures d'or. *Soc. Chimie Belge Bull.* **57**, 432-445.

Bloom, M. and Seward, T.M. (1989) Work in progress.

Braunstein, P. and Clark, R.J.H. (1973) *J. Chem. Soc. Dalton Trans.* 1845-1848.

D'yachkova, I.B. and Khodakovskiy, I.L. (1968) Thermodynamic equilibria in the systems, S–H_2O, Se–H_2O and Te–H_2O in the 25–300°C temperature range and their geochemical interpretations. *Geochem. Internat.* **5**, 1108-1125.

Ellis, A.J. (1967) The chemistry of some explored geothermal systems. In *Geochemistry of Hydrothermal Ore Deposits* (1st edn.), ed. Barnes, H.L. Holt, Rinehart and Winston, New York, 465-514.

Gadet, M. and Pouradier, J. (1972) Hydrolyse des complexes de l'or(I). *C.R. Acad. Sci. Paris* **275**, 1061-1064.

Giggenbach, W.F. (1971) Optical spectra of highly alkaline sulfide solutions and the second dissociation constant of hydrogen sulphide. *Inorg. Chem.* **10**, 1333-1338.

Giggenbach, W.F. (1974) Kinetics of the polysulphide-thiosulphate disproportionation up to 240°C. *Inorg. Chem.* **13**, 1730-1733.

Giggenbach, W.F. (1980) Geothermal gas equilibria. *Geochim. Cosmochim. Acta* **44**, 2021-2032.

Glyuk, D.S. and Khlebnikova, A.A. (1982) Gold solubility in water, HCl, HF and sodium and potassium chloride, fluoride, carbonate and bicarbonate solutions at a pressure of 1000 kg/cm^2. *Dokl. Akad. Nauk SSSR* **254**, 190-194.

Goguel, R.L. (1988) Ultratrace metal analysis of New Zealand geothermal waters by ICP-MS. In *Trace Elements in New Zealand: Environmental, Human and Animal, Proc. NZ Trace Elements Group Conf., Canterbury, Dec. 1988*, 263-270.

Grigor'yeva, T.A. and Sukneva, L.S. (1981) Effects of sulphur and of antimony and arsenic sulphides on the solubility of gold. *Geochem. Internat.* **18**, 153-158.

Hannington, M.D., Peter, J.M. and Scott, S.D. (1986) Gold in seafloor polymetallic sulfide deposits. *Econ. Geol.* **81**, 1867-1883.

Hedenquist, J.W. and Henley, R.W. (1985) Hydrothermal eruptions in the Waiotapu geothermal system, New Zealand: Their origin, associated breccias and relation to precious metal mineralisation. *Econ. Geol.* **80**, 1640-1668.

Helgeson, H.C. (1969) Thermodynamics of hydrothermal systems at elevated temperatures and pressures. *Amer. J. Sci.* **267**, 729-804.

Henley, R.W. (1973) Solubility of gold in hydrothermal chloride solutions. *Chem. Geol.* **11**, 73-87.

Henley, R.W. (1985) Personal communication.

Henley, R.W., Truesdell, A.H., Barton, P.B. Jr and Whitney, J.A. (1984) *Fluid Mineral Equilibria in Hydrothermal Systems. Reviews in Economic Geology,* Vol.1, Society of Economic Geologists, 267pp.

Hyland M.M. and Bancroft G.M. (1989) An XPS study of gold deposition at low temperatures on sulphide minerals: Reducing agents. *Geochim. Cosmochim. Acta* **53**, 367-372.

Jean, G.E. and Bancroft, G.M. (1985) An XPS and SEM study of gold deposition at low temperature on sulphide mineral surfaces: Concentration of gold by adsorption/reduction. *Geochim. Cosmochim. Acta* **49**, 979-987.

Johnson, P.R., Pratt, J.M. and Tilley, R.I. (1978). Experimental determination of the standard reduction potential of the gold(I) ion. *J. Chem. Soc. Chem. Comm.* 606-607.

Jorgensen, C.K. and Pouradier, J. (1970) Un nouveau type de stabilisation du 'champ des ligandes' dans les complexes lineaires de cuivre(I), de l'argent(I) et de l'or(I). *J. Chim. Phys.* **67**, 124-127.

Kakovsky, I.A. and Tyurin, N.G. (1962) Behaviour of gold in polysulphide solutions at elevated temperatures and pressures. *lzv. Vyssh. Ucheb. Zaved, Tsvet. Metallurgiya*, 104-111 (in Russian).

Krauskopf, K.B. (1951) The solubility of gold. *Econ. Geol.* **46**, 858-870.

Krupp, R.E. and Seward, T.M. (1987) The Rotokawa geothermal system, New Zealand: An active epithermal ore-depositing environment. *Econ. Geol.* **82**, 1109-1129.

Latimer, W.M. (1952) *The Oxidation States of the Elements and their Potentials in Aqueous Solutions.* Prentice-Hall, Englewood Cliffs, N.J. 392pp.

Leary, K. and Bartlett, N. (1972) A new oxidation state of gold: The preparation and some properties of $[AuF_6]^-$ salts. *J. Chem. Soc. Chem. Comm.* 903-904.

Lebedev, L.M. (1972) Minerals of contemporary hydrotherms of Cheleken. *Geochem. Internat.* **9**, 485-504.

Lehner, V. (1912) The transport and deposition of gold in nature. *Econ. Geol.* **7**, 744-750.

Linder, J.L. and Gruner, J.W. (1939) Action of alkali sulphide solutions on minerals at elevated temperatures. *Econ. Geol.* **34**, 537-560.

Nekrasov, I. Ya., Konyshok, A.A. and Sorokin, V.I. (1982) Form of gold(I) in antimony-bearing sulphide solutions. *Dokl. Akad. Nauk* **264**, 207-210 (Engl. transl.).

Nekrasov, I. Ya. and Konyshok, A.A. (1982) Heteropolynucleate gold complexes in antimony-bearing sulphide solutions. *Dolk. Akad. Nauk SSSR* **266**, 185-188 (Engl. transl.).

Nikolaeva, N.M., Yerenburg, A.M. and Antinina, V.A. (1972) Temperature dependence of the standard potential of halide complexes of gold. *lzvest. Sib. Otd. Akad. SSSR, Ser. Khim.* **4**, 126-129.

Ogryzlo, S.P. (1935) Hydrothermal experiments with gold. *Econ. Geol.* **30**, 440-424.

Panson, A.J. (1963) Polarography of the ditelluride ion. *J. Phys. Chem.* **67**, 2177-2180.

Panson, A.J. (1964) A study of the telluride ion system. *J. Phys. Chem.* **68**, 1721-1724.

Paramonova, V.I., Vartenev, S.A., Klemina, A.M., Zhuravleva, Ye. L. and Zhuchko, V. Ye (1988) Extraction and concentration of microamounts of gold from Cheleken brines by the sorption method. *Geochem. Internat.* **25**, 124-127.

Parish R.V., Parry O. and McAuliffe C.A. (1981) Characterisation of two-, three- and four-coordinate gold(I) complexes by ^{197}Mössbauer and ^{31}P– (^{1}H) nuclear magnetic resonance spectroscopy. *J. Chem. Soc. Dalton Trans.* 2098-2104.

Parish, R.V. (1984) Gold-197 Mössbauer spectroscopy in the characterization of gold compounds. In *Modern Inorganic Chemistry*, Vol.1, *Mössbauer Spectroscopy Applied to Inorganic Chemistry*, ed. Long, G.J. Pergamon Press, 577-617.

Peshchevitsky, B.I., Yerenberg, A.M., Belevantsev, V.I. and Kazakov, V.P. (1970) Stability of gold complexes in aqueous solutions. *Izv. Sib. Otd. Akad. Nauk SSSR, Ser. Khim.* 75- 83 (in Russian).

Pouradier, J., Gadet, M.C. and Chateau, H. (1965) Électrochimie des sels d'or. 1. Acides auro et aurichlorhydrigues et sels correspondants. *J. Chim. Phys.* **62**, 203-216.

Pryor, W.A. (1960) The kinetics of the disproportionation of sodium thiosulphate to sodium sulfide and sulfate. *J. Amer. Chem. Soc.* **82**, 4794-4797.

Puddephatt, R.J. (1978) *The Chemistry of Gold*. Elsevier, Amsterdam. 274pp.

Pyykkö, P. (1988). Relativistic effects on structural chemistry. *Chem. Rev.* **88**, 563-594.

Renders, P.J. and Seward, T.M. (1989a) The stability of hydrosulphido- and sulphido-complexes of Au(I) and Ag(I) at 25°C. *Geochim. Cosmochim. Acta* **53**, 245-253.

Renders, P.J. and Seward, T.M. (1989b) The adsorption of thio gold(I) complexes by amorphous As_2S_3 and Sb_2S_3 at 25 and 90°C. *Geochim. Cosmochim. Acta* **53**, 255-267.

Ryabchikov, I.D. Baranova, N.N., Zotov, A.V. and Orlova, G.P. (1985) The stability of $Au(OH)^0$ in supercritical water and the metal contents of fluids in equilibrium with a granite magma. *Geochem. Internat.* **22**, 116-117.

Rytuba, J.J. and Dickson, F.W. (1977) Reaction of pyrite + pyrrhotite + quartz + gold with NaCl–H_2O solutions, 300–500°C, 500 to 1500 bars and genetic implications. In *Problems of Ore Deposition, 4th IAGOD Symposium, Varina, 1974*, Vol.II. Bulgarian Acad. Sciences, Sofia.

Schwerdtfeger, P. (1989) Relativistic effects in gold chemistry. II. The stability of complex halides of gold(III). *J. Amer. Chem. Soc.* **111**, 7261-7262.

Schwerdtfeger, P., Dolg, M., Schwarz, W.H.E., Bowmaker, G.A. and Boyd, P.D.W. (1989) Relativistic effects in gold chemistry. I. Diatomic gold compounds. *J. Chem. Phys.* **91**, 1762-1774.

Seward, T.M. (1973) Thio complexes of gold and the transport of gold in hydrothermal ore solutions. *Geochim. Cosmochim. Acta* **37**, 379-399.

Seward, T.M. (1984) The transport and deposition of gold in hydrothermal systems. In *Gold'82: The Geology, Geochemistry and Genesis of Gold Deposits*, ed. Foster, R.P. Balkema Press, Rotterdam, 165-181.

Seward, T.M. (1989) The hydrothermal chemistry of gold and its implications for ore formation: boiling and conductive cooling as examples. *Econ. Geol.Mono* **6**, 398-404.

Shenberger, D.M. (1985) Gold solubility in aqueous sulphide solutions. M.Sc. thesis, Pennsylvania State University. 102pp.

Shenberger, D.M. and Barnes, H.L. (1989) Solubility of gold in aqueous sulphide solutions from 150 to 350°C. *Geochim. Cosmochim. Acta.* **53**, 269-278.

Skibsted, L.H. and Bjerrum, J. (1974) Studies on gold complexes. II. The equilibrium between gold(I) and gold(III) in the ammonia system and the standard potentials of the couples involving gold, diamminegold(I) and tetramminegold(III). *Acta Chem. Scand.* **A28**, 764-770.

Skinner, B.J., White, D.E., Rose, H.J. and Mays, R.E. (1967) Sulfides associated with the Salton Sea geothermal brine. *Econ. Geol.* **62**, 316-330.

Smith, F.G. (1944) The alkali sulphide theory of gold deposition. *Econ. Geol.* **38**, 561-590.

Starling, A., Gilligan, J.M., Carter, A.H.C., Foster, R.P. and Saunders, R.A. (1989) High-temperature hydrothermal precipitation of precious metals on the surface of pyrite. *Nature* **340**, 298-300.

Subzhiyeva, T.M. and Volkov, I.I. (1982) Thiosulfates and sulfites in thermal and hydrothermal waters. *Geochem. Internat.* $\mathbf{19}_1$, 94-98.

Sunder, W.A., Wayda, A.L., Distefano, D., Falconer, W.E. and Griffiths, J.E. (1979) Syntheses and Raman spectra of nitrosyl fluorometallate salts. *J. Fluorine Chem.* **14**, 299-325.

Takano, B. (1987) Correlation of volcanic activity with sulfur oxyanion speciation in a crater lake. *Science* **235**, 1633-1635.

Takano, B. and Watanuki, K. (1988) Quenching and liquid chromatographic determination of polythionates in natural water. *Talanta* **35**, 847-854.

Timakov, A.A., Prusakov, V.N. and Drobysheskii, Yu.V. (1986) Gold heptafluoride. *Dolk. Akad. Nauk SSSR* **291**, 125-128.

Uematsu, M. and Franck, E.U. (1980) The static dielectric constant of water and steam. *J. Phys. Chem. Ref. Data* **9**, 1291-1301.

Vilor, N.V. (1973) Laboratory and theoretical data on gold and silica in hydrochemical process. Candidate thesis, Inst. Zemnoy Kory SO AN SSSR, Irkutsk.

Webster, J.G. (1987) Thiosulphate in surficial geothermal waters, North Island, New Zealand. *Appl. Geochem*, **2**, 579- 584.

Webster, J.G. (1989) The solubility of As_2S_3 and speciation in dilute sulphide-bearing fluids at 25 and 90°C. *Geochim. Cosmochim. Acta* **53** (in press).

Weissberg, B.G. (1970) Solubility of gold in hydrothermal alkaline sulphide solutions. *Econ. Geol.* **65**, 551- 556.

Weissberg, B.G., Browne, P.R.L. and Seward, T.M. (1979) Ore metals in active geothermal systems. In *Geochemistry of Hydrothermal Ore Deposits* (2nd edn.), ed. Barnes, H.L. Wiley-Interscience, New York, 738-780.

Wood, S.A., Crerar, D.A. and Borcsik, M.P. (1987) Solubility of the assemblage pyrite–pyrrhotite–magnetite–sphalerite–galena–gold–stibnite–bismuthinite–argentite–molybdenite in H_2O–NaCl–CO_2 solutions from 200 to 350°C. *Econ. Geol.* **82**, 1864- 1887.

Zotov, A.V., Baranova, N.N., Dar'yina, T.G., Bannykh, L.N. and Kolotov, V.P. (1985) The stability of $AuOH^0$ in water at 300–500°C and 500–1500 atm. *Geochem. Internat.* **22**, 156-161.

Zotov, A.V. and Baranova, N.N. (1989) Themodynamic feature of the aurochloride solute complex, $AuCl_2^-$, at temperatures of from 350 to 500°C and under pressure of from 500 to 1500 bar. *Sciences Géologigues* (in press).

Zviagincev, O.E. and Paulsen, I.A. (1940) Contribution to the theory of formation of vein gold deposits. *Dokl. Akad. Nauk SSSR* **26**, 647-651.

3 Archaean lode gold deposits

D.I. GROVES and R.P. FOSTER

3.1 Introduction

3.1.1 *Global distribution and economic significance*

Lode gold deposits are one of the most characteristic features of Archaean greenstone belts within granitoid-greenstone terranes, with major deposits situated in most major cratonic areas (e.g. Australia, Brazil, Canada, India, South Africa, Zimbabwe: Foster, 1984; Macdonald, 1986; Ho and Groves, 1987). Even relatively small cratons typically contain some Archaean gold deposits (e.g. Tanzania; van Straaten, 1984), and the Archaean belts of China contain major gold deposits, although they are most likely of Mesozoic, not Archaean, age (e.g. Sang and Ho, 1987). Only the major cratons of Finland and the USSR (e.g. Baltic, Ukrainian and Siberian Shields) are virtually devoid of significant Archaean gold mineralization.

The widespread distribution of Archaean lode gold mineralization is reflected in its high contribution to total world gold production (Woodall, 1988). The Superior Province of the Canadian Shield, alone, has produced more than 4500 t (tonnes) (50 million oz) of gold (e.g. Colvine *et al.*, 1988), and other major producers include Western Australia (> 2350 t; Groves and Ho, in press), Zimbabwe (> 2200 t; Foster, 1989) Brazil (1890 t) and India (850t; data for Brazil and India from Milling-Stanley and Green, 1986; Bache, 1987; Milling-Stanley, 1988). Woodall (1988) has emphasized that there were two major periods of lode gold mineralization (i.e. excluding the giant Witwatersrand and modern placers), namely the late Archaean mesothermal deposits and the Mesozoic to Quarternary mesothermal to porphyry-related and epithermal deposits (Figure 3.1). Thus, the Archaean lode gold deposits described here are an extremely important group.

3.1.2 *Genetic concepts*

The genesis of the Archaean lode gold deposits has been the subject of considerable controversy. Prior to the 1960s, most authors discussing the deposits emphasized their structural control and favoured an epigenetic hydrothermal model with granitic intrusions providing the heat source and/or fluid source for fluid circulation: see Boyle (1979) for summary. In the 1960s there was successful application of syngenetic interpretations to a variety of ore-deposit types, including the often structurally complex volcanogenic massive sulphide deposits previously interpreted to be replacement deposits (see, for example, Sangster and Scott, 1976). This led to the common belief that Archaean gold deposits hosted by metasedimentary rocks (e.g.

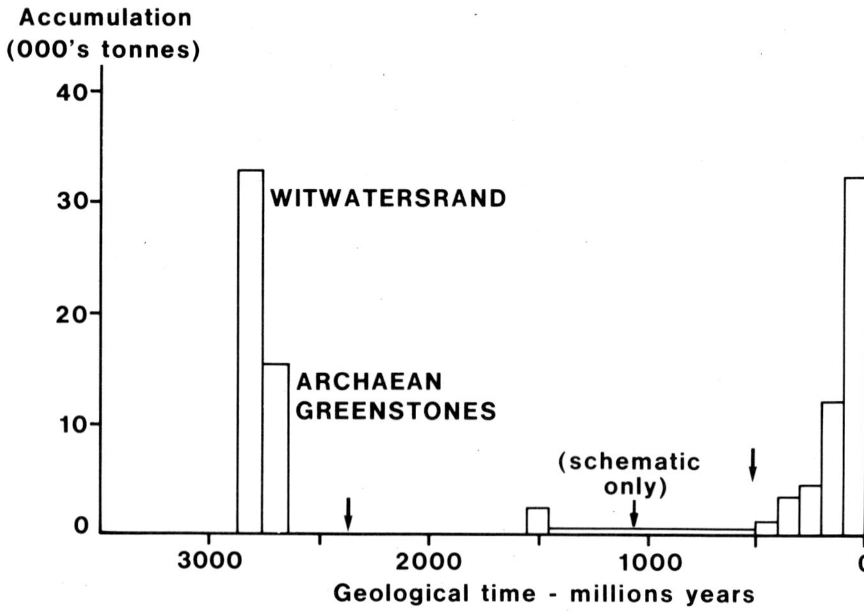

Figure 3.1 Somewhat schematic (due to uncertainties in both production data and age of mineralization) distribution of gold mineralization with time. (adapted from Woodall, 1988.)

oxide facies BIF) were synsedimentary or volcanogenic, with or without subsequent local redistribution of gold (e.g. Anhaeusser, 1976; Hutchinson, 1976), or that synvolcanic or synsedimentary gold concentrations acted as gold-enriched source beds during subsequent 'remobilization' events (e.g. Fripp, 1976a,b; Boyle, 1979). There were proponents of epigenetic hydrothermal models at this time (e.g. Kerrich and Fryer, 1979, 1981), but it was not until the mid-1980s that the importance of such models was generally re-established (e.g. Phillips and Groves, 1983; Foster, 1984; Colvine *et al.*, 1984; Macdonald, 1986; Ho and Groves, 1987). However, while epigenetic models are currently in vogue there are major controversies concerning the source of ore fluids and ore components, with metamorphic (upper or lower crust), magmatic (granitic, felsic porphyry or lamprophyre intrusions) and mantle sources all being invoked (see Groves *et al.*, 1988b for summary). There is also considerable discussion on the precise tectonic setting of gold mineralization. These matters are discussed in more detail below.

3.2 Nature of deposits

3.2.1 *Introduction*

The major features of Archaean lode gold deposits are described below, and are summarized in Tables 3.1 and 3.2 with emphasis on the larger deposits.

3.2.2 *Size and grade*

Highly mineralized greenstone belts within a major Archaean province characteristically contain several hundred to several thousand individual gold deposits or prospects. The majority of these have production ± reserves of less than 1 t Au, but there are commonly several large deposits, often with one or more giant gold districts or camps, in any terrane. For example, in the Superior Province of Canada, there are 25 deposits that contain more than 45 t Au and the Timmins (Porcupine) district has contributed more than 1530 t Au (Colvine *et al.*, 1988). Similarly, in the Yilgarn Block of Western Australia, there are 14 deposits that have produced more than 20 t Au, 25 with current reserves of greater than 10 t Au, with the Golden Mile at Kalgoorlie having produced about 1150 t Au with further reserves of more than 120 t Au (Groves and Ho, in press). Even the relatively small craton in Zimbabwe hosts two deposits that have yielded in excess of 100t Au and a further 26 deposits that have produced more than 10 t Au. A marked concentration of small and medium sized deposits (0.1–10 t Au) is particularly evident around the Globe and Phoenix deposit (production = 120 t Au) in the central part of Zimbabwe (Foster *et al.*, 1986).

The grade of the deposits, as mined, varies widely both historically, from a single mine or group of mines, and between mines in current operation. For example, the average grade of the Archaean lode gold deposits in the Yilgarn Block has fallen from in excess of 40 g/t Au in the 1890s to less than 5 in 1988 (Groves and Ho, in press), and in Zimbabwe fell from more than 220 g/t in 1890 to 5–6 g/t in the 1980s (Foster, 1982): similar trends are evident from other cratons. Most deposits currently exploited by underground mining have mining grades in the range 4 to 8 g/t, but many small deposits and a few large deposits (e.g. Teck-Corona at Hemlo; Burk *et al.*, 1986) have grades of 10–15g/t. Of particular current interest have been the advances in technology which have allowed open-pit mining of weathered ore zones with grades typically below 5 g/t (cut off grades below 0.5 g/t), in the Yilgarn Block of Western Australia. The largest of these deposits, Boddington (total resource 80 t Au), sees current mining in a lateritic profile at grades between 1.5 and 2 g/t (Symons *et al.*, 1988). In Zimbabwe, the Royal Family Mine (10 000t of ore per month) is profitably recovering gold from oxidized ore containing an average of only 1.2 g/t (Graham *et al.*, 1987).

3.2.3 *Structural styles*

Structure is the single most important factor controlling the distribution of gold deposits and the form (geometry) of gold ore shoots. On the regional scale, major gold deposits are commonly sited adjacent to transcraton shear zones (Figures 3.2, 3.3) and on the district scale they are usually sited in shorter strike-length, geometrically related, smaller scale structures. These structures hosting the gold deposits appear to largely reflect the movement on the transcraton shear zones, with different structural styles of mineralization resulting from variations in the orientation of the regional stress field and the strength of the host rocks (or contrasts in strength between adjacent host rocks).

Thus, gold mineralization may occur in cross-cutting or layer- parallel shear zones, strike-extensive laminated quartz veins (more rarely saddle reefs), extensional quartz-vein arrays and/or breccias. The variety of structural styles of mineralization is

Table 3.1 Characteristics of some well-documented, large to giant Archaean lode gold deposits from greenschist facies domains. Data for Western Australia largely derived from Groves *et al.* (1988a), from Canada from Colvine *et al.* (1988) and from Zimbabwe from Foster *et al.* (1986)

Region/Deposit	Gold production	Host rock	Lode type	Alteration in ore zone	Ore minerals
Yilgarn, Australia					
Golden Mile, Kalgoorlie	~1200 t	Tholeiitic dolerite sill	Steeply dipping brittle-ductile shear zones and brittle fracture sets; some breccias	Muscovite + ankerite + pyrite; some silicification and quartz veining	Gold + pyrite + minor scheelite, arsenopyrite and anhydrite; late tellurides
Sons of Gwalia, Leonora	>90 t	Tholeiitic to high Mg basalts	Steeply dipping major shear zone with local boudinaged quartz veins	Muscovite + biotite + ankerite + pyrite; quartz veins	Gold + pyrite + arsenopyrite ± pyrrhotite; minor chalcopyrite
Mararoa-Crown, Norseman	>70 t	Tholeiitic basalts and ultramafic intrusions (?)	Folded laminated quartz veins	Biotite + amphibole ± chlorite ± ankerite – dolomite	Gold + pyrite ± galena ± tellurides ± scheelite
Mt Charlotte, Kalgoorlie	>70 t	Granophyric unit in tholeiitic dolerite sill	Quartz vein set in hydraulic fractures	Sericite + ankerite + pyrite ± albite; some silicification	Gold ± pyrite ± scheelite
Abitibi, Canada					
Hollinger, Timmins	>600 t	Mafic flows; minor felsic flows and pyroclastics; quartz-feldspar porphyries	Quartz veins and stockworks; steeply dipping shear zones	Sericite + ankerite ± chlorite ± calcite; quartz and albite	Gold + pyrite + pyrrhotite ± galena ± sphalerite

Kerr Addison, Larder Lake	>320 t	High-Mg basalts, tholeiitic basalts; felsic porphyry to syenite dykes; clastic sediments	Stockworks ladder veins	Ankerite/dolerite–albite–muscovite	Gold + pyrite + scheelite + arsenopyrite
Sigma, Val d' Or	>110 t	Intrusive dolerite plug with andesite flows; minor feldspar porphyry dykes	Subvertical veins and breccias in brittle-ductile shear zones, flat veins	Calcite + white mica	Gold + pyrite + pyrrhotite; minor chalcopyrite, sphalerite, galena, tellurides, scheelite
Zimbabwe					
Cam and Motor, Kadoma	>145 t	Tholeiitic basalt/andesite; high-Mg basalts; dolerite intrusions; minor clastic sediments	Steeply dipping veins, shear-zone vein arrays, and stockworks	Quartz + ankerite; serpentine in high-Mg rocks	Gold + pyrite + arsenopyrite + stibnite ± sphalerite ± scheelite
Phoenix, Kwekwe	>105 t	Dunite-peridotite intrusive complex	Quartz veins; minor stockworks and silicification	Magnesite ± fuchsite ± talc	Gold + pyrite + stibnite + arsenopyrite
Dalny, Kadoma	>50 t	Tholeiitic basalt flows	Steeply dipping brittle-ductile shear zone	Ankerite + white mica + chlorite + albite	Gold + pyrite + arsenopyrite ± chalcopyrite ± galena ± tetrahedrite ± scheelite ± sphalerite

Table 3.2 Characteristics of some medium and large Archaean lode gold deposits from amphibolite and granulite facies domains

Region/Deposit	Gold production and reserves	Host rock	Lode type	Alteration in ore zone	Ore minerals	Reference
Amphibolite						
Big Bell, Murchison Western Australia	>90 t	Tholeiitic basalt or dolerite	Steeply dipping shear zones	Pyrite + quartz + muscovite; Pyrite + quartz + K-feldspar; minor sillimanite and rutile	Gold + pyrite ± arsenopyrite ± pyrrhotite ± stibnite ± chalcopyrite ± galena ± sphalerite ± magnetite	Phillips (1985)
Hemlo, Canada	>300 t	Probable andesitic to rhyolitic volcanics or pyroclastics: some clastic sediments	Steeply dipping ductile to brittle-ductile shear zones	Quartz + muscovite (some V-rich) ± biotite ± phlogopite ± microcline (some Ba-rich) ± barite ± titanite	Gold + pyrite + molybdenite ± sphalerite ± arsenopyrite ± stibnite; a large variety of minor Pb-, Cu-, Sb-, Hg-, Tl- and Te-bearing phases	Harris (1986); Kuhns *et al.* (1986); Burk *et al.* (1986)
Giant, Yellownife, Canada	>177 t	Pillowed and massive basalt	Steeply dipping brittle-ductile shear zone	Sericite ± ankerite ± chlorite ± albite	Gold ± pyrite + stibnite + sulphosalts + sphalerite + galena	Boyle (1955); Henderson and Brown (1966); Padgham (1979)
Kolar, India	>790 t	Mainly basaltic rocks; subsidiary ultramafic rocks and minor chemical sediments	Quartz veins and sheeted veinlets related to brittle-ductile shear zones; minor replacements of BIF	Quartz + diopside ± hornblende ± biotite; minor sericite and chlorite near veins; calcite and tourmaline	Gold in quartz ± pyrite ± pyrrhotite ± arsenopyrite ± sphalerite ± galena	Hamilton and Hodgson (1986)
Granulite						
Renco[1], Zimbabwe	1.5 t p.a. Reserves not published	Hypersthene–quartz–feldspar–granulite (enderbite)	Folded ductile shear zones with later brittle deformation	Serpentine + biotite + sericite ± epidote	Gold + pyrrhotite + chalcopyrite ± pyrite ± bismuth ± bismuth telluride ± maldonite	Tabeart (1987)

[1]Structural relationships indicate that Renco is of late Archaean age (C.F. Tabeart, personal communication, 1989).

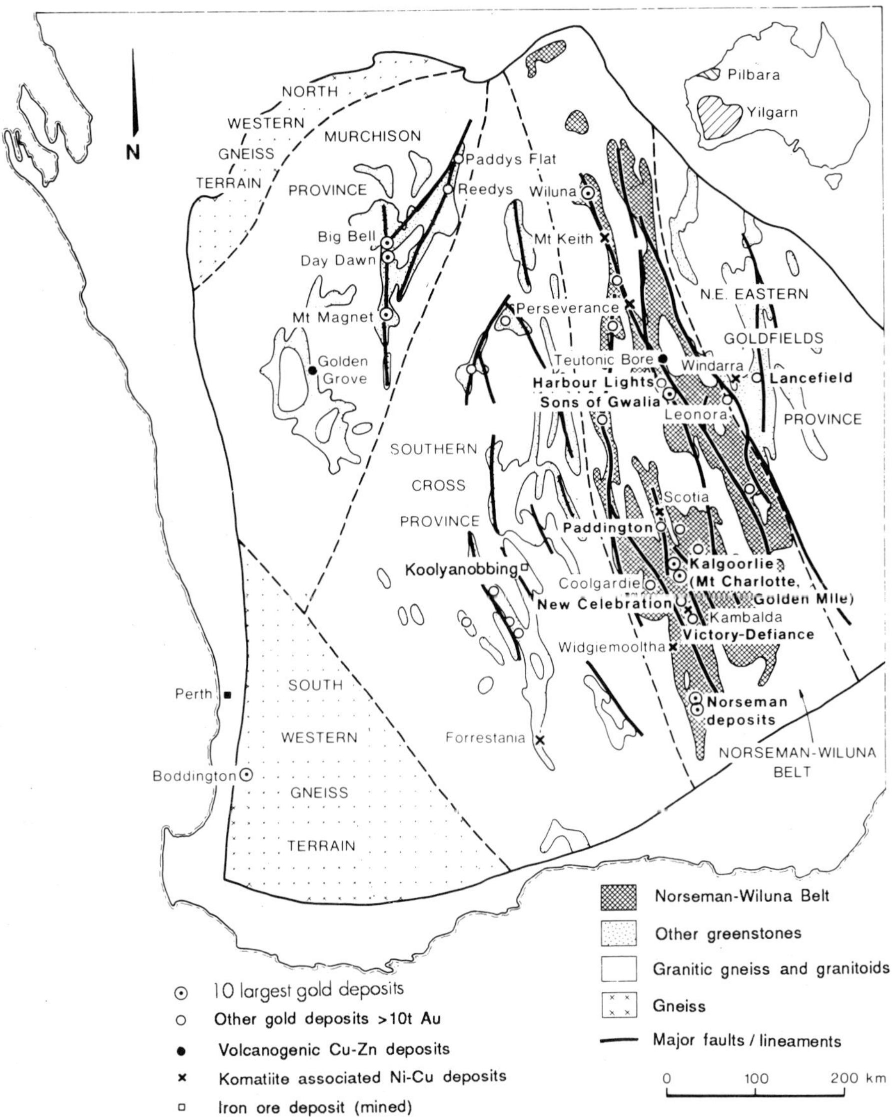

Figure 3.2 Generalized geological map of the Yilgarn Block, Western Australia, showing association of major gold deposits (> 10 t Au) with transcraton deformation zones. Similar associations are shown in Canada (see figures in Colvine *et al.*, 1984, 1988, and Hodgson, 1986) and Southern Africa (see summary figures in Vearncombe *et al.*, 1988).

shown schematically in Figurc 3.4, with individual examples shown in Figure 3.5. Depending on the scale of controlling zones and homogeneity and thickness of host rocks, an individual mine may be dominated by one deposit style or may contain mineralization of several styles. The importance of strength (competency) contrasts is shown by the occurrence of abundant mineralization in shear zones along lithological

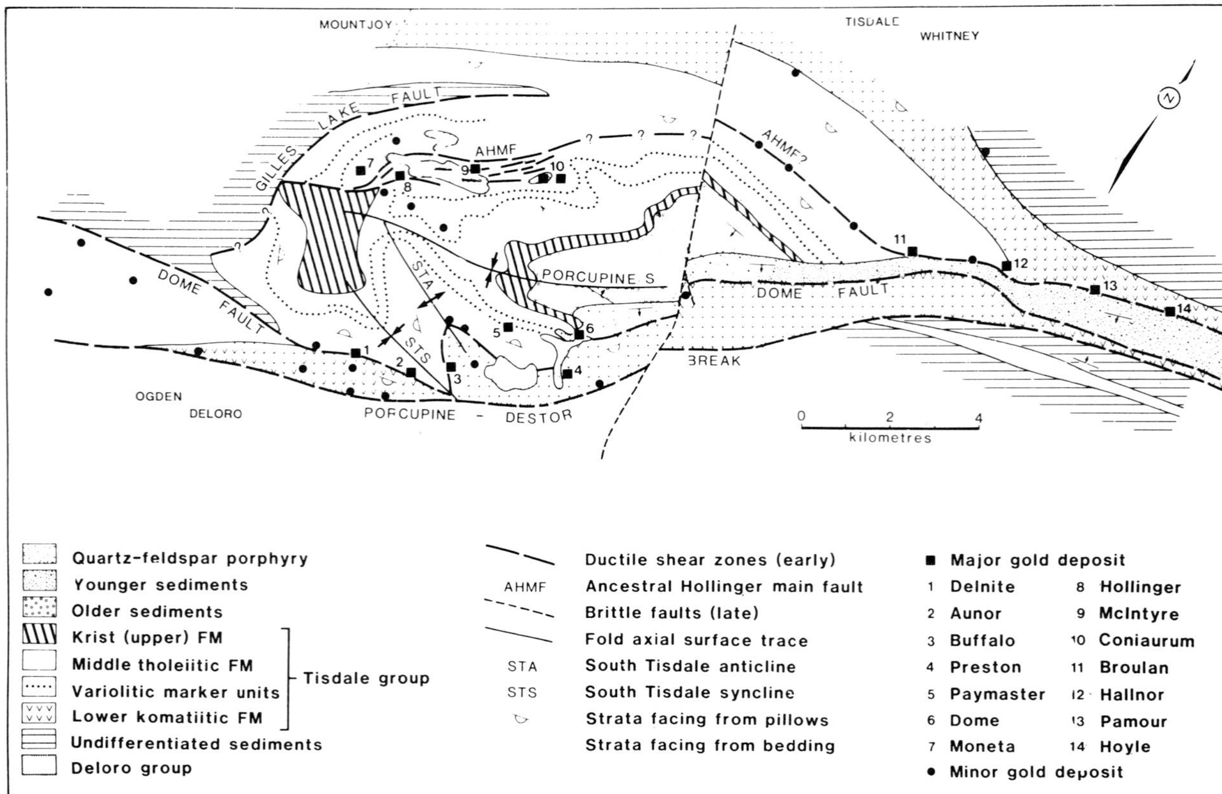

Figure 3.3 Generalized geological map of the Porcupine gold camp, Timmins area, showing relationship between a transcraton deformation ('break') and smaller-scale structures hosting gold deposits in the vicinity. (From Hodgson, 1986 with permission of Instn. Min. Met.)

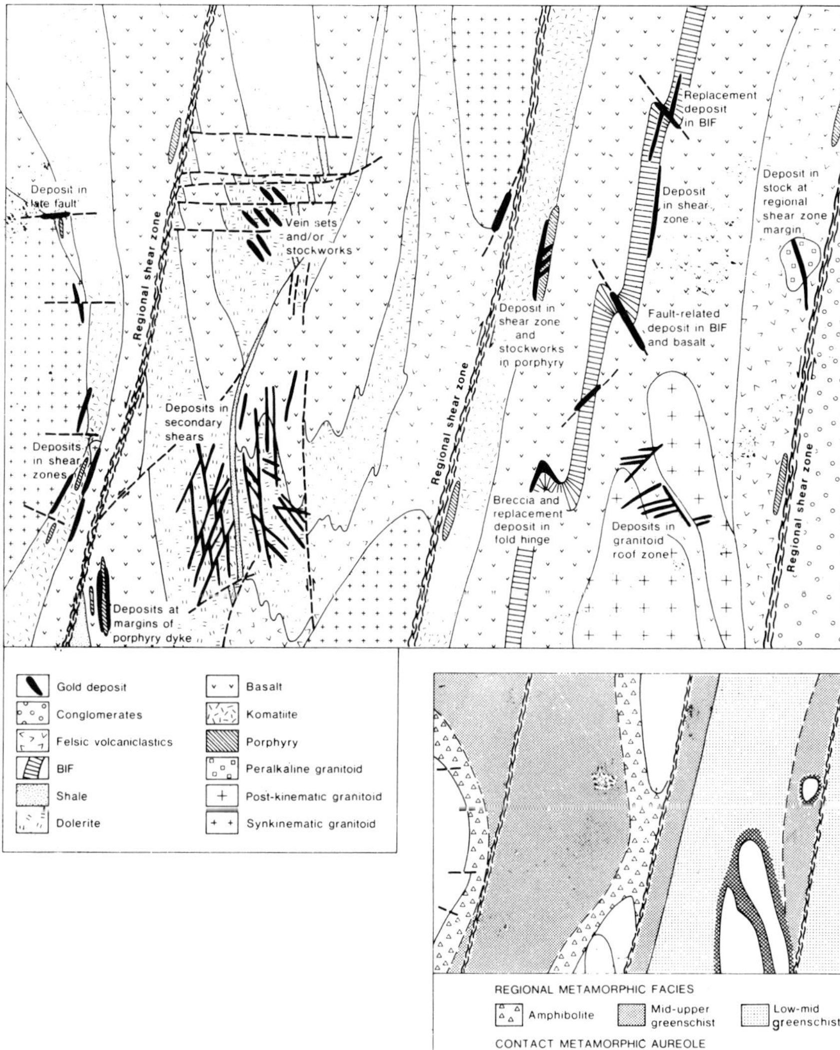

Figure 3.4 Schematic representation of the nature of Archaean gold mineralization showing variable structural styles, host rocks and metamorphic setting. The deposit settings are based largely on Western Australian examples, but are applicable to deposits from other cratons. (From Groves *et al.*, 1988b.)

contacts and the common restriction of discordant vein-style mineralization to specific lithologies (e.g. dolerite sills, felsic porphyry intrusions). Thus, many orebodies have a grossly 'stratabound' appearance, contributing to the epigenetic vs. syngenetic controversy concerning the deposits. This is well demonstrated by the highly mineralized Kalgoorlie district of the Yilgarn Block, Western Australia (Figure 3.6), where three styles of mineralization are evident: shear-zone-hosted gold

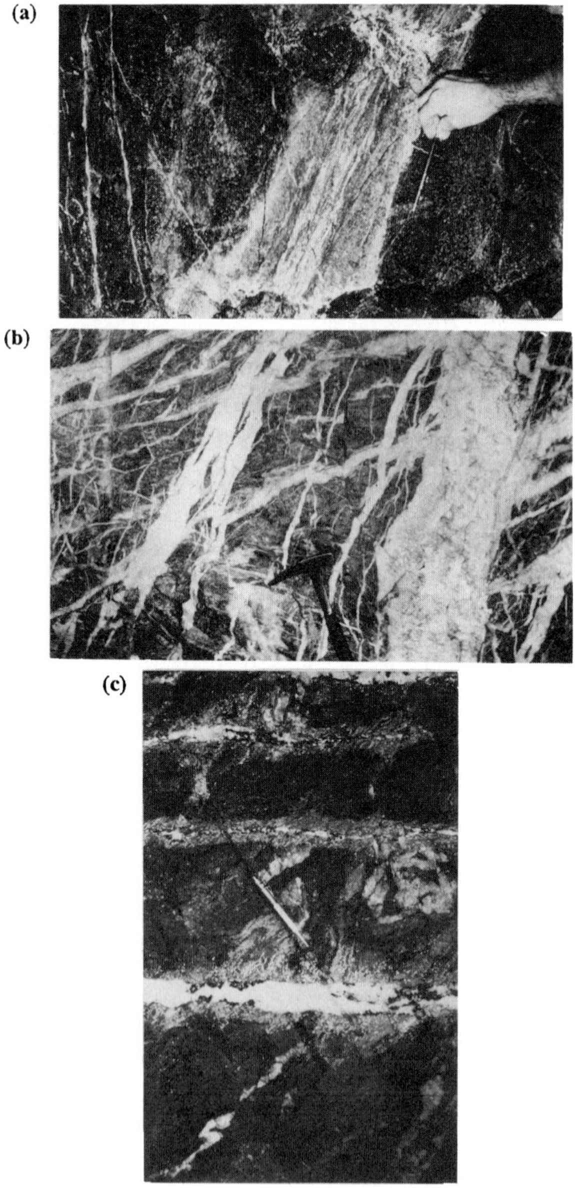

Figure 3.5 Photographs illustrating a variety of structural styles exhibited by Archaean lode gold deposts. (a) Small-scale mineralized shear zone; Perseverance Mine, Kalgoorlie, Australia. (b) Quartz vein arrays in fuchsitic carbonate rock; 12 level, Phoenix section, Globe and Phoenix Mine, Zimbabwe. (c) Quartz veins and sulphidation haloes in Kapai Slate; Victory Mine, Kambalda, Australia. (d) Quartz-cemented breccia; 21 level, Bulawayo section,

(d)

(e)

(f)

Commoner Mine, Zimbabwe. (e) Telluride and gold mineralization in fractures in massive quartz (middle) and lower quartz breccia. Black matter is carbonaceous schist; 21 level, Commoner Mine, Zimbabwe. (f) Quartz vein with lower laminated section, central massive quartz, and upper re-cemented quartz breccia; 4 level, Sunace Mine, Zimbabwe.

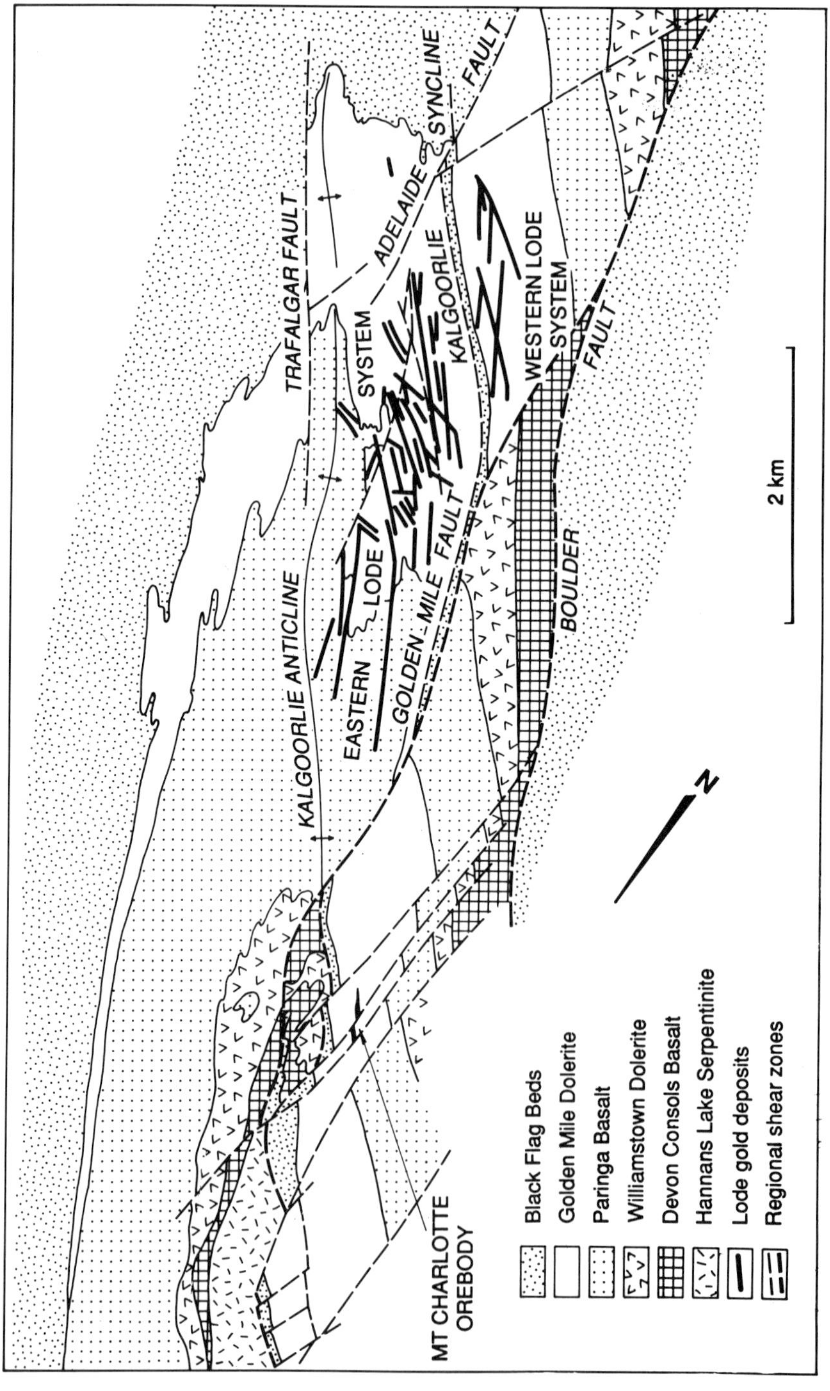

Figure 3.6 Geological map of the Kalgoorlie area, Yilgarn Block, Western Australia, showing the virtual confinement of various styles of mineralization to the Golden Mile Dolerite or its contact with Paringa Basalt.

deposits (Golden Mile) are best developed within the Golden Mile Dolerite, a differentiated tholeiitic dolerite sill; quartz-vein arrays at Mt. Charlotte are virtually restricted to the granophyric Unit 8 of the Golden Mile Dolerite; and high-grade gold-telluride breccias are most commonly developed at or adjacent to the contact between the Golden Mile Dolerite and Paringa Basalt. In the special case where banded iron formations (BIFs) or other Fe-rich metasedimentary rocks are hosts to gold mineralization, the ores may be even grossly stratiform in places due to selective sulphidation of Fe-rich layers (e.g. Phillips *et al.*, 1984; Fuchter and Hodgson, 1986; Gilligan and Foster, 1987).

A variety of structures are mineralized, with strike-, oblique-, reverse-, and normal-slip shear zones all recorded as controlling gold deposits (e.g. Vearncombe *et al.*, 1988). Steep reverse shear zones may be particularly important ore-controlling structures (see Sibson *et al.*, 1988). In many cases, the ore shoots have a marked plunge subparallel to mineral-elongation lineations in the host rocks, suggesting a fundamental structural link between stretching and the development of conduits which focused fluid-flow. In other deposits, the ore shoots plunge parallel to the intersections between shear zones and other planar structures (e.g. lithological contacts, other faults).

Most structures that control gold mineralization show features typical of brittle–ductile deformation, but there is a complete spectrum from control by ductile shear zones to brittle faults and fracture zones. Colvine *et al.* (1988) have proposed that there is a broad depth zonation of structures controlling gold mineralization, with ductile structures dominating in amphibolite facies domains whereas brittle structures are prevalent in sub-greenschist facies domains. This may have validity as a generalization, but there are numerous examples where ductile and brittle structures occur in the same district (e.g. Golden Mile, Kalgoorlie) at the same metamorphic grade. In these cases, strength contrasts between lithologies and changes in strain rate may be as important as the P–T conditions of the rocks at the time of deformation and mineralization. In addition, it is to be expected that physico-chemical conditions will vary during mineralization due to the progressive evolution of the controlling shear zones and faults and the likely uplift associated with reverse and oblique-slip deformation zones (e.g. Sibson *et al.*, 1988).

3.2.4 *Host rocks*

On a world and craton scale, all lithologies in Archaean greenstone belts may be mineralized, and at the mine scale several different rock-types may host economic gold mineralization. There are two ways in which the relative importance of specific lithologies as host rocks to mineralization have been viewed.

First, the contribution to total gold production from deposits in particular host rocks can be simply assessed (Figure 3.7). This approach shows that mafic volcanic and intrusive rocks are by far the most important host rocks in several regions (e.g. Yilgarn Block, Groves and Phillips, 1987; Zimbabwe Craton, Foster, 1985; India, Hamilton and Hodgson, 1986), and are the dominant, or one of two dominant, host rocks in several others (e.g. Canadian Shield, Colvine *et al.*, 1988; Brazil, Thomson, 1986). In the Yilgarn Block, in particular, not only are mafic rocks the dominant hosts but they also contain most of the larger gold deposits, with several authors (see summary in Groves and Phillips, 1987) stressing the importance of their competent nature and their high Fe : (Fe+Mg) ratios; the significance of the latter is described below.

An alternative approach is to assess the contribution from particular host rocks in terms of their relative abundance in the mineralized greenstone belts (e.g. Colvine *et al.*, 1988). In this case, the only rock-type which is consistently more important as a host to gold deposits relative to its overall abundance is BIF where it is present (e.g. Foster, 1985; Groves and Phillips, 1987; Colvine *et al.*, 1988). It is noticeable that, like the mafic host rocks, BIF is Fe-rich and has markedly different strength to surrounding lithologies. In the Abitibi Belt, the late felsic plutons that intrude the greenstone belts are the most selectively mineralized lithology (Colvine *et al.*, 1988), but there are few large deposits in such rocks in other areas (e.g. Western Australia and southern Africa). Although granitoid batholiths are the dominant component of most granitoid-greenstone terranes, they are relatively poorly mineralized (Figure 3.7).

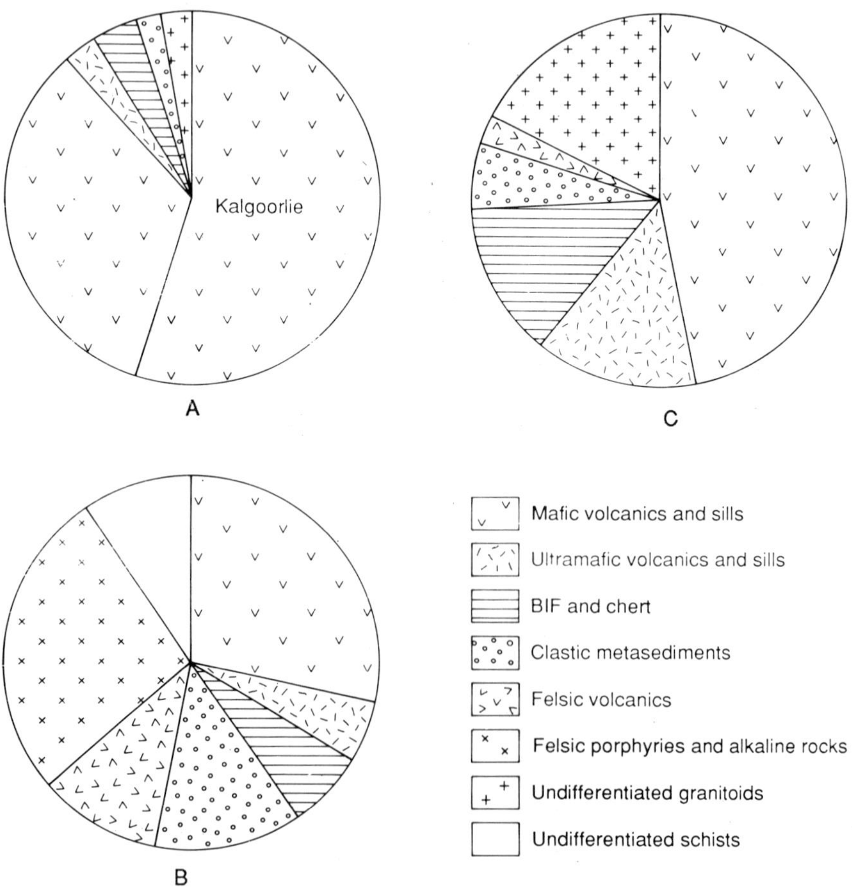

Figure 3.7 Host rocks to lode gold deposits in: (A) the Yilgarn Block, Western Australia, (B) the Abitibi Belt, Canada, and (C) Zimbabwe. (*Source of data*: Groves and Ho (in press), Colvine *et al.* (1988), Roberts *et al.* (in press).

3.2.5 *Mineralization and wallrock alteration*

Although free gold occurs in rich quartz-vein lodes, much gold mineralization is an integral part of wallrock alteration, and, even in the veins, minerals characteristic of the alteration may be present.

The ore mineralogy is generally rather simple with free gold and/or gold sited in pyrite ± pyrrhotite ± arsenopyrite, with a heterogeneous distribution of accessory scheelite, tellurides, stibnite, galena, sphalerite, chalcopyrite, magnetite, haematite and anhydrite: all may be locally abundant in specific ore shoots. Rarely, the gold deposits contain more complex assemblages of sulphides, sulpharsenides and sulphosalts (e.g. Hemlo; Harris, 1986). Multiple phases of mineralization are common, although these probably represent a fluid continuum within an evolving brittle–ductile shear zone rather than temporally discrete pulses of hydrothermal activity.

Wallrock alteration normally involves massive introduction of CO_2, K_2O, S and H_2O, with either introduction or redistribution of SiO_2 and more localized introduction of Na_2O; Rb, Li and Ba are also commonly enriched. There is a distinctive suite of immobile elements (e.g. Al, Ti, V, Y, Zr) and relatively immobile elements (Fe, Mg, Cr, Ni, Sc), and there is commonly a volume increase in mineralized zones. The mineralogical expression of this metasomatism varies both with host rock lithology and metamorphic grade: see summary in Colvine *et al.* (1988) and Table 3.3.

At sub-amphibolite facies grades in most host rocks, there is commonly a wide zone of pervasive carbonate alteration with a central core of ankerite, ferroan dolomite or dolomite (less commonly magnesite) grading out into outer fringes rich in calcite, commonly with associated chlorite. In the strongly mineralized zone, there are normally K-micas, biotite, sericite (less commonly V- or Cr-bearing micas), but these may be absent in Al-poor rocks such as BIF. In addition, albite and less commonly K-feldspar may be present in volcanic or intrusive host rocks in specific deposits. Quartz veining and silicification are variably developed, and tourmaline may be locally abundant. High fluid/rock ratios are indicated by the intensity of alteration (e.g. Phillips, 1986).

In amphibolite facies domains, higher grade metamorphic minerals such as chloritoid, staurolite, cordierite, garnet, andalusite, sillimanite, kyanite and diopside-hedenbergite, together with a variety of amphiboles, may be present (e.g. Mathieson and Hodgson, 1984; Phillips, 1985; Colvine *et al.*, 1988). There may also be alteration minerals more typical of sub-amphibolite domains (e.g. Harris, 1986; Tabeart, 1987) (Table 3.2), almost certainly as a result of retrogression of higher grade alteration minerals. Mueller (1988) has suggested that some calc-silicate alteration assemblages from amphibolite facies domains in the Southern Cross Province of Western Australia are similar to those of Phanerozoic skarn deposits.

The timing of wallrock alteration relative to peak metamorphism is highly controversial in amphibolite facies domains (e.g. Phillips, 1985; Phillips and De Nooy, 1988; Burk *et al.*, 1986; Kuhns *et al.*, 1986). However, in greenschist facies settings, several careful studies (e.g. Phillips and Groves, 1984; Clark *et al.*, 1986; Skwarnecki, 1987) have shown that alteration minerals overprint peak- metamorphic

Table 3.3 Typical wallrock alteration in and adjacent to lode gold deposits in greenschist facies domains. In part adapted from Colvine *et al.* (1988)

Host rock	Main ore zone	Proximal alteration	Distal alteration
Basalt and dolerite	Quartz–carbonate veining; silicification; ankerite + biotite or sericite, some albite; rare green mica; abundant Fe-sulphides ± arsenopyrite; stibnite locally abundant	Ankerite-dolomite + biotite or sericite ± high-Mg chlorite; some Fe-sulphides	Calcite + chlorite + amphibole + albite; minor Fe-sulphides
Komatiites and ultramafic sills	Quartz–carbonate veining; silicification; ankerite-magnesites + phlogopite or green (V or Cr) mica; abundant Fe-sulphides and arsenopyrite; stibnite locally abundant	Dolomite-ankerite-magnesite + talc ± phlogopite; minor Fe-sulphides	Chlorite + amphibole + talc-dolomite- calcite; serpentine ± chlorite ± talc
Banded iron formation	Quartz–carbonate veining; silicification; sulphidation of magnetite or siderite to Fe-sulphides ± arsenopyrite	Partial sulphidation of magnetite or siderite to Fe-sulphides; some chlorite	No obvious alteration
Granitoids and felsic porphyries	Quartz veining; intense silicification; biotite or sericite + calcite; K-feldspar or albite; Fe-sulphides ± arsenopyrite; chalcopyrite, sphalerite, galena, molybdenite and tourmaline locally abundant	Sericite ± K-feldspar ± albite; calcite + titanite ± tourmaline; minor Fe-sulphides	Sericite + calcite; commonly limited alteration

assemblages and hence that wallrock alteration post-dated peak metamorphism. Despite this, the overall complementary variation shown by alteration assemblages and metamorphic grade, together with the common occurrence of biotite in upper greenschist facies domains compared to sericite in lower greenschist facies domains, suggests that wallrock alteration and gold mineralization occurred broadly within a regional tectono-metamorphic event.

3.2.6 *Metal associations*

Archaean gold deposits are commonly termed 'gold-only' deposits (e.g. Hodgson and MacGeehan, 1982) because of the extreme enrichment of Au relative to other metallic elements (e.g. Cu, Pb, Zn, Ag) that commonly accompany Au in other deposit-types (e.g. porphyry Cu-Au, epithermal deposits, Au-rich VMS deposits). The Archaean gold deposits show a characteristic association of Au with As, W, Ag, Sb, Te and B, with generally low base-metal and Mo contents (e.g. Kerrich, 1983; Phillips and Groves, 1983): Figure 3.8 shows the contrast between metal enrichments in some large Archaean gold deposits and volcanogenic massive sulphide deposits. Gold/silver ratios are usually around 10 : 1. Some deposits, however, may be considerably enriched in one or more of Sb, Cu, Pb, Mo and Ag.

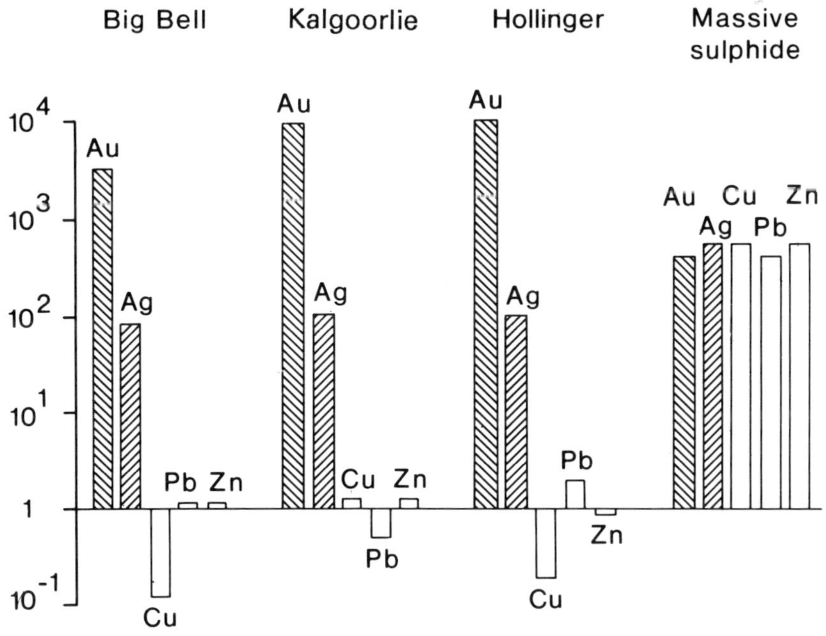

Figure 3.8 Comparison between enrichment factors (above Clarke values) for precious and base-metals in Archaean gold and volcanogenic massive sulphide deposits. (From Groves and Barley, 1988.)

3.3 Regional distribution

3.3.1 *Heterogeneous distribution*

Gold deposits are heterogeneously distributed within greenstone belts, with some large terranes being almost devoid of economic deposits whereas others contain numerous large deposits. Foster (1985) and Groves *et al.* (1987) presented data on the productivity of various greenstone terranes in terms of kg Au per km^2 greenstone. On this basis, the Zimbabwe greenstone belts are the most productive (59 kg/km^2), perhaps partly due to extensive precolonization mining of low grade gold deposits, followed by the areally restricted Barberton belt (50 kg/km^2), the Abitibi Belt (43 kg/km^2) and the Yilgarn Block (25 kg/km^2) including the Norseman-Wiluna Belt (35 kg/km^2).

Where there are sufficiently large granitoid-greenstone terranes for mineralization patterns to be significant (e.g. Canadian and Western Australian Shields), the most highly mineralized greenstone terranes, that also contain the giant (> 500 t Au) deposits, are those dominated by late-Archaean (*ca.* 2.7 Ga) greenstones forming broadly linear belts that are typified by anastomosing crustal-scale deformation zones. The best documented examples are the Abitibi Belt of Superior Province (e.g. Colvine *et al.*, 1988) and the Norseman- Wiluna Belt of the Yilgarn Block (e.g. Groves *et al.*, 1988). Foster (1985) has noted that *ca.* 2.7 Ga greenstone belts of the highly productive Zimbabwe Craton are also the most highly mineralized, although they do not show the same degree of linearity as their Canadian and Western Australian counterparts. In the Superior Province, Yilgarn Block and Zimbabwe Craton, 3.0–2.9 Ga greenstone belts may be highly mineralized within specific areas and contain large gold deposits (e.g. Red Lake, Canada; Mount Magnet-Meekatharra, Western Australia), although available evidence suggests that late tectonism, peak metamorphism and mineralization all post-dated this *ca.* 3.0 Ga greenstone belt volcanism by > 200 Ma.

The Pilbara Block, Western Australia, is a particularly poorly mineralized terrane. It is typified by ovoid granitoid batholiths and stellate 3.5–3.35 Ga greenstone belts with sparse, widely spaced crustal-scale shear zones (Krapez and Barley, 1987); the structural pattern appears to be similar to that of some of the less well mineralized provinces in the west and north-west of the Superior Province. Age need not, however, be a significant factor, because part of the 3.5–3.3 Ga greenstone terrane of the Barberton Mountain Land is anomalously mineralized where there is a high fault/shear-zone density (e.g. Anhaeusser, 1976).

3.3.2 *Structural setting*

As discussed above, the most obvious regional control on the distribution of Archaean gold deposits is structure. Gold mineralization is clearly better developed in greenstone terranes dominated by anastomosing sets of crustal-scale deformation (shear) zones, and within these terranes major gold deposits show a close spatial correlation with these shear zones, commonly resulting in broadly linear arrays of major gold deposits or even gold camps (Figure 3.2). Gold deposits are effectively located in and adjacent to high-strain zones within otherwise generally low-strain greenstone successions.

The best documented examples of such relationships are from the Abitibi Belt and the Norseman-Wiluna Belt. In the Abitibi Belt, large gold deposits occur adjacent to crustal-scale deformation zones (or 'breaks'), which are interpreted to have controlled late-Archaean sedimentation ('Timiskaming-type') and the emplacement of syn- to late-tectonic, intermediate to felsic, silica-saturated and alkaline (including lamprophyres) dykes and stocks (e.g. Colvine *et al.*, 1988). Similarly, in the Norseman-Wiluna Belt, gold deposits occur near crustal-scale, strike- to oblique-slip shear zones that may in part be reactivated synvolcanic faults (Groves and Batt, 1984) and which are also sites for the emplacement of suites of lamprophyric to felsic porphyry intrusions (e.g. Groves *et al.*, 1988a,b; Perring *et al.*, 1988). At the present, there are generally insufficient data on the relative timing of major shear zones and their kinematic history to confidently predict craton-scale stress fields at the time of deformation, although local stress fields may be well defined (e.g. Vearncombe *et al.*, 1988). Colvine *et al.* (1988) suggested NNW-directed compression in the Superior Province could have produced the observed major shear zones, but such interpretations are probably oversimplistic in terms of the variable age of gold mineralization in different terranes of the Province and in view of evidence for more than one movement vector on such major shear zones where well studied in Western Australia (e.g. Mueller *et al.*, 1988; Vearncombe *et al.*, 1988) and Zimbabwe (Carter, 1988).

As discussed above, most of the gold deposits do not actually occur in the crustal-scale shear zones but in subsidiary structures that are geometrically related to them (Figures 3.2, 3.3). The significance of the relationships is discussed below.

3.3.3 *Metamorphic setting*

Gold deposits occur in a variety of settings from sub-greenschist to middle-amphibolite facies. Although some major deposits occur in mid- to upper- amphibolite facies domains (e.g. Hemlo; Big Bell; Kolar), most of the gold production has come from middle- to upper-greenschist facies domains which also contain the giant deposits such as Timmins (Porcupine) and the Golden Mile (Kalgoorlie). This metamorphic P–T regime corresponds broadly to the brittle–ductile transition between higher P–T domains in which ductile mechanisms dominate and lower P–T domains where brittle behaviour is more common (e.g. Sibson, 1987).

3.3.4 *Spatial relationship to intrusive rocks*

Although some gold deposits occur within or adjacent to the contact of regional granitoid batholiths (e.g. Zimbabwe; Foster, 1985), most attention has been given to the spatial association between relatively minor intrusions of intermediate to felsic porphyries and lamprophyres and Archaean gold mineralization.

This association is most pronounced in the Superior Province where intrusions spatially associated with gold deposits range from quartz-bearing types such as granodiorites and monzonite (e.g. Hollinger-McIntyre; Burrows and Spooner, 1986) to silica-undersaturated types such as syenite (e.g. Macassa; Kerrich and Watson, 1984), monzonite (Canadian Arrow; Cherry, 1983) and lamprophyre (e.g. Hollinger-McIntyre; Burrows and Spooner, 1988): the last are abundant in gold camps (Boyle, 1979). Hodgson and Troop (1988) showed that over 70% of gold deposits hosted by other lithologies still have small felsic to mafic porphyritic intrusions in the mine environment. Further, Colvine *et al.* (1988) estimated that felsic intrusions host gold deposits that constitute about 25% of total gold production from the Abitibi Belt whereas they represent less than 4% of the belt as a whole. There are also some specific associations between ore minerals such as molybdenite, scheelite, galena, tellurides and tourmaline and particular types of porphyritic intrusion (Hodgson and Troop, 1988). Such specific associations have led to suggestions of a magmatic source for the gold deposits (e.g. Burrows and Spooner, 1985; Burrows *et al.*, 1986; Hodgson and Troop, 1988).

The situation is much less clear in other cratons where such minor intrusions may be absent or rare in the mine environment (e.g. southern Africa). There is certainly a broad association between major shear zones, porphyritic intrusions and gold deposits in the Yilgarn Block, although there are no major deposits hosted in the porphyritic rocks. Importantly, the lamprophyres may have high gold contents (e.g. Rock and Groves, 1988a,b), whereas the intermediate to felsic porphyries are rarely enriched in gold. In Zimbabwe, there is a close spatial association between gold mineralization and intrusive stocks ranging in composition from trondhjemitic through tonalitic to granodiorite (Mann, 1984; Foster and Wilson, 1984), although again they do not host any major deposits. Quartz- and plagioclase-phyric felsic rocks occur in many deposits as narrow (0.5–5 m) sills and dykes and often form either the hanging wall or footwall of the lode (Maufe, 1913; Foster *et al.*, 1986). Lamprophyric rocks are rare but lamprophyre dykes have been reported from the Phoenix and Gaika Mines in the Kwekwe area (Macgregor, 1932).

Clearly the spatial association between a variety of minor intrusions and gold mineralization has to be accommodated in any genetic model for the gold deposits.

3.3.5 *Timing of mineralization*

In most cratons, the timing of mineralization can only be determined relative to deformational, metamorphic and intrusive events, for which there are no precise absolute ages. In any one terrane, gold mineralization appears to broadly represent a single event related to late-deformation accompanying or outlasting peak metamorphism and post-dating the intrusion of most granitoids, felsic porphyries and lamprophyres. Lead model ages on gold-related sulphides are consistent with a broadly synchronous gold mineralization event late in the evolution of any one terrane (e.g. Browning *et al.*, 1987), but are too imprecise and model-dependent to accurately define the age of mineralization.

Precise U–Pb in zircon geochronology has been used to constrain the age of Archaean gold mineralization through dating of intrusive rocks that are both cut by and cut the gold mineralization in the Canadian Shield (summarized by Colvine *et al.*, 1988). These studies confirm the late timing of the mineralization, and importantly suggest that it significantly post-dated (10–15 Ma; Marmont and Corfu, 1988) the abundant felsic intrusions that are selectively mineralized in the belt. Available data suggest that gold mineralization was broadly contemporaneous with the emplacement of late alkaline intrusions that include syenites and lamprophyres. Attempts to directly date mineralization by $^{40}Ar/^{39}Ar$ dating of micas or U–Pb in rutile and monazite studies in the Canadian Shield provide young ages the significance of which is not entirely clear. The one attempt to date mineralization in the Yilgarn Block by U–Pb in rutile studies (Clark *et al.*, 1988) provided an age that is *ca.* >30 Ma later than the metamorphic peak and 30 Ma older than minor porphyritic intrusions. As in the Superior Province, lamprophyres are closest in timing to gold mineralization as they are both cut by and cut mineralized structures (Rock *et al.*, 1988).

A major late gold event in Zimbabwe is indicated by the occurrence of lodes in basic volcanics of the late Archaean Upper Greenstones which have been bracketed at approximately 2.7 Ga (Wilson *et al.*, 1978; Wilson, 1979).

3.3.6 *Peak mineralization age*

As discussed briefly above, the most highly mineralized terranes in any craton are normally the *ca.* 2.7 Ga greenstone belts. Available data suggest that the late-Archaean (2.65 ± 0.5 Ga), the time of final stabilization of several of the ancient cratons, was a period of intense gold mineralization exceeded only by post-Mesozoic gold deposition (Woodall, 1988). This observation needs to be addressed in any genetic models.

3.4 Constraints on genetic models

3.4.1 *Introduction*

Those parameters that constrain genetic models are discussed below and relevant data are summarized in Table 3.4.

3.4.2 *Nature of ore fluids*

The nature of the ore fluids has been deduced from fluid inclusion studies of quartz veins interpreted to be synchronous with, or slightly post-dating, the main stage of gold mineralization (e.g. Ho *et al.*, 1985; Ho, 1987; Smith *et al.*, 1984; Robert and Kelly, 1987). These studies imply that similar ore fluids were responsible for Archaean gold mineralization in greenschist facies domains in different cratons.

The fluid was a $H_2O - CO_2$ fluid, normally with 10–25 mole% CO_2, low salinity (typically < 2 wt.% NaCl equivalent), and moderate density (0.9 g/cm^3). The fluid was near-neutral to slightly alkaline and normally reducing, although oxidized fluids have been recorded (e.g. Phillips *et al.*, 1986; Cameron and Hattori, 1987). Minimum gold depositional conditions were 200–400°C (normally 250–350°C) and 0.5–4.5 kb (normally 1–3 kb). There is no evidence for systematic temperature gradients in any of the deposits. Significant methane is present in fluids where they interacted with carbonaceous sedimentary rocks (Ho, 1987). Variable degrees of phase separation are recorded (e.g. Ho, 1987), but it is only rarely that saline aqueous fluids are recognized; for example, up to 11 wt.% NaCl equivalent at the Dalny Mine, Zimbabwe (Carter, 1988), and up to 34 wt.% NaCl equivalent in the Sigma Mine, Quebec (Robert and Kelly, 1987).

The similar ranges of stable isotope (C, O, H, S) compositions from Archaean gold deposits world wide (summarized by Colvine *et al.*, 1988) also support generation of the primary ore fluid from a large reservoir of similar nature in all cratons.

3.4.3 *Transport and deposition of gold*

The nature of the ore fluid, the ore–element association, high gold:base-metal ratios, and commonly intimate association of gold with Fe-sulphides all suggest gold transport as reduced sulphur complexes (e.g. Phillips and Groves, 1983), an interpretation which is consistent with available experimental data (Seward, 1973, 1984).

In many of the larger deposits, where the bulk of the gold is intimately associated with Fe-sulphides in host rocks with high Fe/(Fe + Mg) ratios, it is likely that wallrock sulphidation reactions induced instability of the reduced sulphur complexes and caused gold precipitation (e.g. Phillips and Groves, 1984). Such mechanisms are, however, not universal because rock-types with low Fe/(Fe + Mg) ratios are mineralized, and it is likely that other fluid–wallrock reactions induced changes in pH and fO_2 that initiated gold deposition; carbonaceous wallrocks are important in this respect in some instances (Springer, 1985). An alternative mechanism is the later adsorption of aqueous gold-bearing species on to earlier-formed sulphides, due to their conductivity, and then autocatalytic reduction of the aqueous species to metallic gold (Jean and Bancroft, 1985; Hyland and Bancroft, 1989; Starling *et al.*, 1989).

Fluid–wallrock reactions are thus probably most important in producing large, average-grade gold deposits. The formation of high-grade gold shoots (10s–100s g/t) containing free gold is more problematical, but there is increasing evidence that phase separation with formation of coexisting CO_2-rich and H_2O-rich fluids may change the concentrations of H^+, CH_4 and H_2S *within* the fluid and cause gold precipitation (e.g. Spooner *et al.*, 1987; Sibson *et al.*, 1988). This phase separation may be due to lowering of lithostatic pressure or change from lithostatic to hydrostatic pressure during regional uplift and erosion (e.g. Groves *et al.*, 1984; Clark *et al.*, 1988) related to compressional tectonics (e.g. Vearncombe *et al.*, 1988), or more locally due to rupturing of previously sealed conduits (Foster, 1988). Such a mechanism could explain the abrupt downward termination of rich gold shoots in otherwise continuous quartz-vein deposits in Zimbabwe (Foster *et al.*, 1986).

Although local temperature gradients do not appear to be important, the occurrence of the majority of the gold deposits in greenschist facies domains and their restricted depositional temperature range suggests that there was a broad temperature window over which gold deposition occurred.

3.4.4 *Fluid focusing*

The structural style of most Archaean greenstone belts in which relatively narrow high-strain shear zones cut much larger volumes of low-strain rocks appears ideal for the strong focusing of ore fluids required to produce large hydrothermal ore deposits (Boulter *et al.*, 1987). Gold deposits and attendant alteration zones are clearly restricted to zones of structurally induced permeability in the greenstone succession, in which there was dissolution of material and syntectonic mass transfer of dissolved species in a migrating fluid (e.g. Vearncombe *et al.*, 1988). As discussed above, many mineralized structures are geometrically related to crustal-scale shear zones, clearly zones of enhanced heat and fluid flux, but are of an order-of- magnitude smaller scale. An important question arising from this observation is whether there was a fluid continuum between the crustal-scale and smaller ore-bearing structures, carrying with it the implication of a potential deep source for the ore fluid, or whether only the smaller-scale structures had access to the ore fluid.

3.4.5 *Source of fluid and ore components*

The low salinity, $H_2O - CO_2$ auriferous fluids are unlike those of basinal brines, porphyry Cu–Au or epithermal gold systems. They are most compatible in composition with metamorphic fluids from amphibolite facies terranes (e.g. Crawford, 1981), but could also be produced by CO_2 saturation in H_2O- and CO_2-bearing silicate magmas (e.g. Holloway, 1976) or possibly by mantle degassing and subsequent crustal oxidation. In an attempt to resolve the source of ore components and the fluids themselves, stable and radiogenic isotope and incompatible-element ratio studies have been carried out. These are summarized by Perring *et al.* (1987) and Colvine *et al.* (1988), tabulated in Table 3.4, and are briefly discussed below.

There are abundant $\delta^{18}O$ data on quartz veins (e.g. Golding and Wilson, 1987) and few δD data for hydrous silicates (e.g. Kerrich, 1986a). These give calculated $\delta^{18}O$ and δD ranges for the H_2O component of the ore fluid from +2.5 to +10.0‰ and 0 to –70‰, respectively (Colvine *et al.*, 1988). These data are equally consistent with a magmatic or metamorphic fluid source, but exclude a meteoric or sea water fluid source unless there has been such extensive interaction with rocks along fluid pathways that the original isotopic signature of the fluid has been destroyed.

Carbon isotope data for carbonates associated with gold mineralization implicate a $\delta^{13}C$ for CO_2 in the ore fluid of -5 ± 2‰ (see summaries by Golding *et al.*, 1987; Colvine *et al.*, 1988). This is broadly compatible with a juvenile origin for the carbon, with variations between individual deposits (referred to as 'provinciality' by Kerrich, 1986a; Perring *et al.*, 1987) perhaps related to redox changes or mixing with local carbon reservoirs. Several authors have suggested that the $\delta^{13}C$ data are compatible with a magmatic origin for the CO_2 (e.g. Colvine *et al.*, 1984; Burrows *et al.*, 1986), whereas others have suggested metamorphic dissolution of mantle-derived carbonation zones along regional shear zones that also host the gold mineralization (e.g. Groves *et al.*, 1988a,b). The data preclude a source purely derived by metamorphic devolatilization of carbonate related to seafloor alteration of volcanics in the greenstone successions.

Most $\delta^{34}S$ values cluster around 0‰ (e.g. Lambert *et al.*, 1984; Spooner *et al.*, 1985). Sulphur with this composition could equally be derived directly from a magmatic source or via metamorphic devolatilization or dissolution of dominantly volcanic terranes such as the greenstone belts. Negative $\delta^{34}S$ values are recorded from some deposits, implicating relatively oxidized ore fluids attributed either to exsolution

Table 3.4 Physicochemical properties of ore fluid and environment of deposition for Archaean lode gold deposits: modified slightly after Perring *et al.* (1987)

Parameter	Value	Data source	References
T, °C	200–450 (generally 250-350)	Fluid inclusions, oxygen isotopes, thermodynamics	Safonov *et al.* (1978); Kerrich and Fryer (1979); Ho *et al.* (1985); Neall (1985); Smith *et al.* (1984); Ho (1987); Golding *et al.* (in press)
P_{fluid}, kb	1–4.5 (generally 1–2)	Fluid inclusions	Ho *et al.* (1985); Smith *et al.* (1984); Brown, (1986); Ho (1987)
[CO_2], mol.%	3–25 (generally 10–15)	Fluid inclusions, thermodynamics	Ho *et al.* (1985); Smith *et al.* (1984); Neall, (1985); Wood *et al.* (1984); Ho (1987)
Salinity, wt.% NaCl equiv.	<6 (generally <2)	Fluid inclusions	Wood *et al.* (1984); Ho *et al.* (1985); Smith *et al.* (1984); Pretorius *et al.* (1986); Ho (1987); Robert and Kelly (1987); Carter (1988)
Molar Na/K	10–89	Thermodynamics, fluid inclusions	Neall (1985); Ho (1987)
Log fO_2, bars	–33 to –29.7 (reducing)	Thermodynamics, alteration chemistry	Phillips and Groves (1983); Neall (1985)
Log fH_2S, bars	–1.33 to at least –0.6	Thermodynamics	Neall (1985, 1987)
pH	near-neutral to alkaline	Thermodynamics, alteration chemistry, fluid inclusions	Phillips and Groves (1983); Ho *et al.* (1985); Neall (1985)
ρ_{fluid}, g/cm^3	0.8–1.0 (generally 0.9)	Fluid inclusions	Ho *et al.* (1985); Smith *et al.* (1984); Brown (1986); Pretorius *et al.* (1986); Ho (1987)
$\delta^{34}S_{pyrite}$, ‰	+0.8 to +8.1 (generally +1 to +5; G. Mile –9.5 to –1.6)	Stable isotopes	Golding and Wilson (1983); Lambert *et al.* (1984); Lavigne and Crocket (1983); Spooner *et al.* (1985); Phillips *et al.* (1986); Wood *et al.* (1986)
$\delta^{34}S_{fluid}$, ‰	+1 to +3	Stable isotopes	Golding *et al.* (in press)
$\delta^{13}C_{carbonate}$, ‰	–10.0 to +2.2 (generally –8.5 to –2.5)	Stable isotopes	Fyon *et al.* (1982); Golding and Wilson (1983); Spooner *et al.* (1985); Burrows *et al.*, (1986); Kerrich (1986b); Wood *et al.* (1986); Golding *et al.* (1987, in press)

Table 3.4 *Continued*

Parameter	Value	Data source	References
$\delta^{13}C_{fluid}$, ‰	–7 to +1 (generally –4 to –2)	Stable isotopes	Golding and Wilson (1983); Colvine *et al.* (1984); Wood *et al.* (1986); Golding *et al.* (1987, in press)
$\delta^{18}O_{quartz}$, ‰	+8 to +16 (generally +11 to +14)	Stable isotopes	Fyon *et al.* (1982); Golding and Wilson (1983); Fyfe and Kerrich (1984); Kerrich and Watson (1984); Kishida (1984); Kerrich (1986b); McNeil and Kerrich (1986); Golding *et al.* (in press); Foster (1989)
$\delta^{18}O_{fluid}$, ‰	+2.5 to +10 (generally +5 to +8)	Stable isotopes	Golding and Wilson (1983); Fyfe and Kerrich (1984); Kerrich and Watson (1984); Kishida (1984); Kerrich (1986b); McNeil and Kerrich (1986); Wood *et al.* (1986); Golding *et al.* (in press); Foster (1989)
$\delta D_{silicate}$, ‰	–87 to –50	Stable isotopes	Kerrich and Watson (1984); Golding and Wilson (1987); Golding *et al.* (in press)
δD_{fluid}, ‰	–86 to +6 (generally –70 to –30)	Stable isotopes	Fyon *et al.* (1982); Kerrich and Watson (1984); Kerrich (1986a, b); Golding and Wilson (1987); Golding *et al.* (in press)
Au enrichment (relative to background)	10^3 to 10^4	Alteration and vein chemistry	Kerrich (1986a)
Cu enrichment (relative to background)	10^{-1} to 10	Alteration and vein chemistry	Kerrich (1986a)
$^{87}Sr/^{86}Sr$	0.7010–0.7023	Radiogenic isotopes	Kerrich (1986a)
Pb–Pb	μ of mineralization is greater and more variable than μ of country rocks	Radiogenic isotopes	Dahl *et al.* (1987)

from magmatic intrusions (e.g. Cameron and Hattori, 1987) or extensive fluid: wallrock reactions (e.g. Phillips *et al.*, 1986). Phase separation may also change the oxidation state of the fluid, and this mechanism warrants further study.

Variable strontium isotope ratios, including some high $^{87}Sr/^{86}Sr$ initial ratios, from ore-related minerals in the Canadian Shield have been used to demonstrate provinciality of ore fluids and to implicate a lower crustal component in the ore fluid (Kerrich *et al.*, 1987). There are few published studies elsewhere. Data from Kambalda (McNaughton *et al.*, 1988) implicate either a greenstone or granulite-facies basement source.

Lead isotope studies of galenas from greenstone terranes (e.g. Browning *et al.*, 1987; Franklin *et al.*, 1983) confirm the isotopic provinciality of ore fluids and indicate more radiogenic lead than that in contemporaneous mantle for some deposits. In Western Australia, more radiogenic compositions are coincident with areas (e.g. Norseman) with known older sialic basement (e.g. Compston *et al.*, 1986) or granitoids that are derived from such basement but are older than gold mineralization (e.g. Hill and Compston, 1987). In Zimbabwe, there are differences in lead isotope compositions of sulphides from gold deposits in laterally equivalent, but contrasting, greenstone sequences (Kramers and Foster, 1984). These data (cf. Sr data) implicate a non-greenstone source for at least some of the lead in the ore fluid.

Incompatible element ratios (e.g. K/Rb = 200–400; Groves and Phillips, 1987) are generally non-definitive, being typical of crustal values. They do, however, suggest less fractionation than that normally recorded from alteration zones of generally accepted magmatic-hydrothermal mineral deposits. Similarly, a volumetrically reasonable slab of almost any potential source rock could produce the observed metasomatic introduction of Au, CO_2, H_2O, K and S in the ore environments (e.g. Phillips *et al.*, 1987).

In summary, most isotopic data are equivocal. They are broadly compatible with either a metamorphic or magmatic fluid source. Hydrogen isotope data appear to preclude a significant meteoric or seawater contribution. Although broadly comparable between deposits in different cratons, C, S, Pb and Sr isotope data indicate provinciality within specific terranes, suggesting that radiogenic and stable isotope reservoirs were not uniform and/or that upper crustal pathways modified stable isotope compositions. Some Pb and Sr isotope data appear to implicate lower crust as a source for some components of the fluid.

3.5 Genetic models

It is now possible to reconstruct most of the gold mineralizing system (see summaries by Perring *et al.*, 1987; Colvine *et al.*, 1988; Groves *et al.*, 1988b). Auriferous, low-salinity H_2O-CO_2 fluids were channelled up ductile transcrustal shear zones during the late stages of greenstone belt deformation and metamorphism, normally just pre-dating cratonization. P–T conditions at or below the amphibolite/greenschist metamorphic boundary favoured establishment of brittle–ductile and brittle structures in which gold deposits were selectively sited. It is probable that physical gradients (P, T) between the crustal-scale ductile structures and more brittle subsidiary structures generated transient, strongly focused, fluid flow into the latter where lower P and T and fluid–wallrock reactions promoted gold deposition (e.g. Eisenlohr *et al.*, 1989). Gold was precipitated in dilational zones within both discordant and contact-parallel shear zones and in extensional, commonly hydraulic, fractures in low tensile strength lithologies. In lithologies with high Fe/(Fe + Mg) ratios (e.g. tholeiitic basalt and dolerite, BIF), sulphidation reactions destabilized gold reduced-sulphur complexes causing gold precipitation. In other lithologies, fluid–wallrock reactions resulted in changes in fO_2 and pH, causing gold deposition. It is probable that phase separation also induced changes in fO_2, pH and concentration of reduced sulphur species, causing precipitation of gold-rich ores. P, T conditions ranged from 0.5 to 4.5 kb and 250 to 400°C . The high gold/base metal ratios of the deposits reflect the relatively restricted potential for chloride complexing of the base metals in the low-salinity fluids (see Henley, 1985) rather than fluid/rock ratios in the source reservoirs (Kerrich and Fryer, 1981). Only higher salinity fluids derived by convective circulation of seawater through oceanic crust or by dissolution of evaporites have the potential for generating the base metal-rich volcanogenic massive sulphide deposits (Figure 3.8).

While there is general consensus on these aspects of the model, there is still considerable debate on the source of ore fluids and ore components. Proposed models include: (a) metamorphic derivation of fluids by greenstone-belt devolatilization (e.g. Kerrich and Fryer, 1979; Phillips and Groves, 1983; Foster, 1985), crustal outgassing (e.g. Kerrich, 1986a; Perring *et al.*, 1987) or granulitization (Colvine *et al.*, 1988), (b) magmatic models involving release of fluids by minor felsic intrusions (e.g. Burrows and Spooner, 1985; Hodgson and Troop, 1988) or by lamprophyres (e.g. Rock *et al.*, 1987; Rock and Groves, 1988a,b), and (c) direct mantle degassing (see Perring *et al.*, 1987, for discussion). Models are summarized in Figure 3.9.

This diversity of conceptual models is explained by the conjunction of a number of associations on a transcraton scale. Gold mineralization is spatially related to transcraton shear zones, mantle-derived carbonation zones, granitoids, felsic porphyries and lamprophyres. These associations, combined with the overall similarity of isotope data from different cratons, suggest that gold mineralization was related to a single, at least crustal-scale, process. This probably involved a deep-seated tectono-metamorphic event incorporating melting of mantle at depths to 150 km (to produce lamprophyres), melting of the upper mantle or base of the crust (porphyries) and lower crust (granitoids), mantle degassing (regional carbonation zones), interaction of fluids with lower crust (Pb and Sr isotope data), and metamorphic fluids at upper crustal levels (provinciality of most isotope data).

The stage at which gold and other ore components were added to the hydrothermal system is still to be resolved. Certainly, gold deposition commonly occurred during specific events within long-lived shear zones, and it is not clear whether this was due purely to structural controls or to introduction of gold from a specific source at a particular time during tectonism. Also unresolved is whether the ore fluids were mainly magmatic, mainly metamorphic or a combination of both. It does appear in the light of radiogenic isotope studies that gold and ore fluids were not simply derived from devolatilization of greenstone belts. Similarly, field relationships and absolute age data argue against the involvement of magmatic fluid from presently exposed porphyries and granitoids. As pointed out by Rock and Groves (1988a,b), lamprophyres may show primary enrichment of gold, have similar element associations (e.g. high K, Rb, Ba and CO_2 and moderate S) to gold-related alteration, and are the intrusive rocks most similar in timing to gold mineralization in some cratons. If a special gold source is involved, they remain the preferred candidates, although it is most likely that they would have lost their gold to metamorphic fluids

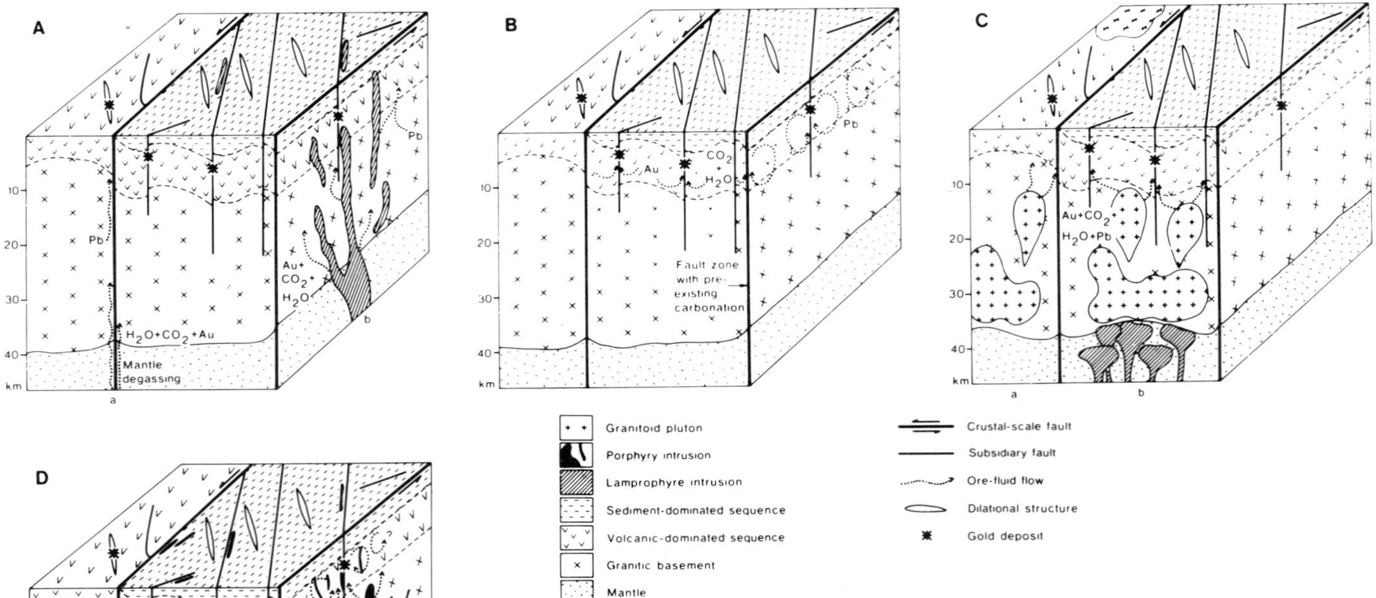

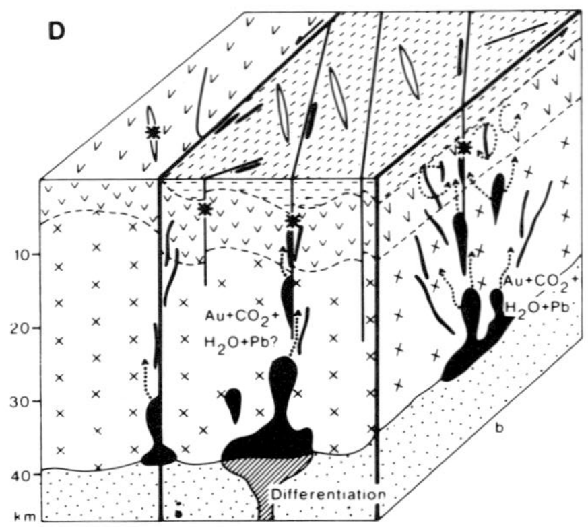

Figure 3.9 Schematic diagrams showing end-member models for: (i) auriferous fluids and ore-component sources, and (ii) ore-fluid flow in Archaean granitoid-greenstone terranes. Intrusive bodies not directly related to a particular model are omitted for the sake of clarity. (A) Mantle-degassing model (H_2O–CO_2± Au) showing lower crustal contamination (Pb); (a) direct degassing; (b) delivery of mantle fluids to upper crust via lamprophyric melts. (B) Metamorphic model showing derivation of H_2O–CO_2–Au fluids from devolatilization of amphibolite-facies greenstones and fluid convection in fault zones that penetrate granitic basement (Pb source?). (C) Magmatic-granitoid model showing granitoids, generated (a) above a mantle plume and (b) above ponded lamprophyric magmas, evolving H_2O–CO_2–Au–Pb fluids throughout diapiric ascent. (D) Magmatic-porphyry model showing porphyry magmas, generated by (a) differentiation of lamprophyric magmas and (b) partial melting at the base of the crust, with porphyry intrusion into fault zones and evolving H_2O–CO_2–Au ± Pb fluids which circulate within the plane of the fault. (Perring *et al.* 1987.)

via interaction with the continental crust rather than exsolving gold at their intrusion site.

It is evident that further data are required to completely resolve current genetic controversies. One promising approach is to examine the tectonic setting of Archaean gold deposits in terms of modern counterparts, as briefly discussed below.

3.6 Tectonic setting of gold mineralization

It is difficult to make valid generalizations about the tectonic setting of Archaean gold mineralization because there has been variable emphasis on this topic in different cratons. Perhaps the best indication that tectonic setting *is* important is provided by the temporal distribution of gold deposits (Woodall, 1988), with major peaks of hydrothermal gold mineralization in the late Archaean and post-Late Palaeozoic (Figure 3.1). The younger deposits are almost entirely restricted to convergent plate boundaries, where epithermal and porphyry-hosted deposits form during subduction-related magmatism in volcanic arcs and marginal basins, and deeper level mesothermal deposits form during deformation in continental-margin orogenic belts; the last (e.g. Mother Lode) are strikingly similar to the Archaean deposits in terms of structural control, ore–element association and wallrock alteration (e.g. Nesbitt *et al.*, 1986, and this volume; Landefield, 1988).

Recent studies of the highly mineralized *ca.* 2.7 Ga Norseman-Wiluna Belt (Barley and Groves, 1988; Perring *et al.*, 1988; Barley *et al.*, 1989) suggest that it represented a convergent-margin setting. A western tectono- stratigraphic association of tholeiites, komatiites and sulphidic shales with isolated rhyolites is interpreted to have formed in a marginal basin setting, whereas an eastern tectono-stratigraphic association of tholeiites, calc-alkaline volcanic and feldspathic sedimentary rocks is interpreted to represent a volcanic arc facies (Figure 3.10). The structural style, in which structural domains are bounded by a regional network of oblique-slip shear zones, and in which recumbent folds and thrusts, upright folds and faults with a dominant oblique- or strike-slip movement are related to oblique crustal shortening (e.g. Vearncombe *et al.*, 1988), is also compatible with a convergent-margin orogen. Similarly, the nature and distribution of granitoids (Figure 3.10) is that expected in such a setting (Perring *et al.*, 1988), and the geochemistry of lamprophyres (LILE enrichment and Nb depletion) is typical of magmatism at Phanerozoic convergent margins (Rock and Groves, 1988a,b). The relatively short time span (<50 Ma) between volcanism and deformation/metamorphism is comparable to that for the opening and closure of younger marginal basins (e.g. Aguire and Offler, 1985).

It is thus probable that gold mineralization in the Norseman- Wiluna Belt was related to oblique closure of a volcanic arc and marginal basin in a convergent margin orogen. Mineralization in older greenstone sequences in adjacent provinces of the Yilgarn Block is of similar age and the host structures were presumably developed in response to the tectonic activity that accompanied closure of the Norseman-Wiluna Belt.

The major transcrustal 'breaks' of the Abitibi Belt, Canada, were initiated as sites of extension and consequent komatiitic to tholeiitic volcanism (Hodgson, 1986). Strike-slip dislocation was dominant during the late stages of volcanism, and sub-aerial and shallow marine clastic sediments accumulated in relatively localized pull-apart basins (Colvine *et al.*, 1988). However, a subsequent shift to compressional

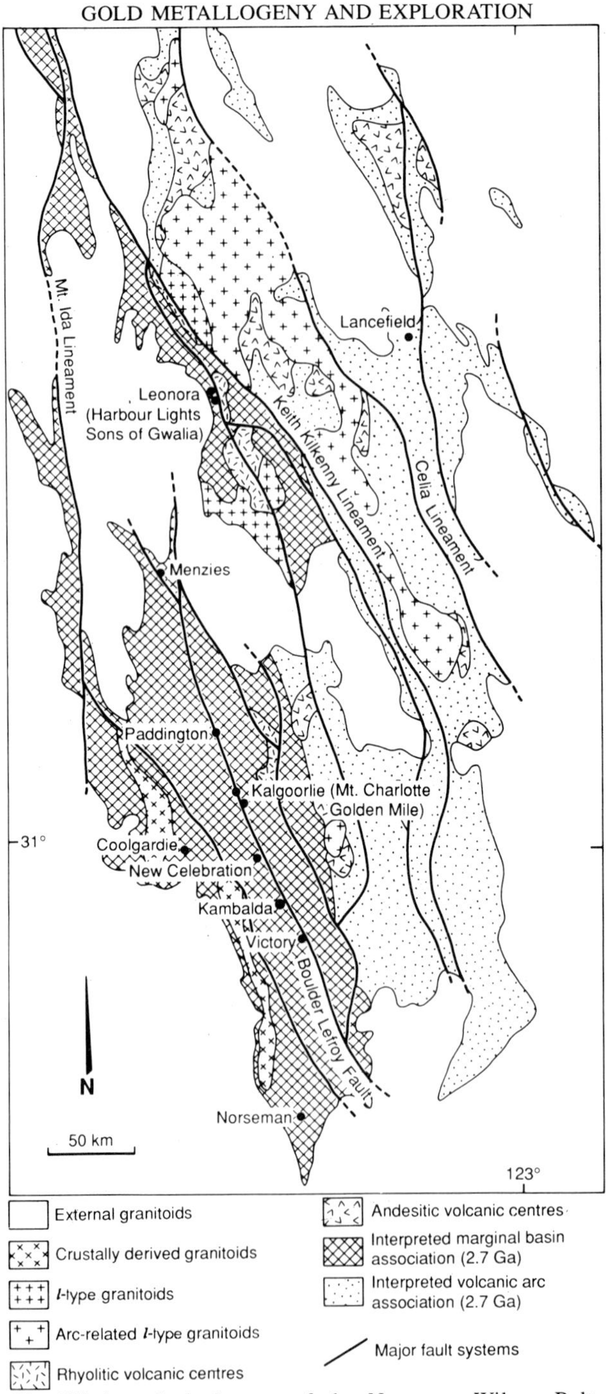

Figure 3.10 Simplified geological maps of the Norseman–Wiluna Belt showing major tectono-stratigraphic associations, granitoid-types, major deformation zones and some of the larger gold deposits. (From Barley *et al.*, 1989.)

tectonism resulting from external shortening, possibly related to subduction and possibly from the internal upwelling of calc-alkaline batholiths, generated oblique-slip zones which acted as channelways for late auriferous fluids (Hodgson, 1986). Again, the gross tectonic environment can be ascribed to a convergent margin (Ludden *et al.*, 1986; Wyman and Kerrich, 1988).

Vearncombe *et al.*, (1988) suggested that gold mineralization in the Sutherland and Murchison greenstone belts of South Africa was related to compressional tectonics during formation of the Limpopo Belt. The mineralized reverse and thrust fault regimes are interpreted to have been generated either during westward emplacement of the Limpopo Central Zone (McCourt and Vearncombe, 1987) or during Himalayan-style continental collision between Kaapvaal and Zimbabwe Cratons (Van Reenen *et al.*, 1987).

Within the Zimbabwe Craton, the auriferous lodes exhibit an important component of reverse movement (Foster, 1989), but the major part of the Limpopo thrusting post-dated the late Archaean (*ca.* 2.6 Ga) adamellites (see review by Wilson, in press) and hence post-dated the bulk of gold mineralization. Some broad correspondence with the Abitibi Belt is evident, with extensional tectonics, komatiitic to tholeiitic to calc-alkaline volcanism and siliciclastic sedimentation culminating in tonalite-dominated intrusive magmatism, regional metamorphism, compressional tectonism and hydrothermal activity. Closure of one or more back-arc basins could be invoked, but the evidence is equivocal. Certainly the structural evolution of the small stellate greenstone belts and the more significant mineralized shear zones was partly dictated by the presence of pre-existing domal granitoid-gneiss terranes (Foster, 1985, 1989; Nutt *et al.*, 1988).

Nevertheless, if such tectonic models are valid, then Archaean gold mineralization is perhaps best viewed as a result of high lithospheric-scale fluid and heat flux. This would certainly explain the apparently conflicting evidence for metamorphic, magmatic and mantle fluid sources and for greenstone, lower crustal and mantle or magmatic sources of ore components.

3.7 Potential exploration significance

Most deposits have been discovered as the result of geochemical and/or geophysical exploration combined with basic field mapping. Certainly, in Western Australia most currently mined deposits were discovered either as a result of drilling of exposed prospects known to early prospectors or drilling of surface or near-surface geochemical (Au ± other elements) anomalies. In the future, geological models are likely to become more important in guiding exploration; this is already evident in some cases (e.g. Roberts, 1988). Some of the most important parameters of mineralization from an exploration viewpoint are summarized below.

(i) The most prospective greenstone belts, and those containing the giant gold deposits, appear to be *ca.* 2.7 Ga in age, although restricted areas of older greenstone belts may also be well mineralized.

(ii) Large to giant deposits are concentrated adjacent to transcraton deformation zones (largely shear zones) in belts typified by a high density of such zones, commonly in anastomosing sets.

(iii) Individual gold deposits are sited in a variety of less extensive structures that are geometrically related to the transcraton shear zones.
(iv) The geometry of gold lodes shows a strong structural control, commonly with a marked plunge of ore shoots subparallel to stretching lineations or to intersection of shear zones with lithological contacts.
(v) In at least some cratons, gold deposits show a strong spatial correlation with zones of enhanced frequency of minor intrusions of intermediate to felsic porphyries and/or lamprophyres.
(vi) Basalts and dolerites are the most common host rocks, and BIF and minor intrusions may have high gold production relative to their abundance in the greenstone terrane. Rocks with high Fe/(Fe + Mg) ratios appear particularly prospective.
(vii) Wallrock alteration assemblages, ore/element ratios and fluid inclusions are normally highly distinctive, thus allowing easy recognition of mineralization.
(viii) The provinciality of isotope data means that major gold mineralization in any one district may be distinctive and hence distinguished from other types of alteration or insignificant gold deposits.
(ix) Within shear zones, ore shoots are marked by volume increase which can be recognized by 'dilution' of immobile elements (e.g. Al, Ti, Zr, Y) even if there is no visible veining.
(x) Fluid phase separation may be important in generating ore shoots rich in free gold. If so, evidence for late uplift during deformation would be important.

In the future, these (and other) parameters are likely to be formalized by generation and statistical manipulation of computer databases. Recent examples are those described by Hodgson and Troop (1988) and Rock *et al.* (in press).

3.8 Brief summary

Lode gold deposits ranging from <1t Au to > 1500 t Au at grades between 50 and 2g Au t^{-1} are a characteristic feature of Archaean granitoid-greenstone terranes. They have accounted for a high percentage of world gold production in the past, and contain a high proportion of known world gold resources.

The deposits are consistently sited in greenstone belts or immediately adjacent granitoids, are structurally controlled, display a wide range of structural styles ranging from shear zones to extensional veins and breccias, and are commonly elongate parallel to stretching lineations or structural intersections. Although mineralization may occur in a variety of host rocks at variable metamorphic grade, many of the large to giant deposits occur in rocks with high Fe/(Fe + Mg) ratios (basalts, dolerites or BIF) at mid- to upper-greenschist facies in structures of the brittle–ductile transition; minor felsic intrusions are important host rocks in the Canadian Shield and are associated with many lodes in the Zimbabwe Craton. Gold mineralization is associated with laterally zoned, wallrock alteration haloes that involved volume increase related to metasomatic addition of SiO_2, K_2O, CO_2, H_2O, and Au. The deposits are characteristically 'gold-only', with Au associated with As, Ag, W, Sb, Te and B with variable, but commonly low, Cu, Pb, Zn and Mo.

Archaean granitoid-greenstone terranes show a characteristic heterogeneous distribution of gold mineralization, both between and within cratons. The best mineralized terranes are those with a high density of anastomosing, oblique- to strike-slip transcraton shear zones which control the geometry of adjacent, subsidiary mineralized structures. Such high-strain zones focused fluid flow in otherwise regionally low-strain greenstone belts. There is commonly a conjunction of these transcraton structures, regional carbonation, emplacement of swarms of minor intrusions of intermediate to felsic porphyries and lamprophyres, and gold mineralization. Available timing constraints suggest that mineralization normally occurred late in greenstone-belt deformation and metamorphism, immediately prior to cratonization, and post-dated most minor porphyritic intrusions. The closest temporal relationship is with lamprophyres and/or silica-undersaturated felsic intrusions in some cratons or regions within them.

Gold mineralization resulted from highly focused flow of low-salinity H_2O–CO_2 fluids along ductile transcraton shear zones with transient fluid flow into adjacent brittle–ductile structures as a result of lowering of pressure during failure. Gold precipitation, normally at $300 \pm 50°C$ and 1–3 kb, occurred as a result of fluid–wallrock reactions, particularly sulphidation of Fe-rich host rocks, or phase separation within the fluid. Stable isotopic data are generally equivocal, supporting either magmatic or metamorphic models for fluid derivation but apparently excluding significant meteoric water or seawater influx. Limited radiogenic isotope data suggest that fluid conduits involved the lower crust, or intrusions derived from it. Collectively, these data indicate that neither exsolution of magmatic fluids from exposed felsic intrusions nor metamorphic devolatilization of greenstone successions alone are valid genetic models.

Hydrothermal gold mineralization peaked in the late Archaean (2.7–2.6 Ga) and again in the Mesozoic to Quaternary where virtually all deposits occur in convergent margin settings. Evidence is accumulating that the late-Archaean gold deposits may be equivalents of the mesothermal deposits found in more modern settings; that is, they are related to closure of *ca.* 2.7 Ga volcanic arc/marginal basin complexes in convergent margin settings. If so, mineralization is best viewed as the result of high lithospheric fluid and heat flux with fluids and ore components being added from a variety of sources and modified by upper crustal fluid pathways. A gold-enriched 'special' source rock is probably not required, although lamprophyres and related magmas potentially fulfil this role in terms of their variable but commonly high Au content and their broad contemporaneity with gold mineralization.

Deposit models for Archaean gold deposits will become increasingly important for development of exploration philosophy as near-surface deposits become exhausted. Computer databases are already being developed to assist in this phase of exploration.

Acknowledgements

D.I.G. acknowledges that many of the data and many of the interpretations presented here were generated within the Archaean Gold Group (now Key Centre for Strategic Mineral Deposits) at the University of Western Australia. He particularly acknowledges Mark Barley, Sue Ho, Neal McNaughton, Neil Phillips, Nick Rock, Julian Vearncombe as well as numerous Ph.D. students who have made important contributions. The research on which the chapter is based was funded by WAMPRI (now MERIWA), ARC and UWA and many mining companies provided access to mines, financial support, encouragement and useful suggestions and discussions: WMC are particularly acknowledged in this respect. This chapter represents a contribution to IGCP Project 247.

R.P.F. acknowledges the continuing logistical and personal support of staff of the Institute of Mining Research, the Geological Survey, and many mining companies in Zimbabwe. Discussion with Jim Wilson (University of Zimbabwe) and Jim Andrews (Southampton), and the lively input of NERC-supported Ph.D. students Alan Carter, Jon Gilligan, Tony Starling and Jamie Wilkinson provided welcome stimulation. The specific comments of Clive Boulter and Alan Carter were appreciated by both authors.

References

Aguire, L. and Offler, R. (1985) Burial metamorphism in the Western Peruvian Trough: its relation to Andean magmatism and tectonics. In *Magmatism at a Plate Edge: the Peruvian Andes*, eds. Pitcher, W.S., Atherton, M.P., Cobbing, E.J. and Beckinsale, R.D. Wiley, New York, 59-71.

Anhaeusser, C.R. (1976) The nature and distribution of Archaean gold mineralization in Southern Africa. *Minerals Science Engng* **8**, 46-84.

Bache, J.J. (1987) *World Gold Deposits: A Geological Classification*. North Oxford Academic, London.

Barley, M.E. and Groves, D.I. (1988) Geological setting of gold mineralization in the Norseman-Wiluna Belt, Eastern Goldfields Province, Western Australia. Geol. Dept. Univ. Extension. Univ. of W. Australia, Pub. 14, 17-46.

Barley, M.E., Eisenlohr, B., Groves, D.I., Perring, C.S. and Vearncombe, J.R. (1989) Late Archaean convergent margin tectonics and gold mineralization: a new look at the Norseman-Wiluna Belt, Western Australia. *Geology* **17**, 826-829.

Boulter, C.A., Fotios, M.G. and Phillips, G.N. (1987) The Golden Mile, Kalgoorlie; a giant gold deposit localized in ductile shear zones by structurally induced infiltration of an auriferous metamorphic fluid. *Econ. Geol.* **82**, 1661-1678.

Boyle, R.W. (1955) The geochemistry and origin of the gold- bearing quartz veins and lenses of the Yellowknife gold deposits, N.W.T., Canada. *Econ. Geol.* **50**, 51-66.

Boyle, R.W. (1979) The geochemistry of gold and its deposits (together with a chapter on geochemical prospecting for the element). Geol. Surv. Can. Bull. 280, 584.

Brown, P.E. (1986) Fluid inclusions, fluids and lode ores, Wabigoon subprovince, NW Ontario, Canada. In *Geocongress '86*, Extended Abstracts Volume. Geol. Soc. South Africa, Johannesburg, 279-282.

Browning, P., Groves, D.I., Blockley, J.R. and Rosman, K.J.R. (1987) Lead isotopic constraints on the age and source of gold mineralization in the Archean Yilgarn Block, Western Australia. *Econ. Geol.* **82**, 971-986.

Burk, R., Hodgson, C.J. and Quartermain, R.A. (1986) The geological setting of the Teck-Corona Au–Mo–Ba deposit, Hemlo, Ontario Canada. In *Proc. of Gold '86, an International Symposium on the Geology of Gold*, ed. Macdonald, A.J. Konsult International Inc., Toronoto, 311-326.

Burrows, D.R. and Spooner, E.T.C. (1985) Generation of an Archean H_2O–CO_2 fluid enriched in Au, W and Mo by fractional crystallization in the Mink Lake intrusion, NW Ontario. *Geol. Soc. Amer. Abstr. Programs* **17**, 536.

Burrows, D.R. and Spooner, E.T.C. (1986) The McIntyre Cu-Au deposit, Timmins, Ontario, Canada. In *Proc. of Gold '86, an International Symposium on the Geology of Gold*, ed. Macdonald, A.J. Konsult International Inc., Toronto, 23-39.

Burrows, D.R. and Spooner, E.T.C. (1988) Relationships between Archaean Au-quartz vein/shear zone mineralization and igneous intrusions in the Val D'Or and Timmins areas, Abitibi Subprovince, Canada; the geochemical characteristics of Au related magmas. In *Bicentennial Gold 88*. Extended Abstracts Oral Programme. *Geol. Soc. Aust. Inc. Abstr.* **22**, 302-306.

Burrows, D.R., Wood, P.C. and Spooner, E.T.C. (1986) Carbon isotope evidence for a magmatic origin for Archean gold-quartz vein ore deposits. *Nature* **321**, 851-854.

Cameron, E.M. and Hattori, K. (1987) Archean gold mineralization and oxidized hydrothermal fluids. *Econ. Geol.* **82**, 1177-1191.

Carter, A.H.C. (1988) The transport and deposition of gold within a major Archaean shear zone; Dalny Mine, Zimbabwe. Southampton Dept. Geol., Rep. C12. 52pp.

Cherry, M.E. (1983) The association of gold and felsic intrusions–examples from the Abitibi belt. In *The Geology of Gold in Ontario*, ed. Colvine, A.C. Ontario Geol. Surv. Misc. Paper 110, 48-55.

Clark, M.E. Archibald, N.J. and Hodgson, C.J. (1986) The structural and metamorphic setting of the Victory gold mine, Kambalda, Western Australia. In *Proc. of Gold '86, an International Symposium on the Geology of Gold*, ed. Macdonald, A.J. Konsult International Inc., Toronto, 243-254.

Clark, M.E., Carmichael, D.M. and Hodgson, C.J. (1988) Metasomatic processes and T-XCO_2 conditions of wall-rock alteration, Victory Gold Mine, Kambalda, Western Australia. In Bicentennial Gold 88 Extended Abstracts Oral Programme. *Geol. Soc. Aust. Inc. Abstr.* **22**, 230-234.

Colvine, A.C., Andrews, A.J., Cherry, M.E., Durocher, M.E., Fyon, A.J., Lavigne, M.J., Macdonald, A.J., Marmont, S., Poulsen, K.H., Springer, J.S. and Troop, D.G. (1984) An integrated model for the origin of Archean lode gold deposits. Ontario Geol. Surv., Open- File Report **5524**. 98pp.

Colvine, A.C., Fyon, J.A., Heather, K.B., Marmont, S., Smith, P.M. and Troop, D.G. (1988) Archean lode gold deposits in Ontario. Ontario Geol. Surv. Misc. Paper **139**. 136pp.

Compston, W., Williams, I.S., Campbell, I.H. and Gresham, J.J. (1986) Zircon xenocrysts from the Kambalda volcanics: Age constraints and direct evidence for older continental crust below the Kambalda-Norseman greenstones. *Earth Planet. Sci. Lett.* **76**, 299-311.

Crawford, M.L. (1981) Fluid inclusions in metamorphic rocks–low and medium grade. In *Short Course on Fluid Inclusions: Applications to Petrology*, eds. Hollister, L.S. and Crawford, M.L. Mineralogical Assoc. of Canada, Vol. 6, 157- 181.

Dahl, N., McNaughton, N.J. and Groves, D.I. (1987) A lead-isotope study of sulphides associated with gold mineralization in selected deposits from the Archaean Eastern Goldfields of Western Australia. Geol. Dept. Univ. Extension, Univ. of Australia, Publ. 11, 189-201.

Eisenlohr, B., Groves, D.I. and Partington, G.A. (1989) Crustal-scale shear zones and their significance to Archean gold mineralization. *Mineral Deposita* **24**, 1-8.

Foster, R.P. (1982) Gold and silver in Zimbabwe. In *Gold '82, Abstracts and Information*, ed. Johnson, A.J. Argosy, Harare, 39-54.

Foster, R.P. ed. (1984) *Gold '82: The Geology, Geochemistry and Genesis of Gold Deposits*. Balkema, Rotterdam. 753pp.

Foster, R.P. (1985) Major controls of Archaean gold mineralization in Zimbabwe. *Trans. Geol. Soc. S. Afr.* **88**, 109-133.

Foster, R.P. (1988) Archaean gold mineralization in Zimbabwe: implications for metallogenesis and exploration. In Bicentennial Gold 88 Extended Abstracts Oral Programme. *Geol. Soc. Aust. Inc. Abstr.* **22**, 62-72.

Foster, R.P. (1989) Archaean gold mineralization in Zimbabwe: implications for metallogenesis and exploration. *Econ. Geol. Monograph.* **6**, 54-70.

Foster, R.P., Mann, A.G., Stowe, C.W. and Wilson, J.F. (1986) Archaean gold mineralization in Zimbabwe. In *Mineral Deposits of Southern Africa*, Vol.1, eds. Anhaeusser, C.R. and Maske, S. Geol. Soc. of South Africa, 43-112.

Foster, R.P. and Wilson, J.F. (1984) Geological setting of Archaean gold deposits in Zimbabwe. In *Proc. of Gold '82: The Geology, Geochemistry and Genesis of Gold Deposits*, ed. Foster, R.P. Balkema, Rotterdam, 521-551.

Franklin, J.M., Roscoe, S.M., Loveridge, W.D. and Sangster, D.F. (1983) Lead isotope studies in Superior and Southern provinces. Geol. Surv. Can. Bull. 351. 60pp.

Fripp, R.E.P. (1976a) Stratabound gold deposits in Archean banded iron-formation, Rhodesia. *Econ. Geol.* **71**, 58-75.

Fripp, R.E.P. (1976b) Gold metallogeny in the Archaean of Rhodesia. In *The Early History of the Earth*, ed. Windley, B.F. Wiley, London, 455-466.

Fuchter, W.A.H. and Hodgson, C.J. (1986) Gold deposits in the North Western Mining Camp, Gwanda greenstone belt. In *Proc. of Gold '86, an International Symposium on the Geology of Gold*, ed. Macdonald, A.J. Konsult International Inc., Toronto, 255-269.

Fyfe, W.S. and Kerrich, R. (1984) Gold: natural concentration processes. In *Gold '82: The Geology, Geochemistry and Genesis of Gold Deposits*, ed. Foster, R.P. Balkema, Rotterdam, 99-127.

Fyon, J.A., Schwarcz, H.P., Crocket, J.H. and Knyf, M. (1982) Gold exploration potential using oxygen, carbon and hydrogen stable isotope systematics of carbonatized rock and quartz veins. Ontario Geol. Surv. Misc. Paper 103, 59-64.

Gilligan, J.M. and Foster, R.P. (1987) Gold mineralization in iron-formation: the importance of contrasting modes of deformation at the Lennox Mine, Zimbabwe. In *African Mining*. Inst. of Mining and Metallurgy, London, 127–138.

Golding, S.D., Clark, M.E., Keele, R.A., Wilson, A.F. and Keays, R.R. (in press) Geochemistry of Archaean epigenetic gold deposits in the Eastern Goldfields Province, Western Australia. Geol. Dept. Univ. Extension, Univ. of W. Australia, Publ. 23.

Golding, S.D. and Wilson, A.F. (1983) Geochemical and stable isotope studies of the No. 4 Lode, Kalgoorlie, Western Australia. *Econ. Geol.* **78**, 438-450.

Golding, S.D., Groves, D.I., McNaughton, N.J., Barley, M.E. and Rock, N.M.S. (1987) Carbon isotopic compositions of carbonates from contrasting alteration styles and gold deposits in supracrustal rocks of the Norseman-Wiluna Belt, Yilgarn Block, Western Australia: Their significance to the source of Archaean auriferous fluids. Geol. Dept. Univ. Extension, Univ. of W. Australia, Publ. 11, 215-213.

Golding, S.D. and Wilson, A.F. (1987) Oxygen and hydrogen isotope relationships in Archaean gold deposits of the Eastern Goldfields Province, Western Australia: Constraints on the source of Archaean gold-bearing fluids. Geol. Dept. Univ. Extension, Univ. of W. Australia, Publ. 11, 203-214.

Graham, N.J.G., Haynes, D., Martin, A., and Storey, M.J. (1987) Development of heap leaching at Royal Family Mine, Zimbabwe. In *African Mining*. The Inst. Mining and Metallurgy, London, 153-159.

Groves, D.I. and Barley, M.E. (1988) Gold mineralization in the Norseman-Wiluna Belt, Eastern Goldfields Province, Western Australia. In *Bicentennial Gold 88, Excursion Guidebook*. Geol. Dept. Univ. Extension, Univ. of W. Australia, Publ. 14, 47-66.

Groves, D.I., Barley, M.E., Ho, S.E. and Hopkins, G.M.F. (1988a) Western Australian gold deposits, Bicentennial Gold 88 Excursion Guidebook. Geol. Dept. Univ. Extension, Univ. of W. Australia, Publ. 14. 152pp.

Groves, D.I. and Batt, W.D. (1984) Spatial and temporal variations of Archaean metallogenic associations in terms of evolution of granitoid-greenstone terrains with particular emphasis on the Western Australian Shield. In *Archaean Geochemistry*, eds. Kröner, A., Hanson, G.N. and Goodwin, A.M. Springer-Verlag, Berlin, 73-98.

Groves, D.I. and Ho, S.E. (in press) Archaean gold deposits in the Yilgarn Block, Western Australia. In *Economic Geology of Australia and Papua/New Guinea*, ed. Hughes, F. Australasian Inst. Mining and Metallurgy, Melbourne.

Groves, D.I., Ho, S.E., McNaughton, N.J., Mueller, A.G., Perring, C.S., Rock, N.M.S. and Skwarnecki, M.S. (1988b) Genetic models for Archaean lode gold deposits in Western Australia. Geol. Dept. Univ. Extension, Univ. of W. Australia, Publ. 12, 1-22.

Groves, D.I. and Phillips, G.N. (1987) The genesis and tectonic control on Archaean gold deposits of the Western Australian Shield: a metamorphic-replacement model. *Ore Geol. Rev.* **2**, 287-322.

Groves, D.I., Phillips, G.N., Ho, S.E., Henderson, C.A., Clark, M.E. and Wood, G.M. (1984) Controls on distribution of Archaean hydrothermal gold deposits in Western Australia. In *Gold '82: The Geology, Geochemistry and Genesis of Gold Deposits*, ed. Foster, R.P. Balkema, Rotterdam, 689-712.

Groves, D.I., Phillips, G.N., Ho, S.E., Houstoun, S.M. and Standing, C.A. (1987) Craton-scale distribution of Archaean greenstone gold deposits: predictive capacity of the metamorphic model. *Econ. Geol.* **82**, 2045-2058.

Hamilton, J.V. and Hodgson, C.J. (1986) Mineralization and structure of the Kolar Gold Field, India. In *Proc. of Gold '86, an International Symposium on the Geology of Gold*, ed. Macdonald, A.J. Konsult International Inc., Toronto, 270-283.

Harris, D.C. (1986) Mineralogy and geochemistry of the Main Hemlo gold deposit, Hemlo, Ontario, Canada. In *Proc. of Gold '86, an International Symposium on the Geology of Gold* ed. Macdonald, A.J. Konsult International Inc., Toronto, 297-310.

Henderson, J.F. and Brown, I.C. (1966) Geology and structure of the Yellowknife greenstone belt, district of Mackenzie. Geol. Surv. of Can., Bull. 44. 87pp.

Henley, R.W. (1985) The geothermal framework of epithermal deposits. In *Geology and Geochemistry of Epithermal Systems*, eds. Berger, B.R. and Bethke, P.M. Reviews in Economic Geology, Soc. of Econ. Geol., Vol.2.

Hill, R.I. and Compston, W. (1987) The age of granite emplacement, southeastern Yilgarn Block, Western Australia. Aust. Nat. Univ. Res. School Earth Sci., Ann. Rept. 1986, 70-71.

Ho, S.E. (1987) Fluid inclusions: Their potential as an exploration tool for Archaean gold deposits. Geol. Dept. Univ. Extension, Univ. of W. Australia, Publ. 11, 239-264.

Ho, S.E. and Groves, D.I., eds. (1987) *Recent Advances in Understanding Precambrian Gold Deposits*. Geol. Dept. Univ. Extension, Univ. of W. Australia, Publ. 11. 368pp.

Ho, S.E., Groves, D.I. and Phillips, G.N. (1985) Fluid inclusions as indicators of the nature and source of ore fluids and ore depositional conditions for Archaean gold deposits of the Yilgarn Block, Western Australia. *Trans. Geol. Soc. S. Afr.* **88**, 149-158.

Hodgson, C.J. (1986) Place of gold ore formation in the geological development of Abitibi greenstone belt, Ontario, Canada. *Trans. Instn Min. Metall.* **95B**, 183-194.

Hodgson, C.J. and MacGeehan, P.J. (1982) A review of the geological characteristics of 'Gold Only' deposits in the Superior Province of the Canadian Shield. In *Geology of Canadian Gold Deposits*. Can. Inst. of Mining and Metallurgy, Spec. Vol. **24**, 211-228.

Hodgson, C.J. and Troop, D.G. (1988) A new computer-aided methodology for area selection in gold exploration: a case study from the Abitibi greenstone belt. *Econ. Geol.* **83**, 952-977.

Holloway, J.R. (1976) Fluids in the evolution of granitic magmas: Consequences of finite CO_2 solubility. *Geol. Soc. Amer., Bull.* **87**, 1513-1518.

Hutchinson, R.W. (1976) Lode gold deposits: the case for volcanogenic distribution. *Proc. Pacific Mining and Metals Conf.*, Oregon Department of Mineral Industries, 64-105.

Hyland, H.M. and Bancroft, G.M. (1989) An XPS study of gold deposition at low temperatures on sulphide minerals: reducing agents. *Geochim. Cosmochim. Acta* **53**, 367-372.

Jean, G.E. and Bancroft, G.M. (1985) An XPS and SEM study of gold deposition at low temperature on sulphide mineral surfaces: concentration of gold by adsorption/reduction. *Geochim. Cosmochim. Acta* **49**, 979-987.

Kerrich, R. (1983) Geochemistry of gold deposits in the Abitibi greenstone belt. Can. Inst. of Mining and Metallurgy Special Vol.27, 48-71.

Kerrich, R. (1986a) Archaean lode deposits of Canada, Part I: Characteristics of the hydrothermal systems and models of origin. Econ. Geol. Research Unit, Univ. of Witwatersrand, Johannesburg, Inform. Circ. 183. 34pp.

Kerrich, R. (1986b) Fluid infiltration into fault zones: chemical, isotopic and mechanical effects. *Pure Appl. Geophys.* **124**, 225-244.

Kerrich, R. and Fryer, B.J. (1979) Archean precious-metal hydrothermal systems, Dome Mine, Abitibi greenstone belt. II. REE and oxygen isotope relations. *Can. J. Earth Sci.* **16**, 440-458.

Kerrich, R. and Fryer, B.J. (1981) The separation of rare elements from abundant base metals in Archean lode gold deposits: implications for low water/rock source regions. *Econ. Geol.* **76**, 160-166.

Kerrich, R., Fryer, B.J., King, R.W., Willmore, L.M. and Vanhesse, E. (1987) Crustal outgassing and LILE enrichment in major lithosphere structures, Archaean Abitibi greenstone belt: Evidence on the source reservoir from strontium and carbon isotope tracers. *Contrib. Mineral. Petrol.* **97**, 156- 168.

Kerrich, R. and Watson, G.P. (1984) The Macassa Mine Archean lode gold deposit, Kirkland Lake, Ontario: geology, patterns of alteration and hydrothermal regimes. *Econ. Geol.* **78**, 1104-1130.

Kishida, A. (1984) Hydrothermal alteration zoning and gold concentration at the Kerr-Addison Mine, Ontario, Canada. Ph.D. thesis, Univ. of W. Ontario, London (unpubl.).

Kramers, J.D. and Foster, R.P. (1984) A reappraisal of lead isotope investigations of gold deposits in Zimbabwe. In *Gold '82: The Geology, Geochemistry and Genesis of Gold Deposits*, ed. Foster, R.P., Balkema, Rotterdam, 569-582.

Krapez, B. and Barley, M.E. (1987) Archaean strike-slip faulting and related ensialic basins: evidence from the Pilbara Block, Australia. *Geol. Mag.* **124**, 555-567.

Kuhns, R.J., Kennedy, P., Cooper, P., Brown, P., Mackie, B., Kusins, R. and Friesen, R. (1986) Geology and mineralization associated with the Golden Giant Deposit, Hemlo, Ontario,

Canada. In *Proc. of Gold '86, an International Symposium on the Geology of Gold*, ed. Macdonald, A.J. Konsult International Inc., Toronto, 327-339.

Lambert, I.B., Phillips, G.N. and Groves, D.I. (1984) Sulphur isotope compositions and genesis of Archaean gold mineralization, Australia and Zimbabwe. In *Gold '82: The Geology, Geochemistry and Genesis of Gold Deposits*, ed. Foster, R.P. Balkema, Rotterdam, 373-387.

Landefield, L.A. (1988) The geology of the Mother Lode gold belt, Sierra Nevada Foothills Metamorphic Belt, California. In *Bicentennial Gold 88*, Extended Abstracts Oral Programme. *Geol. Soc. Aust. Inc. Abstr.* **22**, 167-172.

Lavigne, M.J. and Crocket, J.H. (1983) Geology of the East South C ore zone, Dickenson Mine, Red Lake. In *The Geology of Gold in Ontario*, ed. Colvine, A.C. Ontario Geol. Surv., Misc. Paper. 110, 141-158.

Ludden, J., Hubert, C. and Gariepy, C. (1986) The tectonic evolution of the Abitibi greenstone belt of Canada. *Geol. Mag.* **123**, 153-166.

Macdonald, A.J., ed. (1986) *Proc. of Gold '86, an International Symposium on the Geology of Gold.* Konsult International Inc., Ontario. 517pp.

Macgregor, A.M. (1932) The geology of the country around Que Que, Gwelo District. Geol. Surv. of S. Rhodesia, Bull. 20. 113pp.

Mann, A.G. (1984) Gold mines in Archaean granitic rocks in Zimbabwe. In *Gold '82: The Geology, Geochemistry and Genesis of Gold Deposits*, ed. Foster, R.P. Balkema, Rotterdam, 553-568.

Marmont, S. and Corfu, F. (1988) Timing of gold introduction in the Late Archaean tectonic framework of the Canadian Shield: evidence for U-Pb zircon geochronology of the Abitibi Belt. In Bicentennial Gold 88 Extended Abstracts Oral Programme. *Geol. Soc. Aust. Inc. Abstr.* **22**, 45-50.

Mathieson, N.A. and Hodgson, C.J. (1984) Alteration, mineralization, and metamorphism in the area of the East South 'C' ore zone, 24th level of the Dickenson Mine, Red Lake, northwestern Ontario. *Can. J. Earth Sci.* **21**, 35-52.

Maufe, H.B. (1913) The association of gold deposits with acid igneous rocks in Southern Rhodesia. Geol. Surv. of Rhodesia. Bull. 2.

McCourt, S. and Vearncombe, J.R. (1987) Shear zones bounding the central zone of the Limpopo mobile belt, Southern Africa. *J. Structural Geol.* **9**, 127-137.

McNaughton, N.J., de Laeter, J.R. and Groves, D.I. (1988) Constraints on the source of strontium and alkalis in auriferous alteration zones at Hunt mine, Kambalda, Norseman-Wiluna Belt. Geol. Dept. Univ. Extension, Univ. of W. Australia, Publ. **12**, 209-215.

McNeil, A.M. and Kerrich, R. (1986) Archean lamprophyre dykes and gold mineralization, Matheson, Ontario: the conjunction of LILE-enriched mafic magmas, deep crustal structures and Au concentration. *Can. J. Earth Sci.* **23**, 324-343.

Milling-Stanley, G. (1988) *Gold 1986.* Consolidated Gold Fields plc. London.

Milling-Stanley, G. and Green, T. (1986) *Gold 1989* Consolidated Gold Fields plc. London.

Mueller, A.G. (1988) Archaean gold-silver deposits with prominent calc-silicate alteration in the Southern Cross greenstone belt, Western Australia: analogues of Phanerozoic skarn deposits. Geol. Dept. Univ. Extension, Univ. of W. Australia. Publ. 12, 141-163.

Mueller, A.G., Harris, L.B. and Lungan, A. (1988) Structural control of greenstone-hosted gold mineralization by transcurrent shearing: a new interpretation of the Kalgoorlie Mining District, Western Australia. *Ore Geol. Rev.* **3**, 359-387.

Neall, F.B. (1985) Application of thermodynamics to the study of two Archaean gold deposits in Western Australia. Ph.D. thesis, Univ. of W. Australia, Nedlands (unpubl.).

Neall, F.B. (1987) Sulphidation of iron-rich rocks as a precipitation mechanism for large Archaean gold deposits in Western Australia: thermodynamic confirmation. In *Recent Advances in Understanding Precambrian Gold Deposits*, eds. Ho, S.E. and Groves, D.I. Geol. Dept. Univ. Extension, Univ. of W. Australia, Pub. 11, 265-269.

Nesbitt, B.E., Murowchick, J.B. and Muehlenbachs, K. (1986) Dual origins of lode gold deposits in the Canadian Cordillera. *Geology* **14**, 506-509.

Nutt, T.H.C., Treloar, P.J., Thorpe, R.I. and Cummings, G.L. (1988) The geology and geochemistry of gold-antimony mineralization in the Kwekwe goldfield, Zimbabwe. In *Bicentennial Gold 88*, Extended Abstracts Poster Programme. *Geol. Soc. Aust. Inc. Abstr.* **23**, 103-106.

Padgham, W.A. (1979) Gold deposits of the Northwest Territories. In: R.D. Morton (ed.), *Proc. Gold Workshop, Yellowknife, N.W.T., December, 1979*, 174-201.

Perring, C.S., Barley, M.E., Cassidy, K.F., Groves, D.I., McNaughton, N.J., Rock, N.M.S., Bettenay, L.F., Golding, S.D. and Hallberg, J.A. (1988) The conjunction of linear mobile belts, mantle-crustal magmatism and gold mineralization in the Archaean greenstone belts of Western Australia. In *Bicentennial Gold 88*, Extended Abstracts Oral Programme. *Geol. Soc. Aust. Inc. Abstr.* **22**, 296-301.

Perring, C.S., Groves, D.I. and Ho, S.E. (1987) Constraints on the source of auriferous fluids for Archean gold deposits. Geol. Dept. & Univ. Extension, Univ. of W. Australia, Pub. 11, 287-306.

Phillips, G.N. (1985) Interpretation of Big Bell/Hemlo-type gold deposits: Precursors, metamorphism, melting and genetic constraints. *Trans. Geol. Soc. S. Afr.* **88**, 159-173.

Phillips, G.N. (1986) Geology and alteration of the Golden Mile, Kalgoorlie. *Econ. Geol.* **81**, 779-808.

Phillips, G.N. and DeNooy, D. (1988) High-grade metamorphic processes which influence Archaean gold deposits, with particular reference to Big Bell, Australia. *J. Met. Geol.* **6**, 95-114.

Phillips, G.N. and Groves, D.I. (1983) The nature of Archaean gold-bearing fluids as deduced from gold deposits of Western Australia. *Geol. Soc. Aust. J.* **30**, 25-39.

Phillips, G.N. and Groves, D.I. (1984) Fluid access and fluid- wall rock interaction in the genesis of the Archaean gold-quartz vein deposit at Hunt mine, Kambalda, Western Australia. In *Gold '82: The Geology, Geochemistry and Genesis of Gold Deposits*, ed. Foster, R.P. Balkema, Rotterdam, 389-416.

Phillips, G.N., Groves, D.I. and Brown, I.J. (1987) Source requirements for the Golden Mile, Kalgoorlie: Significance to the metamorphic replacement model for Archaean gold deposits. *Can. J. Earth Sci.* **24**, 1643-1651.

Phillips, G.N., Groves, D.I. and Martyn, J.G. (1984) An epigenetic origin for Archean banded iron-formation hosted gold deposits. *Econ. Geol.* **79**, 162-171.

Phillips, G.N., Groves, D.I., Neall, F.B., Donnelly, T.H. and Lambert, I.B. (1986) Anomalous sulfur isotopic compositions in the Golden Mile, Kalgoorlie. *Econ. Geol.* **81**, 2008- 2015.

Pretorius, A.I., Van Reenen, D.D. and Barton, J.M.J. (1986) Fluid inclusion studies of gold mineralization at the Fumani gold mine, Sutherland greenstone belt. In *Geocongress '86*, Extended Abstracts Volume. Geol. Soc. of South Africa, Johannesburg, 383- 386.

Robert, F. and Kelly, W.C. (1987) Ore-forming fluids in Archean gold-bearing quartz veins at the Sigma Mine, Abitibi greenstone belt, Quebec, Canada. *Econ. Geol.* **82**, 1464-1482.

Roberts, D.E. (1988) Geology of Kambalda gold deposits. *Bicentennial Gold 88*, Excursions 6A and 6B. Western Mining Corporation Ltd, Kambalda.

Roberts, S., Foster, R.P., and Nesbitt, R.W. (in press) Mineralization associated with early Precambrian basic magmatism. In *Early Precambrian Basic Magmatism*, eds. Hall, R.P. and Hughes, D.J. Blackie, Glasgow.

Rock, N.M.S., Duller, P., Haszeldine, R.S. and Groves, D.I. (1987) Lamprophyres as potential gold exploration targets: some preliminary observations and speculations. Geol. Dept. Univ. Extension, Univ. of W. Australia, Pub. 11, 271-286.

Rock, N.M.S. and Groves, D.I. (1988a) Do lamprophyres carry gold as well as diamonds? *Nature* **332**, 253-255.

Rock, N.M.S. and Groves, D.I. (1988b) Can lamprophyres resolve the genetic controversy over mesothermal gold deposits? *Geology* **16**, 538-541.

Rock, N.M.S., Hallberg, J.A., Groves, D.I. and Mather, P.J. (1988) Archaean lamprophyres in the goldfields of the Yilgarn Block, Western Australia: new indications of their widespread distribution and significance. Geol. Dept. Univ. Extension, Univ. of W. Australia, Pub. 12, 245-275.

Rock, N.M.S., Shellabear, J.N., Wheatley, M.R., Poulinet, R. and Groves, D.I. (in press) Microcomputers in mineral exploration: a database for modelling gold deposits in the Yilgarn Block of Western Australia. In *Microcomputer Applications in Geology II*, eds. Hanley, T. and Merriam, D.F. Pergamon Press, New York.

Safonov, Y.G., Radhakrishna, B.P., Rao, B.K., Vasudev, V.N., Raju, K.K., Nosik, L.P. and Pashkov, Y.N. (1978) Mineralogical and geochemical features of endogene gold and copper deposits of south India. *J. Geol. Soc. India* **21**, 365-378.

Sang, J. and Ho, S.E. (1987) A review of gold deposits in China. Geol. Dept. Univ. Extension, Univ. of W. Australia, Pub. 11, 307- 320.

Sangster, D.F. and Scott, S.D. (1976) Precambrian stratabound, massive Cu-Zn-Pb sulfide ores of North America. In *Handbook of Stratabound and Stratiform Ore Deposits*, ed. Wolf, K.H. Elsevier, Amsterdam, 129-222.

Seward, T.M. (1973) Thio complexes of gold in hydrothermal ore solutions. *Geochim. Cosmochim. Acta* **37**, 379-399.

Seward, T.M. (1984) The transport and deposition of gold in hydrothermal systems. In *Gold '82: The Geology, Geochemistry and Genesis of Gold Deposits*, ed. Foster, R.P. Balkema, Rotterdam, 165-181.

Sibson, R.H. (1987) Earthquake rupturing as a hydrothermal mineralizing agent. *Geology* **15**, 701-704.

Sibson, R.H., Robert, F. and Poulsen, K.H. (1988) High-angle reverse faults, fluid-pressure cycling, and mesothermal gold- quartz deposits. *Geology* **16**, 551-555.

Skwarnecki, M.S. (1987) Controls on Archaean gold mineralization in the Leonora area, Western Australia. Geol. Dept. Univ. Extension, Univ. of W. Australia, Pub. 11, 109-136.

Smith, T.J., Cloke, P.L. and Kesler, S.E. (1984) Geochemistry of fluid inclusions from the Mclntyre-Hollinger gold deposit, Timmins, Ontario, Canada. *Econ. Geol.* **79**, 1265- 1285.

Spooner, E.T.C., Bray, C.J., Wood, P.C., Burrows, D.R. and Callan, N.J. (1987) Au–quartz vein and Cu–Ag–Mo–anhydrite mineralizations, Hollinger-Mclntyre Mine, Timmins, Ontario: $\delta^{13}C$ values (Mclntyre), fluid inclusion gas geochemistry, pressure (depth) estimation, and H_2O–CO_2 phase separation as a precipitation and dilation mechanism. Ontario Geol. Surv., Misc. Paper 136, 35-56.

Spooner, E.T.C., Wood, P.C., Burrows, D.R., Thomas, A.V. and Noble, S.R. (1985) Geological, fluid inclusion and isotopic (carbon and sulfur) studies of Au–quartz–carbonate–pyrite–scheelite vein mineralization and intrusion-hosted Cu–(Au–Mo) mineralization in the Hollinger–Mclntyre systems, Timmins, Ontario. Ontario Geol. Surv. Misc. Paper 127, 229-246.

Springer, J.S. (1985) Carbon in Archean rocks of the Abitibi belt (Ontario-Quebec) and its relation to gold distribution. *Can. J. Earth Sci.* **22**, 1945-1951.

Starling, A., Gilligan, J.M., Carter, A.H.C., Foster, R.P. and Saunders, R.A. (1989) High-temperature hydrothermal precipitation of precious metals on the surface of pyrite *Nature* **340**, 298-300.

Symons, P.H., Anderson, G., Beard, T.J., Hamilton, L.H., Reynolds, G.D., Robinson, J.M. and Staley, R.W. (1988) The Boddington gold deposit. In *Bicentennial Gold 88*, Extended Abstracts Oral Programme. *Geol. Soc. Aust. Inc. Abstr.* **22**, 56-61.

Tabeart, C.F. (1987) Controls to deposition of gold, copper and bismuth at Renco Mine, Zimbabwe. In *African Mining*. Inst. of Mining and Metallurgy, London, 347-363.

Thomson, M.L. (1986) Petrology of the Crixas gold deposit, Brazil: evidence for gold associated with hydrothermal alteration, subsequent to metamorphism. In *Proc. of Gold '86*, an International Symposium on the Geology of Gold, ed. Macdonald, A.J. Konsult International Inc., Toronto, 284-296.

Van Reenen, D.D., Barton, J.M., Roering, C., Smith, C.A. and Van Schalkuryk, J.F. (1987) Deep crustal response to continental collision: the Limpopo belt of Southern Africa. *Geology* **15**, 11-14.

Van Straaten, H.P. (1984) Gold mineralization in Tanzania–a review. In *Gold '82: The Geology, Geochemistry and Genesis of Gold Deposits*, ed. Foster, R.P. Balkema, Rotterdam, 673-685.

Vearncombe, J.R., Barley, M.E., Eisenlohr, B., Groves, D.I., Houstoun, S.M., Skwarnecki, M.S., Grigson, M.W. and Partington, G.A. (1988) Structural controls on mesothermal gold mineralization: Examples from the Archean terrains of southern Africa and Western Australia. In Bicentennial Gold 88 Extended Abstracts Oral Programme. *Geol. Soc. Aust. Inc. Abstr.* **22**, 19-23.

Wilson, J.F. (1979) A preliminary re-appraisal of the Rhodesian basement complex. *Spec. Publ. Geol. Soc. S. Afr.* **5**, 1-23.

Wilson, J.F. (in press) A craton and its cracks: some of the behaviour of the Zimbabwe block from the late Archaean to the Mesozoic in response to horizontal movements, and the significance of some of its mafic dyke fracture systems. *J. African Earth Sci.*

Wilson, J.F., Bickle, M.J., Hawkesworth, C.J., Martin, A., Nisbet, E.G. and Orpen, J.L. (1978) Granite-greenstone terrains of the Rhodesian Archaean craton. *Nature* **271**, 23-27.

Wood, P.C., Burrows, D.R., Thomas, A.V. and Spooner, E.T.C. (1986) The Hollinger-Mclntyre Au-quartz vein system, Timmins, Ontario, Canada: geologic characteristics, fluid properties and light stable isotope geochemistry. In *Proc. of Gold '86, an International Symposium on the Geology of Gold*, ed. Macdonald, A.J. Konsult International Inc., Ontario, 56-80.

Wood, P.C., Thomas, A.V., Burrows, D.R., Macdonald, A.J., Noble, S.R. and Spooner, E.T.C. (1984) CO_2-bearing low-moderate salinity fluids in Archaean gold–quartz– carbonate–(W, Mo) vein deposits and magmatically derived Mo, W, Ta and Sn mineralization. *Geol. Assoc. Amer. Abstr. Programs*, **16**, 700.

Woodall, R. (1988) Gold in 1988. *Bicentennial Gold 88* Extended Abstracts Oral Programme, *Geol. Soc. Aust. Inc. Abstr.* **22**, 1-12.

Wyman, D. and Kerrich, R. (1988) Alkaline magmatism, major structures, and gold deposits: Implications for greenstone belt metallogeny. *Econ. Geol.* **83**, 454-461.

4 Phanerozoic gold deposits in tectonically active continental margins

B.E. NESBITT

4.1 Introduction

Phanerozoic gold deposits currently account for approximately 25% of the non-communist world's annual gold production or 53% of annual production, if the deposits of the Witwatersrand are excluded (data from Woodall, 1988). With the anticipated increase in production from newly discovered gold deposits of the southern Pacific and western US, this percentage should increase substantially in the coming years. Consequently, the understanding of and exploration for gold deposits in Phanerozoic terranes merit a substantial effort from both the research and industrial communities.

In contrast to the relatively uniform characteristics of gold deposits of the Precambrian, particularly Archaean deposits (Groves and Phillips, 1987; Colvine *et al.*, 1984; Groves and Foster, this volume), there is a great heterogeneity of deposit types within the classification of gold deposits of the Phanerozoic. Fortunately, in this book, several of the important types of Phanerozoic deposits are dealt with in separate chapters on epithermal, intrusion-related, Carlin-type and auriferous massive sulphide deposits. However, this leaves a large and apparently disparate group of gold deposits spanning all seven continents and all of the Phanerozoic. The bulk of these deposits are representatives of what have been previously classified as turbidite-hosted, greenstone-hosted, metamorphic vein or mesothermal vein deposits (Boyle, 1979; Hutchinson, 1987; Nesbitt and Muehlenbachs, 1989). As will become apparent through the comparative analysis presented in this chapter, these deposits are not as genetically disparate as previously thought and probably constitute a distinct, internally consistent, class of gold deposit, which has strong genetic similarities to many other types of gold deposits, especially those found in Archaean terranes. Throughout the chapter, this class of deposits will be referred to as mesothermal deposits, as it is a more general and less genetically specific term.

The chapter is composed of three segments. The chapter begins with an overview of the distribution in space and time of gold deposits of the Phanerozoic. Following this overview is a review of geological and geochemical characteristics of the mesothermal deposits of the Phanerozoic. Concluding the chapter is a review of genetic models for these deposits and a comparison of the characteristics of the deposits to other deposit types.

4.2 Distribution of Phanerozoic lode gold deposits in space and time

The objective of this section is to provide an overview of the location of Phanerozoic gold deposits, their general tectonic setting and general temporal relations to sedimentation, metamorphism, plutonism and deformation. A thorough compilation of data on many of these aspects has already been provided by the encyclopaedic work of Boyle (1979). Consequently the emphasis in this section will be on a concise review of the distribution of the deposits, on a continent by continent basis, as well as an update of the reference base incorporated in Boyle (1979).

4.2.1 *North America*

Significant quantities of gold have been recovered from two discrete districts on the eastern coast of North America. The northern district is composed of gold mineralization in the Meguma Group of Nova Scotia and the recently discovered deposits of southern Newfoundland. The Meguma Group deposits, generally characterized as turbidite-hosted gold deposits, comprise middle Devonian age, quartz-vein mineralization in Cambro-Ordovician age greywackes (Graves and Zentilli, 1982; Smith and Kontak, 1988a). In contrast, the Chetwynd deposit of Newfoundland comprises disseminated Au–Cu mineralization in silicified and argillically altered Cambrian metasediments and metavolcanics (Yule, 1987).

Gold mineralization in the southern Appalachians is dispersed across a large region composed of portions of the states of Virginia, North and South Carolina and Georgia. The principal deposits in production or development today are volcanic-hosted gold deposits of the late Proterozoic to early Palaeozoic Carolina Slate Belt (Feiss, 1988). These deposits comprise disseminated gold mineralization in sheared silicified, propylitically and argillically altered volcanics and volcaniclastics (Feiss, 1988; Tomkinson, 1988). In many respects, these deposits bear a striking similarity to the Chetwynd deposit of Newfoundland and in some respects to the epithermal deposits of the western U.S.

One of the most areally extensive and economically significant belts of Phanerozoic gold deposits in the world is found in the North American Cordillera stretching from Central America to the western tip of Alaska. Throughout the entire belt, Phanerozoic deposits comprise a complex series of epithermal, porphyry-associated, Carlin-type and mesothermal systems. In this chapter, emphasis will be placed on the last category, since the discussion of deposits in the first three categories will be covered in other chapters. In California, the mesothermal deposits in the Cordillera consist of the western Sierra Nevadan quartz-vein deposits, which include the Mother Lode system, as well as similar deposits in the Klamath Mountains of northern California (Clarke, 1970; Albers, 1981). The Sierra Nevadan gold deposits are hosted by accreted Palaeozoic and Mesozoic units with an age of ore emplacement of approximately mid- to early-Cretaceous (Bohlke and Kistler, 1986). Further north, significant mesothermal gold mineralization is present in Oregon, Washington and Idaho (Boyle, 1979).

Mesothermal gold deposits are widely distributed in the accreted oceanic and island arc terranes of the Canadian Cordillera (Nesbitt and Muehlenbachs, 1989). The deposits are hosted by a variety of Palaeozoic and Mesozoic igneous and sedimentary rock types and span a range in age of ore deposition from early Cretaceous to Tertiary.

A prominent association is noted between the distribution of Au, Sb and Hg deposits and major strike-slip faults in the Cordillera (Nesbitt and Muehlenbachs, 1988). Studies of various deposits of these three metals indicate a vertical mineralization sequence ranging from gold deposition at 10 ± 2 km to Sb mineralization at intermediate depths to Hg deposition near the surface (Nesbitt *et al.*, in press).

Gold mineralization in accreted terranes of Alaska bears a strong resemblance to deposits in other portions of the Cordillera described above. The principal lode gold mining districts are located in the Pacific Border Ranges and Coast Mountains of southern Alaska (Goldfarb *et al.*, 1987). These deposits are composed of gold-bearing quartz veins hosted by greenschist facies sedimentary or plutonic units. The deposits generally appear to have been emplaced following metamorphism in the late Cretaceous or Tertiary (Goldfarb *et al.*, 1987).

4.2.2 *South America*

While the bulk of gold produced in South America is derived from epithermal and/or intrusion-related deposits, significant mesothermal gold production has been recorded in Colombia, Bolivia and Argentina (Boyle, 1979; Utter, 1984; Lehrberger, 1988). The Colombian deposits are hosted by a complex series of metamorphosed sedimentary and igneous rock types ranging in age from Precambrian to Tertiary. The deposits are composed of quartz veins and silicified shear zones and are generally believed to be Mesozoic to Tertiary in age (Utter, 1984). The Bolivian deposits are closely associated with late Palaeozoic Sb-vein deposits in low grade metamorphic rocks (Lehrberger, 1988).

4.2.3 *Australia–New Zealand*

The Tasman Fold System of eastern Australia and Tasmania hosts a wide variety of styles of gold mineralization including Au-rich, volcanogenic massive sulphide, epithermal, intrusion-related, and mesothermal deposits. In New South Wales the bulk of gold production has been derived from Permian to Triassic mesothermal quartz–Au ± Sb, W, As, Ag veins. The location of these veins is strongly controlled by faults and fractures with a general spatial association with S-type granitoids. In addition, important epithermal Au ± Ag deposits and possibly syngenetic, auriferous chert mineralization are present in Carboniferous to Permian units (Degeling *et al.*, 1986; Gilligan and Barnes, 1988).

Gold deposits of Queensland cover a broad range of geological and chronological settings. The bulk of current and past production has been derived from mesothermal Au–quartz veins, similar to the major producers in New South Wales and Victoria. Recently, major interest has been directed towards large tonnage, epithermal and intrusion-related deposits of Permo-Carboniferous age. In addition, substantial past and current production has been derived from deposits closely related to Devonian plutons, which may be transitional between mesothermal and intrusion-related styles of mineralization (Murray, 1986; Morrison, 1988).

By far the most historically important Au-bearing sub-province of the Phanerozoic of eastern Australia is mineralization in the Victoria gold belt. In contrast to the diversity of styles of gold mineralization noted in the northern Australian states, gold mineralization in the Victoria belt is uniformally mesothermal in character (Sandiford and Keays, 1986). The deposits of the Victoria gold belt are regarded as classic examples of turbidite-hosted gold mineralization and are characterized as quartz veins and pods in post-metamorphic brittle structures. The ores are hosted by highly deformed, metamorphosed Cambrian to Silurian sedimentary and volcanic units, with the formation of the ores occurring shortly after a major, early Devonian plutonic event (Sandiford and Keays, 1986; Ramsay and Vandenberg, 1986). In Tasmania the principal sources of gold are by-product gold from volcanogenic massive sulphides of the Mount Read volcanic belt and mesothermal lode deposits in eastern and northern Tasmania (Collins and Williams, 1986; Large *et al.*, 1988).

Phanerozoic gold in New Zealand encompasses deposits of both epithermal and mesothermal styles of mineralization. The epithermal deposits are situated in Tertiary to Recent volcanics of the North Island. Mesothermal deposits are concentrated in Palaeozoic to Mesozoic metasedimentary and plutonic rocks of the South Island and show a prominent association with tungsten and antimony mineralization (Paterson, 1986; Brathwaite, 1988).

Over the last few years, numerous, very economically significant gold discoveries have been made in Phanerozoic terranes of the western island arcs of the Pacific (Sillitoe, 1988). However, since these deposits constitute important examples of both intrusion-related and epithermal deposits, their description and characterization will be covered in the appropriate chapters.

4.2.4 *Asia*

Phanerozoic gold mineralization in Malaysia and Thailand of south-east Asia is composed of mesothermal, epithermal and intrusion-related styles of mineralization. In Malaysia, the principal gold districts comprise mesothermal quartz veins in metamorphosed Ordovician to Jurassic, sedimentary, plutonic and volcanic units. The ores are believed to have been emplaced late in the orogenic history of the area (Yeap Ee, 1988). In Thailand, mesothermal quartz veins in metamorphosed Palaeozoic greywackes have provided the bulk of historical gold production (Tate, 1988).

In China, the majority of the gold production has been derived from mesothermal veins, though locally in eastern China epithermal deposits are economically significant (Wang and Cheng, 1988). The major gold-producing districts in China are found within the North China Craton, which is composed largely of Precambrian metasedimentary and metavolcanic rocks. The gold ores comprise quartz veins and silicified zones, which are closely associated with late Mesozoic granitic plutons of the Yan Shan orogeny (Sang and Ho, 1987). Elsewhere in China, economically significant mesothermal gold–quartz veins are found in metamorphosed Palaeozoic and Mesozoic sedimentary units (Wang and Cheng, 1988). Mesothermal gold deposits in Korea have a geologic setting similar to those of northern China with a close spatial association of quartz vein systems to Jurassic granites in a highly metamorphosed Precambrian terrane (Shelton *et al.*, 1988).

The bulk of the gold produced in the USSR is derived from deposits of Phanerozoic age with mesothermal, intrusion-related, and epithermal being the most significant deposit types (Boyle, 1979; Smirnov, 1977). In the Urals, Palaeozoic, mesothermal gold–quartz veins and silicified zones are hosted by a wide variety of igneous and sedimentary rock types. A second major Palaeozoic gold province in the USSR spans much of the USSR–China border. These deposits are largely composed of mesothermal gold–quartz veins, skarns and intrusion-related deposits (Boyle, 1979; Smirnov, 1977). In Siberia, gold mineralization is largely composed of epithermal deposits of Cenozoic age (Boyle, 1979).

4.2.5 *Europe*

Significant, historic gold production in Britain and Ireland has been recorded from north-eastern Ireland, Scotland and Wales. The majority of this production has been derived from mesothermal gold–quartz veins in metamorphosed, lower Palaeozoic sedimentary units (Steed and Morris, 1986; Fitches, 1987). In the Hercynian Belt of France, important gold mineralization is found in major shear zones with an age of mineralization of late Carboniferous (Cathelineau and Boiron, 1988). Gold mineralization in Czechoslavakia is localized in the Bohemian Massif in mesothermal veins occupying fractures in metamorphosed Proterozoic to Palaeozoic volcano-sedimentary units (Moravek and Pouba, 1987). The genesis of these deposits is believed to be closely related to mid-Palaeozoic regional metamorphism (Moravek and Pouba, 1987). A large belt of Phanerozoic gold-producing districts extends from southern Europe east into central Asia. The bulk of these deposits are either epithermal or intrusion-related in character (Boyle, 1979).

4.2.6 *Africa*

Phanerozoic gold deposits are widely scattered throughout the African continent with principal concentrations of deposits occurring in Morocco, Nigeria, Zambia and the Cape System of South Africa (Boyle, 1979; Andrews-Speed *et al.*, 1984; Woakes and Bafor, 1984; Akande and Fakorede, 1988). In Zambia the distribution of gold–quartz veins appears to be closely linked to the distribution of granitic plutons and major zones of strike-slip faulting formed during the Pan African tectonic event (Andrews-Speed *et al.*, 1984). Gold–quartz veins in Nigeria are similarly a product of the Pan African orogenic process (Woakes and Bafor, 1984).

4.3 Geological and geochemical characteristics

Presented in this section and Table 4.1 is a synthesis of the geological and geochemical characteristics of Phanerozoic mesothermal lode gold deposits, since this subject is not covered in other chapters. One of the principal objectives of this review is to evaluate to what degree the various characteristics of the deposits are similar from region to region and to what degree they differ. The resulting information aids significantly in developing models of fundamental processes involved in ore formation and generally applicable exploration guidelines.

4.3.1 *Host rocks*

A wide variety of rock lithologies host Phanerozoic mesothermal lode gold deposits. Probably the most common host unit is a greywacke-shale assemblage formed in continental margin, turbidite settings (Boyle, 1986). Prime examples of gold deposits in such units are found in the Nova Scotia (Canada) and Victoria (Australia) gold districts. The Meguma deposits of Nova Scotia are hosted by an exotic, 5–9 km thick, flyschoid sequence of shales and quartz wackes. The sediments appear to have been derived from a cratonic provenance with little or no volcanic contribution (Graves and Zentilli, 1982). Similarly, the deposits of the Victoria gold belt are hosted by greywackes and argillites with some volcanosedimentary units in the eastern portion of the belt (Sandiford and Keays, 1986). Other good examples of greywacke-shale-hosted mineralization include deposits in parts of the North American Cordillera, particularly Alaska, and mesothermal deposits of New Zealand (Goldfarb *et al.*, 1987; Paterson, 1986).

Table 4.1 General geological and geochemical characteristics of Phanerozoic mesothermal deposits

Tectonics	Typically in accreted, deformed and metamorphosed continental margin or island arc terranes.
Size and grade	Generally several hundred thousand to a few million tonnes, typically 5–25 g/t
Host lithology	Widely variable; greywackes-pelites, chemical sedimentary units, volcanics, plutons, ultramafics
Metamorphism	Typically sub- to upper-greenschist; occasionally host terranes were metamorphosed to higher grade prior to mineralization
Relations to plutons	Variable; some areas close spatial and probable genetic link; other districts no evidence of plutonic activity
Structure	Varies from fold to fault control; where fault controlled, mineralization is generally confined to second-order faults related to major structures
Timing	Late in orogenic sequence; subsequent to principal deformation and metamorphism
Ore morphology and textures	Thick quartz veins, typically banded, occasionally vuggy with high-grade ore shoots; vertically continuous mineralized zones; occasional stockwork and disseminated mineralization
Mineralogy and paragenesis	Early phases: quartz, Ca–Mg–Fe carbonates, arsenopyrite, pyrite, albite, sericite, chlorite, scheelite, stibnite, pyrrhotite, tetrahedrite, chalcopyrite, tourmaline. Late phases: gold, galena, sphalerite, tellurides
Hydrothermal alteration	Carbonatization, albitization, sericitization, silicification, sulphidation, chloritization; listwanite development
Zoning and elemental geochemistry	Au:Ag typically >1; associated elements: Ag, Sb, As, W, Hg, Bi, Mo, Pb, Zn, Cu, Ba. Zoned from high-temperature Au ± Ag, As, Mo, W to Sb ± Au, Hg, W to Hg ± Sb
Fluid inclusions	H_2O–CO_2 inclusions, typical XCO_2 0.05 to 0.2; <5 equiv.wt.% NaCl; *T* (homogenization) 250-350°C, *P*> 1000 bars
Stable isotopes	Typical values: $\delta^{18}O$ (Qz) = + 11 to + 18‰; $\delta^{13}C$ (Cb) = − 25 to −3‰; δD (fluid inclusions) = − 160 to − 30‰; $\delta^{34}S$ −10 to + 10‰
Radiogenic isotopes	Indicate heterogeneous crustal sources for Sr, Pb and Nd

Significant Phanerozoic gold mineralization is also found in volcanosedimentary settings, which essentially constitute a 'greenstone' environment similar to the principal host lithologies of the Archaean deposits (Hutchinson, 1987). A good example of this setting is present in the Mother Lode district of California, where mineralization is hosted by serpentinites, volcanics, and plutonic units in addition to sedimentary units (Landefeld, 1988). Similar host-rock lithologies are present in several of the mesothermal gold deposits of the Canadian Cordillera and Colombia, South America (Utter, 1984; Nesbitt and Muehlenbachs, in press). Faulted and altered granitic to intermediate plutons are a common site for gold mineralization in the Chinese deposits, as well as portions of the Sierra Nevadan and Canadian Cordilleran regions (Bohlke and Kistler, 1986; Sang and Ho, 1987; Nesbitt and Muehlenbachs, in press). A fourth host-rock lithology is intermediate to high metamorphic grade Precambrian gneiss, which hosts important Palaeozoic and Mesozoic mesothermal gold deposits in northern China and Korea (Sang and Ho, 1987; Shelton *et al.*, 1988).

In contrast to the diversity of host-rock lithologies, the metamorphic grade of the host rocks is strikingly uniform from district to district, varying from sub- to upper-greenschist facies. In regions where higher metamorphic grades are present, such as the Meguma terrane of Nova Scotia, the areas of gold mineralization are generally restricted to greenschist facies units (Taylor and Schiller, 1966). The principal exception to this generality is found in areas such as northern China or portions of the Mother Lode, where the host units had been exposed to higher metamorphic grades prior to the mineralization event (Sang and Ho, 1987; Bohlke and Kistler, 1986).

Felsic to intermediate plutons are present in most of the terranes containing Phanerozoic mesothermal gold deposits and frequently host important gold deposits. However, the genetic significance of the plutons appears to vary substantially from district to district. In some districts, as in Queensland and northern China, there appears to be a close spatial, temporal and genetic link between gold mineralization and granitic plutons. These deposits generally have anomalously high base-metal concentrations in the gold ores and are probably gradational to intrusion-related deposits (Sang and Ho, 1987; Morrison, 1988). More typical is the relationship observed in Victoria, Nova Scotia and parts of the North American Cordillera where felsic to intermediate plutons are broadly associated with the mineralization in space and time but little evidence exists for a direct genetic link between mineralization and plutonism (Sandiford and Keays, 1986; Graves and Zentilli, 1982). At the other extreme, in several districts in the Sierra Nevada and Canadian Cordillera, plutons are not present at all or age dating has shown that the plutonic units are either substantially younger or older than the ores (Bohlke and Kistler, 1986; Nesbitt and Muehlenbachs, in press).

In summary, Phanerozoic mesothermal gold deposits do not appear to be closely associated with any particular host-rock lithology; however, they do appear frequently to be restricted to greenschist facies metamorphic units. A ubiquitous genetic association with felsic plutons is not indicated but a general regional association is common.

4.3.2 *Structure*

The regional tectonic setting for many of the districts considered here is one of ore formation in accreted, often exotic, terranes. This is well documented for deposits on both coasts of North America, as well as the deposits of eastern Australia (Graves and Zentilli, 1982; Coney *et al.*, 1980; Degeling *et al.*, 1986).

The terranes hosting the ore are characteristically highly deformed by one or more folding or faulting events. The degree to which regional folding or faulting events control the location of mineralization varies from region to region. In the Meguma deposits of Nova Scotia, the principal controlling structural features are large scale, doubly plunging, en echelon anticlines and synclines. Mineralized zones are generally concentrated on the anticlinal domes or adjacent hinge zones (Mawer, 1986) (Figure 4.1). In addition, recent results indicate that strike-parallel, ductile shear zones play an important role in controlling the location of mineralization (Kontak and Smith, 1988).

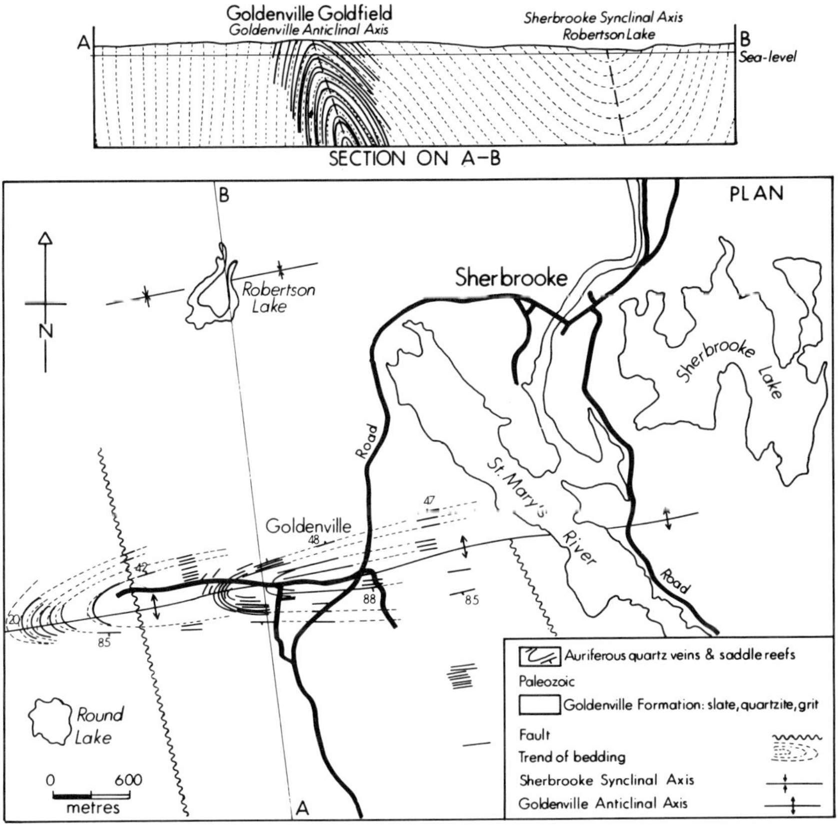

Figure 4.1 Distribution of gold lodes in anticlinal noses at Goldenville, Nova Scotia. (Modified from Boyle, 1979.)

In the Victoria gold belt, pre-mid-Devonian folding produced upright, north-trending folds. Subsequent to the folding event, major, steep, north-trending, strike-slip faults disrupted the folded terrane. Gold mineralization was to some extent localized by this strike-slip system (Sandiford and Keays, 1986) (Figure 4.2). Two types of ore morphologies, saddle reefs and fault reefs, are common in the Victoria gold belt. Fault reef mineralization occurs in second-order reverse faults, which are

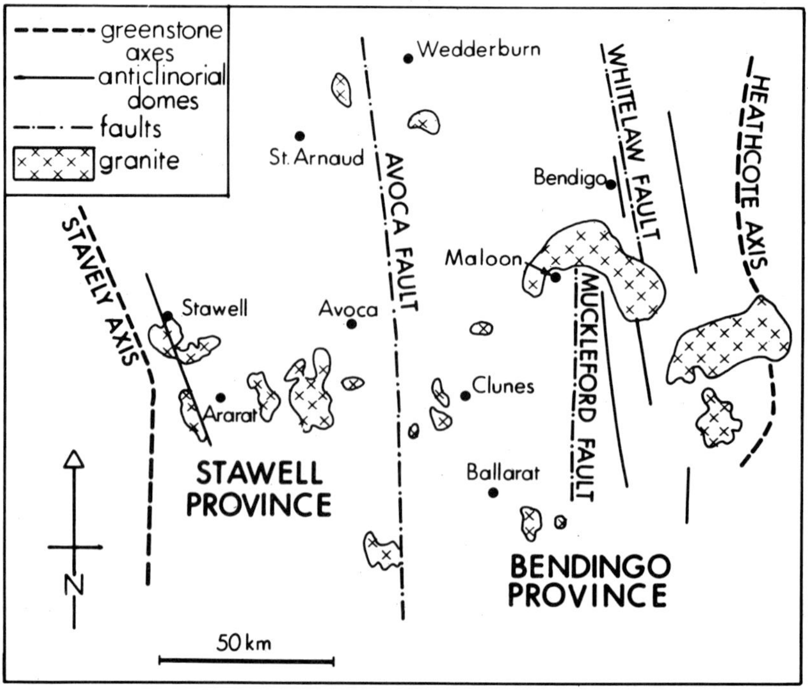

Figure 4.2 Distribution of principal gold lodes in the Ballarat slate belt, Victoria, emphasizing the relation between faults and mineralization. (After Sandiford and Keays, 1986.)

closely related to major but barren strike-slip faults. Saddle reefs comprise mineralized zones in the hinges of anticlines (Sandiford and Keays, 1986), which closely resemble the principal style of mineralization in the Meguma deposits. However, in the Victoria gold belt, the saddle reefs invariably form at the intersection of a reverse fault with an anticline (Sandiford and Keays, 1986) (Figure 4.3).

In the North American Cordillera and northern China, the location of mineralization was dominantly controlled by faults, which had a major component of strike-slip motion. A good example of the link between faulting and mesothermal ores is evident in the spatial association between the Melones fault zone and gold mineralization of the Sierra Nevada district (Figure 4.4). The veins and mineralized zones are located in a series of second-order faults (Figure 4.5), which adjoin the major, generally barren Melones fault zone (Landefeld, 1988). A second example of this relationship is found in the Jia Pi Gou gold district of northern China (Figure 4.6). In the Canadian Cordillera, a similar relationship is noted between strike-slip faulting and gold mineralization. Of particular significance is the observation that mineralization is principally confined to second-order fault structures related to the main, but barren strike-slip fault (Nesbitt and Muehlenbachs, 1988).

In general, mesothermal gold mineralization events appear to occur very late in the sequence of deformation, metamorphism and plutonism which affects the mineralized terranes. This is a surprisingly consistent observation from most of the principal Phanerozoic districts (Sandiford and Keays, 1986; Nesbitt and Muehlenbachs, 1988; Bohlke and Kistler, 1986; Paterson, 1986; Mawer, 1986; Kontak *et al.*, 1988a; Stuwe,

1986) and raises significant questions as to the source of ore fluids since the peak of metamorphism often preceded the formation of the ores.

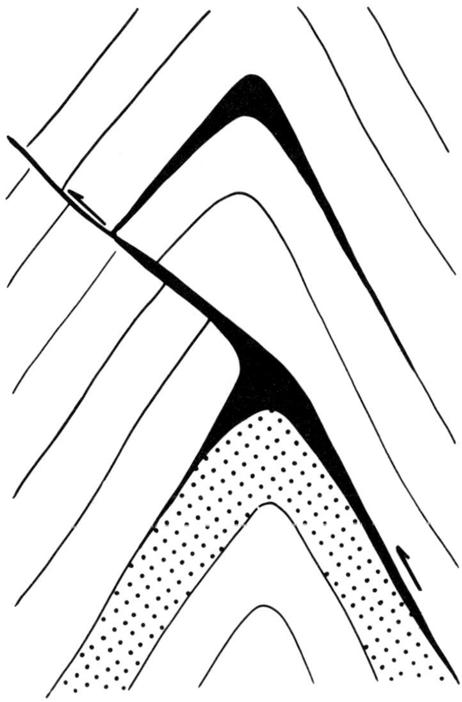

Figure 4.3 Generalized distribution of quartz–Au lodes (in black) in saddle reefs at Bendigo, Victoria. Note the close relation between saddle reef and fault reef mineralization. Stipple pattern indicates psammitic lithologies. (After Sandiford and Keays, 1986.)

In summary, principal host structures for Phanerozoic mesothermal gold deposits vary from faults to folds, though there is usually some evidence of fault-controlled fluid pathways. Almost universally, the faults which host the ores are second-order structures, subsidiary to large, often barren, strike-slip faults. This relationship between mineralized second-order and barren first-order faults is a common product of ore formation in fault-controlled systems (Sibson, 1987). The timing of ore formation, late in the orogenic history of a region, is also relatively universal between principal mineralized districts.

4.3.3 *Ore morphology and textures*

The dominant style of mineralization in most Phanerozoic mesothermal gold deposits is thick, vertically extensive quartz veining. The veins vary considerably in thickness with a typical range of 0.5 to 3 m and often exhibit significant vertical and longitudinal continuity. In several regions, such as the Sierra Nevada, Canadian Cordillera and Victoria, mineralized zones have often been mined to depths of over 1000 m. In most mines, high-grade gold zones are discontinuous longitudinally along

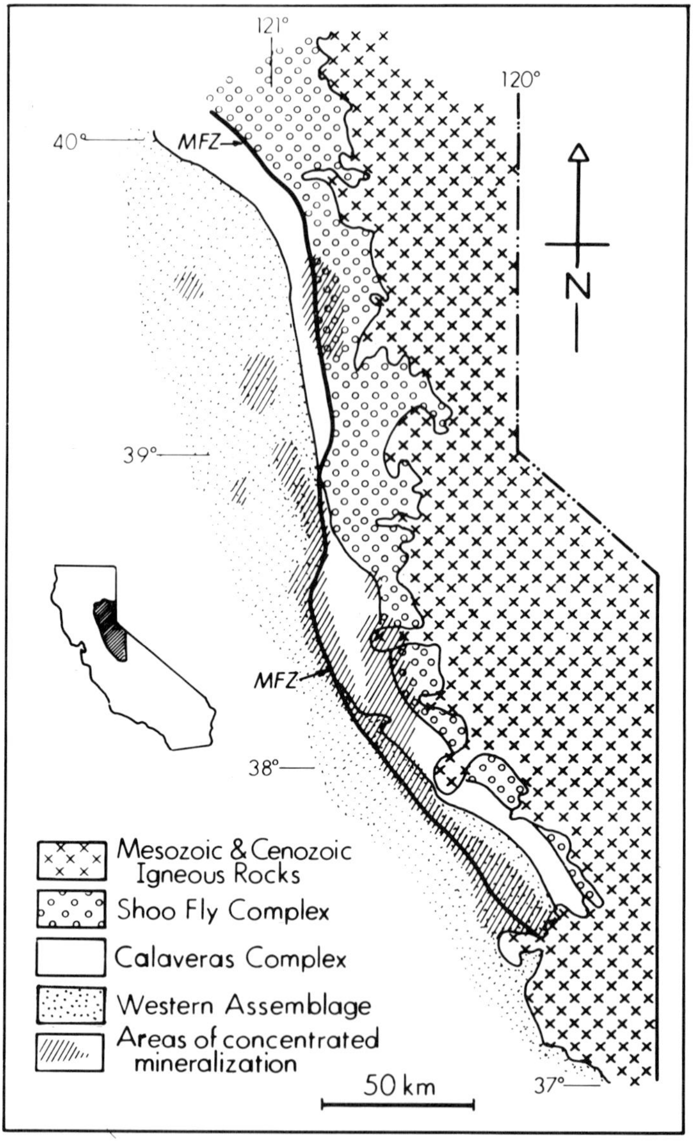

Figure 4.4 Distribution of principal gold-mining districts in the Sierra Nevada region of California. Note the close spatial correlation of much of the mineralization with the Melones Fault Zone (MFZ). (Modified from Bohlke and Kistler, 1986.)

the veins, producing a typical ore shoot structure consisting of sub-vertical interspersed zones of ore grade and sub-ore grade vein material (Knopf, 1929; Cairnes, 1937; Bowen and Whiting, 1975).

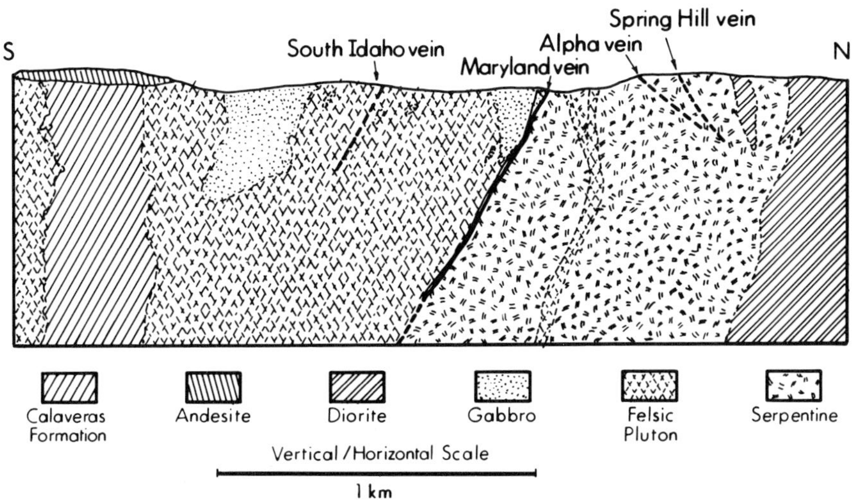

Figure 4.5 Diagram illustrating fault control and the typical distribution of veins in the Mother Lode district at the southern end of the mineralized region shown in Figure 4.4. (Modified from Knopf, 1929.)

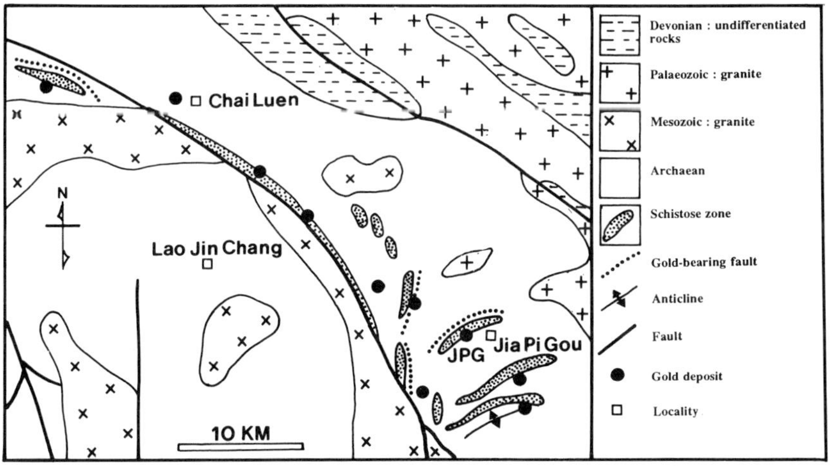

Figure 4.6 Distribution of major gold deposits in the Jia Pi Gou district, China. Note the close spatial link between the major fault and mineralization. (Modified from Sang and Ho, 1987.)

Texturally, the quartz veins are typically milky, occasionally vuggy, and often laminated or banded (Figure 4.7). The laminations comprise fine grained sheet

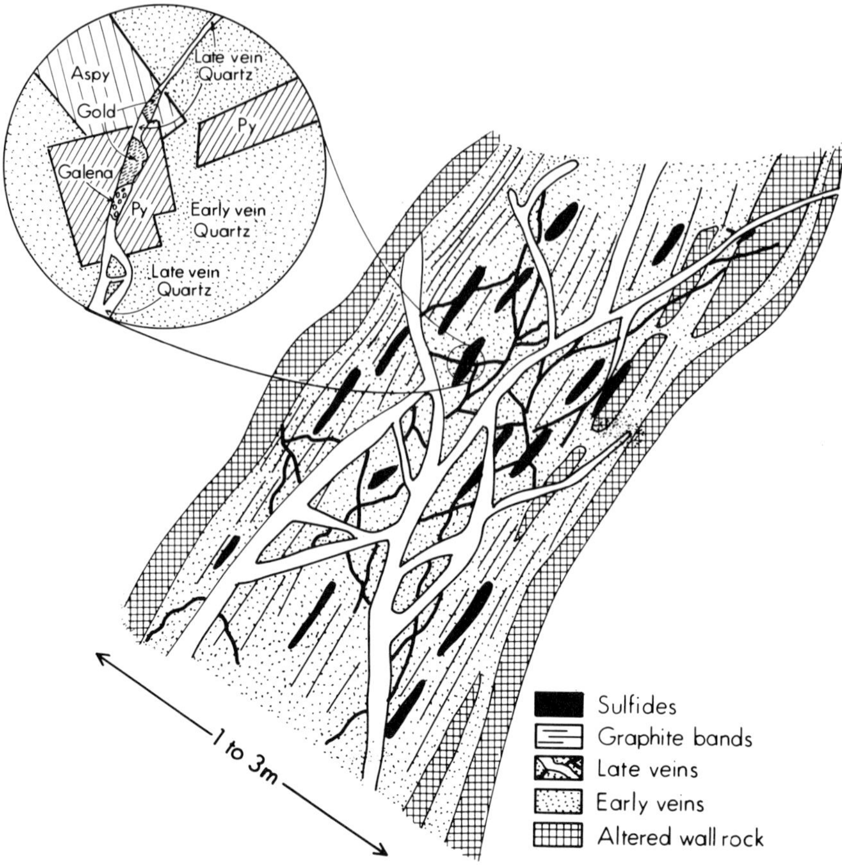

Figure 4.7 Schematic illustration of a cross-section of a mesothermal vein. The blow-up illustrates the late fracture-filling and/or replacement textures observed in most Phanerozoic mesothermal gold deposits.

silicates, sulphides and carbonaceous material and have been variously ascribed to wallrock material incorporated during dilation or a product of wallrock replacement (Sandiford and Keays, 1986; Smith and Kontak, 1988b). In well-mineralized zones there is often evidence for several generations of quartz veining and indications that the majority of the gold was introduced by the late veining events (Knopf, 1929; Dussel, 1986; Smith and Kontak, 1988b) (Figure 4.7).

In addition to quartz veining, stockwork and disseminated mineralization constitute important styles of mineralization in parts of most districts (Landefeld, 1988; Nesbitt and Muehlenbachs, 1989; Bowen and Whiting, 1975). These typically high-tonnage, low-grade mineralized zones are becoming increasingly popular as exploration targets.

4.3.4 *Mineralogy and paragenesis*

A very consistent mineralogy and paragenesis is noted between all of the major districts (Figure 4.8). The mineralized veins are invariably dominated by quartz, typically ranging from 70 to 95 vol.% of the vein. Other important, common vein-forming phases are Ca–Fe–Mg carbonate, albite, sericite, chlorite, pyrite, and arsenopyrite. Occurring less frequently or in trace amounts are graphite, pyrrhotite, scheelite, stibnite, chalcopyrite, tourmaline and tetrahedrite. All of the phases listed above form relatively early in the typical paragenesis. Wherever detailed petrographic studies have been conducted, gold has been shown to be paragenetically late, commonly associated with late quartz, carbonate, galena, sphalerite ± tellurides and typically concentrated in fractures in early sulphides (Weir and Kerrick, 1987; Nesbitt and Muehlenbachs, 1989; Graves and Zentilli, 1982; Green *et al.*, 1982; Smith and Kontak, 1988b) (Figure 4.8).

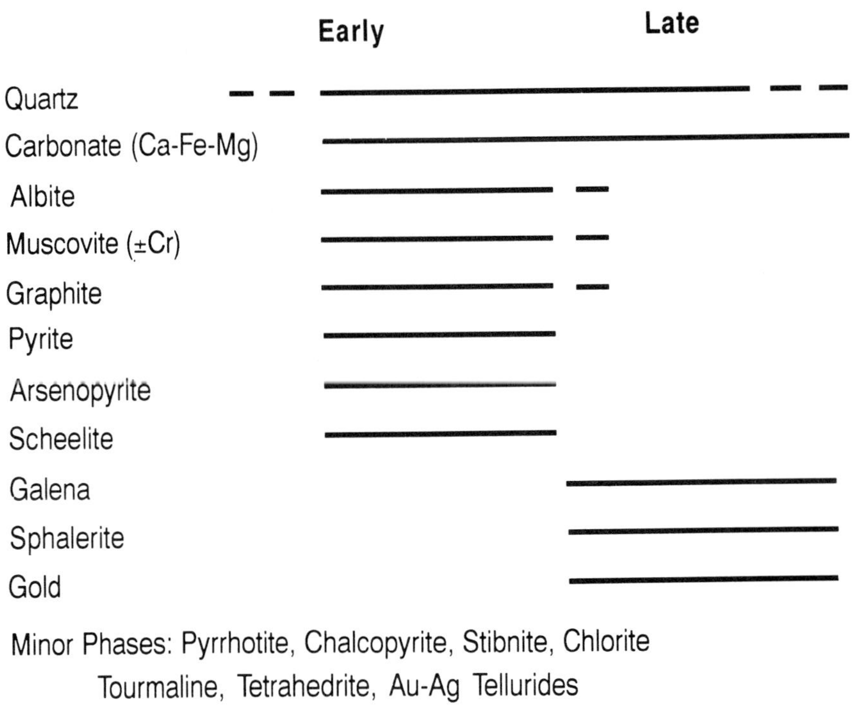

Figure 4.8 Generalized mineralogy and paragenesis of mesothermal gold deposits.

4.3.5 *Hydrothermal alteration*

In contrast to the remarkable consistency in vein mineralogy, paragenesis and textures, the mineralogy and geochemistry of the hydrothermal alteration zones varies considerably both within districts and between districts. The apparent principal control on the variations in alteration style is a difference in host-rock lithology. Hydrothermal alteration associated with ores hosted by arenaceous wackes is generally minimal with minor disseminated sulphides, sericite, and carbonate (Sandiford and Keays, 1986), though occasionally pervasive cryptic silicification is observed (Smith and Kontak, 1988a). In more feldspathic to pelitic units, alteration can be more pronounced as exemplified by the Coquihalla deposits in southern British Columbia, where extensive albitization of the host rocks has occurred (Ray *et al.*, 1986).

In intermediate to mafic volcanics and ultramafics, hydrothermal alteration is much more intense and pervasive. The dominant style of alteration in such units is carbonatization by Ca–Mg–Fe carbonates (Bohlke and Kistler, 1986; Nesbitt and Muehlenbachs, 1989; Landefeld, 1988). In addition to carbonates, chlorite, pyrite, sericite, graphite and talc are common in these alteration zones. In ultramafic units, the green mica, fuchsite, is common, giving rise to the distinctive speckled green, listwanites, which are often associated with the ores (Dussell, 1986). Alteration of felsic plutonic rocks varies from minor to pervasive albitization, silicification, chloritization and carbonatization (Sang and Ho, 1987; Shelton *et al.*, 1988).

4.3.6 *Elemental geochemistry and zoning*

One of the most consistent features of Phanerozoic, mesothermal gold deposits is a relatively high Au/Ag ratio. The ratio is nearly always greater than 1, typically around 10, and in many cases the deposits are essentially gold-only deposits (Boyle, 1979).

A relatively typical suite of elements in anomalous concentrations associated with Au includes Ag, Sb, As, W, Hg ± Bi, Mo, Pb, Zn, Cu, Ba (Haynes, 1986; Paterson, 1986; Bowen and Whiting, 1975; Boyle, 1979). It is interesting to note that this assemblage of elements is similar to the assemblage associated with epithermal gold deposits (Henley, this volume) and consequently epithermal and mesothermal gold systems cannot be distinguished on the basis of their respective suites of associated elements (Nesbitt, 1988a). In Australia, New Zealand, the Canadian Cordillera, Alaska and Bolivia, stibnite concentrations are occasionally sufficient to justify the extraction of antimony and often a transition from Au to Sb ± Au to Sb deposits is noted (Henley *et al.*, 1976; O'Shea and Pertzel, 1988; Nesbitt and Muehlenbachs, 1989). Studies of gold mineralization in New Zealand indicate that in addition to antimony, tungsten is often deposited in such settings (Paterson, 1986). The general indication from fluid inclusion and other data is that formation of antimony and tungsten mineralization occurs at somewhat lower temperatures and probably shallower depths than gold mineralization (Paterson, 1986). This concept has been extended further in studies of Canadian Cordilleran deposits to include a vertical Au to Sb ± W to Hg zonation with Hg deposition occurring at temperatures of 150°C or lower and within 0 to 2 km of the surface (Nesbitt *et al.*, in press).

4.3.7 *Fluid inclusions*

Fluid-inclusion studies of Phanerozoic mesothermal gold deposits (Table 4.2) range from relatively thorough in the North American Cordillera and New Zealand to relatively limited for deposits in Nova Scotia and Australia (Leach *et al.*, 1987; Nesbitt and Muehlenbachs, 1989; Weir and Kerrick, 1987; Coveney, 1981; Graves and

Zentilli, 1982; Sang and Ho, 1987; Shelton *et al.*, 1988; Paterson, 1986; Kontak *et al.*, 1988a). Based on the data that are available, there appears to be relatively good agreement in terms of inclusion characteristics between districts. Primary inclusions generally contain both H_2O and CO_2. Mole percentages of CO_2 typically range between 5 and 20%. Secondary H_2O or H_2O–CO_2 inclusions are common, which is to be expected given the multiple stages of deformation and veining that affected many of these systems (Roedder, 1984). Salinity determinations indicate low salinities for the fluids, typically <5 wt.% NaCl equivalent. Daughter minerals are rare. Heating tests have produced a large range in temperatures of homogenization but typically the values cluster between 250 and 350°C. Pressure estimates from the inclusion studies generally indicate values greater than or equal to 1 kb. Studies by Leach *et al.*(1987) have identified minor amounts of N_2 and CH_4 in addition to H_2O and CO_2 in fluid inclusions from the Alaskan mesothermal deposits.

Table 4.2 Fluid inclusion data from Phanerozoic mesothermal gold deposits

Area	Temp. (homog.) (°C)	Est. mole% CO_2	Salinity (wt.% NaCl equiv.)	References
Victoria	160–330	~ 5	1–9	1
Nova Scotia	200–400	13–14	<4–6	2
California	150–280	<11	<2	3, 4
Canadian Cordillera	200–350	5–20 (up to 100)	2–8	5
Alaska	150–310	5–20 (up to 50)	<5	6
Korea	290–375	15–25	~4	7
China	100–350	< 5	~4	8

References: 1. Green *et al.* (1982); 2. Kontak, MacDonald and Smith., (1988); 3. Weir and Kerrick (1987); 4. Coveney (1981); 5. Nesbitt and Muehlenbachs (in press) 6. Goldfarb *et al.* (1988); 7. Shelton *et al.* (1988); 8. Sang and Ho (1987).

4.3.8 *Stable isotopes*

$\delta^{18}O$ values for vein quartz have been obtained from most of the major Phanerozoic mesothermal districts. The $\delta^{18}O$ values are relatively enriched in ^{18}O with a general range of values from +12 to + 19‰ (SMOW) (Goldfarb *et al.*, 1988; Nesbitt *et al.*, in press; Shelton *et al.*, 1988; Taylor, 1986; Weir and Kerrick, 1987; Wilson and Golding, 1988; Kontak *et al.*, 1988c) (Table 4.3). These results, in terms of both their absolute values and their relatively restricted range of values, are similar to results obtained from Archaean, mesothermal gold deposits (Kerrich, 1987). In some Phanerozoic deposits, there appears to be evidence for control of $\delta^{18}O$ values of vein material by the $\delta^{18}O$ characteristics of the host rocks (Bohlke and Kistler, 1986). However, in most districts the $\delta^{18}O$ values of the veins appear to reflect regional lithologic controls (Nesbitt *et al.*, in press; Wilson and Golding, 1988). Calculated $\delta^{18}O$ values for the quartz-depositing fluids range from + 6 to + 11‰.

$\delta^{18}O$ values for carbonates (ankerite, dolomite, calcite, and magnesite) are in general 2–4‰ lighter in $\delta^{18}O$ than the vein quartz (Table 4.3), producing a typical range of values of +12 to +17 ‰ (SMOW) (Goldfarb *et al.*, 1987; Kontak, *et al.*, 1988c; Nesbitt *et al.*, in press; Wilson and Golding, 1988). Results of quartz-carbonate isotope thermometry vary considerably. In some Sierra Nevadan deposits, relatively good agreement was obtained between stable isotope thermometry and other temperature estimates (Bohlke and Kistler, 1986). However, carbonate-quartz isotope temperatures generally have a wide range in values and are often 50–100°C lower than temperatures estimated from fluid inclusions (Weir and Kerrick, 1987; Green *et al.*, 1982). Consequently, calculated $\delta^{18}O$ values for H_2O in equilibrium with the vein carbonates are similar to or slightly lower than the values calculated from the analyses of vein quartz.

Table 4.3 Stable isotope data from Phanerozoic mesothermal lode gold deposits (in ‰).

Area	$\delta^{18}O$	$\delta^{13}C$	δD	$\delta^{34}S$	References
Victoria	+14 to +21 (Qz) +11 to +18 (Cb)	–20 to –3 (Cb)		–10 to +10 (Py, Po, Apy)	1, 2
Nova Scotia	+12 to +18 (Cb)	–25 to –14 (Cb)			3
California	+14 to +23 (Qz) +13 to +19 (Cb)	–6 to +7 (Cb) –5 to –2 (CO_2) –25 to –23 (CM)	–100 to –30 (FI) –60 to –40 (Sil)	–4 to +6 (Py, Po, Gl, Spl)	4, 5, 6
Canadian Cordillera	+13 to +18 (Qz) +13 to +17 (Cb)	–10 to –2 (Cb) –17 to –8 (CO_2) –26 to –25 (CM)	–160 to –100 (FI)	+8 to +13 (Py, Apy)	7
Alaska	+14 to +18 (Qz)		–100 to –120 (FI) ~ –60 (Sil)		8, 9
Korea	+11 to +15 (Qz) +6 to +9 (Cb)	–6 to –5 (Cb)	–110 to –80	+4 to +12 (Py, Po, Gl, Spl)	10
China	+6 to +13 (Qz) +6 to +9 (Cb)	–8 to –3 (Cb)	–140 to –60 (FI)	–8 to +16 (Py)	11, 12

Abbreviations: Qz, quartz; Cb, carbonate; FI, fluid inclusions; CM, carbonaceous material; Sil, silicate; Py, pyrite; Apy, arsenopyrite; Po, pyrrhotite; Gl, galena; Spl, spahlerite

References: 1. Wilson and Golding (1988); 2. Gulson *et al.* (1988); 3. Kontak, Smith, and Kerrich, (1988); 4. Taylor (1986); 5. Bohlke and Kistler (1986); 6. Weir and Kerrick (1987); 7. Nesbitt *et al.* (in press); 8. Goldfarb *et al.* (1988); 9. Mitchell *et al.* (1981); 10. Shelton *et al.* (1988); 11. Sang and Ho (1987); 12. Huan-Zhang Lu (1988).

The interpretation of the genetic significance of the calculated $\delta^{18}O$ values for the ore fluids is difficult since at least three fluids of distinctly different origins may have $\delta^{18}O$ values in this range (Taylor, 1979). Both metamorphic and magmatic fluids are believed to have $\delta^{18}O$ values which would reasonably fit the range of observed values, though in some locations the calculated $\delta^{18}O$ values are somewhat enriched in $\delta^{18}O$ relative to 'typical' magmatic values. A third possible fluid source is isotopically evolved meteoric water. This is meteoric water which, due to deep circulation and interaction with rocks at low water:rock ratios, evolved isotopically to $\delta^{18}O$ values similar to the observed values (Nesbitt *et al.*, 1986). Based on $\delta^{18}O$ data alone, none of these three potential fluid sources can be excluded.

$\delta^{13}C$ values obtained from carbonates in mesothermal deposits generally fall in the range of 0 to –10‰ (PDB) (Nesbitt *et al.*, in press; Taylor, 1986; Wilson and Golding, 1988) with the exception of the Meguma deposits where $\delta^{13}C$ values for carbonates range to values as low as –25‰ (Kontak *et al.*, 1988c). In the Canadian Cordillera, in addition to $8^{13}C$ data from carbonates, $\delta^{13}C$ values have been obtained from fluid-inclusion CO_2 and from carbonaceous material in the veins. The $\delta^{13}C$ values for fluid-inclusion CO_2 range from − 8 to −17‰, whereas the $\delta^{13}C$ values of carbonaceous material are even more depleted, −25‰ to −26‰ (Nesbitt *et al.*, in press).

Considerable discussion has occurred in the literature on the significance of $\delta^{13}C$ data in interpreting the origin of Archaean, mesothermal ore fluids. Burrows *et al.*(1986) argued that a tight range of $\delta^{13}C$ values with an average value of − 3.1 ± 1.3‰ observed in the Timmins district of Ontario indicated a magmatic origin for the ore fluids. However, a review by Kerrich (1987) of Archaean mesothermal deposits of Ontario and Quebec documented a greater range in $\delta^{13}C$ values and defined regional differences, all of which are interpreted as indicating a metamorphic origin for the ore fluids. As is shown in Table 4.3, a large range in $\delta^{13}C$ values is reported for carbonates associated with Phanerozoic mesothermal gold deposits. Following the lines of argument used in the discussion on Archaean deposits, this large range in values appears to eliminate an igneous source for the carbon. However, due to the complexities of the interpretation of carbon isotope systematics (Ohmoto and Rye, 1979), this conclusion is equivocal.

$\delta^{34}S$ values for sulphides from Phanerozoic mesothermal gold deposits range from − 10 to + 25‰ (CDM) with most values falling between 0 and +10‰ (Gulson *et al.*, 1988; Nesbitt, 1988b; Stuwe *et al.*, 1988; Shelton *et al.*, 1988; Taylor, 1986; Kontak *et al.*, 1988c). The interpretation of the genetic significance of this range of values is complex (Ohmoto and Rye, 1979). In general, values near 0‰ are considered to be indicative of an igneous source for the sulphur, whereas more positive values are likely to reflect a sedimentary source (Ohmoto and Rye, 1979). Results from studies in both the Victoria gold belt and the Canadian Cordillera indicate that $\delta^{34}S$ values may have some utility as an indicator of zoning around ore veins (Gulson *et al.*, 1988; Nesbitt, 1988b; Stuwe *et al.*, 1988).

δD values have been obtained by the analysis of hydrous silicates and fluid inclusion fluids from many deposits of the North American Cordillera, Korea and China (Nesbitt *et al.*, in press; Mitchell *et al.*, 1981, Bohlke and Kistler, 1986; Shelton *et al.*, 1988; Taylor, 1986; Huan-Zhang, 1988; Goldfarb *et al.*, 1988). Values obtained from inclusion fluids range from − 30 to − 160‰ (SMOW). Agreement between the fluid inclusion analyses and the δD values calculated from δD analyses of hydrous silicates from the deposits varies from good to poor, with the hydrous silicates in general indicating more D-rich values (Goldfarb *et al.*, 1988; Nesbitt *et al.*, in press; Bohlke and Kistler, 1986). Two interpretations of these results are presented in the literature:

(i) The analyses of fluid inclusion fluids reflect the values of ore fluids whereas the hydrous silicates retain an earlier, metamorphic imprint (Nesbitt *et al.*, 1987).

(ii) The values calculated from hydrous silicate data reflect the ore-fluid values whereas the fluid inclusion data reflect contamination by later secondary fluids (Goldfarb *et al.*, 1988).

There are several arguments favouring the first interpretation. As pointed out by Roedder (1984) and earlier in this review, gold is paragenetically late in these systems and hosted by small fractures in the earlier phases. Such small fractures also contain secondary inclusions indicating that at least some of the secondary inclusions in the quartz are gold-related. Secondly, even in samples which have high primary: secondary inclusion ratios, there is no indication of a significant component of fluid more enriched in D (Shelton *et al.*, 1988; Nesbitt *et al.*, in press). Thirdly, there *is* good agreement at some locations between both $\delta^{18}O$ and δD values calculated from values for hydrous silicates and the values from vein quartz (Nesbitt *et al.*, in press). Finally, the results of analyses of noble gases of fluids from laser opened, primary inclusions are consistent with the first interpretation (Bohlke *et al.*, in press).

In the Sierra Nevada deposits, δD values range from – 40 to – 70‰, which is similar to the range in Archaean mesothermal gold deposits (Kerrich, 1987; Groves and Foster, this volume). The genetic significance of this range of values is difficult to determine, since the range is typical of magmatic, metamorphic or California meteroic water. However, in the Canadian and Alaskan Cordillera and in the Chinese and Korean deposits, fluid-inclusion δD values range from – 100 to – 160‰ and decrease with increasing latitude (Mitchell *et al.*, 1981; Nesbitt *et al.*, in press; Shelton *et al.*, 1988; Huan-Zhang, 1988). Such low values and their latitudinal dependence unambiguously indicate the involvement of meteoric water in the formation of these deposits.

4.3.9 *Applications of Sr, Pb and Nd isotope ratios*

Sr, Pb and Nd isotope systematics have been applied to a variety of Phanerozoic, mesothermal gold deposits as tracers of the origin of vein components. In the Sierra Nevada deposits, Bohlke and Kistler (1986) found a range in $^{87}Sr/^{86}Sr$ initial ratios from 0.704 to 0.718, indicating that either the Sr in the deposits was derived from diverse sources or re-equilibrated with the differing rock units which host the ores. Similar results indicating heterogeneous basement sources have been obtained from $^{87}Sr/^{86}Sr$ studies of the Meguma deposits (Kontak, Smith and Kerrich, 1988c). In the Cassiar deposits of northern British Columbia, using $^{87}Sr/^{86}Sr$ and $^{143}Nd/^{144}Nd$ results, Nesbitt *et al.* (1989) document the equilibration of the ore fluids with autochthonous, Palaeozoic sedimentary units prior to the influx of the fluids into an allochthonous, oceanic basaltic package.

Lead isotope data from the Bridge River district in British Columbia indicate that Pb in the ores is a mixture of primitive mantle-type lead and more radiogneic lead (Leitch *et al.*, 1988). In the Victoria gold belt, lead isotope data appear to indicate derivation of the lead from greenstone units underlying the host sedimentary units (Gulson *et al.*, 1988). In the Chinese deposits, Pb data are consistent with a model of derivation of ore-lead from adjoining and underlying older terranes (Sang and Ho, 1987). Consequently, the general picture derived from the application of Sr, Pb and Nd isotope systematics is one of large-scale systems with the ore components being derived from underlying, often heterogeneous, sources.

4.4 Genetic models

Due to the efforts of a large number of geologist and geochemists over many years, a significant data base now exists of the characteristics of Phanerozoic mesothermal gold deposits, and a summary of the geological and geochemical characteristics of these deposits is presented in Table 4.1. In several of the categories of Table 4.1, there is considerable heterogeneity among various districts, especially in host rock-type relation to plutons, and hydrothermal alteration. However, surprisingly there are other characteristics, particularly metamorphism, timing of ore emplacement, ore morphology and textures, mineralogy and paragenesis, fluid inclusions, and stable isotopes, for which there are considerable similarities across the principal districts. The existence of this degree of uniformity for many characteristics suggests that certain aspects of the genesis of these deposits must be common to all deposits of this type. Consequently, it appears probable that there is a common genetic process among the various districts, with differences between the districts being a product of local geologic heterogeneities. A similar concept of a common genetic process with local variations has frequently been applied to Archaean, mesothermal gold deposits (Groves and Phillips, 1987; Kerrich, 1987; Colvine *et al.*, 1984).

Theories of origin for Phanerozoic mesothermal gold deposits are nearly as numerous as the number of deposits and the number of geologists who have studied them. The theories span a broad range including syngenetic, metamorphic, magmatic and meteoric modes of origin, with numerous variations within each of the categories. A thorough, historical review of theories of origin has been compiled by Boyle (1979, 1986) and consequently this section will focus only on more recent genetic concepts.

Syngenetic hypotheses for the origin of these deposits generally have few adherents, due to the striking epigenetic characteristics present in most districts. However, in studies of the Meguma deposits, where the ores are more stratabound than in other districts, there is some support for a syngenetic origin. Haynes (1986) proposed that the stratabound deposits were the result of seafloor hot spring activity and that veins which cross-cut bedding represent feeder conduits for fluids which formed the seafloor deposits.

Genetic models for Phanerozoic mesothermal deposits, which involve felsic plutons, vary from models where magmatic fluids are the principal hydrothermal fluid source to models where the plutons provide only heat to drive the systems. There is little support amongst current researchers for the concept that magmatic fluids are the primary, ore-forming fluid. This is largely the result of the lack of consistent and close spatial and temporal relationships between felsic plutons and the mesothermal gold deposits (Bohlke and Kistler, 1986). In addition, the large ranges in δD and $\delta^{13}C$ values are inconsistent with derivation of the ore fluids from a magmatic source. On the other hand, there is considerable support for the idea that felsic plutons regionally provided heat to drive the ore-forming systems. Such a process appears to have been important in the Victoria and Nova Scotia gold districts and the deposits of northern China and Korea, where there is a close age-relation between late tectonic granites and the ores (Sandiford and Keays, 1986; Shelton *et al.*, 1988; Sang and Ho, 1987; Kontak and Smith, 1988). In the Alaska and Sierra Nevada deposits, the link between plutonism and gold mineralization is less direct, with the thermal drive for mineralization being ascribed to a general, post-metamorphic, subduction-plutonism event (Bohlke and Kistler, 1986; Landfeld, 1988; Goldfarb *et al.*, 1988). Yet in other areas such as parts of the Canadian Cordillera, there is good evidence that plutonic activity either significantly preceded or post-dated mineralization (Sketchley *et al.*, 1986).

The most generally accepted genetic model for the formation of Phanerozoic mesothermal gold deposits is one of derivation from metamorphic processes and metamorphic fluids (Graves and Zentilli, 1982; Sandiford and Keays, 1986; Goldfarb *et al.*, 1988; Kontak and Smith, 1988). In this model, high geothermal gradients with or without plutonic heat cause regional metamorphism and devolatilization. This metamorphic fluid plus any remnant pore fluid acquires CO_2, Si, S, and Au, as the fluid migrates through the metamorphic units toward major structures, often strike-slip faults. Fluid-migration in the vicinity of major structures is aided by hydrofracturing in response to fluid pressures exceeding lithostatic (geostatic) pressure plus the tensile strength of the rock. Chemical data from several districts indicate that this migration process occurs on a large scale with many of the elements being derived from units significantly (1 km or greater) below the ore-hosting units. As the fluids rise, they cool, effervesce CO_2 and lose sulphur to the gas phase or in response to sulphidation of the host rocks. The aggregate effect of these processes is to bring about vein formation, gold precipitation, and wallrock alteration.

The metamorphic model is consistent with the bulk of the geologic and geochemical data. It clearly accounts for the link to metamorphic grade, the characteristics of the fluid inclusions, the source of the metals, and the $\delta^{18}O$, $\delta^{13}C$ and $\delta^{34}S$ data. However, it is often difficult to account for the general post-peak metamorphic timing of vein formation, since the bulk of the devolatilization occurs during prograde metamorphism (Stuwe, 1986; Kontak *et al.*, 1988b). This problem is generally ignored, ascribed to a second, deeper metamorphic event, or explained by complex trapping mechanisms for ore fluids during metamorphism (Bohlke and Kistler, 1986; Paterson, 1982). A second problem is that the model is inconsistent with δD data obtained from many of the deposits in the Canadian Cordillera, Alaska, Korea and China.

The newest model for the genesis of Phanerozoic mesothermal gold deposits is one that invokes deep convection of meteoric water in the brittle continental crust as the principal process in ore formation. The convection process is believed to be driven by heat from either the regional geothermal gradient or deep plutons (Nesbitt *et al.*, 1986; Shelton *et al.*, 1988). Temperatures in excess of 350°C are attained, indicating that the fluids interacted with rocks undergoing greenschist facies metamorphism. During convection, the fluids acquire CO_2, Si, S and Au, and, due to water–rock interactions at relatively low water/rock ratios, the $\delta^{18}O$ values are strongly enriched in ^{18}O (Nesbitt and Muehlenbachs, in press). As a result of the significant degree of water–rock interaction, the meteoric fluids chemically resemble metamorphic fluids in many respects. δD values of the fluids do not change appreciably due to the low hydrogen content of many rock units. As the fluids rise along relatively permeable, major fault zones, they deposit the ore in response to cooling and loss of CO_2 and sulphur. Studies in the Canadian Cordillera indicate that at shallower levels and cooler temperatures these fluids form Sb ± Au mineralization and, near the surface, Hg ± Sb deposits (Nesbitt *et al.*, in press).

The convection process requires permeability values in excess of 10^{-17} m^2 and fluid pressures approximately equal to hydrostatic pressure values (Nesbitt, 1988a). Values for permeability equal to or greater than 10^{-17} m^2 are common in fractured and jointed rocks of the brittle rheological regime (Brace, 1984). A probable lower limit for such permeability values in the crust most likely occurs at the brittle–ductile transition at 350–450°C (Sibson, 1983; Kusznir and Park, 1983), which is consistent with the probable upper limit of temperature in the ore fluids. The requirement that the fluid pressure approximates the hydrostatic head generally precludes hydrofracturing from being an important phenomenon in vein formation, which is consistent with the results of permeability studies of Brace (1980; 1984) and the Russian deep-drilling programme (Kozlovsky, 1984). However, locally hydrofracturing could result from fluid pressures within wallrocks exceeding fluid pressures within an opening extensional fracture in a zone of dilation (Sibson, 1986). In addition, under hydrostatic conditions, 1000 bar of fluid pressure is equivalent to approximately 10 000 m of depth as opposed to approximately 3300 m of depth under lithostatic conditions. Consequently, the depths at which the gold ores form are probably much deeper than previously believed.

The advantage of the meteoric water model is that it is consistent with all of the data generally cited in support of the metamorphic model as well as being in agreement with the δD data. In addition, the late timing of vein formation in many districts probably reflects the influx of meteoric water into still warm, metamorphic units during uplift after maximum burial.

In summary, the two principal genetic models, metamorphic and meteoric, are similar in many respects, such as the importance of regional structure and the chemistry of the ore deposits, and differ in the source and migration paths of the fluids. In spite of the differences, recent studies involving either model are consistent with two fundamental conclusions. First, mesothermal gold deposits are recognized as volumetrically small, but economically important, parts of large-scale, fluid-flow systems in the crust. There is increasing evidence that the vein constituents and ore fluids migrated over large distances and that the ores are not products of local remobilization or concentration. Secondly, as a corollary to the first point, there is a recognizable and regular association of the deposits with crustal scale, generally subvertical structures. These structures provided fluid pathways for the ore fluids, sites for ore deposition and, consequently, were essential to the formation of the deposits.

Future research will concentrate on the verification and amplification of many of the concepts discussed above. Additional testing using a variety of geochemical techniques is required to resolve the conflicts over fluid source between the metamorphic and meteoric models. Further structural studies documenting the relations between large scale structures and ore-hosting structures will be useful in defining fluid-flow paths and ore-deposition controls. Computer modelling of fluid flow and chemical evolution will be useful in determining the constraints on gold mineralization within the larger scale fluid system. Finally, more detailed investigations of the structure and chemistry of ore zones are necessary to determine the exact controls on gold deposition.

4.5 Comparisons of Phanerozoic mesothermal deposits to other types of gold mineralization

As pointed out by a variety of authors (Boyle, 1979; Kerrich, 1987; Nesbitt *et al.*, 1986; Hutchinson, 1987; Sang and Ho, 1987), there are a number of similarities between Phanerozoic and Archaean mesothermal gold deposits. In fact, since most characteristics summarized in Table 4.1 are common to both Phanerozoic and Archaean mesothermal deposits, the deposits of both age groups should more appropriately be considered as part of one age-independent class of mesothermal gold deposits. The recognition of the similarities between Archaean and Phanerozoic deposits is important, since the results of studies of the younger deposits, which generally have undergone less post-ore deformation and metamorphism, may be pertinent to the Archaean deposits as well.

One significant difference between the Archaean and Phanerozoic deposits is the overall larger accumulation of gold in Archaean mesothermal deposits (Boyle, 1979; Woodall, 1988; Hutchinson, 1987). This is especially significant given the larger area of exposure of Phanerozoic tectonic belts. This difference in abundance of gold cannot be simply ascribed to the greater productivity of gold camps in the Archaean, since at least two Phanerozoic districts, the Mother Lode of California and Bendigo-Ballarat in Victoria, compare favourably with the richest Archaean districts (Boyle, 1979). Consequently, it appears that there is simply a greater density of lode gold deposits per unit area in the Archaean relative to the Phanerozoic. Many authors have noted this disparity but no satisfactory explanation has been obtained (Boyle, 1979). Hutchinson (1987) noted that the end of the major era of gold enrichment in the Precambrian coincided with the oxygenation of the Earth's atmosphere and hydrosphere. For this factor to be important it would require a significant involvement of the hydrosphere in ore generation. An alternative possibility may be linked to a higher regional geothermal gradient in the Archaean, which may have stimulated either greater meteoric water convection or shallower metamorphic devolatilization. As yet no generally accepted answer has been obtained and the differences in gold abundance between the Phanerozoic and Archaean terranes remain enigmatic.

Of the other three major types of Phanerozoic deposits (intrusion-related, Carlin, and epithermal), the Phanerozoic mesothermal deposits only have significant similarities and possible overlap with the first two. As described above, many Phanerozoic deposits have close spatial and possibly genetic links to igneous plutons, which may indicate a transition from mesothermal to intrusion-related deposits. However, many characteristics of mesothermal deposits, in particular ore morphology, mineralogy, paragenesis, hydrothermal alteration, Au/Ag ratios, and fluid-inclusion characterisitics, differ substantially from the corresponding characterisitics of intrusion-related deposits. The most probable explanation for the differences between the two styles of pluton associated mineralization is one of different depths of emplacement ranging from the relatively shallow, intrusion-related deposits to the deeper mesothermal deposits. Mesothermal gold deposits and Carlin-type deposits share some similarities, especially in terms of structural controls on mineralization and characteristics of associated elements. At least one model has been proposed which attempts to explain the relationships between Phanerozoic mesothermal and Carlin-type deposits based on depth of meteoric water convection in the genesis of the deposits (Nesbitt, 1988a). In this model, the depth of meteoric water convection is linked to the depth of the brittle-ductile transition in both settings. In high heat flow areas such as the extensional regime of Nevada, which hosts the Carlin and related deposits, the depth of convection, and hence mineralization, is shallower than in the case of the mesothermal deposits. This leads to distinctly different chemical properties of the ore fluids and subsequently to differing characteristics of mineralization and hydrothermal alteration (Nesbitt, 1988a).

4.6 Conclusions

The metallogeny of gold in the Phanerozoic is distinctive relative to other aeons in its wide variety of styles of economic gold mineralization. With epithermal, intrusive-related, Carlin-type, placer, Au-rich massive sulphide and mesothermal deposits all being important styles of mineralization, no other aeon offers such a variety of potential gold targets. In addition, with the recent, major discoveries in many of these categories, the exploration potential in Phanerozoic terranes is clearly documented.

This chapter has focused on the geological and geochemical characteristics of mesothermal gold deposits of the Phanerozoic. One of the principal observations which has arisen from this review is that in many characteristics, especially in timing of ore formation, structure, metamorphism, ore morphology, textures, associated elements, mineralogy, paragenesis, fluid inclusions, and stable and radiogenic isotopes, there is a considerable degree of similarity among the major districts, world wide. The existence of this degree of similarity suggests that in many respects the genetic processes are similar as well. The principal differences between districts, largely in host-rock type, hydrothermal alteration, and relation to plutons, most likely reflect local geological heterogeneities.

One of the surprising conclusions arising from this synopsis is the close similarity between Archaean and Phanerozoic mesothermal deposits. Recognition and documentation of this similarity here and elsewhere (Hutchinson, 1987; Kerrich, 1987) support a classification scheme of a single general category of mesothermal lode gold deposits encompassing mesothermal ores ranging from Archaean to Tertiary in age.

Acknowledgements

I would like to thank P. Lhotka for a helpful review of an earlier version of this manuscript.

References

Akande, S.O. and Fakorede, O. (1988) Gold mineralisation in the Nigerian Schist Belts. In *Bicentennial Gold 88*, eds. Goode, A.D.T., Smyth, E.L., Birch, W.D. and Bosma, L.I. *Geol. Soc. Aust. Abstr.* **23**, 140-142.

Albers, J.P. (1981) A lithologic-tectonic framework for the metallogenic provinces of California. *Econ. Geol.* **76**, 765-790.

Andrews-Speed, C.P., Sliwa, A. and Unrug, R. (1984) Primary gold occurrences in Zambia and their geological controls. In *Gold '82* ed. Foster, R.P. *Proceedings of the Symposium Gold '82.* Geol. Soc. of Zimbabwe, Special Pub. No.1, 493-505.

Bohlke, J.K. and Kistler, R.W. (1986) Rb-Sr, K-Ar, and stable isotope evidence for the ages and sources of fluid components of gold-bearing quartz veins in the Northern Sierra Nevada Foothills Metamorphic Belt, California. *Econ. Geol.* **81**, 296- 322.

Bohlke, J.K., Kirschbaum, C. and Irwin, J.J. (in press) Simultaneous analyses of noble gas isotopes and halogens in fluid inclusions in neutron-irradiated quartz veins using a laser-microprobe noble gas spectrometer. In *New Frontiers in Stable Isotope Research: Laserprobes,*

Ion Probes and Small Samples Analysis, eds. Sharks, W.C., and Criss, R.E. US Geol. Surv., Bull.

Bowen, K.G. and Whiting, R.G. (1975) Gold in the Tasman Geosyncline, Victoria. In *Economic Geology of Australia and Papua New Guinea*, ed. Knight, C.L. The Australasian Inst. of Mining and Metallurgy, Monograph Series No.5, 647-659.

Boyle, R.W. (1979) The geochemistry of gold and its deposits. Geol. Surv. of Can., Bull. 280. 584pp.

Boyle, R.W. (1986) Gold deposits in turbidite sequences: Their geology, geochemistry and history of the theories of their origin. In *Turbidite-hosted Gold Deposits*, eds. Keppie, J.D., Boyle, R.W. and Haynes, S.J. Geol. Assoc. of Can., Special Paper 32, 1-13.

Brace, W.F. (1980) Permeability of crystalline and argillaceous rocks. *Internat. J. Rock Mech. Min. Sci. Geomech. Abstr.* **17**, 241-251.

Brace, W.F. (1984) Permeability of crystalline rocks: New insitu measurements. *J. Geophys. Res.* **89**, 4327-4330.

Brathwaite, R.L. (1988) The tectonic setting and control of gold deposits in New Zealand. In *Bicentennial Gold 88*, eds. Goode, A.D.T. Smyth, E.L., Birch, W.D. and Bosma, L.I. *Geol. Soc. Aust. Abstr.* **22**, 191-196.

Burrows, D.R., Wood, P.C. and Spooner, E.T.C. (1986) Carbon isotope evidence for a magmatic origin for Archaean gold-quartz vein ore deposits. *Nature* **321**, 851-854.

Cairnes, C.E. (1937) Geology and mineral deposits of the Bridge River mining camp, British Columbia. Geol. Surv. of Can., Memoir 213. 140pp.

Cathelineau, M. and Boiron, M.C. (1988) Fluid-mineral equilibria in French hydrothermal gold veins. In *Bicentennial Gold 88*, eds. Goode, A.D.T., Smyth, E.L., Birch, W.D. and Bosma, L.I. *Geol. Soc. Aust. Abstr.* **23**, 428-430.

Clarke, W.B. (1970) Gold districts of California. California Div. of Mines and Geol., Bull. 193. 182pp.

Collins, P.L.F. and Williams, E. (1986) Metallogeny and tectonic development of the Tasman Fold Belt system in Tasmania, *Ore Geol. Rev.* **1**, 153-201.

Colvine, A.C., Andrews, A.J., Cherry, M.E., Durocher, M.E., Fyon, A.J., Lavingne, M.J. Jr., Macdonald, A.J., Marmont, S., Poulsen, K.H., Springer, J.S. and Troop, D.G. (1984) An integrated model for the origin of Archean lode gold deposits. Ontario Geol. Surv., Open-File Report 5524. 85pp.

Coney, P.J., Jones, D.L. and Monger, J.W.H. (1980) Cordilleran suspect terranes. *Nature* **288**, 329-333.

Coveney, R.M.Jr. (1981) Gold quartz veins and auriferous granite at the Oriental Mine, Alleghany District, California. *Econ. Geol.* **76**, 2176-2199.

Degeling, P.R., Gilligan, L.B., Scheibner, E. and Suppel, D.W. (1986) Metallogeny and tectonic development of the Tasman Fold Belt System in New South Wales. *Ore Geol. Rev.* **1**, 259-313.

Dussell, E. (1986) Listwanites and their relationship to gold mineralization at Erickson Mine, British Columbia, Canada. Unpublished M.Sc. thesis, Western Washington Univ. Bellingham, Washington. 90pp.

Feiss, P.G. (1988) Gold mineralization in the southern Appalachian Piedmont–Pardee and Park redivivus. In *Proc. of North American Conf. on Tectonic Control of Ore Deposits and the Vertical and Horizontal Extent of Ore Systems*, eds. Kisvarsanyi, G. and Grant, S.K. Univ. of Missouri-Rolla, 491- 500.

Fitches, W.R. (1987) Aspects of veining in the Welsh lower Palaeozoic Basin. In *Deformation of Sediments and Sedimentary Rocks*, eds. Jones, M.E., and Preston, R.M.F. Geol. Soc. Spec. Publ. 29, 325-342.

Gilligan, L.B. and Barnes, R.G. (1988) Gold in the New England Fold Belt in New South Wales. In *Bicentennial Gold 88*, eds. Goode, A.D.T., Smyth, E.L., Birch, W.D. and Bosma, L.I. *Geol. Soc. Aust. Abstr.* **23**, 215-218.

Goldfarb, R.J., Nelson, S.W., Berg, H.C. and Light, T.D. (1987) Distribution of mineral deposits in the Pacific Border Ranges and Coast Mountains of the Alaskan Cordillera. In *Geoexpo/86–Exploration in the North American Cordillera*, eds. Elliott, I.L. and Smee, B.W. The Assoc. of Exploration Geochemists, 19-41.

Goldfarb, R.J., Leach, D.L. and Pichthorn, W.J. (1988) Application of fluid inclusion and stable isotope techniques to the study of Tertiary mesothermal gold deposits, Southern Alaska, U.S.A.

In *Bicentennial Gold 88*, eds. Goode, A.D.T., Smyth, E.L., Birch, W.D. and Bosma, L.I. *Geol. Soc. Aust. Abstr.* **23**, 439-441.

Graves, M.C. and Zentilli, M. (1982) A review of the geology of gold in Nova Scotia. In *Geology of Canadian Gold Deposits*, eds. Hodder, R.W. and Petruk, W. Can. Inst. of Mining and Metallurgy, Special Vol. 24, 233-242.

Green, A.H., Donnelly, T.H., Jahnke, F.M. and Keays, R.R. (1982) Evolution of gold-bearing veins in dykes of the Woods Point dyke swarm, Victoria. *Mineral. Deposita* **17**, 175-192.

Groves, D.I. and Phillips, G.N. (1987) The genesis and tectonic control on Archaean gold deposits of the Western Australian Shield–a metamorphic replacement model. *Ore Geol. Rev.* **2**, 287-322.

Gulson, B.L., Andrew, A.S., Mizon, K.J., Keays, R.R. and Stuwe, K. (1988) Source of gold in Ballarat Slate Belt deposits and potential exploration applications. In *Bicentennial Gold 88*, eds. Goode, A.D.T., Smyth, E.L., Birch, W.D. and Bosma, L.I. *Geol. Soc. Aust. Abstr.* **22**, 331-337.

Haynes, S.J. (1986) Geology and chemistry of turbidite-hosted gold deposits, greenschist facies, Eastern Nova Scotia, Canada. In *Turbidite-hosted Gold Deposits*, eds. Keppie, J.D., Boyle, R.W. and Haynes, S.J. Geol. Assoc. of Can., Special Paper 32, 161-177.

Henley, R.W., Norris, R.J. and Paterson, C.J. (1976) Multistage ore genesis in the New Zealand Geosyncline, a history of postmetamorphic lode emplacement. *Mineral. Deposita* **11**, 180-196.

Huan-Zhang Lu. (1988) Fluid inclusion studies on different types of Chinese gold deposits. In *Bicentennial Gold 88*, eds. Goode, A.D.T., Smyth, E.L., Birch, W.D. and Bosma, L.I. *Geol. Soc. Aust. Abstr.* **23**, 442-444.

Hutchinson, R.W. (1987) Metallogeny of Precambrian gold deposits: Space and time relationships. *Econ. Geol.* **82**, 1993- 2007.

Kerrich, R. (1987) The stable isotope geochemistry of Au–Ag vein deposits in metamorphic rocks. In *Stable Isotope Geochemistry of Low Temperature Fluids*, ed. Kyser, T.K. Mineralogical Assoc. of Can., Short Course 13, 287-336.

Knopf, A. (1929) The Mother Lode system of California. US Geol. Surv. Prof. Paper 157. 88pp.

Kontak, D.J. and Smith, P.K. (1988) Meguma Gold Studies VI: An integrated model for the genesis of Meguma-hosted lode gold deposits. In *Mines and Minerals Branch Report of Activities 1987, Part B*, ed. MacDonald, D.R. and Brown, Y. Nova Scotia Dep. of Mines and Energy, Report 88-1, 111-119.

Kontak, D.J., MacDonald, D. and Smith, P.K. (1988a) Fluid inclusion study of the Beaver Dam Gold Deposits, Meguma Terrane, Nova Scotia. Nova Scotia Dep. of Mines and Energy, Mineral Resources Division Report of Activities 1988, Part A, Report 88- 3, 63-69.

Kontak, D.J., Smith, P.K. and Reynolds, P.H. (1988b) Meguma Gold Studies III. Timing of mineralization at the Beaver Dam Gold Deposit, Eastern Meguma Terrane, Nova Scotia: Implications for Meguma gold deposits. In *Mines and Minerals Branch Report of Activities 1987, Part B*, ed. MacDonald, D.R. and Brown, Y. Nova Scotia Dep. of Mines and Energy, Report 88-1, 77-83.

Kontak, D.J., Smith, P.K. and Kerrich, R. (1988c) Meguma Gold Studies V: Carbon, strontium, oxygen and sulphur isotopic compositions of vein minerals: Implications for source reservoirs and fluid:rock ratios attending vein formation. In *Mines and Minerals Branch Report of Activities 1987, Part B*, ed. MacDonald, D.R. and Brown, Y. Nova Scotia Dept. of Mines and Energy, Report 88-1, 101-110.

Kozlovsky, Y.A. ed. (1984) *The Superdeep Well of the Kola Peninsula*. Springer-Verlag.

Kusznir, N.J. and Park, R.G. (1983) A mathematical model of the brittle-ductile transition within the lithosphere. *Geol. Soc. London J.* **140**, 839-842.

Landefeld, L.A. (1988) The geology of the Mother Lode Gold Belt, Sierra Nevada Foothills Metamorphic Belt, California. In *Bicentennial Gold 88*, eds. Goode, A.D.T. Smyth, E.L., Birch, W.D. and Bosma, L.I. *Geol. Soc. Aust. Abstr.* **22**, 167-172.

Large, R., Huston, D., McGoldrick, P., McArthur, G. and Ruxton, P. (1988) Gold distribution and genesis in Palaeozoic volcanogenic massive sulphide systems, Eastern Australia. In *Bicentennial Gold 88*, eds. Goode, A.D.T. Smyth, E.L., Birch, W.D. and Bosma, L.I. *Geol. Soc. Aust. Abstr.* **22**, 121-126.

Leach, D.L., Goldfarb, R.J. and Light, T.D. (1987) Fluid inclusion constraints on the genesis of the Alaska-Juneau Gold Deposit. In *Geoexpo/86–Exploration in the North American Cordillera*, eds. Elliott, I.L. and Smee, B.W. The Assoc. of Exploration Geochemists, 150-159.

Lehrberger, G. (1988) Gold-antimonite deposits in marine sediments of the Eastern Cordillera of the Bolivian Andes. In *Bicentennial Gold 88*, eds. Goode, A.D.T., Smyth, E.L., Birch, W.D. and Bosma, L.I. *Geol. Soc. Aust. Abstr.* **23**, 319-321.

Leitch, C.H.B., Dawson, K.M. and Godwin, C.I. (1988) Late Cretaceous–Early Tertiary gold mineralisation: A galena lead isotope study of the Bridge River Mining Camp, Southwestern British Columbia, Canada. In *Bicentennial Gold 88*, ed. Goode, A.D.T., Smyth, E.L., Birch, W.D. and Bosma, L.I. *Geol. Soc. Aust. Abstr.* **23**, 448-451.

Lindgren, W. (1933) *Mineral Deposits*, (4th edn.). McGraw- Hill, New York, 191-202.

Mawer, C.K. (1986) The bedding-concordant gold-quartz veins of the Meguma Group, Nova Scotia. In *Turbidite-hosted Gold Deposits*, eds. Keppie, J.D., Boyle, R.W. and Haynes, S.J. Geol. Assoc. of Can., Special Paper 32, 135-148.

Mitchell, P.A., Silberman, M.L. and O'Neil, J.R. (1981) Genesis of gold vein mineralization in an Upper Cretaceous turbidite sequence, Hope-Sunrise District Southern Alaska. US Geol. Surv., Open-File Report 81-355, 34-49.

Moravek, P. and Pouba, Z. (1987) Precambrian and Phanerozoic history of gold mineralization in the Bohemian Massif. *Econ. Geol.* **82**, 2098-2114.

Morrison, G.W. (1988) Palaeozoic gold deposits of Northeast Queensland. In *Bicentennial Gold 88*, eds. Goode, A.D.T. Smyth, E.L., Birch, W.D. and Bosma, L.I. *Geol. Soc. Aust. Abstr.* **22**, 91-101.

Murray, C.G. (1986) Metallogeny and tectonic development of the Tasman Fold Belt System in Queensland. *Ore Geol. Rev.* **1**, 315-400.

Nesbitt, B.E. (1988a) Gold deposit continuum: A genetic model for lode Au mineralization in the continental crust. *Geology* **16**, 1044-1048.

Nesbitt, B.E. (1988b) Reconnaissance study of ^{34}S values of sulphides from mesothermal gold deposits of the Eastern Canadian Cordillera. In *Geological Fieldwork 1987–A summary of field activities and current research.* Mineral Resources Division, Geological Survey Branch, B.C. Ministry of Energy, Mines and Petroleum Resources, Paper 1988-1, 543-545.

Nesbitt, B.E., Cavell, P.A. and Muehlenbachs, K. (1989) Isotopic characteristics of the McDame gold-silver deposits, Cassiar District, British Columbia. In *Current Research, Part E*, Can. Paper 89-1E, 5-16.

Nesbitt, B.E. and Muehlenbachs, K. (1988) Genetic implications of the association of mesothermal gold deposits with major strike- slip fault systems. In Proc. North American Conf. on Tectonic Control of Ore Deposits and the Vertical and Horizontal Extent of Ore Systems, eds. Kisvarsanyi, G. and Grant, S.K. Univ. of Missouri-Rolla, 57-66.

Nesbitt, B.E. and Muehlenbachs, K. (1989) Geology, geochemistry and genesis of mesothermal lode gold deposits of the Canadian Cordillera: Evidence for ore formation from evolved meteoric water. In *Economic Geology Monograph*, eds. Keays, R.R., Groves, D.I. and Ramsay, R. **6**, 552-563.

Nesbitt, B.E., Muehlenbachs, K. and Murowchick, J.B. (in press) Genetic implications of stable isotope characteristics of mesothermal Au deposits and related Sb and Hg deposits in the Canadian Cordillera. *Econ. Geol.*

Nesbitt, B.E., Murowchick, J.B. and Muehlenbachs, K. (1986) Dual origins of lode gold deposits in the Canadian Cordillera. *Geology* **14**, 506-509.

Nesbitt, B.E., Murowchick, J.B. and Muehlenbachs, K. (1987) Reply to a comment on 'Dual origins of lode gold deposits in the Canadian Cordillera'. *Geology* **15**, 472-473.

Ohomoto, H. and Rye, R.O. (1979) Isotopes of sulfur and carbon. In *Geochemistry of Hydrothermal Ore Deposits* (2nd edn.), ed. Barnes, H.L. Wiley, 509-567.

O'Shea, P.J. and Pertzel, B.A. (1988) The Brunswick gold-antimony mine, Costerfield, Victoria. In *Bicentennial Gold 88*, eds. Goode, A.D.T., Smyth, E.L., Birch, W.D. and Bosma, L.I. *Geol. Soc. Aust. Abstr.* **23**, 282-284.

Paterson, C.J. (1986) Controls on gold and tungsten mineralization in metamorphic-hydrothermal systems, Otago, New Zealand. In *Turbidite-hosted Gold Deposits*, eds. Keppie, J.D., Boyle, R.W. and Haynes, S.J. Geol. Assoc. Can., Special Paper 32, 25-39.

Ramsay, W.R.H. and Vandenberg, A.H.M. (1986) Metallogeny and tectonic development of the Tasman Fold Belt System in Victoria. *Ore Geol. Rev.* **1**, 213-257.

Ray, G.E., Shearer, J.T. and Niels, R.J.E. (1986) The geology and geochemistry of the Carolin gold deposit, southwestern British Columbia, Canada. In *Proc. of Gold '86 Symposium, Toronto*, ed. Macdonald, A.J., 470-487.

Roedder, E. (1984) Fluid inclusions. In *Reviews in Mineralogy*, Vol. 12, ed. Ribbe, P.H. Mineralogical Soc. of Amer. 644pp.

Sandiford, M. and Keays, R.R. (1986) Structural and tectonic constraints on the origin of gold deposits in the Ballarat Slate Belt, Victoria. In *Turbidite-hosted Gold Deposits*, eds. Keppie, J.D., Boyle, R.W. and Haynes, S.J. Geol. Assoc. of Can., Special Paper 32, 15-24.

Sang, J. and Ho, S.E. (1987) A review of gold deposits in China. Geol. Dept. Univ. Extension, Univ. of W. Australia, Pub. No.11, 307-320.

Shelton, K.L., So, C-S. and Chang, J-S. (1988) Gold-rich mesothermal vein deposits of the Republic of Korea: Geochemical studies of the Jungwon Gold Area. *Econ. Geol.* **83**, 1221-1237.

Sibson, R.H. (1983) Continental fault structure and the shallow earthquake source. *Geol. Soc. London J.* **140**, 741- 767.

Sibson, R.H. (1986) Brecciation processes in fault zones: inferences from earthquake rupturing. *Pure Appl. Geophys.* **124**, 159-175.

Sibson, R.H. (1987) Earthquake rupturing as a mineralizing agent in hydrothermal systems. *Geology* **15**, 701-704.

Sillitoe, R.H. (1988) Environments, styles and origins of gold deposits in western Pacific island arcs. In *Bicentennial Gold 88*, eds. Goode, A.D.T., Smyth, E.L., Birch, W.D. and Bosma, L.I. *Geol. Soc. Aust. Abstr.* **22**, 127-138.

Sketchley, D.A., Sinclair, A.J. and Godwin, G.I. (1986) Early Cretaceous gold-silver mineralization in the Sylvester allochthon, near Cassiar, north central British Columbia. *Can. J. Earth Sci.* **23**, 1455-1458.

Smirnov, V.I. ed. (1977) Gold deposits. In *Ore Deposits of the U.S.S.R.*, III. Pitman, 3-81.

Smith, P.K. and Kontak, D.J. (1988a) Meguma gold studies I: Generalized geological aspects of the Beaver Dam gold deposit, Eastern Meguma Zone, Nova Scotia. In *Mines and Minerals Branch Report of Activities 1987, Part B*, eds. MacDonald, D.R. and Brown, Y. Nova Scotia Dept. of Mines and Energy, Report 88-1, 45-59.

Smith, P.K. and Kontak, D.J. (1988b) Meguma gold studies II: Vein morphology, classification and information, a new interpretation of 'crack-seal' quartz veins. In *Mines and Minerals Branch Report of Activities 1987, Part B*, eds. MacDonald, D.R. and Brown, Y. Nova Scotia Dept. of Mines and Energy, Report 88-1, 61-76.

Steed, G.M. and Morris, J.H. (1986) Gold mineralization in Ordovician greywackes at Clontibret, Ireland. In *Turbidite- hosted Gold Deposits*, eds. Keppie, J.D., Boyle, R.W. and Haynes, S.J. Geol. Assoc. of Can., Special Paper 32, 67-86.

Stuwe, K. (1986) Structural evolution of the Port Wells gold mining district, Prince William Sound, south central Alaska: Implications for the origin of the gold lodes. *Mineral. Deposita* **21**, 288-295.

Stuwe, K., Keays, R.R. and Andrew, A. (1988) Wall rock alteration around gold quartz reefs at the Wattle Gully Mine, Ballarat Slate Belt, Central Victoria. In *Bicentennial Gold 88*, eds. Goode, A.D.T., Smyth, E.L., Birch, W.D. and Bosma, L.I. *Geol. Soc. Aust. Abstr.* **23**, 478-480.

Tate, N.M. (1988) Styles and distribution of gold mineralisation in Thailand. In *Bicentennial Gold 88*, eds. Goode, A.D.T., Smyth, E.L., Birch, W.D. and Bosma, L.I. *Geol. Soc. Aust. Abstr.* Series Number **23**, 236-237.

Taylor, B.E. (1986) Origin and isotopic characteristics of Mother Lode hydrothermal fluids and gold deposits with comparison to Archaean analogues. In *Gold '86*, ed. Chater, A.M. An Internat. Symp. on the Geology of Gold Deposits, Toronto, 1986, 148-150.

Taylor, F.C. and Schiller, E.A. (1966) Metamorphism of the Meguma Group of Nova Scotia. *Can. J. Earth Sci.* **3**, 959- 974.

Taylor, H.P.Jr. (1979) Oxygen and hydrogen isotope relationships in hydrothermal mineral deposits. In *Geochemistry of Hydrothermal Ore Deposits* (2nd edn.), ed. Barnes, H.L. Wiley, New York, 236-277.

Tomkinson, M.J. (1988) Gold mineralization in phyllonites at the Haile Mine, South Carolina. *Econ. Geol.* **83**, 1392- 1400.

Utter, T. (1984) Geological setting of primary gold deposits in the Andes of Colombia (South America). In *Gold '82*, ed. Foster, R.P. *Proceedings of the Symposium Gold '82*. Geol. Soc. of Zimbabwe, Special Pub. No.1, 731-753.

Wang X. and Cheng J. (1988) Major geological characteristics and origin of gold deposits in China. In *Bicentennial Gold 88* eds. Goode, A.D.T., Smyth, E.L., Birch, W.D. and Bosma, L.I. *Geol. Soc. Aust. Abstr.* **23**, 408-413.

Weir, R.H.Jr. and Kerrick, D.M. (1987) Mineralogic, fluid inclusion, and stable isotope studies of several gold mines in the Mother Lode, Tuolumne and Mariposa Countries, California. *Econ. Geol.* **82**, 328-344.

Wilson, A.F. and Golding, S.D. (1988) Stable isotope constraints on fluid sources for granitoid-and metamorphic-hosted gold-quartz vein deposits in Eastern Australia. In *Bicentennial Gold 88*, eds. Goode, A.D.T., Smyth, E.L., Birch, W.D. and Bosma, L.I. *Geol. Soc. Aust. Abstr.* **23**, 495-499.

Woakes, M. and Bafor, B.E. (1984) Primary gold mineralization in Nigeria. In *Gold '82*, ed. Foster, R.P. *Proceedings of the Symposium Gold '82*. Geol. Soc. of Zimbabwe, Special Publ. No.1, 661-671.

Woodall, R. (1988) Gold in 88–Keynote address. In *Bicentennial Gold 88*, eds. Goode, A.D.T., Smyth, E.L., Birch, W.D. and Bosma, L.I. *Geol. Soc. Aust. Abstr.* **23**, 1-12.

Yeap Ee Beng (1988) Gold mineralisation in the Central Belt of Peninsular Malaysia. In *Bicentennial Gold 88*, eds. Goode, A.D.T., Smyth, E.L., Birch, W.D. and Bosma, L.I. *Geol. Soc. Aust. Abstr.* **23**, 238-242.

Yule, A. (1987) Geology of the Chetwynd Gold Deposit, Newfoundland. *Maritime Sediments and Atlantic Geology*, Abstracts, 110-111.

5 Epithermal gold deposits in volcanic terranes

R.W. HENLEY

5.1 Introduction

Epithermal mineral deposits are the products of large-scale hydrothermal convective systems driven by magmatic heat in the upper 5–10 km of the Earth's crust. The choice of such a broad definition reflects (a) the wide use of the term 'epithermal' in contemporary mineral exploration and (b) the recognition that active geothermal systems in volcanic terranes are the present day *equivalent* of the epithermal systems responsible for ore formation in the geologic past (e.g. White, 1981; Henley and Ellis, 1983).

The term 'epithermal' was coined by Lindgren (see Lindgren, 1922, 1933, p. 210) in an attempt to systematize hydrothermal mineral deposits on a basis similar to that beginning to be applied to metamorphic rocks. Pressure (or depth) and temperature were recognized as the controlling factors for metamorphic assemblages and both appeared able to be estimated for ore assemblages based on available data for mineral stabilities and textures. Some authors (e.g. Schmitt, 1950) have attempted to use the term 'epithermal' to refer to or to infer a specified depth range; since in most cases depths are derived from temperature estimates, the restriction of the term to a temperature range commensurate with that of explored geothermal fields (50 to ~350°C) is preferred here. Such a definition of 'epithermal systems' lacks the accustomed (but generally unwieldy) specificity of many ore-deposit classifications but has the advantage of focusing on crustal environment and crustal process. Deposit types may be subdivided for convenience (e.g. when deriving exploration models) using other criteria depending on the objective of the classification; gold deposits in sedimentary rock overthrust terranes (Carlin-type deposits), for example, are reviewed in this volume by Berger and Bagby. Similarly, volcanogenic massive sulphide deposits are a product of such a hydrothermal environment but are characterized by parageneses and form which reflect seafloor depositional environments. In this chapter, epithermal deposits closely associated in time and space with terrestrial volcanic rocks are considered. Common factors in the origin of the various epithermal deposit-settings are discussed by Berger and Henley (1989).

Lindgren (1922) based his classification of hydrothermal deposits on the concept of a fluid continuum from high, magmatic temperatures to lower temperatures at the surface. This approach and his focus on high-grade vein deposits reflects the mineral prices and mining technology of the time whereas today there is a greater, although not exclusive, emphasis on bulk-tonnage, low-grade deposits. As a result, present-day classification focuses more on the geological and hydrothermal 'environment' of mineral deposit formation and far less on an ad hoc use of formation temperatures. In recognizing the broader scale environment of formation of epithermal deposits, I wish to re-emphasize the concept of Lindgren (see also Sillitoe, 1989, and this volume) that a continuum does indeed exist in ore-forming hydrothermal systems in volcanic terranes from magmatic fluid to almost entirely meteoric-groundwater environments. Thus, porphyry gold and related deposits described elsewhere in this volume relate to a higher temperature hydrothermal-magmatic environment where depositional processes are in part controlled by the depressurization of magmatic vapour and its interaction with surrounding groundwater (Henley and McNabb, 1978).

Epithermal gold deposits have had a major impact on world economics past and present. The epithermal districts of northern Greece and Romania (7–10 million oz), for example, have respectively underpinned the economies of Ancient Greece and Rome (P. Eimon, personal communication). More recently, consequent on the increased value of the metal, epithermal deposits have become the focus of major exploration booms in the south-western Pacific and western United States.

New discoveries such as Lihir Island, Papua New Guinea, and Hishikari, Japan, with published reserves of 6.0×10^8 g Au and 9.8×10^7 g Au respectively, rank with world-class gold deposits of other types; Kalgoorlie's Golden Mile (Archaean shear zone type), for example, has yielded some 10^9 g gold. Together with open-pit mining methods, the use of modern cyanide extraction and heap leach techniques has lowered recovery costs for low-grade deposits so that bulk tonnage epithermal deposits may provide a high rate of return on invested capital. Thus, both high-grade vein and large- to medium-tonnage, low-grade disseminated or breccia-hosted deposits are attractive exploration targets.

Figure 5.1 provides a summary of ore reserves and grades for 'epithermal' deposits and comparison with porphyry-related Cu–Au deposits in volcanic terranes.

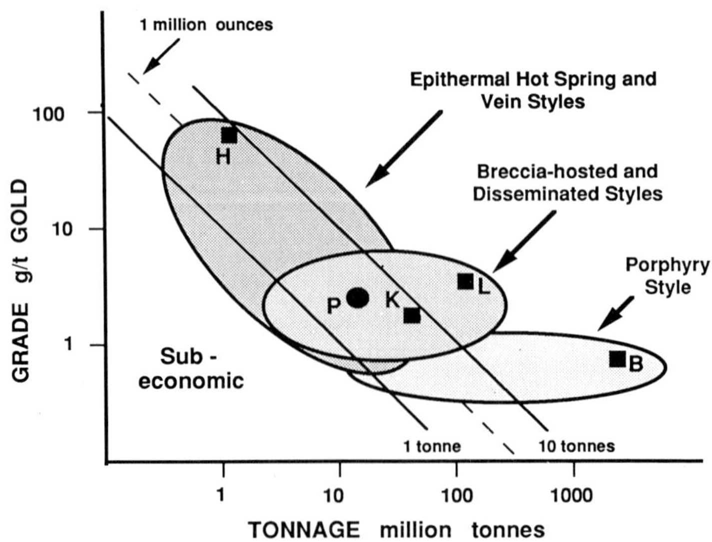

Figure 5.1 Summary of grade and tonnage data for epithermal- and porphyry-style gold deposits. The subdivisions reflect more the economics of gold recovery than any intrinsic geological discontinuity between deposit styles. H=Hishikari, B=Bougainville, L=Ladolam, K=Kelian and P=Paradise Peak.

This chapter will first review some exploration case histories and then proceed to discuss the geological and geochemical environment of ore formation, the distribution of 'epithermal' deposits in time and space, and exploration techniques. A comprehensive review of epithermal deposits is not possible in the space available but a number of excellent reviews and special volumes are available elsewhere (e.g. Berger and Bethke, 1985; Hedenquist, White and Siddeley, 1990).

5.2 Exploration case studies

A number of epithermal deposits have been intensively studied and reported in the literature. Here we review the discovery and characteristics of three world-class deposits (~10^8 g contained gold, ≈ 30 million oz) illustrative of the bonanza vein and disseminated bulk-tonnage styles of epithermal gold deposits. The discovery histories of a number of other epithermal deposits have been summarized by Hollister (1985). Specific features of other new discoveries or of well-established epithermal deposits are considered elsewhere in this chapter.

5.2.1 *Hishikari, Japan*

The Hishikari deposit, 45 km north-west of Kagoshima in Kyushu, rates as one of the greatest bonanza discoveries of recent times. Drilling commenced in 1980 following a five-year exploration programme by the Metal Mining Agency of Japan and the first hole encountered 0.15 m at 290 g/t gold plus 167 g/t silver. A series of 18 drillholes further defined the extent of high-grade mineralization and the deposit was then explored from a 2 km decline. Hot water at 65°C was encountered in the ore zone and has required a major pumping programme to lower the water table below the ore zone. Concurrent with production, exploration is continuing on this deposit the published reserves of which currently stand at 1.4 million tonnes with an average grade of 70 g/t gold. Through 1988, Hishikari (at US$50/oz) was the lowest cost gold-producer in the world.

Figure 5.2 shows the geology of the Hishikari deposit and is based largely on Urashima and Izawa (1982) and Urashima *et al.* (1987). Gold (18–30 g/t gold in clay–quartz–calcite veins) was discovered in the Neogene andesite volcanic sequence in the eighteenth century and was worked until 1940. The main host to the deeper bonanza mineralization are shales of the underlying Shimanto group. The vein mineralization consists of an anastomosing sequence of veins over a strike length of at least 1100 m and may be interpreted as a dilational jog in a transcurrent fault system. Syn-mineralization faults (Ishihara *et al.*, 1986) and the presence of large angular clasts of wallrock in some parts of the vein provide evidence of an active tectonic environment during mineralization.

The bonanza veins are up to a few metres wide and consist of symmetrically banded (syn. crustiform) chalcedonic silica and up to 30% adularia sometimes interbanded with bladed calcite. The latter has been subsequently leached out. Gold occurs as electrum (70%Au) grains up to 70 μm in diameter. Other vein minerals consist of silver sulpho-selenides and antimonides, pyrite, minor sphalerite and chalcopyrite, and a late-stage, coarse stibnite. Kaolinite bands also occur as well as bands of fine

(a)

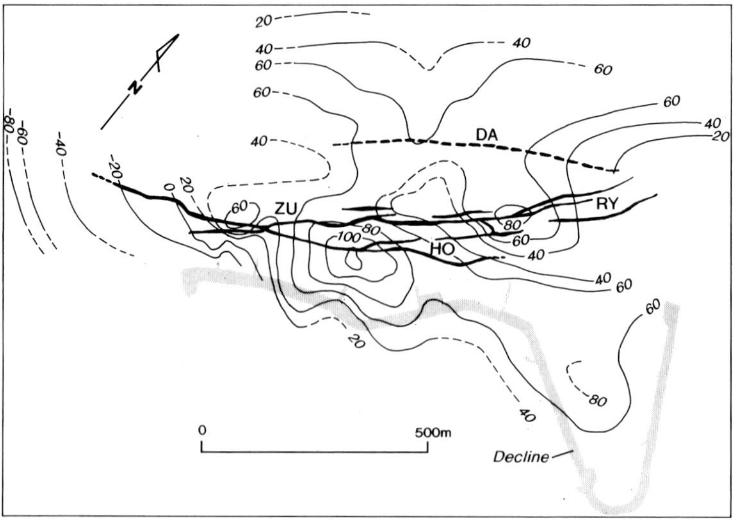

(b)

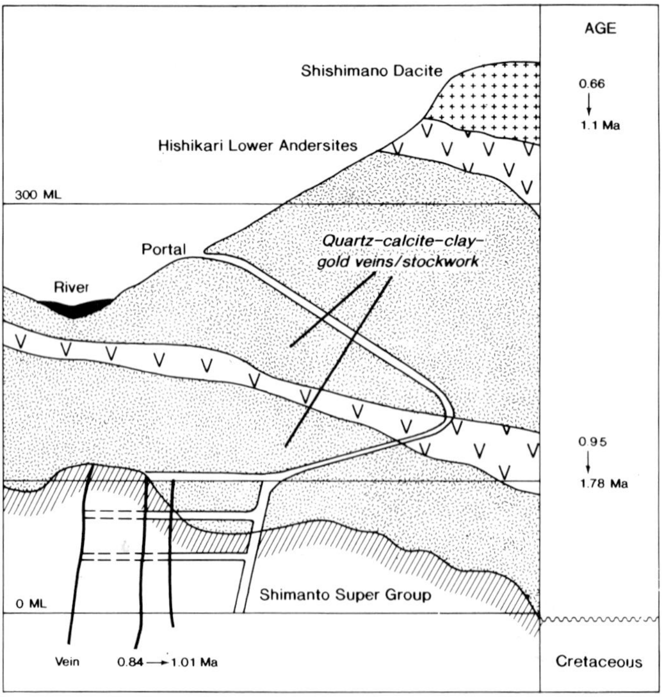

Figure 5.2 Hishikari Deposit, Kyushu, Japan. (a) Plan view of veins projected to the + 70 metre level in relation to the upper surface of the Shimanto Supergroup sediments (shown as metres a.s.l.) and the access decline. Ho=Hosen vein, Da=Daisen vein, Ry=Ryosen vein and Zu=Zuisen vein. (b) General NW-SE cross-section of the Hishikari deposit. (Redrawn with permission, R. Suzuki, Sumitomo Mining Company.)

disseminated hematite. Massive reyerite ($[Na, K]_2 Ca_{14} Al_2 Si_{24} O_{58} (OH)_8, 6H_2O$) is relatively common in the vein system and zeolites occur in veins in the andesite.

Propylitic alteration is most marked in the andesite sequence and consists of K-feldspar, chlorite and mixed layer clays. Shales are only visibly altered within a few centimetres of vein contacts. Fluid inclusion temperatures from vein quartz and calcite generally are 200°C with a range from 150 to 300°C and some high-grade bands at 100°C. K–Ar dating of adularia provides a mineralization age in the range 0.78–1.05 Ma.

The geological setting of the deposit has been discussed by Ishihara *et al.* (1986). Based on sulphur isotope data and a regional study of the gold content of basement rocks, these authors suggested that the deposit is associated with an unexposed porphyry intrusive related to the magnetite-series magmatic arc of southern Japan.

5.2.2 *Kelian, Kalimantan, Indonesia*

The Kelian district was first recognized through local workings of alluvial gold in the Kelian River in the 1950s and early 1960s. The discovery of the Kelian deposit came after a programme of mapping, soil sampling, deep augering and trenching between 1976 and 1979 (Van Leeuwen *et al.*, 1990) followed up by a six-hole drilling programme on the most promising anomalies. This programme realized low-grade mineralization at about 2 g/t. Encouraged by the escalation of the price of gold in the early 1980s, further drilling (a total of 300 holes), together with more surface exploration and a successful I.P. geophysics survey, has presently outlined a bulk-tonnage deposit of 40 Mt at an average gold grade of 2 g/t (Van Leeuwen *et al.*, op. cit.). Mineralization has been proven well below the 200 m reserve drilling.

Figure 5.3 shows the geology of the Kelian deposit; the data are from Van Leeuwen *et al.* (op. cit.). The deposit is associated with Late Oligocene–Early Miocene andesites, Upper Eocene(?) pyroclastics and minor rhyolites, and some Plio-Pleistocene basalts. It lies on a north-easterly regional trend which also contains significant epithermal mineralization at Mt. Muro and Masuparia.

At the deposit scale, mineralization occurs in the margins of a set of andesite bodies intrusive into the Eocene pyroclastic unit. Mineralization is hosted by fractured andesite and tuff and by a variety of breccias ranging from primary pyroclastic and intrusion breccias to clast-supported hydrothermal breccias. Alteration is intense. A chlorite–carbonate–sericite assemblage is preserved in the cores of the larger andesite bodies and, in the East and West Prampus ore zones (Figure 5.3), alteration occurred in the following stages: sericite–pyrite, quartz–sericite–adularia–pyrite, carbonate–pyrite–base metal sulphide. This alteration is overprinted by a widespread kaolin +Fe–Mn carbonate assemblage. The adularia has been dated at 20.2 ± 0.3 Ma.

Mineralization is closely associated with pyrite, which makes up a few per cent of the ore, and is disseminated or is present as a fine stockwork throughout the orebodies. Other sulphides include sphalerite, galena and minor chalcopyrite, tennantite–tetrahedrite, cinnabar and arsenopyrite. Gold has been observed commonly in polished sections and is associated with the carbonate–base metal sulphide assemblage as inclusions or on grain boundaries. However, much of the gold is submicroscopic, possibly associated with pyrite.

Fluid inclusions from quartz, sphalerite and carbonate give temperatures in the range 270–310°C and generally low salinities, from 0.5 to 4.2 wt.% NaCl equivalent.

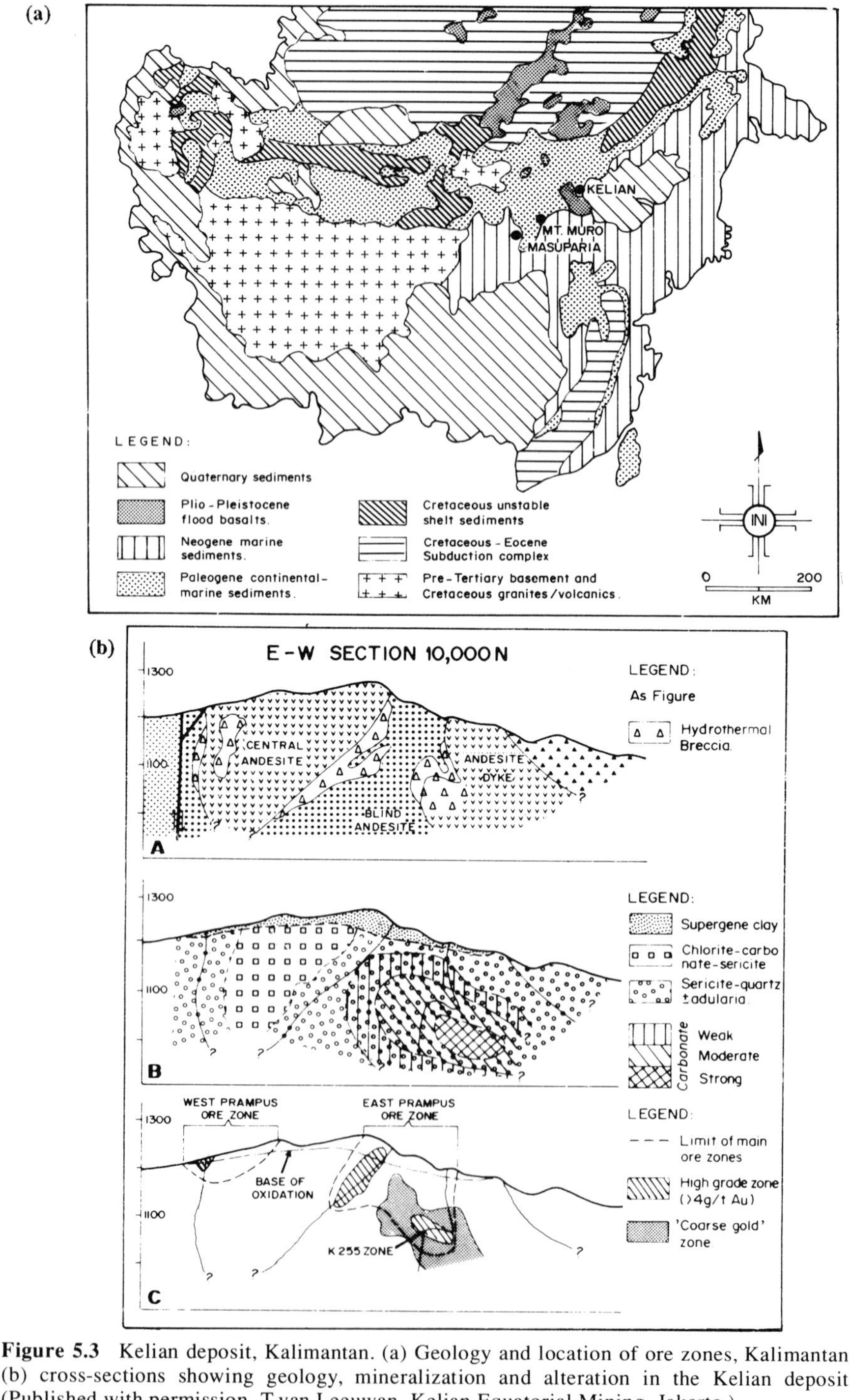

Figure 5.3 Kelian deposit, Kalimantan. (a) Geology and location of ore zones, Kalimantan, (b) cross-sections showing geology, mineralization and alteration in the Kelian deposit. (Published with permission, T.van Leeuwan, Kelian Equatorial Mining, Jakarta.)

5.2.3 *Ladolam deposit, Lihir Island, Papua New Guinea*

The Ladolam deposit was discovered in 1982 as part of a regional gold reconnaissance programme of the Tabar-Feni island chain to the north-east of New Ireland, Papua New Guinea (Moyle *et al.*, 1990). Earlier geological reconnaissance had recognized hydrothermal activity and alteration but there was no history of gold workings on Lihir or adjacent islands. The discovery outcrops consist of a set of silicified and pyritized rocks adjacent to hot-spring discharges on the beach of Luise Harbour (Figure 5.4). Mapping and trenching plus soil and rock geochemistry followed in 1983 and identified the Coastal, Kapit and Lienetz mineralized zones. Surface work and diamond drilling proceeded through late 1983 to 1984 and outlined a large zone of mineralization. Drilling on vegetation anomalies in 1985 further defined the extent of the orebodies and identified a fourth ore zone, the Minifie area. By 1987, 153 diamond holes (41 000 m) and 420 reverse circulation holes (14 000 m) had been completed, identifying 167 million tonnes of ore at 3.43 g/t gold. About 5% of this is oxide ore.

Lihir Island lies north-east of New Ireland, within the Tabar-Feni volcanic chain which extends to the south towards Bougainville Island, location of the Panguna porphyry Cu–Au deposit (Fountain, 1972; Eastoe, 1978). Lihir itself lies at the intersection of this chain with a north-north-east-trending zone of seismic activity. The Tabar-Feni Islands are alkaline in composition and may, as suggested by Johnson (1987), be the result of partial melting of mantle lithosphere modified by earlier subduction. The Ladolam deposit lies within the Luise collapse caldera (Figure 5.4) inside the former Luise volcano. The deposit itself underlies an area of about 3 km^2. The Lienetz, Minifie and Coastal Zone orebodies are subhorizontal, breccia-hosted deposits in volcanics and monzonitic intrusives. As at Kelian, the ore deposit is hosted by very extensive breccias derived from volcanic, intrusive and hydrothermal events. One of these is a unique open-space hydrothermal breccia, consisting of subrounded, potassic-altered intrusive and volcanic clasts, which occurs at 50–200 m below surface in the Lienetz and Coastal areas. This breccia, which typically contains gold grades in excess of 5 g/t, grades downward into a more weakly mineralized breccia in which the interstices are filled with anhydrite and calcite (G. Ballantyne, personal communication).

Alteration consists of an early porphyry-style event with potassic and propylitic assemblages. This is overprinted by advanced argillic, argillic and phyllic alteration associated with gold mineralization. The former is very widespread and occurs to about 100 m and locally much deeper. Alteration ages range from 0.9 Ma for biotite-altered volcanics, through 0.33 Ma for biotite–anhydrite veins, to 0.15 Ma for alunite (Davies and Ballantyne, 1987).

Mineralization consists of pyrite–marcasite with minor base metal sulphides, arsenopyrite, sulphosalts and gold–silver tellurides. Free gold (50–100 μm, 990 fineness) occurs in the oxide zone but is rare in the sulphide zone where gold occurs as < 20 μm particles in pyrite (Moyle *et al.*, 1990).

Fluid inclusion homogenization temperatures show two maxima of 140–150°C and 200–220°C. They have salinities averaging 3.8 wt.% NaCl equivalent and have relatively high gas contents (Moyle *et al.*, 1990).

In addition to its setting on an oceanic island, the Ladolam deposit has a number of important features. The Ladolam ore zones lie within an active geothermal system with boiling hot springs, solfatara, and submarine discharge in Luise Harbour

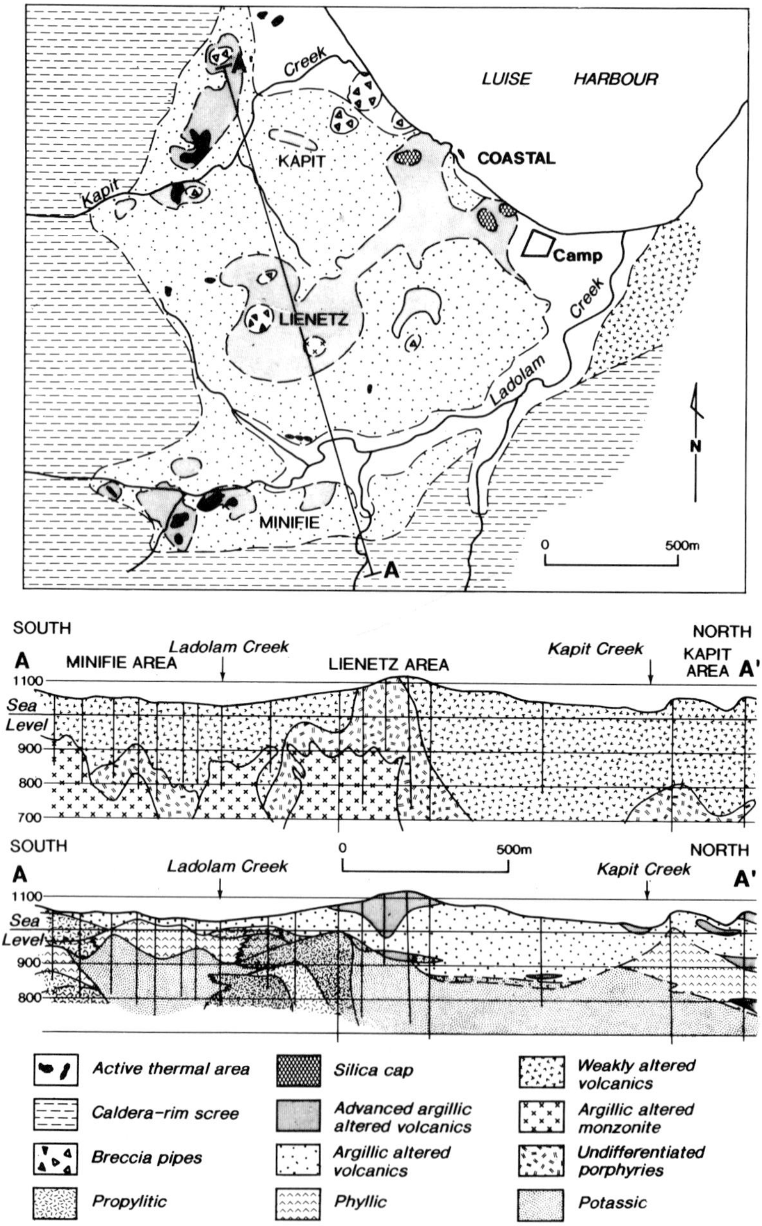

Figure 5.4 Ladolam deposit, Lihir Island, Papua New Guinea. Geology, alteration and locations of the principal ore zones (Coastal Lienitz, Minifie and Kapit). From Moyle *et al.*, (1990); redrawn with permission, A.J. Moyle and Kennecott Explorations (Australia) Ltd. and the Australasian Institute of Mining and Metallurgy.

emphasizing the relation between ore deposition and present-day active processes. The current hydrothermal activity represents a late evolutionary stage in the formation of this unique deposit which commenced with subvolcanic porphyry-style alteration and mineralization and evolved to the present style following caldera collapse. This has resulted in the telescoped alteration zoning and overprinting of assemblages and emphasizes the continuum from magmatic-hydrothermal to meteoric water dominated hydrothermal environments.

5.3 Environment of alteration and mineralization

5.3.1 *Volcanic association*

In contrast to the 'Carlin-type' gold deposits described by Berger and Bagby in Chapter 7 (this volume), the deposits described here all show a close temporal and spatial association with volcanism. Detailed descriptions of the geology and geological environment of individual deposits abound in the recent literature (e.g. Hedenquist, White and Siddeley, 1990, proceedings of conferences such as Pacrim [Aust. I.M.M., Gold Coast, Queensland, 1987], Gold '88 [Geol. Soc. Australia, Melbourne, 1988], Bulk Mineable Precious Metal Deposits of the Western United States (Schafer *et al.*, 1988) and many field excursion guides). Reviews are given by Tooker (1985) and Berger and Bonham (1990).

Cenozoic epithermal gold deposits occur in convergent plate margin settings and are a consequence of alkalic, andesitic, or felsic volcanism (Berger and Henley, 1989; Berger and Bonham, 1990; Sillitoe, 1989, and this volume). The latter relates to anomalous heat flow due to intrusions from the lithosphere in back arc environments such as the Taupo Volcanic Zone of New Zealand (Henley and Hoffman, 1987) or analogous extensional environments such as the Northern Great Basin, Nevada (Noble *et al.*, 1988).

Mineralization in the San Juan Volcanic Field (Colorado) exemplifies the common time and space association of epithermal deposits with caldera development (Steven and Lipman, 1976; Lipman *et al.*, 1976; Lipman, 1987) and Sillitoe and Bonham (1984) have discussed deposit-style as a function of volcanic landforms such as calderas, stratovolcanoes and domes. Such styles are the consequence of the hydrology induced by the volcanic structure and the relative level of magmatism (cf. Berger and Henley, 1989). Sillitoe *et al.* (1984) and Sillitoe and Bonham (1984) have also recognized an association of gold mineralization with large maar-diatremes in volcanic fields and cite examples such as Wau (New Guinea), Cripple Creek (Colorado) and Baguio (Phillippines). This association is not restricted to any particular volcanic composition.

While most epithermal deposits are associated with calc-alkaline volcanics and intrusives, some show a close association with alkalic volcanics. The Ladolam deposit and the Cripple Creek district, Colorado, are examples and it may be argued that the potential for such rocks to act as sources of gold is dependent on magma evolution, especially with respect to the saturation state of sulphides (Keays and Scott, 1976; Wyborn, 1988)

5.3.2 *Structural controls*

The structural environment of formation of epithermal deposits has received relatively little attention in the recent literature. Most information derives from earlier mining which focused on high-grade vein deposits such as at Tonopah, Nevada (see for example Wisser, 1960, and McKinstry, 1941, 1948).

Both vein and bulk-tonnage deposits are often closely associated with major regional transcurrent fault structures such as the Walker Lane in Nevada. In many cases, high-grade ore shoots formed within dilational transcurrent fault jogs as discussed by Sibson (1987). Examples include Hishikari (Japan), Camp Bird (Colorado) and Pajingo (Queensland) (see Porter, 1988). At Hishikari (and at Golden Cross, New Zealand) low-grade vein sets, formed within a few hundred metres of the palaeosurface in poorly consolidated volcanics, give way at depth to major, high-grade, repetitively mineralized shear zones. This may be controlled by effective rock-stress and hydro-fracturing as discussed by Sibson (1981). The regional scale structural control of ore-forming systems may relate to a variety of structures associated with reactivation of deep crustal shears in a manner analogous to the flower structures associated with fault-reactivation beneath sedimentary basins (Harding, 1985). The Kelian deposit is also related to a major regional trend as are the Porgera and Ladolam deposits, Papua New Guinea. The latter are hosted by extensive zones of phreato-magmatic and hydrothermal breccias the formation of which may well be related to interaction of active faults with cooling magma bodies.

5.3.3 *Wallrock alteration*

From a review of published data, Hayba *et al.* (1985) and Heald *et al.* (1987) have subdivided volcanic-hosted epithermal deposits into two classes according to their predominant rock-alteration assemblage: the 'adularia–sericite' style and the 'acid–sulphate' style. In this paper (and see Berger and Henley, 1989) the *mineralogical* term 'alunite–kaolinite' style is preferred for the latter. The characteristics of the two styles are shown in Table 5.1. Their distinctive mineral assemblages reflect formation under quite different chemical conditions relating to the near-neutral and highly acidic pH of the ambient fluid at certain stages in their formation.

Adularia–sericite-type deposits are exemplified by the three case histories outlined above. Figure 5.5 shows the general distribution of mineralization and alteration in relation to the hydrological structure of the parent hydrothermal system. Their alteration corresponds to the intermediate argillic and propylitic assemblages defined by Meyer and Hemley (1967), although, where erosion has not occurred for more than a few tens of metres below the palaeosurface, a surficial advanced argillic alteration assemblage (due to oxidized steam condensate) may be preserved. In the field, the mappable zonation of alteration is largely dependent on alteration intensity and primary rock composition, and involves minerals which in the field may not be readily identified. However, in volcanic rocks, selective alteration of phenocrysts (e.g. pyroxene to chlorite) may aid field recognition of alteration zonation. Perhaps more important in exploration is the recognition of alteration-overprinting which may relate to specific phases of ore deposition.

The temperature-dependence of common alteration minerals in the propylitic–argillic alteration assemblage may be empirically calibrated from geothermal drilling data (Figure 5.6) so that the careful determination of clay mineralogy can provide a useful exploration guide to temperature and, used carefully in conjunction with an appropriate boiling point vs depth curve (Figure 5.7), to depth of formation. For example, at Golden Cross, zonation of clay minerals is closely related to ore zones (De Ronde, 1986; De Ronde and Blattner, 1988).

Table 5.1 Characteristics of adularia–sericite and kaoline-alunite-style epithermal deposits (modified with permission from Hayba *et al.*, 1985)

Characteristic	Adularia–sericite	Alunite-kaolinite
Structural setting	Structurally complex volcanic environments, commonly in calderas	Intrusive centers, 4 out of the 5 studied related to the margins of calderas
Size length : width ratio	Variable; some very large usually 3 : 1 or greater	Relatively small equidimensional
Host rocks	Silicic to intermediate and alkalic volcanics	Rhyodacite typical
Timing of ore and host	Similar ages of host and ore	Similar ages of host and ore (< 0.5 m.y.)
Mineralogy	Argentite, tetrahedrite, tennantite, native silver and gold, and base-metal sulphides. Chlorite common, selenides present, Mn gangue present, no bismuthinite	Enargite, pyrite, native gold, electrum, and base-metal sulphides. Chlorite rare, no selenides, Mn minerals rare, sometimes bismuthinite
Production data	Both gold- and silver-rich deposits variable base metal production	Both gold- and silver-rich deposits, noteworthy Cu production
Alteration	Propylitic to argillic Supergene alunite, occasional kaolinite, abundant adularia	Advanced argillic to argillic (± sericitic) Extensive hypogene alunite, major hypogene kaolinite, no adularia
Temperature	100–300°C	200–300°C[1]
Salinity	0–13 wt.% Nacl equiv.[3]	1–24 wt.% NaCl equiv.[2]
Source of fluids	Dominantly meteoric	Dominantly meteoric, possibly significant magmatic component
Source of sulphide sulphur	Deep-seated magmatic or derived by leaching wallrocks deep in system	Deep-seated, probably magmatic
Source of lead	Precambrian or Phanerozoic rocks under volcanics	Volcanic rocks or magmatic fluids

1. Limited data, possibly unrelated to ore.
2. Salinities of 5–24 wt.% NaCl equiv. are probably related to the intense acid-sulphate alteration which preceded ore deposition.
3. Note: a lower salinity range applies more to Au–Ag deposits than to base-metal Ag ± Au deposits (see text).

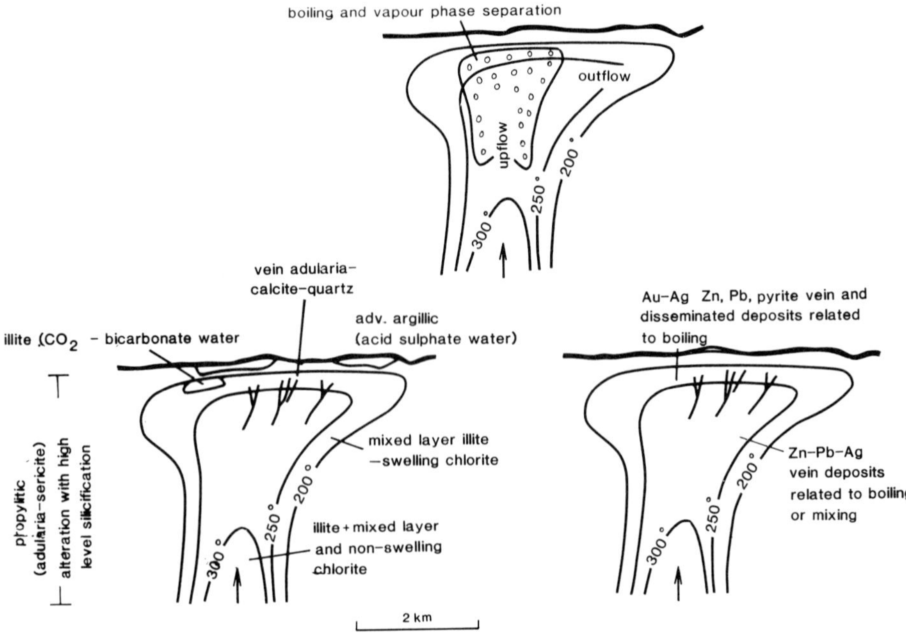

Figure 5.5 Schema for the environments of formation of epithermal gold deposits in relation to hydrology, thermal structure, and rock alteration. Note that this 'mushrooming' commonly drawn in such schema reflects *dispersion in the plane of major fracture systems* and a much narrower thermal anomaly would be seen transverse to such structures. While drawn with reference to a silicic volcanic terrane with deep magmatism, similar patterns apply in other types of volcanic terrane (cf. Henley and Ellis, 1983).

Certain minerals act as indicators of original subsurface conditions. Adularia and bladed (or pseudomorphed) calcite, for example (see below), are reliable indicators of original permeability and the occurrence of boiling conditions (Browne, 1978). A number of adularia–sericite-style epithermal precious metal deposits are closely associated with well-preserved silica-sinter deposits containing casts of plants, mud cracks and other diagnostic features. Berger and Eimon (1983) reviewed such deposits, which may be exemplified by McLaughlin, California (Lehrman, 1986), and by the National District (Vikre, 1985), Buckhorn (Plahuta, 1986) and Sulphur (Wallace, 1987) deposits in Nevada. Silica sinters containing plant fossils and other features common to modern sinters are well preserved in the newly discovered Palaeozoic epithermal deposits of North Queensland (Cunneen and Sillitoe, 1989; White, Wood and Lee, 1989). Such 'hot springs-style' deposits are often associated with hydrothermal eruption breccias and lacustrine sediments (Figure 5.8; cf. Figure 5.5) and are a subset of the wider adularia–sericite-style deposits.

The most comprehensively studied adularia–sericite-style epithermal system is Creede, Colorado. Originally a major silver district (Steven and Eaton, 1975) with production from an extensive vein set, recent exploration has located high-grade gold mineralization 4–5 km north along strike of the base metal–silver veins (Bethke, 1988). Isotope and inclusion studies now show that the gold mineralization relates to

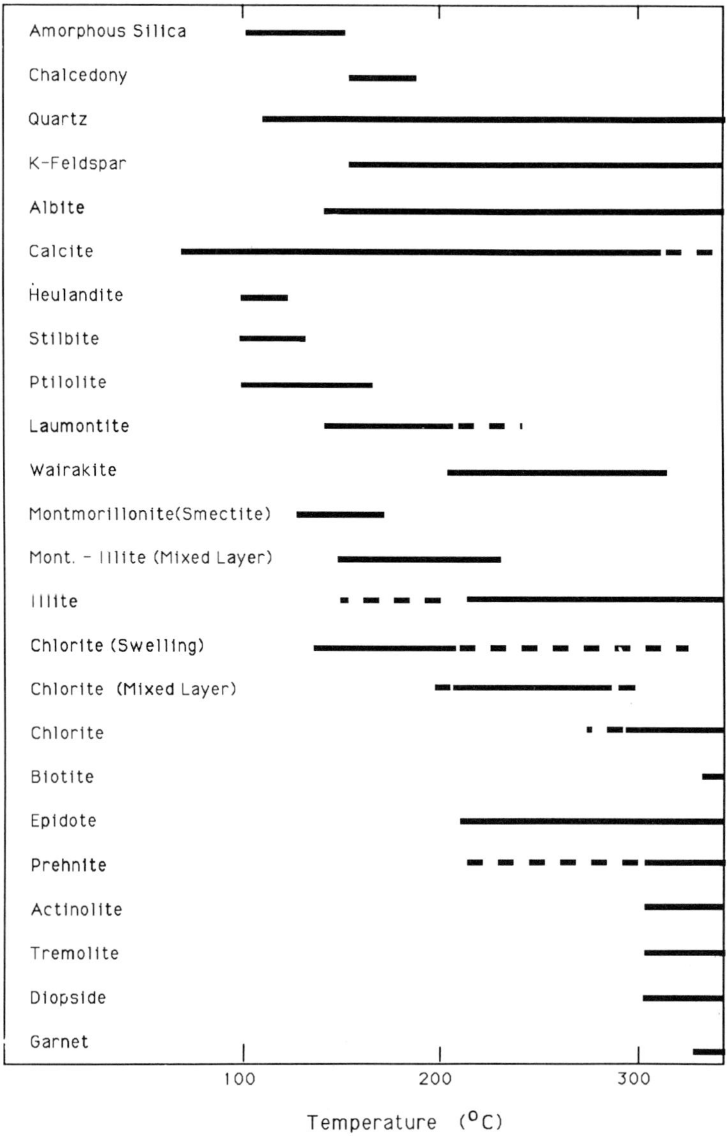

Figure 5.6 Temperature ranges for the occurrence of common alteration minerals in active geothermal systems based on data from drill-core and downhole measurements; note that the occurrences of calcic minerals are dependent both on temperature and the CO_2 content of the ambient fluid. (From Henley and Ellis, 1983.)

the original, low-salinity, convective upflow of the system controlled by magmatism on the margin of the San Luis caldera. The silver mineralization relates to a satellite convection system sourced by higher salinity fluid derived from the moat lake of the Creede caldera to the south (Rye *et al.*, 1988). Similarly at Guanajato, Mexico, base metal–gold mineralization occurs at depth in the Veta Madre silver veins and in a newly discovered vein set cross-cutting the subparallel Villapando vein.

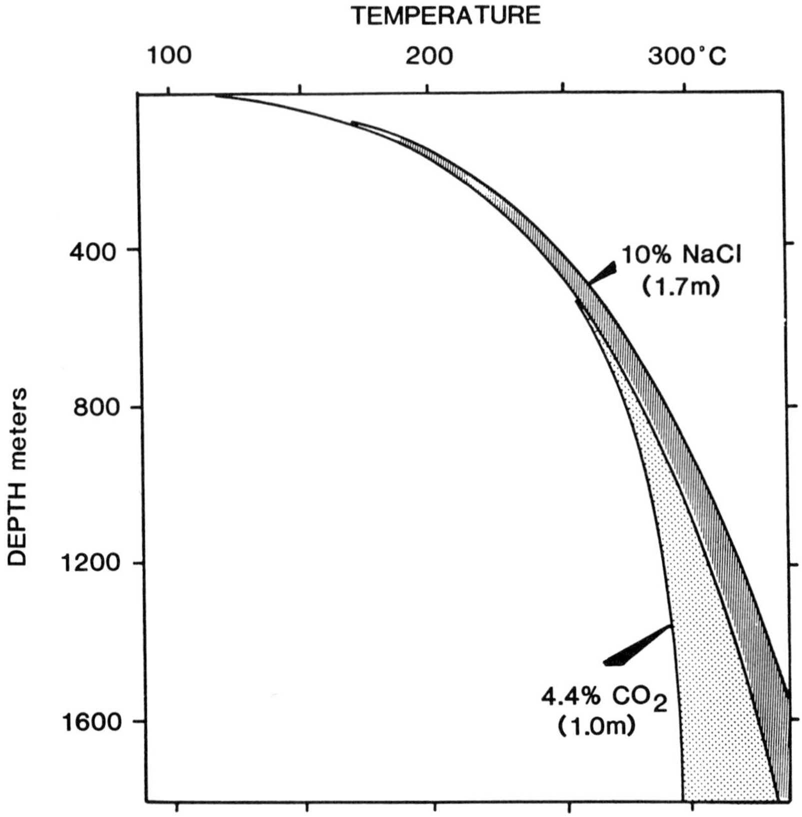

Figure 5.7 Temperature-depth relations for 'boiling fluids containing NaCl and CO_2 in relation to boiling point vs. depth relations for pure water. (From Henley, 1985.)

Alunite-kaolinite-style deposits are exemplified by Goldfield, Nevada (Ransome, 1909), Summitville, Colorado, Rodalquilar, Spain, and Paradise Peak, Nevada (Thomason, 1986), and their general characteristics are shown in Figure 5.9a. Somewhat smaller but otherwise similar deposits occur in Japan and these are the Nansatsu-type deposits (Hedenquist *et al.*, 1988) such as Iwato and Kasuga (Kyushu). Sillitoe *et al.* (1990) have described a similar deposit from Nalesbitan, near Luzon, Phillippines, which shows a close association with a major regional transcurrent fault system. In some deposits, the occurrence of pyrophyllite indicates alteration temperatures above about >280°C (Hemley *et al.*, 1980) and a close genetic association of alunite–kaolinite-style mineralization with magmas and porphyry mineralization is highlighted in a number of deposits, e.g. Guinaoang deposit, Luzon, Phillippinnes (Sillitoe and Angeles, 1985).

At Summitville, an assemblage of covellite–luzonite ± gold passes downward to chalcopyrite–tennantite; gold also occurs with barite and supergene goethite–jarosite

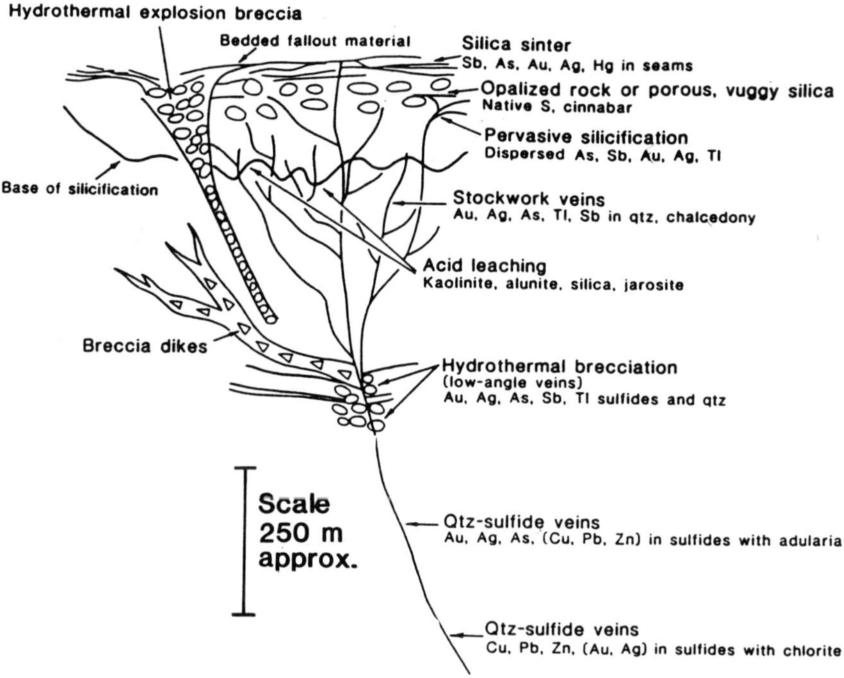

Figure 5.8 Schematic cross-section showing the main features of a hot-springs sub-type epithermal deposit. (Redrawn with permission from Berger and Eimon, 1983.)

near the surface (Stoffregen, 1987). This mineralization is mostly hosted by a zone of vuggy silica developed as pipe-like bodies within a host quartz latite. The quartz-bearing zones (up to 70 m wide) are enclosed by about 30 m of intense alunite–quartz alteration and decrease in thickness downward where they are mantled by quartz–kaolinite with much less alunite (Figure 5.9(a)). At Rodalquilar, where mineralization is associated with the rhyolitic tuffs and domes of a nested caldera in the Miocene Cabo de Gata volcanic field, deep drilling has defined mineralization to more than 860 m. Deep sericitic alteration grades upward into argillic, advanced argillic, and silicic zones and the gold mineralization (averaging 7.8 g/t) is associated primarily with the latter (Arribas *et al.*, 1988; Cunningham *et al.*, 1989). In this example, stable isotope data suggest that seawater was the dominant source of fluid to the deep system.

In many alunite–kaolinite-style deposits, mineralization is capricious but the El Indio deposit in Chile is a spectacular very high-grade example with ore in a more coherent vein system. Bonanza ore grades, averaging 225 g/t Au, 104 g/t Ag and 2.4%

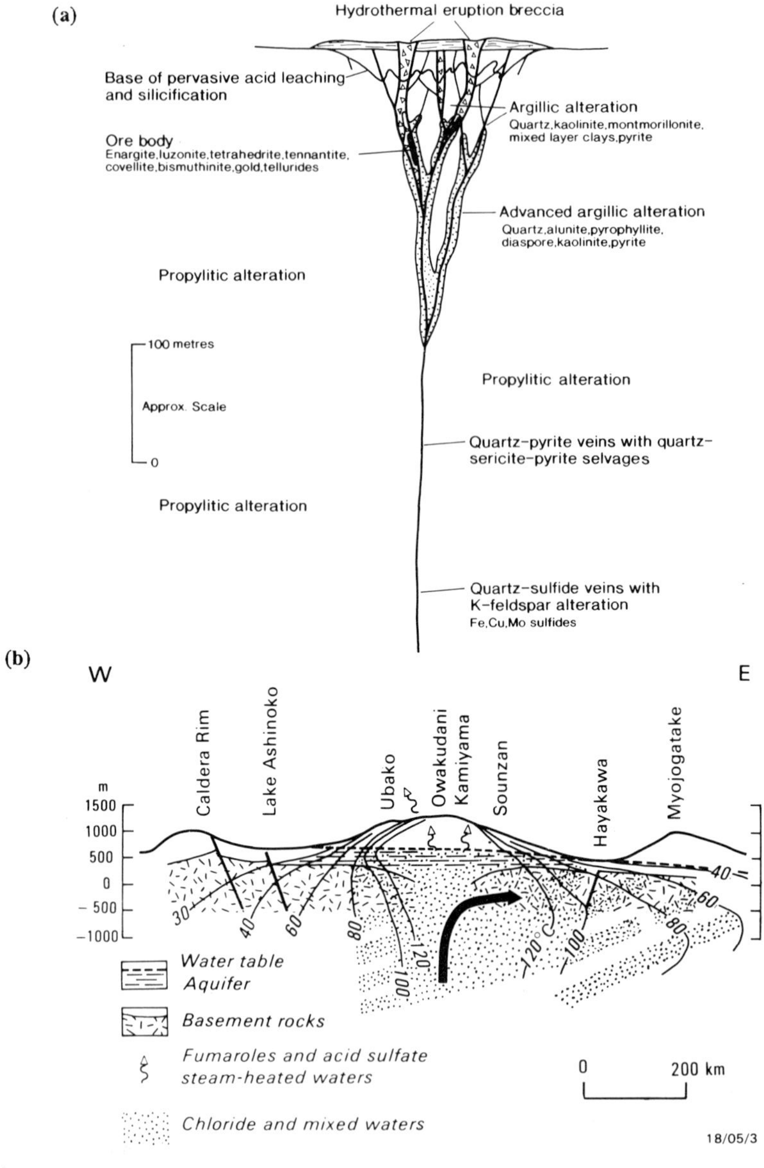

Figure 5.9 Alunite–kaolinite–style epithermal deposits. (a) Schematic cross-section of an alunite–kaolinite-style epithermal deposit showing alteration, mineralogy and general location of ore zones. (Redrawn with permission from Silberman and Berger, 1985.) (b) Hydrology of the active hydrothermal system within the Hakone caldera and in relation to the occurrence of rhyolite domes. The regional hydraulic gradient controls groundater-flow from west to east across the system. In such an environment, high-level magma degassing at an early stage may produce extensive alunite–kaolinite alteration which subsequently becomes the focus for ore deposition when invaded by the groundwater-based adularia-sericite system. Hakone is used here as an analogy of the hydrodynamic setting of some alunite–kaolinite systems without speculation on the possibility of gold mineralization in the system. (Redrawn from Oki, 1983.)

Cu, occur in quartz–gold veins cross-cutting enargite-pyrite veins several metres in width (Siddeley and Araneda, 1986; Jannas *et al.*, 1989). Barite–alunite–quartz vein and breccia deposits occur nearby at El Tambo and perhaps represent a nearer-surface part of the system.

Alunite-kaolinite-style deposits appear to be more closely related to high-level magma degassing environments than do adularia-sericite-style deposits but Berger and Henley (1989) have stressed some common characteristics which suggest that *two* stages of alteration and mineralization are involved in their formation. Stage 1 involves the intense acid alteration of host volcanics and stage 2 is the invasion of these alteration zones by a more-normal near-neutral pH fluid from which gold and associated minerals are precipitated by reaction with the alunite–kaolinite assemblage. Deen *et al.* (1988), using stable isotope data, have demonstrated this dynamic sequence in the Julcani district, Peru. The type of hydrodynamic environment for such mineralization is illustrated in Figure 5.9(b) through analogy with the Hakone geothermal field in Japan.

5.3.4 *Fluid inclusions and light stable isotopes*

The systematics of fluid inclusions and stable isotopes in epithermal deposits are reviewed by Bodnar *et al.* (1985) and by Field and Fifarek (1985) respectively. Fluid inclusion data are sparse for alunite–kaolinite systems; Matsuhisa (personal communication) recorded only low salinities for the Nansatsu alunite–kaolinite-style deposits in Japan despite the close association with a magmatic environment. A consistent pattern, however, emerges for adularia–sericite systems (Hedenquist and Henley, 1985a), showing low salinities for epithermal *gold–silver* deposits and higher (≥ seawater) salinities for epithermal *base metal–silver* ± gold deposits.

In addition to geothermometry using mineral pairs, stable isotope data provide information on the dominant source or sources of water, sulphur and carbon in hydrothermal fluids. In adularia–sericite-style epithermal systems in subaerial volcanic terranes, meteoric water generally dominates throughout the evolution of the system but there is evidence in some deposits for transient input of magmatic fluid or of highly exchanged fluids from some deep crustal source. Ocean water may be shown to contribute to or to substitute for meteoric water in the convective systems developed in epithermal deposits in oceanic island settings (e.g. Drake silver–goldfield, New South Wales, Andrew *et al.*, 1985). Carbon and sulphur isotopes commonly show magmatic signatures. In alunite–kaolinite-style deposits, as noted above, oxygen isotopes show an evolution from magmatic to meteoric sources while sulphur isotopes record the breakdown of primary magmatic sulphur dioxide to sulphate and hydrogen sulphide (Rye *et al.*, 1989).

Gold in a number of epithermal deposits is often associated with tellurium- and selenium-bearing minerals such as calaverite and naumannite. The specific controls on this element association are quite unknown, although it is often assumed that it relates to input from magmatic fluid; examples include Vatukoula, Fiji (Ahmad *et al.*, 1987), Thames, New Zealand (Merchant, 1986), and Bessie G, Colorado (Saunders and May, 1986). The chemistry of fluids responsible for mineralization and alteration lies within the same self-consistent chemical framework as is observed in active geothermal systems. In the next section, the fluid chemistry, alteration and mineralization in the active systems are briefly reviewed. For more comprehensive treatments see, for example, Henley *et al.* (1984) and Henley and Hedenquist (1986).

5.4 Active geothermal systems

It has long been recognized that active geothermal systems in volcanic terranes are the present-day equivalents of those ancient systems responsible for gold- and silver-bearing base-metal mineralization in epithermal mining districts (e.g. Lindgren, 1933; White 1955, 1981; Henley and Ellis, 1983). Weissberg (1969), Weissberg *et al.* (1979) and Hedenquist and Henley (1985b) have discussed the occurrence and chemistry of hot spring precipitates in the Ohaaki Pool at Broadlands and Champagne Pool at Waiotapu, New Zealand, both of which contain around 3 oz/t gold, 6–16 oz/t silver, about 2% arsenic and antimony and, at Ohaaki, 2000 mg/kg mercury. Similar hot spring precipitates have also been described from Steamboat Springs, Nevada (White, 1967). Hot springs such as Champagne Pool represent major discharge points from extensive subsurface geothermal systems and may be depositing gold at rates of the order 0.13 million oz per 1000 years at depth (Hedenquist and Henley, 1985b).

Integrated studies of mineralization and alteration in drill core from active systems and the geochemistry of geothermal fluids have provided a basis for the understanding of the hydrodynamics and the geochemistry of epithermal ore-forming systems. Studies include Browne (1969, 1971) and Browne and Ellis (1970) on Broadlands, Hedenquist and Henley (1985b) on Waiotapu, Christenson (1987) on Kawerau, and Krupp and Seward (1987) on Rotokawa. Although reliable geochemical data are now available for fumarolic gas in active volcanoes (e.g. Giggenbach, 1982,) for self-evident reasons few studies have been published on alteration and mineralization in such extreme environments – the equivalent of the early stage of evolution of alunite–kaolinite-style deposits. Thus the geochemical data reviewed below provide insight to the geochemistry of the adularia–sericite-style epithermal systems and the late stages of the evolution of alunite–kaolinite-style deposits.

Exploitation of geothermal energy involves the large-scale discharge of hot water and steam from geothermal fields. The consequent rapid pressure changes and longer term changes in temperature profiles and discharge chemistry of production wells provide critical information on the hydrodynamics of geothermal systems (Elder, 1981) and the recognition of free convection as the control on the gross heat and mass transfer in such large-scale hydrothermal systems. Geological data and simulation studies have provided contraints of 10^4 to 10^5 years on the duration of such hydrothermal systems. Taken together with geological and geochemical data, these observations provide the basis of an integrated geochemical and hydrodynamic framework for mineralization in epithermal systems (Henley and Ellis, 1983; Henley and Hedenquist, 1986).

The well-studied systems of the Taupo Volcanic Zone in New Zealand are hosted largely by porous ashflows and by well-fractured rhyolites. As with epithermal deposits, structural features control the major flows of hydrothermal fluid and these are marked by both intensity of alteration (particularly the abundance of adularia) and often the presence of hydrothermal breccias (Browne, 1978; Grindley and Browne, 1976). Temperatures in the upper few hundred metres of these systems are controlled by mixing with cool groundwaters or by phase separation ('boiling'). Boiling point vs. depth curves for fluids of differing salinity and gas content have been provided by Haas (1971) and Hedenquist and Henley (1985a) (Figure 5.7). These curves are usually drawn assuming no transfer of groundwater pressure from outside the system, although this certainly occurs in some systems (Fournier, 1987) and leads to higher temperatures at a given depth.

While the bulk of the fluid in a typical system is a near-neutral pH, alkaline chloride water containing CO_2 and H_2S (see below), in the near-surface regime phase separation leads to two types of fluid. Steam-heated (syn. acid sulphate) waters result from the condensation of steam and adsorption of H_2S in regions open to atmospheric oxygen (Henley and Stewart, 1983). The resultant oxidation of sulphide to sulphate leads to low-pH conditions and a consequent distinctive, intense alteration of surficial rocks to alunite–kaolinite–silica assemblages often associated with pyrite and native sulphur. At deeper levels, the adsorption of CO_2-rich vapour into cooler groundwater results in CO_2-rich waters of moderate acidity (Hedenquist and Stewart, 1985). It is these waters which are responsible for many of the illite–kaolinite 'clay caps' seen in many epithermal deposits and which, through incursions consequent on declining pressure in the mineralizing systems, probably account for the late-stage occurrence of kaolinite in banded veins (e.g. Hishikari, Golden Cross) and breccia deposits (e.g. Kelian) which otherwise have mineral assemblages (adularia, calcite, quartz) indicative of near-neutral pH, boiling conditions. The second type of fluid evolved in the near-surface discharge regime, consequent on phase separation at greater depths, is an alkaline chloride water supersaturated with silica. During ascent to and discharge at the surface, the alkaline chloride water precipitates amorphous silica to form the familiar sinter terraces seen in geothermal fields and the silicified zones, containing banded chalcedonic silica seen in hot-spring-type epithermal deposits. Remixing of these water types and mixing with shallow groundwaters results in the wide range of compositions and temperatures of hot springs.

Geothermal systems have been explored by drilling to around 3 km but in the majority of cases only indirect evidence (for example the comparison of trace and major gas contents of geothermal and volcanic gases [Giggenbach, 1986]) provides evidence of deep magmatic fluid-input to these systems. Such an input may be indirectly argued by analogy with porphyry systems and in some cases this is supported by stable isotope (δD, $\delta^{18}O$) data from epithermal deposits. J.W. Hedenquist, (personal communication) has noted that, in the Philippines, one geothermal well is reported to have discharged SO_2 which may have been of volcanic origin.

Table 5.2 provides a summary of the major element composition of a representative suite of geothermal fluids. Apart from fluids in the Salton Trough and a few other cases (e.g. Cessano, Italy), salinities lie in the range 1000–10 000 mg/kg chloride and fluid pH is close to that of neutral for water at the ambient temperature. The major element compositions are controlled by alteration mineral–fluid equilibria (Browne and Ellis, 1970; Giggenbach, 1981, 1984) and, through charge-balance, are indirectly dependent on salinity (Ellis, 1970; Henley *et al.*, 1984).

Carbon dioxide concentrations are largely independent of salinity and reflect input from deep magmatic and/or metamorphic sources (Table 5.2). Isotope data tend to favour the former so that analyses of gases from active volcanic environments (Table 5.3) may be representative of such input – both to the rocks of adularia–sericite systems and, at a higher crustal level, to the early formative stages of alunite–kaolinite-style systems. The concentrations of H_2S in Table 5.2 correlate with CO_2 and Giggenbach (1980) has suggested that the ratio $CO_2 : H_2S$ is strongly controlled by alteration temperature through equilibria involving $CaCO_3$–(FeO)–FeS_2 where (FeO) represents ferrous iron in aluminosilicates such as chlorite, epidote or prehnite.

Table 5.2 Concentrations (mg/kg) of major constituents in fluids discharged from geothermal wells in some representative geothermal fields. Compositions have been recalculated by recombination of liquid- and vapour-phase analyses of samples taken at the wellhead. (For discussion, see Henley and Hedenquist, 1986.)

Geothermal fields	t^0C	Na	K	Ca	Mg	Cl	SO_4	Si_2	B	H_2S	CO_2
Wajrakei, New Zealand	250	926	154	11.8	0.04	1543	22	484	20	9	348
Broadlands, New Zealand	261	705	150	5	0.6	1238	7	556	34	72	4104
Rotokawa, New Zealand	320	265	87	0.5	0.01	520	4	710	15	245	4495
Matilo, Tongonan Philippines	324	5018	1379	122	1.4	9124	18	771	194	85	2945

Table 5.3 Chemical composition of gas discharges from three calc–alkaline volcanoes; note that only the most abundant gases are shown; $x_g$is the anhydrous gas content of the fumarole sample (mmole per mole of H_2O) and n^* is the average oxidation state of sulphur (e.g. S = +4 in SO_2 and –2 in H_2S). From Heinrich *et al.*(1989) based on data from Giggenbach (1982) and Gerlack and Casadevall (1986). Note the abundance of acidic gases and the relative abundance of SO_2 over H_2S relative to the gases in near-neutral pH geothermal fluids – in the latter the acidity of any deep magmatic gas in neutralized by rapid alteration of primary minerals to pyrophyllite, alunite, etc.

	t^0C	x_g	CO_2	S_t	n^*	HCl	HF	NH_3	H_2	N_2
White Island, New Zealand	620	165	617	297	+3.2	59	0.66	0.13	24.1	3.3
Ngauruhoe, New Zealand	520	101	696	179	+1.6	31	2.6	0.31	28.4	142.4
Mt. St. Helens, Washington	860	84	824	67	+0.22	—	—	—	101.4	—

5.5 Metal transport in epithermal systems

The solution chemistry of gold in hydrothermal solutions is now well understood through both experimental studies and direct observation in geothermal systems. These studies demonstrate quite clearly that in the epithermal environment the principal gold-transport species is $Au(HS)_2^-$ (Seward, 1973). Confirmation comes from the observation by Brown (1986) of bonanza-grade (6 wt.% gold) precipitates resulting from boiling of geothermal fluid near the control plates of discharging geothermal wells. These data confirm that the Broadlands fluid is, at depths of a few hundred to a thousand metres, close to saturation with gold as a bisulphide complex. Equivalent gold arsenide species may contribute in a small way to the solubility, but the systematics of gold transport and deposition may be discussed satisfactorily in terms of the dominant bisulphide complex.

Based on solubility reactions of the form,

$$Au + 2H_2S = Au(HS)_2^- + H^+ + 0.5H_2$$

$$ZnS + 2H^+ + 2Cl^- = ZnCl_2 + H_2S$$

for Au, Ag, Cu, Zn, Pb, Henley (1985, 1990) has shown that the geologic solubility of gold and base metals (Ag, Pb, Zn) may be calculated for a wide range of salinities and H_2S contents. These relations are developed from empirical and thermodynamic data relating to pH, f_{H_2}, a_{K^+}/a_{H^+}, a_{H_2S}/f_{H_2} and alteration assemblages in active geothermal systems. The resulting solubility isopleths are shown in Figure 5.10 for a temperature of 250°C. It is important to note that, in constructing these solubility maps, reliable data are not available for the important copper and silver bisulphide complexes, nor for copper–chloride complexes. For gold, the solubility due to $AuCl_2^-$ has been estimated from the thermodynamic extrapolation of Helgeson and Garrels (1986) and experimental data by Bloom and Seward (in prep.) and Henley (unpublished data) and becomes important only under extreme salinity conditions near halite saturation. The chloride complex may become more important for ore transport at higher temperatures (400°C or so) in the porphyry environment.

The important feature to note in these solubility maps is that the solubility of gold in epithermal environments is inversely proportional to salinity and directly proportional to m_{H_2S}. The solubility of chloride-complexed metals is the inverse of this behaviour and this provides the essential chemical distinction between epithermal precious metal and epithermal base metal–silver deposits. This is borne out by fluid inclusion salinity data (Hedenquist and Henley, 1985a). The solubility–salinity data underpin the observation that polymetallic volcanogenic massive sulphide deposits, such as the Kuroko deposits of Japan, are essentially seafloor equivalents of terrestrial epithermal systems. The basic ingredients of their formation are the same but different or additional depositional controls apply due to the overlying seawater. As with epithermal systems, some are gold-enriched and others not, reflecting some fundamental geochemical control in the deep source regime of the systems.

Gold transport is maximized in solutions containing relatively high concentrations of H_2S which, as noted above, is inversely related, through fluid-mineral equilibria, to the concentration of CO_2. Using the thermodynamic and empirical relations of Giggenbach (1980, 1981), the $H_2 : H_2S : CO_2$ ratios may be related to temperature so that the solubility of gold may be expressed in f_{XO_2}–m_{Cl^-}-temperature space as illustrated in Figure 5.10(b). The implication of these data is clear – the solubility of gold in crustal fluids in volcanic terranes is strongly linked to CO_2 concentration *and* to temperature in the alteration-buffered roots of hydrothermal systems. In addition to its role as a source of metals (Au, Cu, etc.), this highlights the importance of magma degassing and magma evolution in developing epithermal ore-forming systems (note that CO_2 /SO_2 /H_2S /H_2 relations in the porphyry environment are controlled both by equilibria in the magma and by progressive high-temperature acid-alteration of country rock).

(a)

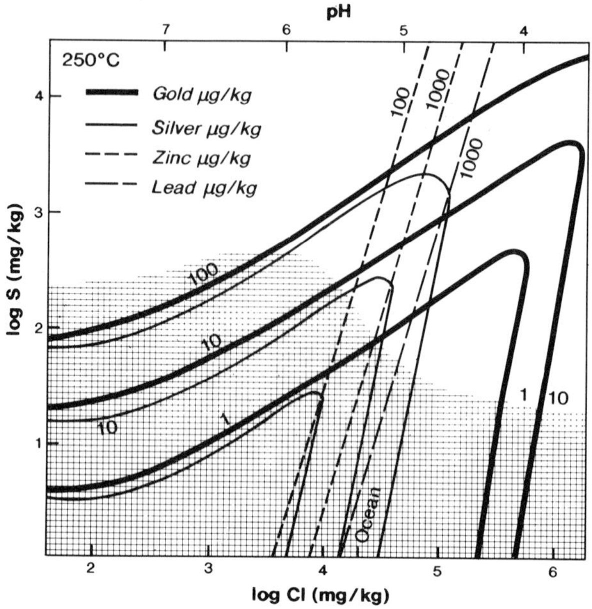

(b)

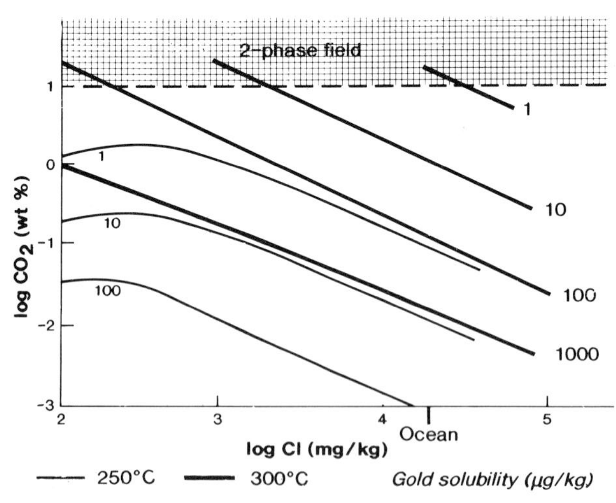

Figure 5.10 (a) Solubility controls of gold, silver, and base metals. (a) Relative solubilities of precious and base metals at 250°C in chloride–sulphide solutions the pH and redox state of which are controlled by common alteration mineral assemblages (see text). The composition range for fluids observed in active geothermal systems is shown by the stipple. H_2S is the dominant sulphide species across most of the diagram but sulphate occurs under the low-salinity, low-sulphur conditions in the bottom left of the figure. Halite saturation is at about 5.54 log Cl. (b) Solubility of gold at 250 and 300°C in chloride–sulphide–CO_2 solutions buffered by common mineral equilibria (see text). The stippled area shows the general region within which two-phase conditions occur, dependent on pressure. (From Berger and Henley, 1989.)

5.6 Physico-chemical conditions in the depositional regime

In discussing the transport chemistry of gold and base metals, it was shown that the principal controlling variables in rock-buffered systems are pH, m_{H_2S}, f_{H_2} and temperature. In the high-level discharge/depositional regime of epithermal hydrothermal systems these factors change rapidly due to phase separation – *boiling* – in response to decreasing pressure and to dilution by near-surface cold waters (Figure 5.5). Pressures and temperatures in the depositional regime are constrained by regional hydrology and by the phase relations of the CO_2–H_2O–X system; X represents total dissolved solids and is a subordinate factor relative to CO_2 in epithermal gold systems. Conductive heat loss is not significant in mineralized systems since the formation of an epithermal deposit itself implies high mass-flow rates and consequent high rates of convective heat transfer.

During phase separation, dissolved gases partition strongly to the vapour phase (Giggenbach, 1980; Henley *et al.*, 1984; Drummond and Ohmoto, 1985). Transfer of CO_2 out of solution in this way leads to a pH increase of 1–2 pH units relative to the original mineral-buffered solution through the reaction,

$$CO_{2(g)} + H_2O = HCO_3^- + H^+$$

Loss of H_2S and H_2 also occurs and these combined effects increase the saturation of gold from initial values close to one to high values. The loss of H_2 may be such as to oxidize reduced sulphur species to sulphate, further increasing gold supersaturation. Thus boiling is one of the most important and most efficient processes whereby gold is deposited as ore but there is no preferred temperature range. Calcite and adularia act as mineralogical indicators of this process (see below) and some textures may give circumstantial evidence of episodic boiling (e.g. banded chalcedonic or opaline silica).

Dilution can also be an important process leading to gold deposition in more saline systems. In such systems, dilution of chloride ion and accompanying temperature drop lead to precipitation of co-transported base metals as sulphides. The consequent removal of H_2S lowers the solubility of gold and leads to the deposition of discontinuous gold-rich shoots in some polymetallic veins (e.g., El Bronces, Chile – Skewes and Camus, 1988).

Deposition due to chemical changes associated with wallrock reactions are less important than in other styles of mineral deposit. Exceptions are the Carlin-type deposits (cf. Berger and Bagby, this volume) which involve dissolution of carbonate and the alunite–kaolinite-style deposits where, as discussed above, gold deposition results from the reaction of later stage, near-neutral fluids with acidic, alunite-rich alteration assemblages developed during earlier, magmatic-stage alteration.

The gangue mineralogies of epithermal deposits record chemical processes in the depositional regime. Loss of CO_2 due to boiling leads to supersaturation with respect to calcite according to,

$$CaCO_3 + CO_{2(g)} + H_2O = Ca^{++} + 2HCO_3^-$$

Due to the retrograde solubility of calcite (Arnorrsson, 1978; Fournier, 1985a), the mass of calcite deposited in this way is (in addition to the original CO_2 or Ca content of the fluid) dependent on the initial and final temperatures. Calcite deposited by boiling has a characteristically massive, bladed texture which is often found replaced by silica or leached following silica precipitation in the matrix. The change in pH attendant on boiling commonly leads to the precipitation of adularia in vugs (e.g. Kelian and Ladolam) or interbanded with silica (e.g. Hishikari, Golden Cross, and McLaughlin). In exploration, it must be remembered that the occurrence of adularia and calcite are indicative of phase separation and are not themselves indicative of the gold potential of a deposit or prospect.

Increasing concentration of silica due to vapour loss leads to the deposition of quartz or chalcedonic silica from fluids initially at about 250–300°C (Fournier, 1985b). The maximum mass of silica deposition occurs where fluids boil adiabatically to atmospheric pressure and the high supersaturations attained lead to amorphous and opaline-chalcedonic silica deposition as observed in silica sinters and shallow veins. Banding records transient pressure changes in the vein environment due to mineral deposition, permeability changes due to phase separation or, near surface, to changes in atmospheric pressure. The solubility of quartz passes through a maximum at about 320°C, so that boiling of fluids at initially higher temperatures leads first to undersaturation with respect to quartz. Thus at Kelian, only small, well-terminated quartz crystals occur sporadically in vugs and there are no major zones of silicification as occur in many other epithermal deposits.

Silica deposition from fluids may occur during dilution by cool groundwaters (Fournier, 1985b) but the saturation levels attained are low; this favours slow deposition and the formation of well-crystallized quartz above about 200°C. Chalcedony may form at lower temperatures but generally lacks the banding characteristic of boiling environments. Massive chalcedonic silica often occurs with alunite–kaolinite deposits; in this case, the high silica saturations relate to the rapid destruction of primary aluminosilicates by acidic fluids.

Over time the various polymorphs of silica recrystallize to quartz but primary deposition textures in banded veins and silica sinters are generally preserved; for example, in the 324 ± 5 Ma Pajingo deposit, Queensland, recrystallization has occurred and destroyed any primary fluid inclusions but vein textures are well preserved (Etminan *et al.*, 1988).

5.7 Epithermal deposits through geologic time

Until recently the majority of epithermal deposits were known from Cenozoic to Recent volcanic terranes. However, the ingredients for their formation – high-level magmatism, brittle fracture systems and groundwater flow – are timeless. For this reason it is likely that epithermal deposits have been formed throughout crustal evolution and the only restriction upon their present distribution relates to preservation from erosion or severe metamorphic overprinting. Newly discovered deposits in north-eastern Queensland (e.g. Pajingo–Etminan *et al.*, 1988; Wirralie–Fellows and Hammond, 1988; Yandan–Wood *et al.*, 1989) are Permo- Carboniferous in age and associated with acid to intermediate volcanic suites. Their preservation has been due to younger cover rocks which are only recently being stripped in response to uplift of the Eastern Highlands. Epithermal deposits of similar age also occur at Drake, New South Wales (Bottomer, 1986) and Cracow, Queensland. Palaeozoic epithermal deposits are recognized in the Lachlan fold belt of south- eastern Australia and in the Appalachians. In both regions, alunite-kaolinite-type deposits are preserved (e.g. Temora, New South Wales–Thompson *et al.*, 1986; Pilot Mountain – Klein and Criss, 1988; Haile, Carolina – Kiff and Spence, 1988). Epithermal deposits have also tentatively been recognized in Archaean rocks in Canada and Australia (e.g. Adams, 1976; Groves, 1988).

5.8 Exploration

Modern exploration for epithermal deposits follows the methodology developed in the 1960s and 1970s for porphyry deposits and volcanogenic massive sulphide deposits. Thus, regional geochemistry and the mapping of alteration in bedrock or float material are of prime importance. Geochemical stream sediment and soil surveys are proving increasingly useful but their utility as a cost-effective exploration method has been dependent on the development of sub-ppm, carbon-rod, atomic absorption techniques. The bulk-cyanide leach technique has also developed as a qualitative and inexpensive tool for regional gold surveys. Arsenic, antimony and mercury are covariant with gold and may be used as additional pathfinders (Silberman and Berger, 1985) and base metals and copper may be useful for some deposit styles. Thallium and tellurium may be useful as discriminants but are generally precluded by the difficulties and cost of analysis.

The majority of large epithermal systems are characterized by extensive intense alteration haloes consequent on widespread feldspar and ferromagnesian mineral destructive reactions, silicification, and pyritization. Other characteristics of epithermal systems are the transfer of potassium into the upper few hundred metres of the system and, due to the presence of reduced sulphur, the widespread demagnetization of host rocks. These features ensure that airborne magnetic, radiometric and resistivity surveys are very useful in locating epithermal systems (Allis, 1990). The former have been very successful in epithermal exploration throughout eastern Australia (Tennison Woods and Webster, 1985; Irvine and Smith, 1990). Such surveys also provide basic geological information and, depending on correct flight line orientation, can locate major regional and related structures which may provide a fundamental control on the location of ore districts. The recognition of possible regional structures is also aided by satellite thematic mapper and SLAR radar imagery. Zones of extensive rock alteration may, in some cases, be detected through gravity survey (Locke and De Ronde, 1987).

At the *prospect* scale, geochemical surveys based primarily on gold are obviously of great importance in locating ore zones. Where outcrop is sufficient or pilot drilling has been completed, alteration mapping (backed by X-ray diffraction and petrography) are both important, although the distinction of hypogene or supergene origin for kaolinite and alunite may often be impossible. Recognition of multiple alteration stages may be very important. Similarly, mapping of breccia types provides an important guide to mineralization style and possible targets. Fluid inclusion thermometry is useful in characterizing deposits but *only* if great care is taken in selecting suitable samples and in interpreting the homogenization data.

Strong variations in electrical properties within hydrothermally altered rocks allow conductive clay–pyrite zones and resistive silicified zones to be mapped in detail by standard resistivity methods (Irvine and Smith, 1990). Controlled Source Audio-

Magnetic-Telluric (CSAMT) surveys have been used with varying success in targeting capricious vein systems at depth and discriminating resistive quartz-rich zones or conductive clay–pyrite zones (Austpac, 1988). In many epithermal districts, recognition of the original geomorphology provides clues to the palaeohydrology which may then aid in developing exploration targets. At Creede, Colorado, for example, the silver–base metal vein district lies on the outflow portion of a much larger epithermal system which upflowed several kilometres to the north near the ring fracture of the San Luis caldera and developed some high-grade gold mineralization.

No single technique provides the key to epithermal exploration. Instead, the use of careful mapping in conjunction with geophysics and geochemistry may lead to the recognition of good prospects. Drilling as always is the ultimate test. The discovery of the Hishikari bonanza deposit, for example, was the result of a five-year combined geological and geophysical programme. Exploration drilling tested a series of helicopter-detected electromagnetic and ground resistivity anomalies coincident with a basement high recognized from gravity survey (Johnson and Fujita, 1985; Urashima *et al.*, 1987). Drilling also tested the complementary concept of depth-extension to the exposed andesite-hosted veins.

5.9 Summary

This chapter has focused on the geochemical character of epithermal gold deposits in volcanic terranes rather than the geology of their occurrence. This is deliberate in that epithermal deposits owe more to the chemistry *imposed* on their host-rock package *by* hydrothermal fluids at high water/rock ratios than they do on their intrinsic geochemistry.

Epithermal gold deposits are the product of large-scale hydrothermal systems in volcanic terranes. Their essential ingredients are a magmatic heat source in the upper few kilometres of the crust, a source of groundwater, metal and reduced sulphur, and zones of brittle fracture. These ingredients have been available throughout crustal history so that there is no restriction on age, only on preservation. Mixing of these ingredients leads to the formation of deposits and deposit- settings which share more common features than differences – one of the primary controls on deposit style is the relative level in the crust where magma degassing occurs.

In the past there has been much controversy on the role of magmas as metal sources. To the author, the growing body of field evidence and geochemical data is unequivocal; high-level magmas are the source of gold in epithermal gold systems and in most cases are also the source of the sulphur required for gold transport. The ability of degassing magmas to supply metals is becoming more firmly established; Meeker (1988), for example, has shown that in December, 1986, Mt. Erebus, Antarctica, discharged daily about 0.1 kg Au together with 0.2 kg Cu, 200t HCl and 56t SO_2 – equivalent to 360t Au per 10 000 years. What differs between deposit styles, as discussed above, may relate largely to the relative depth of intrusion. Thus the alunite–kaolinite-style deposits relate to degassing of high level magmas (e.g. rhyolite domes), with later hydrothermal flow driven by the larger, deeper magma system, whereas adularia–sericite-style systems relate to deeper magma bodies degassing into an overlying groundwater system. Convecting groundwaters serve to disperse the magma fluid. In high-permeability systems, dispersion may be so strong as to prevent the formation of an ore deposit even though a billion grams of gold may be disseminated at low grade in the upper few hundred metres of the system (e.g. Broadlands, New Zealand). In lower permeability host rocks, major structures control groundwater flow and focus fluids to a high-level deposition site (e.g. Hishikari, Japan). Extensive brecciation, due to magmatic processes in some way associated with major regional structures, appears to provide an optimum environment (Kelian, Indonesia, and Ladolam, Papua New Guinea, are examples).

Epithermal gold deposits are attractive exploration targets. Although many questions remain open, the quantitative understanding of epithermal deposits based on geothermal physics and chemistry provides a much stronger, less speculative geological framework than for many other types of deposit. It is this framework, used wisely, that provides the basis for their systematic and successful exploration.

Acknowledgements

In this chapter I have attempted to distil some of the essential essence of epithermal systems and their gold mineralization into a few pages. This in no way does justice to the huge and growing body of literature on this style of mineral deposit. Theo van Leeuwan, Roy Suzuki and Geoff Ballantyne very kindly provided review comments on the Kelian, Hishikari and Lihir case histories respectively and permission to utilize published diagrams on these deposits, and Noel White (BHP) kindly provided some of the grade-tonnage data used in Figure 5.1. Jeff Hedenquist, Julian Hemley, Dick Sillitoe, Barney Berger, Pat Browne, Paul Eimon and Mike Smith kindly provided helpful comments on the manuscript. I thank them and the body of explorationists and researchers through whose efforts and enthusiasm our knowledge of epithermal gold deposits has flourished and will continue to flourish into the future.

References

Adams, G.W. (1976) Precious metal veins of the Berens River mine, Northwest Ontario. Unpublished M.Sc. thesis, Univ. of Western Ontario, London, Ontario.

Ahmad, M., Solomon, M.J. and Walshe, J.L. (1987) Mineralogical and geochemical studies of the Emperor gold telluride deposit, Fiji. *Econ. Geol.* **82**, 345-370.

Allis, R.G. (1990) Geophysical signatures of epithermal systems. *J. Geochem. Explor.* **36**, 339-374.

Andrew, A.S., Scott, K.M. and Bottomer, L.R. (1985) Mineralization and alteration in the Drake silver-gold field. CSIRO, Division of Mineralogy and Geochemistry, Research Review, 44-45.

Arnorsson, S. (1978) Precipitation of calcite from flashed geothermal waters in Iceland. *Contrib. Mineral. Petror.* **66**, 21-28.

Arribas, A. Jr., Rytuba, J.J., Cunningham, C.G., Kelly, W.C., Rye, R.O. and Castroviejo, R. (1988) Rodalquilar Deposits, Spain, First example of caldera related epithermal mineralisation in Europe; Part 2, Ore Genesis. *Geol. Soc. Amer. Abstr. Programs*, Denver, 351.

Austpac, (1988) Annual Report to Shareholders, 1988.

Berger, B.R. and Bethke, P.M. (1985) *Geology and Geochemistry of Epithermal Systems.* Reviews in Economic Geology, Soc. of Econ. Geol., Vol.2, 298.

Berger, B.R. and Bonham, H.F. (1990) Epithermal Au–Ag deposits in the western United States. *J. Geochem. Explor.* **36**, 103-142.

Berger, B.R. and Eimon, P.I. (1983) Conceptual models of epithermal precious metal deposits. In *Cameron Volume on Unconventional Mineral Deposits*, ed. Shanks, W.C. Soc. Mining Engineers of AIME, 191-205.

Berger, B.R. and Henley, R.W. (1989) Recent advances in the understanding of epithermal gold-silver deposits – with special reference to the western United States. *Econ. Geol. Monograph*, **6**, 405-423.

Bethke, P.M. (1988) The Creede, Colorado, ore-forming system; a summary model. US Geol. Surv., Open-File Report 88-403. 29pp.

Bodnar, R.J., Reynolds, T.J. and Kuehn, C.A. (1985) Fluid inclusion systematics in epithermal systems. In *Geology and Geochemistry of Epithermal Systems*, eds. Berger, B.R. and Bethke, P.M. Reviews in Economic Geology, Soc. of Econ. Geol., Vol.2, 73-98.

Bottomer, L.R. (1986) Epithermal silver-gold mineralisation in the Drake area, Northeastern New South Wales: *Aust. J. Earth Sci.* **33**, 457-473.

Brown, K.L. (1986) Gold deposition from geothermal discharges in New Zealand. *Econ. Geol.* **81**, 979-983.

Browne, P.R.L. (1969) Sulphide mineralisation in a Broadlands geothermal drill hole, Taupo Volcanic Zone, New Zealand. *Econ. Geol.* **64**, 156-159.

Browne, P.R.L. (1971) Mineralisation in the Broadlands Geothermal Field, Taupo Volcanic Zone, New Zealand. *Soc. Mining Geol. Japan, Special Issue* **2**, 64-75.

Browne, P.R.L. (1978) Hydrothermal alteration in active geothermal fields. *Ann. Rev. Earth Plan. Sci.* **6**, 229-250.

Browne, P.R.L. and Ellis, A.J. (1970) The Ohaki-Broadlands hydrothermal area, New Zealand: Mineralogy and related geochemistry. *Amer. J. Sci.* **269**, 97-131.

Christenson, B.W. (1987) Fluid-Mineral Equilibria in the Kawerau Hydrothermal System, Taupo Volcanic Zone, New Zealand. Unpublished Ph.D. Dissertation, Univ. of Auckland. 444pp.

Cunneen, R. and Sillitoe, R.H. (1989) Paleozoic hot-spring sinter in the Drummond Basin, Queensland. *Econ. Geol.* **84**, 135-142.

Cunningham, C.G., Arribas, A. Jr., Rytuba, J.J., Rye, R.O., Kelly, W.C., Podwysocki, M.H., Mckee, E.H., Arribas, A. Snr. and Castroviejo, R. (1989) The origin of United States–Spain gold–alunite deposits: the caldera-related epithermal gold/alunite deposits at Rodalquilar, Spain, Internat. Geol. Cong., Washington, D.C., Abstracts–Poster Session, (in press).

Davies, R.G. and Ballantyne, G. (1987) Geology of the Ladolam gold deposit, Lihir Island, Papua New Guinea. *Proc. Pacrim Congress, Gold Coast, Queensland.* Aust, I.M.M., 943-949.

De Ronde, C.E.J. (1986) The Golden Cross gold-silver deposit. In *Guide to Active Epithermal (Geothermal) Systems and Precious Metal Deposits of New Zealand*; eds. Henley, R.W. Hedenquist, J.W. and Roberts, P.J. Gebruder Borntraeger, Monograph Series on Mineral Deposits, 26, 129-146.

De Ronde, C.E.J. and Blattner, P. (1988) Hydrothermal alteration, stable isotopes and fluid inclusions of the Golden Cross Epithermal Gold–Silver Deposit, Waihi, New Zealand. *Econ. Geol.* **83**, 896-917.

Deen, J.A., Rye, R.O. and Drexler, J.W. (1988) Polymetallic mineralization related to magma evolution and magmatic-hydrothermal fluid mixing, *Geol. Soc. Amer. Abstr. Programs*, Denver, A351.

Drummond, S.E. and Ohmoto, H. (1985) Chemical evolution and mineral deposition in boiling hydrothermal systems. *Econ. Geol.* **80**, 126-147.

Eastoe, C.J. (1978) A fluid inclusion study of the Panguna porphyry copper deposit, Bougainville, Papua New Guinea. *Econ. Geol.* **73**, 721-748.

Elder, J. (1981) *Geothermal Systems*. Academic Press, London. 508pp.

Ellis, A.J. (1970) Quantitative interpretation of chemical characteristics of hydrothermal systems. *Geothermics*, Special Issue, **2**, Pt. 1, 516-528.

Etminan, H., Porter, R.G., Hoffman, C.F., Shen Su Sun and Henley, R.W. (1988) Initial hydrothermal alteration, fluid inclusion and stable isotope studies at Pajingo Gold Deposit, North Queensland. *Bicentennial Gold 88,* eds. Goode, A.D.T. Smyth, E.L. Birch, W.D. and Bosma, L.I. Melbourne. *Geol. Soc. Aust. Abstr.* **23**, Australia, 434-435.

Fellows, M.L. and Hammond, J.M. (1988) Wirralie Mine, Mt. Coolon, Queensland, in Core Shed Guidebook, *Proc. Bicentennial Gold 88*, Melbourne, ed. Bloom, M.S. Australia, 55.

Field, C.W. and Fifarek, R.H. (1985) Light stable isotope systematics in the epithermal environment, In *Geology and Geochemistry of Epithermal Systems*, eds. Berger, B.R. and Bethke, P.M. Reviews in Economic Geology, Soc. of Econ. Geol., Vol. 2, 99-128.

Fountain, R.J. (1972) Geological relationships in the Panguna porphyry copper deposit, Bougainville island, Papua New Guinea. *Econ. Geol.* **67**, 1049-1064.

Fournier, R.O. (1985a) Carbonate transport and deposition in the epithermal environment, In *Geology and Geochemistry of Epithermal Systems*, eds. Berger, B.R. and Bethke, P.M. Reviews in Economic Geology, Soc. of Econ. Geol., Vol. 2, 63-72.

Fournier, R.O. (1985b) The behavior of silica in hydrothermal solutions, In *Geology and Geochemistry of Epithermal Systems*, eds. Berger, B.R. and Bethke, P.M. Reviews in Economic Geology, Soc. of Econ. Geol., Vol. 2, 45-62.

Fournier, R.O. (1987) Conceptual models of brine evolution in magmatic-hydrothermal systems. In *Volcanism in Hawaii*. US Geol. Surv., Prof. Paper 1350, 1487-1506.

Gerlach, T.M., and Casadevall, T.J. (1986) Fumarole emissions at Mount St Helens volcano, June 1980 to October 1981: degassing of a magma hydrothermal system. *J. Volc. Geothermal Res.* **28**, 141-160.

Giggenbach, W.F. (1980) Geothermal gas equilibria. *Geochim. Cosmochim. Acta* **44**, 2021-2032.

Giggenbach, W.F. (1981) Geothermal mineral equilibria. Geochim. Cosmochim. Acta 45, 393-410.

Giggenbach, W.F. (1982) The chemical and isotopic composition of gas discharges from New Zealand andesite volcanos. *Bull. Volc.* **45**(3), 253-255.

Giggenbach, W.F. (1984) Mass transfer in hydrothermal alteration systems. *Geochim. Cosmochim. Acta* **48**, 2693-2711.

Giggenbach, W.F. (1986) The use of gas chemistry in delineating the origin of fluids discharged over the Taupo Volcanic Zone. *Proc. 5th Internat. Symp. Volc. Cong., Volcanism, Hydrothermal Systems and Related Mineralisation*, February 1986.

Grindley, G.W. and Browne, P.R.L. (1976) Structural and hydrological factors controlling the permeability of some hot water geothermal fields. *Second United Nations Symp. on the Development and Use of Geothermal Resources, San Francisco*, 377-386.

Groves, I.M. (1988) Epithermal gold and porphyry deposits; end members of a continuum? In *Advances in Understanding PreCambrian Gold Deposits*, Vol. II, eds. Ho, S.E. and Groves, D.I. Geol. Dept., Univ. of W. Australia Pub. 12, 321-334.

Haas, J.L. Jr. (1971) The effect of salinity on the maximum thermal gradient of a hydrothermal system at hydrostatic pressure: *Econ. Geol.* **66**, 940-946.

Harding, T.P. (1985) Seismic characteristics of negative flower structures, positive flower structures and positive structural inversion. Amer. Assoc. of Petrol Geol., Bull. 63, 1016-1058.

Hayba, D.O., Bethke, P.M., Heald, P. and Foley, N.K. (1985) Geologic, mineralogic, and geochemical characteristics of volcanic-hosted epithermal precious-metal deposits. In *Geology and Geochemistry of Epithermal Systems*; eds. Berger, B.R. and Bethke, P.M. Reviews in Economic Geology, Soc. of Econ. Geol., Vol. 2, 129-168.

Heald, P., Foley, N.K. and Hayba, D.O. (1987) Comparative anatomy of volcanic-hosted epithermal deposits: acid sulphate and adularia-sericite types. *Econ. Geol.* **82**, 1-26.

Hedenquist, J.W. and Henley, R.W. (1985a) The importance of CO_2 on freezing point measurements of fluid inclusions: Evidence from active geothermal systems and implications for epithermal ore deposition. *Econ. Geol.* **80**, 1379-1406.

Hedenquist, J.W. and Henley, R.W. (1985b) Hydrothermal eruptions in the Waiotapu geothermal system, New Zealand: Their origin, associated breccias, and relation to precious metal mineralisation. *Econ. Geol.* **80**, 1640-1668.

Hedenquist, J.W. and Stewart, M.K. (1985) Natural steam-heated waters in the Broadlands-Ohaaki geothermal system, New Zealand; their chemistry, distribution and corrosive nature. Trans. Geothermal Resources Council Annual Mtg, Hawaii, Vol. 9, 245- 250.

Hedenquist, J.W., Matsuhisa, Y., Izawa, E., Marumo, K., Aoki, M. and Sasaki, K. (1988) Epithermal gold mineralisation of acid leached rocks in the Nansatsu district of southern Kyushu, Japan. *Bicentennial Gold 88*, Melbourne, Extended Abstracts Oral Program. *Geol. Soc. Aust. Abstr.* **22**, 183-190.

Hedenquist, J.W., White, N.C. and Siddeley, G. (1990) eds. Epithermal gold mineralization of the Circum-Pacific:geology, geochemistry, origin and exploitation. Volumes I and II, *J. Geochem. Explor*. **35** and **36**.

Heinrich, C.A., Henley, R.W. and Seward, T.M. (1989) *Hydrothermal Systems*. Australian Mineral Foundation, Adelaide.

Helgeson, H.C. and Garrels, R.M. (1968) Hydrothermal transport and deposition of gold. *Econ. Geol.* **63**, 622-635.

Hemley, J.J., Montoya, J.W., Marinenko, J.W. and Luce, R.W. (1980) Equilibria in the system Al_2O_3-SiO_2-H_2O and some general implications for alteration/mineralisation processes. *Econ. Geol.* **75**, 210-228.

Henley, R.W. (1985) The geothermal framework for epithermal deposits. In *Geology and Geochemistry of Epithermal Systems*; eds. Berger, B.R. and Bethke, P.M. Reviews in Economic Geology, Soc. of Econ. Geol., Vol. 2, 1-24.

Henley, R.W. (1990) Ore transport and deposition in epithermal environments, In *Stable Isotopes and Fluid Processes in Mineralisation*; eds. Hubert, H., Golding, S. and Ho, S.E., Geol. Dept. and Univ. Extension, Univ. of Western Australia Publ. **23**, 51-69.

Henley, R.W. and Ellis, A.J. (1983) Geothermal systems ancient and modern: A geochemical review. *Earth-Sci. Rev.* **19**, 1-50.

Henley, R.W. and Hedenquist, J.W. (1986) Introduction to the geochemistry of active and fossil geothermal systems, *Guide to the Active Epithermal (Geothermal) Systems and Precious Metal Deposits of New Zealand*, eds. Henley, R.W. Hedenquist, J.W. and Roberts, P.J. Gebruder Borntraeger Monograph Series on Mineral Deposits 26, 1-22.

Henley, R.W. and Hoffman, C.F. (1987) Gold: sources to resources. In *Proc. Pacific Rim Congress 87*, Aust. Inst. Min. Met., 159-168.

Henley, R.W. and McNabb, A. (1978) Magmatic vapour plumes and ground-water interaction in porphyry copper emplacement: *Econ. Geol.* **73**, 1-20.

Henley, R.W. and Stewart, M.K. (1983) Chemical and isotopic changes in the hydrology of the Tauhara geothermal field due to exploitation at Wairakei. *J. Geotherm. Res.* **15**, 285- 314.

Henley, R.W., Trueadell, A.H., Barton, P.B. Jr. and Whitney, J.A. (1984) *Fluid-Mineral Equilibria in Hydrothermal Systems*. Reviews in Economic Geology, Soc. of Econ. Geol., Vol. 1 267pp.

Hollister, V.F. (1985) *Discoveries of Epithermal Precious Metal Deposits*. Soc. of Mining Engineers, A.I.M.M., New York. 169pp.

Irvine, R.J. and Smith, M.J. (1990) Geophysical exploration for epithermal gold deposits. *J. Geochem. Explor.* **36**, 375-412.

Ishihara, S., Sakamaki, Y., Sasaki, A., Teraoka, Y. and Terashima, S. (1986) Role of the basement in the genesis of the Hishikari gold-quartz vein deposit, Southern Kyushu, Japan. *Mining Geol.* **36**, 495-510.

Jannas, R.R., Beane, R., Ahler, B. and Brosnahan, D. (1989) Gold and copper mineralization at the El Indio deposit, Chile Mineralogy, fluid inclusions and phase relations. *J. Geochem. Explor.* **36**, 233-266.

Johnson, I.M. and Fujita, M. (1985) The Hishikari gold deposit: an airborne EM discovery. Can. Inst. of Mining, Bull. 78, 61-66.

Johnson, R.W. (1987) Delayed partial melting of subduction- modified magma sources in western Melanesia: new results from the late Cenozoic. *Proc. Pacrim Congress, Gold Coast, Queensland.* Aust. I.M.M., 211-214.

Keays, R.R. and Scott, R.B. (1976) Precious metals in ocean ridge basalts: Implications for basalts as source rocks for gold mineralisation, *Econ. Geol.* **71**, 705-720.

Kiff, I.T. and Spence, W.H. (1988) Gold mineralisation and associated alteration patterns at the Haile mine, Lancaster County, South Carolina. In *Southeast Geological Excursions*, ed. Secor, D.T. S. Carolina Geol. Survey, 29-39.

Klein, T.L. and Criss, R.E. (1988) An oxygen isotope and geochemical study of meteoric-hydrothermal systems at Pilot Mountain and other selected localities, Carolina Slate Belt. *Econ. Geol.* **83**, 801-821.

Krupp, R.E. and Seward, T.M. (1987) The Rotokawa geothermal system, New Zealand: An active epithermal gold-depositing environment. *Econ. Geol.* **82**, 1109-1129.

Lehrman, N.J. The McLaughlin Mine, Napa and Yolo Counties, California. In *Precious-Metal Mineralization in Hot Springs Systems, Nevada-California*, (1986) eds. Tingley, J.V. and Bonham, H.F. Nevada Bur. of Mines and Geol., Report 41, 85-89.

Lindgren, W. (1922) A suggestion for the terminology of certain mineral deposits, *Econ. Geol.* **17**, 292-294.

Lindgren, W. (1933) *Mineral Deposits* (4th edn.). McGraw-Hill, New York. 930pp.

Lipman, P.W. (1987) Oligocene central San Juan caldera cluster, Colorado [abs.]. *Geol. Soc. Amer. Abstr. Programs* **19**(5), 315.

Lipman, P.W., Fisher, F.S., Mehnert, H.H., Naeser, C.W., Luedke, R.G. and Steven, T.A. (1976) Multiple ages of mid-Tertiary mineralization and alteration in the western San Juan Mountains, Colorado: *Econ. Geol.* **71**, 471-588.

Locke, C.A. and De Ronde, C.E.J. (1987) Delineation of gold-bearing hydrothermally altered rocks using gravity data – a New Zealand example. *Geoexploration* **24**, 471-481.

McKinstry, H.E. (1941) Structural control of ore deposition in fissure veins. *Amer. Inst. Mining Engineers Trans.* **144**, 65-95.

McKinstry, H.E. (1948) *Mining Geology*. Prentice-Hall, New Jersey. 680pp.
Meeker, K. (1988) The emission of gases and aerosols from Mt Erebus volcano, Antarctica. M.Sc. thesis, New Mexico Institute of Mines and Technology. 175pp.
Merchant, R.J. (1986) Mineralisation in the Thames District. In *1986, Guide to the active epithermal* (Geothermal) *systems and precious metal deposits of New Zealand*; eds. Henley, R.W., Hedenquist, J.W. and Roberts, P.J. Gebruder Borntraeger Monograph Series on Mineral Deposits 26, 147-164.
Meyer, C. and Hemley, J.J. (1967) Wall rock alteration. *Geochemistry of Hydrothermal Ore Deposits*, ed. Barnes, H.L. Holt, Rinehart and Winston, New York, 166-235.
Moyle, A.J., Doyle, B.J., Hoogvliet, H and Ware, A.R. (1990) Ladolam gold deposit, Lihir Island. In Geology and Mineral Deposits of Australia and Papua New Guinea. Austr. Inst. Min. Met., in press.
Noble, D.C., McCormack, J.K., McKee, E.H., Silberman, M.L. and Wallace, A.B. (1988) Time of mineralisation in the evolution of the McDermitt caldera complex, Nevada-Oregon, and the relation of middle Miocene mineralisation in the northern Great Basin to coeval regional basaltic magmatic activity. *Econ. Geol.* **83**, 859-863.
Oki, Y. (1983) Geology and hydrothermal system of Hakone volcano and Tanzawa Mountains. Bulletin Hot Springs Research Institute, Kanagawa Prefecture, 15, Special Issue. 68pp.
Plahuta, J.T. (1986) Geology of the Buckhorn Mine, Eureka County, Nevada. In *Precious-Metal Mineralization in Hot Springs Systems, Nevada-California*, eds. Tingley, J.V. and Bonham, H.F. Nevada Bur. of Mines and Geol., Report 41, 103-107.
Porter, R. (1988) The Pajingo Gold Mine. In *Bicentennial Gold 88*, Excursion Guidebook. ed. Morrison, G.W. Economic Geology Research Unit, James Cook Univ., Contribution 29, 23-34.
Ransome, F.L. (1909) Geology and ore deposits of Goldfield, Nevada. US Geol. Surv., Prof. Paper 66. 258pp.
Rye, R.O., Bethke, P.M. and Wasserman, M.D. (1989) Diverse origins of alunite and acid sulphate alteration: stable isotope systematics. US Geol. Surv., Open-File Report 89-5. 33pp.
Rye, R.O., Plumlee, G.S., Bethke, P.M. and Barton, P.B. (1988) Stable isotope geochemistry of the Creede, Colorado, hydrothermal system. US Geol. Surv., Open-File Report 88-356. 40pp.
Saunders, J.A. and May, E.R. (1986) Bessie G: a high grade epithermal gold telluride deposit, La Plata County, Colorado. *Proc. of Gold '86 Symp.*, ed. MacDonald, A.J. Toronto, 436- 444.
Schafer, R.W., Cooper, J.J. and Vikre, P.J., eds. (1988) *Proc. Symp. Bulk Mineable Precious Metal Deposits of the Western United States*. Geol. Soc. Nevada, Reno. 755pp.
Schmitt, H. (1950) Origin of the 'epithermal' mineral deposits, *Econ. Geol.* **45**, 191-200.
Seward, T.M. (1973) Thio complexes of gold and the transport of gold in hydrothermal ore solutions: *Geochim. Cosmochim. Acta* **37**, 379-399.
Sibson, R.H. (1981) Controls on low-stress, hydro-fracture dilatancy in thrust, wrench and normal fault terrains. *Nature* **289**, 665-667.
Sibson, R.H. (1987) Earthquake rupturing as a mineralising agent in hydrothermal systems. *Geology* **15**, 701-704.
Siddeley, G. and Araneda, R. (1986) The El Indio-Tambo gold deposits, Chile. In *Proc. of Gold'86 Symp.*, ed. MacDonald, A.J. Toronto, 445-456.
Silberman, M.L. and Berger, B.R. (1985) Relationship of trace element patterns to alteration and morphology in epithermal precious metal deposits. In *Geology and Geochemistry of Epithermal Systems*, eds. Berger, B.R. and Bethke, P.M. Reviews in Economic Geology, Soc. of Econ. Geol., Vol. 2, 203- 232.
Sillitoe, R.H. (1989) Gold deposits in western Pacific island arcs: the magmatic connection. *Econ. Geol.* (in press).
Sillitoe, R.H. and Angeles, C.A. (1985) Geological characteristics and evolution of a gold-rich porphyry copper deposit at Guinaoang, Luzon, Phillippines. *Asian Mining '85*. Inst. of Mining and Metallurgy, London, 15-26.
Sillitoe, R.H., Angeles, C.A. Jr., Comia, G.M., Antioquia, E.C. and Abeya, R.B. (1990) A Nansatsu-type deposit at Nalesbitan, Luzon, Phillippines. *J. Geochem. Explor.* **35**, 387-412.
Sillitoe, R.H., Baker, E.M. and Brook, W.A. (1984) Gold deposits and hydrothermal eruption breccias associated with a maar volcano at Wau, Papua, New Guinea. *Econ. Geol.* **79**, 638-655.
Sillitoe, R.H. and Bonham, H.F. Jr. (1984) Volcanic landforms and ore deposits: *Econ. Geol.* **79**, 1286-1298.

Skewes, M.A. and Camus, F. (1988) Fluid inclusions and the mechanism of gold deposition in the Chilean epithermal deposit, El Bronce de Petorca. *Geol. Soc. Amer. Abstr. Programs*, Denver, A352.

Steven, T.A. and Eaton, G.P. (1975) Environments of ore deposition in the Creede mining district, San Juan Mountains, Colorado: 1. Geologic, hydrologic and geophysical setting. *Econ. Geol.* **70**, 1023-1037.

Steven, T.A. and Lipman, P.W. (1976) Calderas of the San Juan Volcanic Field, southwestern Colorado. US Geol. Surv., Prof. Paper 958. 35pp.

Stoffregen, R.E. (1987) Genesis of acid-sulphate alteration and Au–Cu–Ag mineralization at Summitville, Colorado. *Econ. Geol.* **82**, 1575-1591.

Tennison Woods, K. and Webster, S.S. (1985) Geophysical signature of gold and porphyry copper mineral deposits in the Lachlan Fold Belt, New South Wales. *Proc. 4th Australian Soc. of Economic Geophysicists Conf.*, 325-331.

Thomason, R.E. (1986) Geology of the paradise peak gold/silver deposit, Nye County, Nevada. In *Precious-Metal Mineralization in Hot Springs Systems, Nevada-California*, eds. Tingley, J.V. and Bonham, H.F. Nevada Bur. of Mines and Geol., Report 41, 90-92.

Thompson, J.F.H., Lessman, J. and Thompson, A.J.B. (1986) The Temora gold-silver deposit: A newly recognized style of high sulphur mineralization in the Lower Paleozoic of Australia. *Econ. Geol.* **81**, 732-738.

Tooker, E.W. (1985) Geologic characteristics of sediment and volcanic hosted disseminated gold deposits – search for an occurrence model. US Geol. Surv., Bull. 1646, 150pp.

Urashima, Y. and Izawa, E. (1982) Hydrothermal alteration observed from drilling into the Hishikari gold deposit. *Proc. Annual Mtg of Mineral Resources-Related Societies of Japan*, G, 13-16. (in Japanese).

Urashima, Y., Ibaraki, K. and Suzuki, R. (1987) The Hishikari gold-silver deposit. In *1987, Gold Deposits and Geothermal Fields in Kyushu*, ed. Urashima, Y. Soc. of Mining Geologists of Japan, Guidebook 2, 27-38.

Van Leeuwen, Leach, T., Hawke, A. and Hawke, M. (1990) The Kelian disseminated gold deposit. East Kalimantan, Indonesia. *J. Geochem. Explor.* **35**, 1-61.

Vikre, P.G. (1985) Precious-metal vein systems in the National district, Humboldt County, Nevada. *Econ. Geol.* **80**, 360-393.

Wallace, A.B. (1987) Geology of the Sulphur District, Nevada. In *Bulk Mineable Precious Metal Deposits of the Western United States*, ed. Johnson, J.L. Guidebook for Field Trips, 165-171.

Weissberg, B.G. (1969) Gold-silver ore-grade precipitates from New Zealand thermal waters: *Econ. Geol.* **64**, 95-108.

Weissberg, B.G., Browne, P.R.L. and Seward, T.M. (1979) Ore metals in active geothermal systems. In *Geochemistry of Hydrothermal Ore Deposits*, (2nd edn.), ed. Barnes, H.L. Wiley Interscience, New York, 738-780.

White, D.E. (1955) Thermal springs and epithermal ore deposits. *Econ. Geol.*, 50th Anniversary Vol. 99-154.

White, D.E. (1967) Mercury and base metal deposits with associated thermal and mineral waters. In *Geochemistry of Hydrothermal Ore Deposits*, ed. Barnes, H.L. Holt, Rinehart and Winston, New York, 575-631.

White, D.E. (1981) Active geothermal systems and hydrothermal ore deposits. *Econ. Geol.*, 75th Anniversary Vol., 392-423.

White, N.C., Wood, D.G. and Lee, M.C. (1989) Epithermal sinters of Paleozoic age in North Queensland, Australia. *Geology* **17**, 718-722.

Wisser, E. (1960) Relation of ore deposition to doming in the North American Cordillera. Geol. Soc. of Amer., Memoir 77.

Wood, D.G., Porter, R.G. and White, N.C. (1990) Geological features of some epithermal occurrences in northeastern Queensland, Australia. *J. Geochem. Explor.* **36**, 413-443.

Wyborn, D. (1988) Ordovician magmatism, gold mineralisation and an integrated tectonic model for the ordovician and silurian history of the Lachlan foldbelt in New South Wales. Bureau of Mineral Resources, Canberra, Australia, *Res. Newsletter* **8**, 13-14.

6 Intrusion-related gold deposits

R.H. SILLITOE

6.1 Introduction

A broad spectrum of gold mineralization styles is found in the epizonal intrusive environment, and a number of these give rise to world-class gold deposits (Figure 6.1). Here the spectrum is subdivided into: (a) intrusion-hosted stockwork/disseminated deposits of both porphyry and non-porphyry types, the former possessing all the essential geological attributes (especially multidirectional stockworks) of typical porphyry copper and/or molybdenum deposits; (b) skarn and non-skarn replacement deposits in carbonate wallrocks; (c) stockwork, disseminated and replacement deposits in non-carbonate wallrocks; (d) wallrock-hosted breccia pipes; and (e) veins in both intrusions and wallrocks. This spectrum of deposits is transitional upwards to the volcanic-hosted, epithermal gold environment, which is mentioned only briefly in this chapter but discussed at length by Henley (this volume).

This chapter documents the variety of intrusion-related gold deposits, and considers their interrelationships and origin. Major, newly discovered deposits of current economic importance are emphasized throughout (Figure 6.1). Deposits in which gold possesses only by-product status are largely omitted even though the tonnage of contained gold may be large (e.g. Bingham porphyry copper–molybdenum deposit, Utah, USA).

6.2 Geotectonic settings

The majority (at least 85%) of the intrusion-related gold deposits used as examples in this review were generated at Phanerozoic convergent plate margins above zones of active subduction. The convergent margins ranged from primitive through mature island arcs to continental margins. Clearly the overall constitution and thickness of the lithosphere are not fundamental controls of intrusion-related gold deposits, although carbonate-hosted gold mineralization is more widespread in continental-margin arcs than in island arcs because of the restriction of shelf-carbonate sequences to the former (Sillitoe, 1989a). However, linear zones of crustal weakness that underwent repeated reactivation localized some of the deposits; Fortitude and McCoy-Cove (Tables 6.3 and 6.4) in the Battle Mountain-Eureka trend of Nevada, USA (Roberts, 1966), and Tai Parit (Table 6.4) in the Bau trend of Sarawak, East Malaysia (Sillitoe and Bonham, 1990), are typical examples.

In the western Pacific region, overriding plates and their volcano-plutonic arcs were commonly subjected to neutral or extensional stress regimes during mid- to

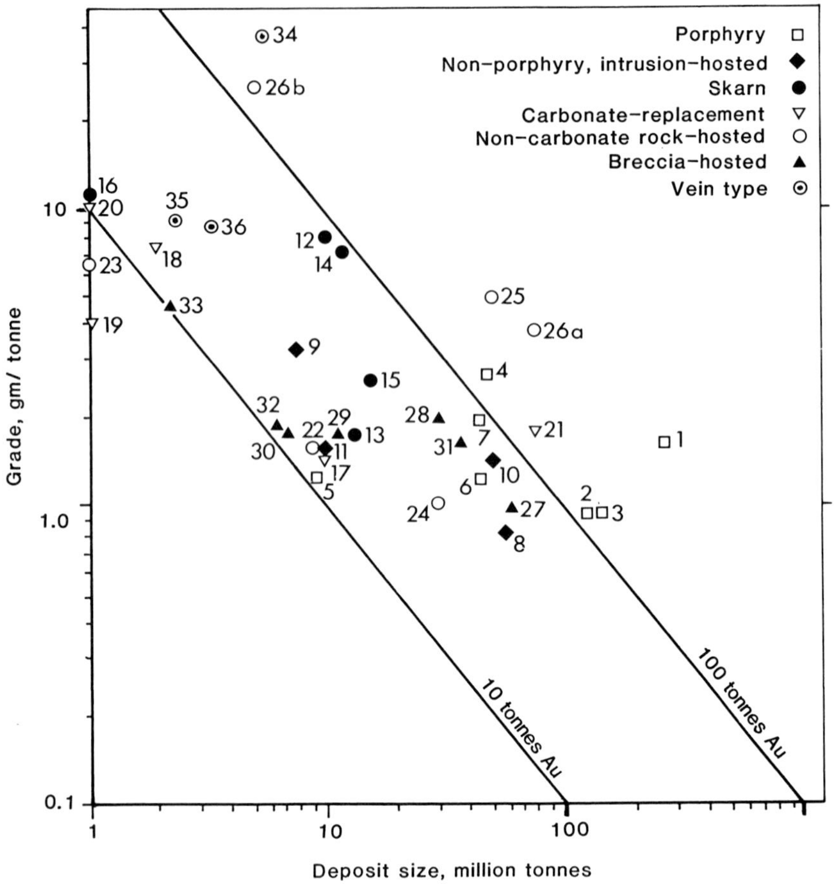

Figure 6.1 Grade–tonnage plot for representative gold deposits reviewed in this chapter. Data are taken from references listed in Tables 6.1 to 6.7 or from the mining press. The total gold tonnages in Tables 6.1 to 6.7 may be greater than those implied by this figure which for some deposits does not include previous production because the grades are unknown. Numbers: 1 Lepanto, 2 Santo Tomas II, 3 Dizon, 4 Ok Tedi, 5 Cuervo, 6 Marte, 7 Boddington, 8 Zortman-Landusky, 9 Salave, 10 Gilt Edge, 11 Kori Kollo, 12 Fortitude, 13 McCoy, 14 Nickel Plate, 15 Red Dome, 16 Thanksgiving, 17 Barney's Canyon, 18 Tai Parit, 19 Star Pointer, 20 Ketza River, 21 Cove, 22 Beal, 23 Quesnel River, 24 Equity Silver, 25 Mount Morgan, 26a Porgera (Waruwari), 26b Porgera (Zone VII), 27 Montana Tunnels, 28 Golden Sunlight, 29 Colosseum, 30 Ortiz, 31 Kidston, 32 Mount Leyshon, 33 Chadbourne, 34 Charters Towers, 35 Los Mantos de Punitaqui, 36 Masara.

late-Cainozoic intrusion and gold mineralization, but more rarely overthrusting (e.g. Ok Tedi, Papua New Guinea; Table 6.1) or strike-slip faulting (e.g. Masara, Philippines; Table 6.7) were prevalent (Sillitoe, 1989b). Extensional conditions were commonly accompanied by bimodal magmatism dominated by silicic products (Sillitoe, 1989a, b), and a comparable geotectonic setting may have characterized northern Queensland, Australia, in the Carboniferous when gold deposits (Kidston,

Red Dome, Mount Leyshon; Tables 6.3 and 6.6) were generated in association with rhyolitic to trachytic intrusions (Morrison, 1988). Alkaline magmatism also marks extensional situations, commonly in back-arc environments, as exemplified by the late Cretaceous to mid-Tertiary intrusions and gold deposits of Montana, South Dakota, Colorado, and New Mexico in the western USA (e.g. Giles, 1983; Mutschler *et al.*, 1985; DeWitt *et al.*, 1986).

Many convergent plate boundaries were constructed by accretion of numerous tectono-stratigraphic terranes, some including parts of volcano-plutonic arcs, which attained their present sites as the combined results of subduction and transform faulting. Such tectonically emplaced allochthonous terranes may contain intrusion-related gold deposits, including the Jurassic deposits at Nickel Plate (Table 6.3) and Quesnel River (Table 6.5) in British Columbia, Canada (McMillan *et al.*, 1987).

The geotectonic settings of several of the older gold deposits discussed below are more obscure. This is the case particularly for deposits like Boddington, Young-Davidson (Table 6.1) and Chadbourne (Table 6.6) that are components of Archaean greenstone belts. Nevertheless, some recent hypotheses draw analogies between the evolution of greenstone belts and convergent plate boundaries, and assign intrusion-related gold deposition to the final stages of subduction, possibly in a transpressional tectonic regime (e.g. Wyman and Kerrich, 1988). However, the Hercynian and Pan-African orogens that host the Salave, Spain (Table 6.2), and the Tiouit, Morocco (Table 6.7), and Navachab, Namibia (Table 6.3), intrusion-related gold deposits, respectively, are generally considered to be the products of convergence that resulted in continental collision. Magmatism and associated gold mineralization may have been triggered by crustal thickening consequent upon collisional tectonics as, for example, in the Damaran part of the Pan-African orogen in Namibia (Miller, 1983).

6.3 Intrusion-hosted stockwork/disseminated deposits

6.3.1 *Porphyry deposits*

Several major porphyry copper–gold deposits (Table 6.1) are (or shortly will be) exploited for their gold contents, with copper commonly constituting either a co- or by-product. However, the leached capping at Ok Tedi (Hewitt *et al.*, 1980) and part of the laterite profile at Boddington (Symons *et al.*, 1988) constitute gold-only deposits above sulphide-bearing copper–gold mineralization. On the basis of the geological characteristics of gold-rich porphyry copper systems, Sillitoe (1979, 1983a) predicted the existence of copper-deficient porphyry gold deposits in Phanerozoic volcano-plutonic arcs. These have now been discovered in Chile and are exemplified by the Marte deposit (Table 6.1; T. Vila, R.H. Sillitoe, J. Betzhold and E. Viteri, in prep.). Small porphyry gold deposits may also be present in some Archaean greenstone belts (Table 6.1; Franklin and Thorpe, 1982). However, most cited examples of porphyry gold mineralization (e.g. Lawrence, 1978; Schmidt, 1985) are not of porphyry type in the strict genetic sense defined below.

Gold-rich porphyry deposits, like those dominated by copper or molybdenum, are centred on stocks, which are commonly composite (Figure 6.2) and subvolcanic in character. The stocks tend to be rather basic in composition but commonly contain quartz and/or K-feldspar (Table 6.1). Gold ore constitutes steeply inclined bodies

Table 6.1 Characteristics of selected porphyry gold and copper-gold deposits

Deposit	Status	Contained Au, tonnes	Age, Ma	Progenitor intrusion	Associated mineralization	Host rocks	Metal association	Key hypogene alteration products	Data source(s)
Lepanto, Philippines	Feasibility stage (underground flotation)	440	3.5	Quartz porphyry stock	Acid-sulphate type enargite-Au deposit	Late Tertiary andesitic volcanics,	Cu–Au–(Mo)	Biotite, anhydrite, magnetie	Gonzalez (1956); R.H. Sillitoe (pers. observ.)
Santo Tomas II, Philippines	Mine: block caving, flotation	200	1.5	Diorite and quartz diorite porphyry stock	None	Mainly Late Cretaceous-Palaeogene andesitic volcanics	Cu–Au	Biotite, actinolite, anhydrite, magnetite	Serafica and Baluda(1977); Sillitoe and Gappe (1984)
Dizon, Philippines	Mine: open pit, flotation	130	2.7	Quartz diorite porphyry	None	Late Tertiary andesitic volcanics	Cu–Au–(Mo)	Sericite, clay, chlorite martitized magnetite	Sillitoe and Gappe (1984); Malihan (1987)
Ok Tedi, Papua New Guinea	Mine: open pit, flotation (formerly CIP)	368	1.2	Monzonite porphyry stock	Cu–Au skarn deposits	Late Cretaceous siltstone, Eocene-mid-Miocene limestone	Cu–Mo–Au	Biotite, K-feldspar	Bamford (1972); Howell *et al.* (1978); Hewitt *et al.* (1980)

Cuervo (Sullivan), Nevada, USA	Feasibility stage	12.4	Jurassic (?)	Microdiorite sill	None	Triassic metavolcanics	Cu–Au	Clay	J.F.H. Thompson (pers. comm. 1989); H.F. Bonham, Jr. (pers. comm. 1989)
Marte, Chile	Mine: open pit, heap leach	53.5	13.3	Quartz diorite porphyry stock	None	Miocene andesitic volcanics	Au–(Cu)	Sericite, clay, chlorite, gypsum (after anhydrite), hematite (partly after magnetite)	T. Vila, R.H. Sillitoe J. Betzhold, and E. Viteri (in prep)
Young-Davidson, Ontario, Canada	Abandoned mines, under exploration	19	*ca.*2700	Syenite stock	Au veins, porphyry Cu–Mo–Au protore	Archaean metavolcanics and metasediments	Au–(Cu–W)	K-feldspar, hematite, tourmaline	Sinclair (1982)
Boddington, Western Australia	Mine: open pit, heap leach	120	2650-2670	Quartz microdiorite bodies	None	Archaean andesitic lava	Au–Cu–Mo–W	Biotite, K-feldspar, actinolite, epidote	Symons *et al.* (1988)

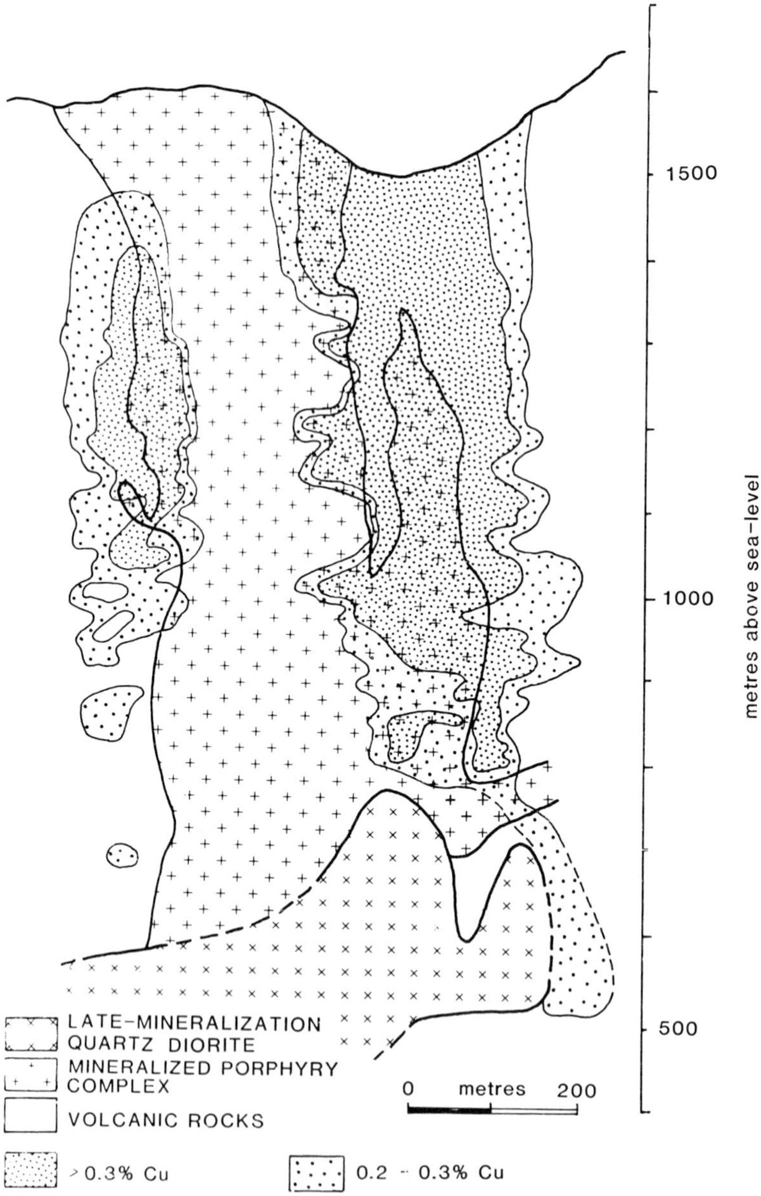

Figure 6.2 Cross-section of the Santo Tomas II porphyry copper–gold deposit, Philippines, after Sillitoe and Gappe (1984). The annular form of the copper–gold ore, approximated by the copper isopleths, is due to the lower grade mineralization present in inter-mineralization porphyry intrusions occupying the core of the cylindrical stock. The stock, K-silicate alteration, and orebody were truncated about 800 m below the present surface by a propylitized and poorly mineralized, late-mineralization, equigranular quartz diorite.

(Figure 6.2) of quartz stockwork and subordinate disseminated mineralization, which at some deposits contain low-grade cores attributable to either inter-mineralization intrusions (Figure 6.2; Serafica and Baluda, 1977; Sillitoe and Gappe, 1984) or bodies of replacement quartz (Hewitt *et al.*, 1980).

Gold was generally introduced as a component of K-silicate alteration in which biotite may be accompanied by K-feldspar and/or actinolite (Table 6.1; Bamford, 1972; Sillitoe and Gappe, 1984; Symons *et al.*, 1988). Cuervo, with argillic alteration reportedly dominant (Table 6.1), is an exception. Anhydrite and its supergene hydration product, gypsum, are commonplace. Gold in hypogene porphyry copper–gold deposits is commonly present in pyrite-poor, chalcopyrite–bornite assemblages, whereas in porphyry gold deposits pyrite is the only significant sulphide (Table 6.1). Magnetite, in part transformed to hypogene hematite, is a common but not ubiquitous associate of the sulphides, in both veinlet and disseminated form (Sillitoe, 1979). At least 8 vol.% iron oxides are present at Santo Tomas II (Sillitoe and Gappe, 1984) and Marte (T. Vila, R.H. Sillitoe, J. Betzhold and E. Viteri, in prep.). The presence of hypogene iron oxides and anhydrite testifies to the highly oxidized condition of the ore fluids.

In porphyry copper–gold deposits, the contents of the two contained metals typically display a positive correlation, a relationship that supports their co-transport and co-precipitation and argues strongly against the introduction of significant amounts of gold later than the K-silicate–copper association. The high-fineness gold is closely related to chalcopyrite and bornite, especially the latter where present, in porphyry copper–gold deposits, but occurs mainly with quartz or pyrite in the porphyry gold deposits.

Gold-bearing K-silicate alteration was overprinted in several shallowly eroded deposits, such as Dizon and Marte, by a sericite–clay–chlorite assemblage, which seems to have introduced some of the pyrite, martitized some of the magnetite, but induced little modification of gold and copper grades (Sillitoe and Gappe, 1984; T. Vila, R.H. Sillitoe, J. Betzhold and E. Viteri, in prep.). It is not clear, however, if the clay at Cuervo was developed at the expense of earlier K-silicate alteration.

These shallowly eroded deposits may also be partly overlain and/or flanked by erosional remnants of extensive zones of volcanic-hosted advanced argillic alteration dominated by the quartz–alunite assemblage and containing bodies of chalcedonic silicification. These zones contain ~10 vol. % pyrite along with high-sulphidation minerals such as enargite and native sulphur. At Lepanto, a major enargite-gold orebody has been mined from a body of massive pyrite and chalcedony which is located at higher elevations than, but marginal to, the porphyry copper–gold deposit (Gonzalez, 1956; Sillitoe, 1983b). It is an acid-sulphate or high-sulphidation-type epithermal deposit in the sense of Bonham (1986) and Heald *et al.* (1987) (see Henley, this volume).

6.3.2 *Other intrusion-hosted stockwork/disseminated deposits*

Several disparate stockwork or disseminated gold deposits are hosted, at least in large part, by stocks (Table 6.2), but none of them is of porphyry-type *sensu stricto* because of the absence of multidirectional stockworks and associated K-silicate alteration. The category includes shallow, low-sulphidation, adularia–sericite-type epithermal

Table 6.2 Characteristics of selected non-porphyry-type stockwork and disseminated gold deposits hosted by intrusions

Deposit	Status	Contained Au, tonnes	Age, Ma	Progenitor intrusion	Associated mineralization	Host rocks	Metal association	Key hypogene alteration minerals	Data source(s)
Zortman-Landusky, Montana, USA	Mine: open pit, heap leach	48	62.6	Syenite and quartz latite porphyry stock cut by trachyte porphyry dykes	Carbonate-replacement Au deposits	Precambrian gneiss, Phanerozoic sediments	Au–Ag–Te–As –(Cu–Mo–Pb–Zn–Bi)	Sericite, kaolinite, orthoclase, fluorite	Rogers and Enders (1982); Hastings (1987, 1988)
Salave, Spain	Trial mining: open pit, heap leach	30	285	Granodiorite stock cut by granodiorite dykes	None	Cambro-Ordovician quartzite and slate	Au–As–Mo–Sb–(Zn)	Sericite, albite, dolomite, calcite	Harris (1980a, b)
Gilt Edge, South Dakota, USA	Mine: open pit, heap leach	87	50–60	Trachyte porphyry stocks	Porphyry Mo protore (in breccia clasts)	Early Protero-zoic metasediments and amphi-bolite, Cam-brian sandstone	Au–Cu–As–Mo –(Pb–Zn–Te)	Sericite, clay, orthoclase, fluorite	Macleod (1987); R.H.Sillitoe (pers. observ., 1980)
Kori Kollo, Bolivia	Mine: open pit, heap leach	15	15.7	Rhyodacite porphyry stock	Veins with Cu–Au–W–Pb–Zn–Ag–Bi–Sb–As	Siluro-Devonian siltstone	Au–Ag–Cu	Sericite	Redwood (1987); Anzoleaga (1988)

deposits (Bonham, 1986; Heald *et al.*, 1987; Henley, this volume) through to deeper-level types.

The Kori Kollo deposit in the La Joya district would be considered epithermal by many investigators, although its stockwork gold–copper mineralization, pervasive sericitic alteration and, in the nearby La Joya stock, the presence of tourmaline combine to suggest a porphyry-type deposit. The rhyodacite porphyry stocks in the La Joya district are locally flow-banded and, in accordance with Redwood's (1987) interpretation, are considered as the roots of lava domes. Such shallow intrusion offers a ready explanation for the combination of epithermal and porphyry-type characteristics. Hastings (1987, 1988) likewise designated the Zortman-Landusky deposits as epithermal because of the vein textures and Hg–Te–As trace-element signature, but emphasized their intimate intrusive connection. The Gilt Edge deposit was also considered to be of epithermal type by Paterson *et al.* (1987).

In contrast, Salave creates the overall impression of being a deeper-seated gold deposit. The dominantly disseminated form of the arsenopyrite-rich ore and its association with pervasive sericite–albite–carbonate alteration (Harris, 1980a, b) suggest that the deposit may be considered as a greisen.

The stocks that host these stockwork and disseminated gold deposits possess a wider compositional range than those that carry the porphyry-type gold deposits (Tables 6.1 and 6.2). The geometries of the ore zones appear to be more varied than those in the porphyry deposits, with structural localization playing an important role at Zortman-Landusky (Figure 6.3; Rogers and Enders, 1982) and Gilt Edge

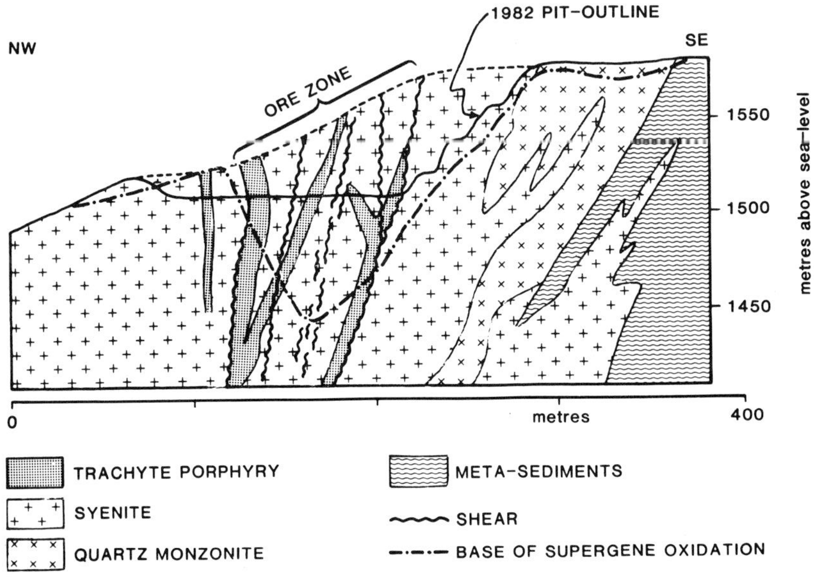

Figure 6.3 Cross-section of the August orebody in the non-porphyry-type intrusion-hosted Zortman-Landusky gold–silver deposit, Montana, USA, after Enders and Rogers (1983) with modifications from Hastings (1988). The orebody is centred on a steeply dipping zone of shearing and trachyte porphyry dykes which, due to enhanced permeability, gave rise to deeper supergene oxidation – a prerequisite for the development of gold ore.

Table 6.3 Characteristics of selected skarn gold deposits

Deposit	Status	Contained Au, tonnes	Age, Ma	Ore-related intrusion	Associated mineralization	Host rocks	Metal Association	Key hypogene alteration products	Data source(s)
Fortitude, Nevada, USA	Mine: open pit, CIP	96	37.2	Porphyritic granodiorite stock and dykes	Porphyry Cu–Mo protore, Cu–Au skarns, Pb–Zn veins	Carboniferous-Permian limestone and calcareous siltstone	Au–Ag–Cu–Zn–Pb–As–(Bi-Te)	Hedenbergite–andradite, actinolite–chlorite–calcite-prehnite–K-feldspar	Wotruba *et al.* (1988); Myers and Meinert (1989)
McCoy, Nevada, USA	Mine: open pit, heap leach	30	39.7	Porphyritic granodiorite stock and dykes	Minor fracture-controlled Au in stock, Cove Au–Ag deposit (Table 6.4)	Triassic limestone	Au–Ag–(Cu)	Garnet, diopside, calcite, orthoclase, chlorite	Kuyper (1988)
Nickel Plate (Hedley), British Columbia, Canada	Mine: open pit, CIP	83.5	Early Jurassic (>200<225)	Diorite and quartz diorite porphyry sills and dykes	None	Triassic calcareous and tuffaceous siltstone	Au–Cu–As–Zn–(Co–Ni–Bi–Mo–Te)	Clinopyroxene-garnet, epidote-chlorite-–clinozoisite–prehnite	Ray *et al.* (1988)

Red Dome, Queensland, Australia	Mine: open pit, heap leach	39	Carboniferous	Rhyolite porphyry dykes	Porphyry Mo protore, Cu–Pb–Ag skarn deposits	Silurian limestone	Au–Cu–As–Zn–(W–Sn–Bi–Te)	Hedenbergite, andradite, wollastonite-andradite-ferrobustamite, chlorite-calcite	Torrey *et al.* (1986); Ewers and Sun (1988)
Thanksgiving, Philippines	Mine: underground, flotation	13	5.5	Diorite porphyry stock and dykes	Kennon porphyry Cu–Au deposit, Au veins	Miocene limestone	Au–Ag–Zn–(Pb–Cu–As–Te)	Garnet, clinozoisite, chlorite, calcite	Callow (1967); Tangonan(1987)
Suan district, North Korea	Mines: underground, flotation	>100	Jurassic	Granite pluton	Au veins	Late Proterozoic-Cambrian limestone and dolomite	Cu–Au–(Zn–Pb–As–Bi)	Garnet-diopside, actinolite, phlogopite, ludwigite, chlorite, talc, tremolite, wollastonite	Watanabe (1943); R.H.Sillitoe (unpub. observ., 1987)
Navachab, Namibia	Mine development: open pit, heap leach	22	Cambro-Ordovician	Leucogranite dykes	None	Late Proterozoic dolomitic marble	Au–(Cu–Pb–Zn–W–Bi–Te–As–Mo)	Garnet-diopside, biotite-tremolite	R.H.Sillitoe (unpub. observ., 1986, 1987)

(MacLeod, 1987), and ponding of ore fluids beneath relatively impermeable wallrock quartzites and slates appearing to have been influential at Salave. Furthermore, at Zortman-Landusky, a transition from deeper narrow veins to shallow, wide, low-grade stockworks, breccias and intensely fractured 'sheeted' zones is observed (Hastings, 1987, 1988).

All the deposits (Table 6.2) underwent gold-introduction along with sericitic ±carbonate alteration. Gold is accompanied by a more varied suite of metals (Table 6.2) than in the porphyry-type deposits, although copper and/or molybdenum are still present in minor quantities in all deposits. Gold is present at least partly in the native state, although gold-bearing tellurides are dominant at Zortman-Landusky (Rogers and Enders, 1982). Appreciable silver contents also characterize Zortman-Landusky and Kori Kollo.

6.4 Deposits in carbonate rocks

6.4.1 *Skarn deposits*

Skarn deposits in which gold occurs as the principal commodity or as an important by-product are widespread (Table 6.3; Orris *et al.*, 1987). Many of these deposits may be assigned to the iron, copper, lead-zinc or porphyry copper skarn types defined by Einaudi *et al.* (1981), whereas others constitute a separate and, as yet, poorly characterized gold skarn category (Table 6.3; Orris *et al.*, 1987; Meinert, 1988). However, the divisions between the gold-bearing skarn types are not clearcut geologically, and some deposits are difficult to categorize. For example, the Thanksgiving deposit (Table 6.3) is porphyry copper-related but classified as a gold-rich zinc skarn by Orris *et al.* (1987).

The gold skarns may be related genetically to either unmineralized intrusions with porphyritic or equigranular textures, or linked to porphyry stocks carrying mineralization of porphyry copper (Fortitude, Thanksgiving) or porphyry molybdenum (Red Dome) type (Table 6.3). The skarns may also be subdivided into proximal, where they abut the progenitor intrusion (McCoy, Red Dome, Thanksgiving and Suan), or distal, where they are observed or inferred to occur at distances of several hundred metres or more from stock contacts (Fortitude, and probably Nickel Plate and Navachab).

Most auriferous skarn is present in the exoskarn environment, although endoskarn also constitutes ore locally at some deposits (e.g. Red Dome; Torrey *et al.*, 1986). The gold-rich exoskarns are normally calcic and developed from limestone and/or calcareous siltstone, but magnesian skarns developed from dolomite dominate at Suan (Watanabe, 1943).

Gold skarns, in common with other skarn types, were localized invariably by a combination of lithological and structural factors. At most large deposits, such as Fortitude (Figure 6.4; Wotruba *et al.*, 1988), Suan (Watanabe, 1943) and McCoy (Kuyper, 1988), auriferous skarn developed as stratabound mantos in favourable carbonate horizons. Steep, pre-mineral faults that united progenitor intrusions and sites of mineralization acted as feeders for the mineralizing fluids at Fortitude (Figure 6.4; Wotruba *et al.*, 1988) and McCoy (Kuyper, 1988). Smaller skarn bodies tend to form as discontinuous rinds along intrusive contacts, as at Red Dome, Thanksgiving and, at depth, McCoy. Even in large auriferous skarns, however, local features, such

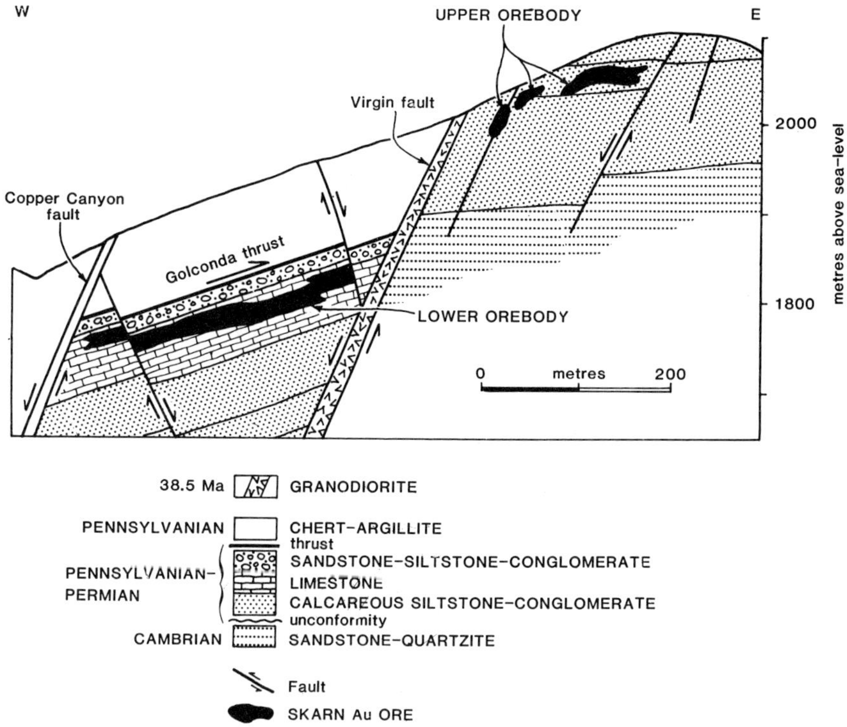

Figure 6.4 Cross-section of the Fortitude skarn gold deposit, Nevada, USA, after Wotruba *et al.* (1988). The principal conduit for mineralizing fluids was the pre-mineralization Virgin fault which was intruded by a granodiorite dyke, an offshoot of the Copper Canyon stock some 800 m farther south. The dyke, in common with the stock, underwent K-silicate alteration. Prograde skarn replaced limestone in the lower orebody and calcareous siltstone and conglomerate in the upper orebody.

as the sills, dykes and small-scale fold hinges at Nickel Plate (Ray *et al.*, 1988), acted as cogent ore controls.

Gold skarns, like those carrying other metals, developed in several stages, which may be subdivided into early, anhydrous, prograde and late, hydrous, retrograde events (Einaudi *et al.*, 1981). Prograde skarn, typically comprising Fe-rich garnet and pyroxene (Table 6.3), originally may have been widespread (6 km^2 at surface in the Nickel Plate area) or restricted to small pods, patches and veins (e.g. Navachab, Thanksgiving). Retrograde events may have almost completely destroyed prograde skarn, but commonly acted only weakly, as at Fortitude (Myers and Meinert, 1989) and Nickel Plate (Ray *et al.*, 1988).

Gold may be precipitated during both prograde and retrograde stages of skarn formation. Proximal skarns commonly display an association between gold and retrograde events (e.g. Red Dome; Torrey *et al.*, 1986), whereas gold in several major distal skarns was introduced during prograde alteration (e.g. Fortitude – Myers and Meinert, 1989; Nickel Plate – Ray *et al.*, 1988). Auriferous skarn is rich in sulphides, principally pyrite and/or pyrrhotite, and, locally, also in magnetite; total sulphide

Table 6.4 Characteristics of selected carbonate-replacement gold deposits

Deposit	Status	Contained Au, tonnes	Age, Ma	Ore-related intrusion	Associated mineralization	Host rocks	Metal association	Key hypogene alteration products	Data source(s)
Barney's Canyon, Utah, USA	Mine: open pit, heap leach	14	38(?)	Monzonite porphyry stock	Bingham porphyry Cu–Mo–Au, skarn Cu–Au + carbonate-replacement Pb–Zn–Ag–(Au) deposits	Carboniferous dolomite	Au–As–Sb	Sanding, minor jasperoid	H.F.Bonham, Jr. (pers. comm., 1988); J.F.H. Thompson (pers. comm; 1989)
Tai Parit, Bau district, Sarawak, East Malaysia	Abandoned open-pit mine	15.6	Mid-late Tertiary	Dacite porphyry stocks and dykes	Porphyry Cu–Mo–Au protore and other sediment-hosted Au deposits	Late Jurassic–Cretaceous limestone and calcareous siltstone	Au–As–Sb	Jasperoid	Geikie (1906); Wolfenden (1965); R.H.Sillitoe (unpub. obs., 1982-1988)
Purísima Concepción, Yauricocha district, Peru	Mine: open pit and underground CIP	Private data	7	Granodiorite to quartz monzonite stock	Porphyry Cu–Mo protore, enargite veins and polymetallic carbonate-replacement deposits	Late Cretaceous impure limestone	Au–As–Sb–Ag–Mn–Tl–Te	Silicification, decalcification, sericite, rhodochrosite	Alvarez and Noble (1988)

Kendall, Montana, USA	Mine: open pit, heap leach	22		66	Syenite porphyry laccolith, sills and dykes	Syenite-hosted breccia with minor Au–Ag–Pb– Cu–Mo	Carboniferous faulted or solution-brecciated limestone	Au–(Pb–Zn–Cu–As–Sb – Hg–Tl)	Sanding, silicification	Lindsey (1985); Linsdsey and Naeser (1985); R.W.Schafer (pers. comm., 1989)
Foley Ridge, South Dakota, USA	Mine: open pit, heap leach	42		50–60	Monzonite sills	Disseminated Au in sills	Cambrian calcareous sandstone and siltstone	Au–As–(Pb–Zn–Te)	Silicification, decalcification, fluorite, barite	Taylor and Bolin (1988); DeWitt *et al.* (1986)
Star Pointer, Robinson district, Nevada, USA	Mine: open pit, CIP	5		109–111	Quartz monzonite porphyry stocks	Ely porphyry Cu deposit and Cu skarns	Permian calcareous sandstone	Au– (Pb–Zn–As–Te–Sb–Tl)	Silicification, decalcification, kaolinite	Smith *et al.* (1988), Durgin (1989)
Ketza River, Yukon, Canada	Mine: underground, CIP	12.5		Mid-Creta-ceous (?)	Concealed stock (hornfels)	Au–bearing veins	Early Cambrian limestone and late Proterozoic argillite	Au–As–Cu– (Pb–Zn–Bi)	Silicification, sericite	Abbott (1986)
Cove, Nevada, USA	Mine: open pit, heap leach (CIP planned)	149		39.5	Granodiorite or quartz monzonite porphyry stock, dykes and sills	McCoy Au skarn (Table 6.3)	Triassic limestone and clastic rocks	Au–Ag–Zn–Pb–Cu–Sn–As	Silicification, jasperoid, sericite, clays	Emmons and Coyle (1988)

contents commonly exceed 10 vol.% and, as in parts of the Fortitude orebody, even 50 vol.% (Wotruba *et al.*, 1988).

Gold in retrograde skarns is usually associated closely with the sulphides, in which copper or, less commonly, zinc occur as the main base metals (Table 6.3). As emphasized recently by Meinert (1988), many, but apparently not all, gold-rich skarns are enriched in arsenic as arsenopyrite, bismuth as the native metal, bismuthinite, hedleyite and other minerals, and tellurium (Table 6.3). At Thanksgiving, tellurides are the principal gold-bearing minerals (Callow, 1967).

6.4.2 *Carbonate-replacement deposits*

Intrusion-related, carbonate-replacement deposits are well-known sources of lead, zinc and silver (e.g. Titley and Megaw, 1985), but are less widely recognized as receptacles for gold. Carbonate-replacement deposits generally occur distally beyond the skarn front, but in some districts skarn is either poorly developed and devoid of significant metallization (Bau, Yauricocha) or absent entirely (Kendall, Ketza River, Foley Ridge).

As reflected by Table 6.4, the carbonate-replacement category is amenable to subdivision into deposits which, like the gold skarns, contain gold in combination with at least small quantities of the common base metals (Star Pointer, Cove, Ketza River) and deposits characterized by a gold–arsenic–(antimony) association (Barney's Canyon, Tai Parit, Purísima Concepción, Foley Ridge, Kendall). Foley Ridge and Kendall do, however, also contain anomalous amounts of base metals. Cove contains tin (Emmons and Coyle, 1988).

Most deposits may be related readily to exposed intrusions, although the Ketza River deposit is only inferred on the basis of hornfelsing to overlie a concealed stock (Abbott, 1986). The stock that is linked to the Barney's Canyon deposit is the focus of the Bingham district, with its porphyry Cu–Mo–Au, skarn Cu–Au and carbonate-replacement Pb–Zn–Ag–(Au) deposits. Barney's Canyon is on the northern fringe of the district, some 8 km from the stock. However, a genetic connection is supported by the presence of similar, structurally controlled gold-arsenic mineralization as an overprint to the other deposit types in the Bingham district, and by the fact that at least some of this gold mineralization pre-dated end-stage molybdenum introduction in the skarn environment (Cameron and Garmoe, 1987). Porphyry-type mineralization, with or without skarn and/or carbonate-replacement deposits, is also present with intrusions related genetically to the Tai Parit (Sillitoe and Bonham, 1990), Purísima Concepción (Alvarez and Noble, 1988) and Star Pointer (Smith *et al.*, 1988) gold deposits, and the Cove gold-silver deposit is only 1.5 km from, and apparently related to, the McCoy gold-skarn deposit (Emmons and Coyle, 1988).

As in the case of the skarn gold deposits, the carbonate-replacement deposits were localized by combinations of lithological and structural features, with steep faults (e.g. Tai Parit; Figure 6.5), steep fractures (e.g. Foley Ridge; DeWitt *et al.*, 1986), lithological contacts (e.g. Ketza River, Tai Parit; Figure 6.5) and regionally extensive aquifers (e.g. Kendall; Giles, 1983) playing prominent roles.

Gold and associated sulphides in the carbonate-replacement deposits are present in a variety of forms. They may be disseminated and fracture-controlled, as in all the gold–arsenic–(antimony) deposits, but locally, as at Ketza River and Cove, more pervasive replacement has taken place. Silicification is commonplace, and jasperoid

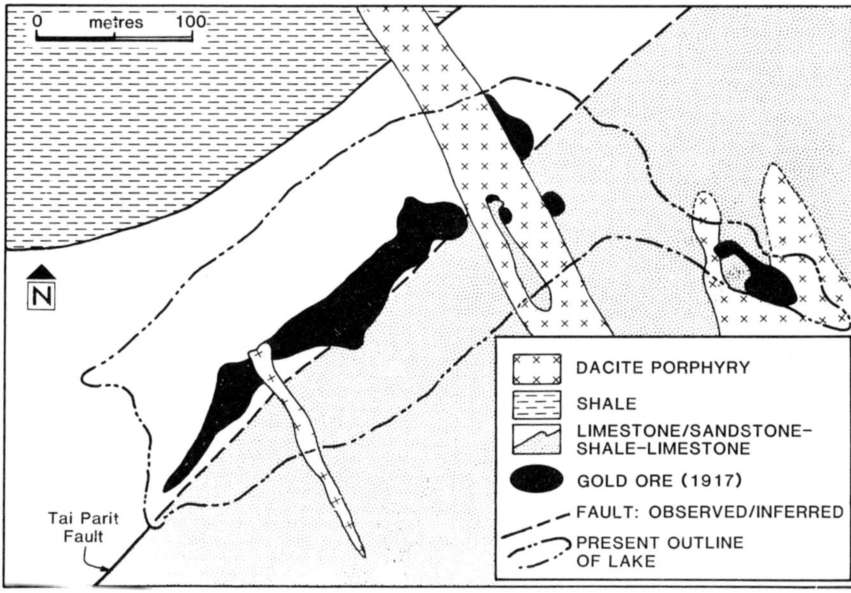

Figure 6.5 Surface map of the Tai Parit carbonate-replacement gold deposit, Sarawak, East Malaysia, after Wolfenden (1965). Gold ore was generated along the Tai Parit normal fault and dacite porphyry dyke-contacts directly beneath a cover of flat-lying shale, which is preserved immediately north-west of the orebody. The dykes emanate from a cluster of porphyry stocks 600 m or more farther south. The ornamental lake coincides roughly with the outline of the old open-pit.

was formed in parts of the Star Pointer (Smith *et al.*, 1988), Cove (Emmons and Coyle, 1988) and Tai Parit deposits. At Tai Parit, the jasperoid is extensively hydrothermally brecciated (Geikie, 1906; Wolfenden, 1965). However, sanding (matrix removal) of the mineralized carbonate rocks was the main alteration feature at Barney's Canyon. At Cove, part of the deeper orebody comprises sulphide stockworks in clastic sedimentary rocks rather than mineralized carbonates (Emmons and Coyle, 1988), thereby making this ore type similar to the gold deposits hosted by non-carbonate rocks. The pipe and manto configuration described for gold ore at Ketza River (Abbott, 1986) is reminiscent of the geometry of many carbonate-replacement lead–zinc–silver deposits.

The overall features of the gold–arsenic–(antimony) deposits, including the paucity of base metals and the frequent occurrence of arsenopyrite, realgar, orpiment and stibnite (or their supergene oxidation products) as accompaniments to decalcification and various degrees of silicification and argillization of the carbonate host rocks led Giles (1983), Lindsey (1985), Alvarez and Noble (1988) and Sillitoe and Bonham (1990) to compare one or more of the deposits summarized in Table 6.4 with sediment-hosted (Carlin-type) gold deposits.

The carbonate-replacement deposit characterized by the gold–base metal association at Cove, in common with three of the gold–arsenic–(antimony) deposits, formed alongside the skarn environment. However, at Star Pointer, reconstruction of district geology reveals that gold mineralization took place at shallower depths than those conducive to skarn development, in an environment where silica-pyrite formed

Table 6.5 Characteristics of selected stockwork, disseminated and replacement gold deposits in non-carbonate rocks

Deposit	Status	Contained Au, tonnes	Age, Ma	Ore-related intrusion	Associated mineralization	Host rocks	Metal association	Key hypogene alteration minerals	Data source(s)
Porgera, Papua New Guinea	Mine development (open pit/ underground, CIP)	420	7–10	Diorite stocks, sills and dykes	Au–bearing stockwork and breccia in some stocks	Late Cretaceous silty shale	Au–Ag–Zn–Pb–As–Te–(Hg)	Sericite-dolomite, roscoelite-quartz, anhydrite	Fleming *et al.* (1986); Handley and Bradshaw (1986); Henry (1988)
Beal, Montana, USA	Mine: open pit, heap leach	13.9	71.8	Diorite stock and dykes, granodiorite pluton	Diorite similarly mineralized	Late Cretaceous sandstone, mudstone and conglomerate: all hornfelsed	Au–Cu–(As–Mo–Zn–Bi–Te)	Silicification, chlorite, sericite, carbonate, clays	Hastings and Harrold (1988)
Muruntau, Uzbekistan, USSR	Mine: open pit, resin in pulp	1000–2500	Late Palaeozoic	Peraluminous syenite and granite dykes. (Buried pluton inferred)	None (?)	Late Proterozoic (?) shale and siltstone	Au–As–W–Cu–Pb–Zn–Sb–Bi–Ag	Silicification, alkali feldspar, chlorite, tourmaline, biotite	Krason (1984); Strishkov (1986); Bloomstein (1987)

Andacollo, Chile	Small-scale mining	33	Early Cretaceous	Granodiorite porphyry stock	Porphyry Cu–Mo–Au deposit and vein Au	Early Cretaceous felsic volcanics	Au– (Cu–Zn –As)	Silicification, alkali feldspar, chlorite	Llaumett (1980); Sillitoe (1983c)
Quesnel River (QR), British Columbia, Canada	Feasibility stage	12.5	201	Diorite-monzodiorite stock and porphyry dykes and sills	Porphyry Cu protore	Late Triassic–early Jurassic basaltic lava, tuff and breccia	Au–Cu–Zn–Pb–As– (Mo –Sb)	Epidote, chlorite, calcite, tremolite	Fox *et al.* (1987); Melling and Watkinson (1987)
Equity Silver, British Columbia, Canada	Mine: open pit, flotation	27	58–59	Quartz monzonite stock	Porphyry Cu–Mo protore	Cretaceous felsic volcanics	Au–Ag–Cu–Sb–As– (Zn –Pb–Mo–W –Bi)	Sericite, andalusite pyrophyllite, tourmaline, scorzalite, corundum, dumortierite, chlorite	Cyr *et al.* (1984); Wojdak and Sinclair (1984)
Mount Morgan, Queensland, Australia	Abandoned mine (open pit, flotation)	280	362	Tonalite pluton	Several small breccia pipes with minor Cu–Au	Middle Devonian felsic volcanics	Au–Cu–(Zn–Bi–Te)	Quartz-pyrite, sericite, chlorite	Cornelius (1969); Taube (1986); Arnold and Sillitoe (1989)

at the expense of carbonate rocks and sericitic and/or advanced argillic rather than K-silicate alteration affected contiguous intrusive rocks (James, 1976; Einaudi, 1982; Durgin, 1989).

6.5 Stockwork, disseminated and replacement deposits in non-carbonate rocks

Fewer intrusion-related stockwork, disseminated or replacement gold deposits are found in non-carbonate than in carbonate rocks, which is presumably a reflection of the less-reactive nature of the former. However, the overall mineralization styles and metal associations (Table 6.5) resemble those characteristic of skarn and carbonate-replacement gold deposits.

The gold deposits in non-carbonate rocks may be in contact locally with the related intrusions as at Mount Morgan (Figure 6.6) and Porgera, or may occur up to several hundred metres away from them (Andacollo, Quesnel River, Equity Silver). Dykes associated with the Muruntau deposit are inferred to overlie a larger, concealed pluton (Krason, 1984). The intrusions carry porphyry copper-type mineralization at Andacollo (Llaumett, 1980), Quesnel River (Fox *et al.*, 1987) and Equity Silver (Cyr *et al.*, 1984), whereas the dioritic stocks at Porgera and Beal are mineralized like, although to a lesser extent than, the sedimentary wallrocks (Fleming *et al.*, 1986; Hastings and Harrold, 1988).

As in the case of gold deposits hosted by carbonate rocks, fault and fold structures and rock permeability and/or reactivity are obvious ore controls. Gold ore is distributed parallel to bedding as grossly manto-like bodies at Andacollo (Llaumett, 1980), Quesnel River (Fox *et al.*, 1987) and Equity Silver (Cyr *et al.*, 1984), whereas control by pre-mineral structures dominates at Porgera (Henry, 1988) and Beal (Hastings and Harrold, 1988). The fault-localized, late-stage Zone VII at Porgera carries very high-grade ore (40 g/t Au) and is epithermal in character (Handley and Bradshaw, 1986; Henry, 1988). In contrast, the annular fractures that delimit the pipe-like Mount Morgan orebody are believed to have been generated as a consequence of alteration and mineralization (Figure 6.6; Arnold and Sillitoe, 1989).

Mineralization is mainly disseminated in form at Beal, Andacollo, Quesnel River and Equity Silver but is largely veinlet-controlled at Porgera and Muruntau. Massive to semi-massive sulphide intergrown with quartz constitutes the major replacement body at Mount Morgan (Figure 6.6; Cornelius, 1969), and small massive sulphide lenses of replacement origin are also reported at Equity Silver (Cyr *et al.*, 1984) and Quesnel River (Fox *et al.*, 1987). It is interesting to note that the two deposits hosted by felsic volcanic rocks – Equity Silver and Mount Morgan – have both been considered as volcanogenic deposits of overall Kuroko type (Ney *et al.*, 1972; Taube, 1986). Rock reactivity was apparently augmented at Porgera by the weakly calcareous nature of some of the shales (Fleming *et al.*, 1986), and at Quesnel River by the presence of pre-mineralization calcite and pyrite in the basaltic host rocks (Fox *et al.*, 1987). Small lenses of sulphide- and magnetite-bearing retrograde skarn after limestone occur alongside and beneath the Mount Morgan orebody (Arnold and Sillitoe, 1989), and traces of skarn are also reported at Porgera (Fleming *et al.*, 1986).

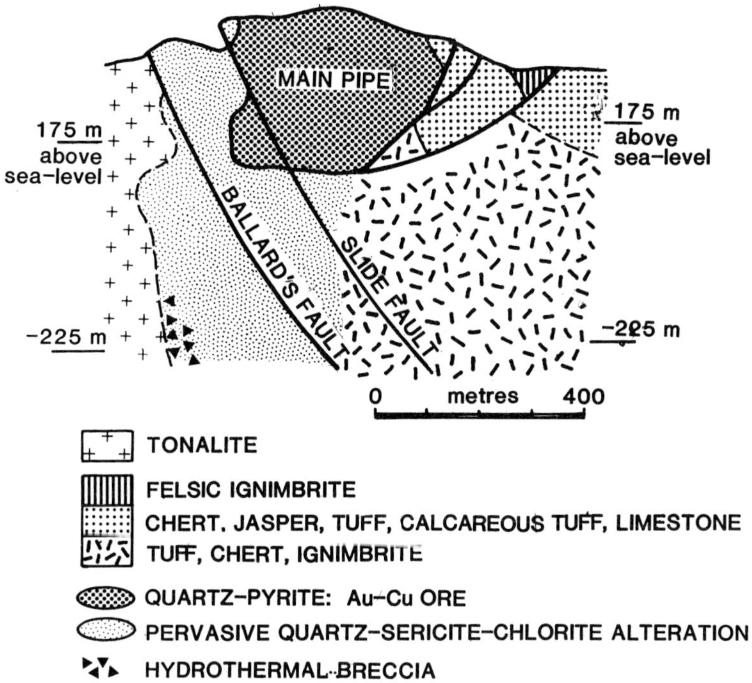

Figure 6.6 Cross-section of the Mount Morgan gold–copper replacement deposit in predominantly non-carbonate rocks, Queensland, Australia, after Arnold and Sillitoe (1989). Post-mineral displacement on the Slide-Ballard's fault system has been restored. Texture-destructive alteration along the pluton contact extended for at least 280 m beneath the orebody, and locally also affected the tonalite itself. The footwall of the Main Pipe (the principal orebody component) is delimited by an annular array of listric faults, which define a down-dropped block largely replaced by quartz–pyrite. Collapse of the ore-bearing block is attributed to volume reduction consequent upon rock replacement. The orebody has been mined out and the open pit is largely filled by water containing cyanide derived from a major tailings retreatment operation.

The introduction of ore-related hydrothermal quartz, sericite and/or chlorite was characteristic of most deposits in this category (Table 6.5). The obvious exceptions are Equity Silver and Quesnel River, where gold deposition was accompanied by aluminium- and boron-rich advanced argillic alteration (Table 6.5; Cyr *et al.*, 1984; Wojdak and Sinclair, 1984) and propylitization (Table 6.5; Fox *et al.*, 1987; Melling and Watkinson, 1988), respectively. These deposits in largely non-carbonate rocks are characterized by the same general gold–base metal association evident in most of the deposits in carbonate host rocks. Arsenic as arsenopyrite is a prominent component and is related intimately to gold at Equity Silver (Cyr *et al.*, 1984). Beal (Hastings and Harrold, 1988), Muruntau (Bloomstein, 1987), and Mount Morgan (Cornelius, 1969) also display the late-stage gold–bismuth–tellurium association, which is so widespread in carbonate-hosted gold deposits (Table 6.3), and bismuth is also reported from Equity Silver (Cyr *et al.*, 1984) and tellurium from Porgera (Handley and Bradshaw, 1986).

6.6 Breccia-hosted deposits

Although hydrothermal breccias are common components of intrusion-related gold deposits, especially those hosted by the intrusions themselves, a discrete class of gold

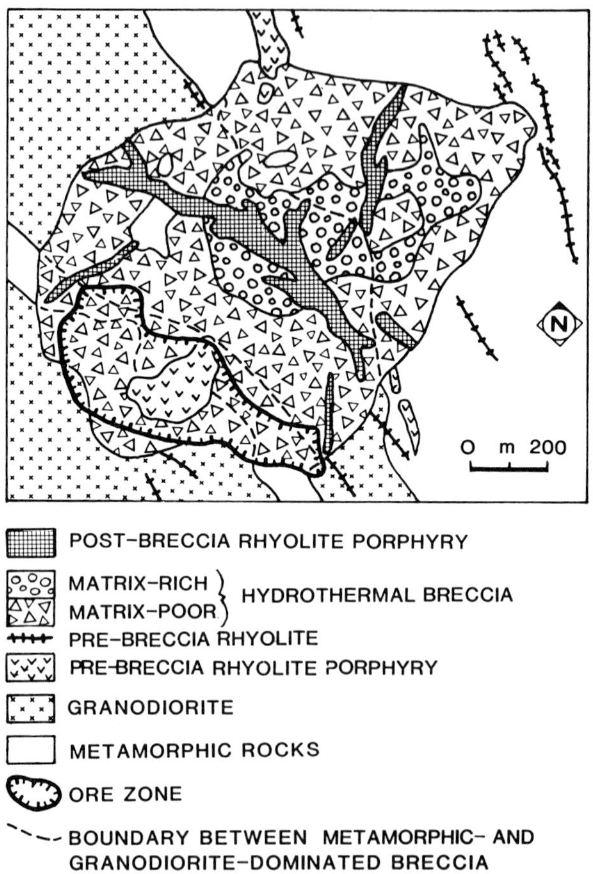

Figure 6.7 Surface plan of the Kidston gold-bearing breccia deposit, Queensland, Australia, after Baker (1987). Emplacement of the magmatic-hydrothermal breccia resulted in little mixing of fragments, and wallrock contacts may be traced across the pipe. Rhyolite porphyry was intruded immediately before and after brecciation, and the porphyry body and associated fragments within the outline of the gold orebody underwent pre-brecciation development of a quartz–molybdenite stockwork. The ore zone is centred on an annular swarm of sheeted veins and veinlets that follow the south-western contact of the pipe.

deposits confined to pipe-like breccias is also evident. These breccias (Table 6.6) generally transect wallrock lithologies, and therefore must be related to concealed intrusions. However, outcropping intrusions, commonly in the form of felsic dykes and/or sills (Figure 6.7; Kidston, Montana Tunnels, Golden Sunlight, Colosseum, Mount Leyshon and probably Ortiz) may be interpreted reasonably to possess genetic and perhaps physical connections with the inferred intrusions at depth. Porphyry copper or molybdenum protore is associated with the auriferous breccia pipes at Kidston (Baker, 1987), Mount Leyshon (Morrison *et al.*, 1988) and Golden Sunlight (R.H. Sillitoe and H.F. Bonham, Jr., personal observations, 1980), but everywhere pre-dated brecciation, and at Kidston and Mount Leyshon was incorporated as clasts in the pipe breccias.

Using the breccia classification outlined by Sillitoe (1985), the gold-bearing pipes may be subdivided into two distinct types: magmatic-hydrothermal breccias at Golden Sunlight (cf. Porter and Ripley, 1985), Colosseum, Ortiz, Kidston (cf. Baker, 1987; Baker and Andrew, 1988) and Chadbourne; and phreatomagmatic breccias (diatremes) at Montana Tunnels (Sillitoe *et al.*, 1985) and, in part, Mount Leyshon (Morrison *et al.*, 1987, 1988). In the case of Ortiz, the auriferous pipe abuts a poorly mineralized vent filled with latitic tuff (Lindquist, 1980), which was interpreted as a magmatic (rather than phreatomagmatic) diatreme by Sillitoe (1985). The magmatic-hydrothermal breccias are characterized by angular to rounded clasts in a matrix of comminuted rock flour and/or hydrothermal products, which in the phreatomagmatic breccias are accompanied by juvenile tuffaceous material.

Table 6.6 shows that the gold and associated mineralization were accompanied by feldspar-destructive alteration dominated by sericite and carbonates. However, at Kidston (Baker, 1987; Baker and Andrew, 1988) and Mount Leyshon (Morrison *et al.*, 1987, 1988), K-silicate alteration was related to pre-breccia porphyry-type mineralization and, at Kidston, also to early brecciation events. Base and other metals accompany the gold in all the deposits considered (Table 6.6), but attain tenors that are exploitable or potentially so only in the case of zinc and lead at Montana Tunnels (Sillitoe *et al.*, 1985) and tungsten at Ortiz (Wright, 1983). Gold ore generally occurs only in restricted portions of the breccia pipes, commonly in sheeted zones along parts of the pipe contacts (e.g. Kidston – Wilson *et al.*, 1986, Figure 6.7; upper levels at Colosseum – Sharp, 1984).

6.7 Vein-type deposits

Numerous gold-bearing quartz veins, many of them insignificant from the standpoint of their gold contents (say, < 1 tonne), are associated, at least spatially, with intrusions. Here a representative selection of larger vein-type gold districts, including three world-class camps (Charters Towers, Segovia, Zhao-Ye), is presented (Table 6.7).

The auriferous veins may be subdivided into two broad categories: massive quartz veins of ‘mesothermal’ type hosted by equigranular plutons or their wallrocks (Charters Towers, Tiouit, Los Mantos de Punitaqui, Segovia, Zhao-Ye); and partly crustified quartz veins of adularia–sericite epithermal type cutting the wallrocks to porphyry stocks, which at both Masara and Paracale are associated with porphyry- and skarn-type base- and precious-metal mineralization (Table 6.7).

Table 6.6 Characteristics of selected breccia-hosted gold deposits

Deposit	Status	Contained Au, tonnes	Age, Ma	Ore-related intrusion	Associated mineralization	Host rocks	Metal association	Key hypogene alteration minerals	Data source(s)
Montana Tunnels, Montana, USA	Mine: open pit, flotation	60	45–50	Quartz latite porphyry dykes	None	Late Cretaceous and middle Eocene andesitic and felsic volcanics	Au–Ag–Zn–Pb–Mn– (Cu)	Sericite, carbonates	Sillitoe *et al.* (1985)
Golden Sunlight, Montana, USA	Mine: open pit, CIP	59	Early Tertiary	Latite porphyry sills	Minor porphyry Cu–Mo protore, Au veins	Proterozoic calcareous shale	Au– (Cu–Pb–Zn–Bi–Te)	Silicification, decalcification	Porter and Ripley (1985)
Colosseum, California, USA	Mine: open pit, CIP	20	100	Rhyolite plug and dykes	Ag veins	Precambrian gneiss	Au– (Zn–Cu– Pb)	Sericite, carbonates	Sharp (1984)
Ortiz, New Mexico, USA	Abandoned mine (open pit, heap leach)	11	34	Monzonite stock, latite porphyry plugs and dykes	Au breccias and veins, Cu–Au skarns	Late Cretaceous quartzite and argillite	Au–W–Cu– (Pb)	Sericite, carbonates	Lindquist (1980); Wright (1983)
Kidston, Queensland, Australia	Mine: open pit, CIP	101	321	Rhyolite plugs and dykes	Porphyry Mo protore	Proterozoic metamorphics and Proterozoic or Palaeozoic granodiorite	Au–Zn–Cu–Mo–Pb– (As–Bi–Te)	Orthoclase–biotite– siderite, sericite–carbonates	Mustard (1986); Wilson *et al.* (1986); Baker (1987); Baker and Andrew (1988)

Mount Leyshon, Queensland, Australia	Mine: open pit, heap leach	26	280	Trachyte plugs and dykes	Porphyry Cu–Mo protore	Cambrian metasediments	Au–Zn–Pb– Cu–(Mo–Bi)	Biotite–K–feldspar, sericite	Morrison *et al.* (1987, 1988)
Chadbourne, Quebec, Canada	Mine: underground, flotation	9	Archaean	Syenite stocks	None	Archaean felsic and mafic metavolcanics	Au– (Cu–Zn)	Sericite, carbonates	Walker and Cregheur (1982)

Table 6.7 Characteristics of selected vein gold deposits related to intrusions

Deposit	Status	Contained Au, tonnes	Age, Ma	Ore-related intrusion	Associated mineralization	Host rocks	Metal association	Key hypogene alteration minerals	Data source(s)
Charters Towers, Queensland, Australia	Abandoned mines, under exploration	211	~400	Granodiorite-tonalite-diorite pluton	None	Devonian intrusive rocks and minor schist	Au–Pb–Zn–(Cu–As–Te)	Sericite, carbonates	Blatchford (1953); Peters and Golding (1987)
Tiouit, Morocco	Mine: underground, flotation	6	Late Proterozoic	Granodiorite pluton	None	Late Proterozoic granodiorite	Au–Ag–Cu–(As–Pb–Zn–Bi)	Chlorite, sericite	Jouravsky (1952), R.H.Sillitoe (unpub. rept., 1987)
Los Mantos de Punitaqui, Chile	Abandoned mine: under exploration	22.5	Early Cretaceous	Granodiorite pluton	None	Cretaceous granodiorite, andesite and limestone	Au–Cu–Hg–(Sb)	Sericite	McAllister *et al.* (1950)
Segovia, Colombia	Mines: underground, cyanidation	>200	Jurassic	Quartz diorite pluton	None	Jurassic quartz diorite	Au–Ag–Pb–Zn–(Cu–As–W)	Sericite	Rodriguez and Pernett (1983)
Zhao–Ye, (Jiaojia), Shandong, China	Mines: underground and minor open pit, flotation (?)	~500	~150	Granite and granodiorite pluton	None	Jurassic granite and granodiorite	Au–Cu–Pb–Zn–Ag–Bi	Sericite, K–feldspar, chlorite, carbonate	Sang and Ho (1987); Zhu (1988)

Masara, Philippines	Mines: underground, flotation	28	4.7	Quartz diorite porphyry stocks	Porphyry Cu–Au deposit and protore, minor Cu–Au skarn	Cretaceous andesitic volcanics and volcaniclastics	Au–Zn–Pb –Cu	Sericite	Mercado *et al.* (1987)
Paracale, Philippines	Mines: open pit and underground	76	17	Diorite porphyry stocks	Larap skarn Fe–(Cu–Mo–Au–W) deposit and porphyry Cu–Mo –Au protore	Pre-Tertiary gneissic granodiorite amphibolite and ultramafics	Au–Zn–Pb–Cu–(Mo–W)	Sericite	Frost (1959)

Both vein categories are confined to brittle structures, which range in scale from district-wide to local (e.g. Tiouit), and in dip from steep to shallow (e.g. Charters Towers, Tiouit, Segovia, upper levels at Los Mantos de Punitaqui). Locally, lithological parameters, such as the contact between gneissic granodiorite and ultramafic rocks at Paracale (Frost, 1959), influenced gold deposition.

Both vein categories are also characterized by an association of gold with base metals, and by a predominance of narrow sericitic alteration selvages around veins (Table 6.7). However, at Los Mantos de Punitaqui, mercurian tetrahedrite and cinnabar – sulphides normally ascribed to epithermal processes – in association with magnetite and specular hematite are major components of the mesothermal vein, from which mercury ore was once exploited (McAllister *et al.*, 1950).

6.8 Deposit interrelationships and metal zoning

The different types of intrusion-related gold deposits reviewed above commonly occur in juxtaposition, and locally exhibit transitions between one another. This may be apparent at the scale of a gold province as well as within individual districts. An example of province-scale associations is provided by the early Tertiary gold province of the northern Black Hills, South Dakota, USA, where deposits of non-porphyry-type intrusion-hosted (e.g. Gilt Edge; Table 6.2), carbonate-replacement (e.g. Foley Ridge; Table 6.4) and breccia-hosted types, as well as gold deposits transitional between these three types, are products of alkaline igneous activity (DeWitt *et al.*, 1986).

At the district scale, one of the pre-eminent features apparent from Tables 6.3 to 6.7 is the intimate association between wallrock-hosted gold deposits and porphyry-type copper–molybdenum or molybdenum mineralization, with or without gold, in the progenitor intrusions. This intrusion-hosted porphyry-type mineralization is commonly developed only weakly (cf. Sillitoe, 1988). In a few districts, two or more wallrock-hosted deposit types occur together: proximal Cu–Au skarns and the Fortitude and other distal gold skarns in the Copper Canyon district, Nevada (Table 6.3); the McCoy skarn and Cove carbonate-replacement deposits in the McCoy district, Nevada (Tables 6.3 and 6.4); skarn Cu–Au, carbonate-replacement Pb–Zn–Ag–(Au) and the Barney's Canyon and one other carbonate-replacement Au–(As–Sb) deposits in the Bingham district, Utah (Table 6.4); carbonate-replacement base metals plus silver and the Purísima Concepción carbonate-replacement Au–(As–Sb) deposit in the Yauricocha district of Peru (Table 6.4); and gold veins and skarn-type deposits in the Larap-Paracale and Masara districts, Philippines (Table 6.7).

Using the information on interrelationships furnished by these and other deposits, a typical intrusion-centred gold district characterized by carbonate wallrocks may be reconstructed (Figure 6.8). Porphyry-type mineralization carrying Cu and/or Mo (± Au) is abutted by a Cu–Au or Au–Cu skarn environment. In the Copper Canyon district, Nevada, an outward zonation from copper to gold is present within the skarn environment (Blake *et al.*, 1984; Myers and Meinert, 1989). Further from the intrusion, skarn gives way typically to carbonate-replacement-type Pb–Zn–Ag mineralization which may or may not carry Cu or Au. Meinert (1987) has documented in detail a transition between skarn and carbonate-replacement deposits. Upwards,

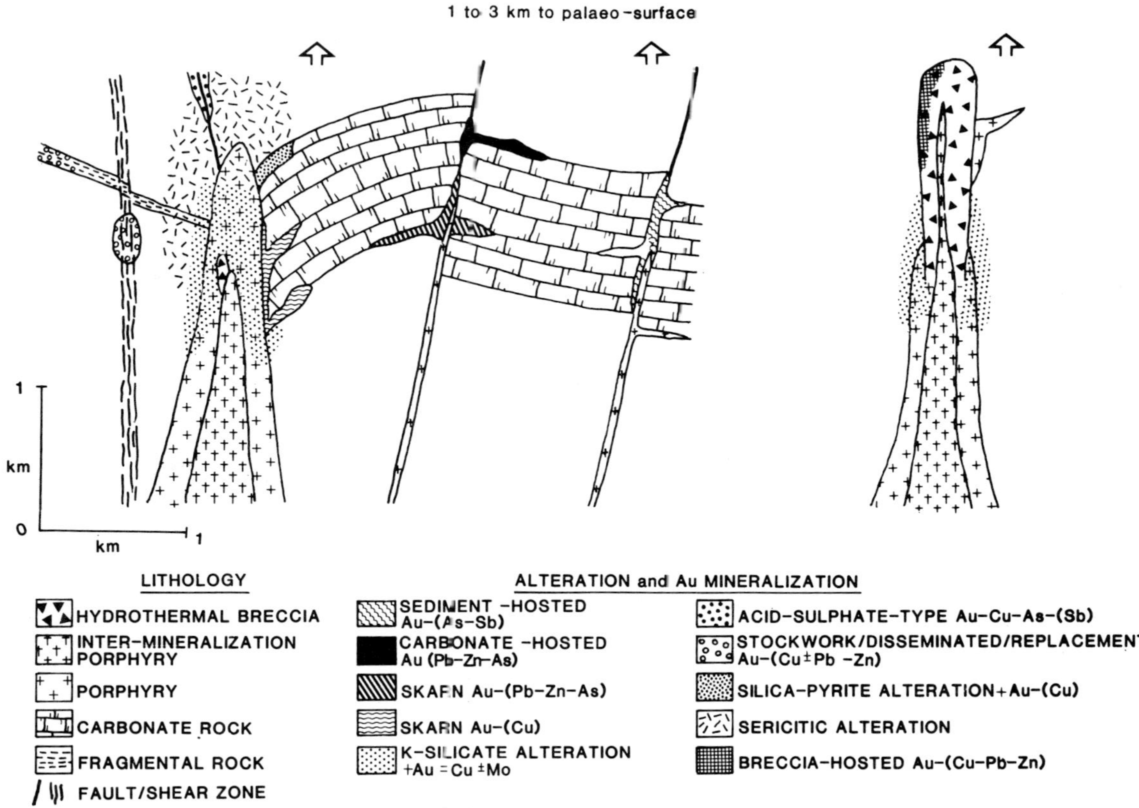

Figure 6.8 Schematized interrelationships between progenitor intrusions and some of the gold mineralization styles described in the text. The presence of all styles in a single district is not necessarily implied. The environments depicted are transitional upwards to the epithermal setting (see Figure 6.10). Depths ranging from 1 to 3 km beneath the palaeosurface are supported by geological evidence (e.g. Lindsey and Fisher, 1985; Moore, 1973; Paterson *et al.*, 1988; Sillitoe and Gappe, 1984) and fluid-inclusion studies (e.g. Bowman *et al.*, 1987; Wojdak and Sinclair, 1984).

skarn gives way to silica-pyrite alteration which is gold-bearing at Star Pointer (Table 6.4). Beyond the base-metal-bearing carbonate-replacement environment, a few districts possess base- metal-depleted carbonate-replacement Au–As–Sb mineralization. Where carbonate wallrocks are absent, district-scale zoning beyond progenitor stocks appears to be less well developed, as seen in the case of the non-carbonate-hosted disseminated, stockwork and replacement gold deposits and the breccia-hosted gold deposits (Figure 6.8).

Only one district, Bingham, displays the complete proximal-to-distal sequence depicted in Figure 6.8. More commonly, only parts of this sequence are developed, and there is complete dominance by a single gold deposit type. The absence of porphyry-type stockworks from some progenitor stocks may therefore be regarded in the same way as a deficiency of, say, carbonate-replacement gold (e.g. Ok Tedi) or skarn gold (e.g. Kendall) in other districts.

Hydrothermal overprinting and/or telescoping are common complicating factors in many intrusion-centred districts, and result in the juxtaposition or superimposition of earlier and later gold deposit types. Thus, gold-bearing porphyry-type mineralization may be immediately overlain by or situated alongside acid–sulphate-type gold deposits associated with advanced argillic alteration (Figure 6.8), as at Lepanto (Table 6.1) and Equity Silver (Table 6.5). Distal Au–As–Sb mineralization may be localized by structures cutting more proximal mineralization types, as at Bingham (see above). Breccia-hosted gold deposits may cut through and partially destroy at least the upper parts of porphyry-type stockworks (Figure 6.8), as at Kidston and Mount Leyshon (Table 6.6). The occurrence of gold deposits of epithermal character within or alongside intrusions, as at Porgera, Zortman-Landusky, Gilt Edge and Kori Kollo (Tables 6.2 and 6.5), may be attributed to the telescoping of environments, which could result from deep penetration of late-stage, shallow-level ore fluids and/or extremely high-level magma emplacement.

It is clear, therefore, that gold has the potential to precipitate in economic concentrations both within, and at various distances away from, progenitor intrusions. Traditional hydrothermal zoning theory (Fe → Ni → Sn → Cu → Zn → Pb →Ag → Au → Sb → Hg; Barnes, 1975) does not explain this behaviour but predicts only distal precipitation of gold along with Sb, Hg, and possibly As. However, more recently, Hemley *et al.* (1987) showed that hydrothermal zoning is dependent mainly on temperature–concentration relationships, and suggested that zoning reversals are due to differences in the relative concentrations of metals. Presumably gold-rich systems possess the potential for gold precipitation in one or more proximal sites, which leads to variations in the zonal position of gold relative to base metals. For example, gold at Fortitude and other distal skarn gold deposits in the Copper Canyon district is located immediately inboard of the main concentration of Pb and Zn (Blake *et al.*, 1984), whereas at Bingham the Barney's Canyon gold deposit occurs several kilometres beyond the main carbonate-replacement Pb-Zn zone.

A surprising number of the wallrock-hosted gold deposits summarized in Tables 6.3 to 6.7, including skarn, carbonate-replacement, non-carbonate-hosted, breccia-pipe and vein types, carry As and small amounts of Bi and/or Te besides the common base metals. It appears, therefore, that the As–Bi–Te association may be typical not only of gold skarns (Meinert, 1988) but of all types of wallrock-hosted gold deposits related to intrusions.

6.9 Genetic considerations

6.9.1 *Magma type*

Ishihara (1981) proposed that all significant copper, molybdenum, lead–zinc and gold–silver deposits in volcano-plutonic arcs, especially porphyry-type deposits, are related genetically to magnetite-series magmas characterized by their relatively oxidized nature (bulk $Fe_2O_3/FeO > 0.5$). All magnetite-series magmas fall within the I-type classification of Chappell and White (1974). Cameron and Hattori (1987) also stressed the relationship between magmas with relatively high oxygen fugacities and the generation of major gold deposits. In apparent contrast, Keith and Swan (1987a) proposed an association of gold deposits with some of the more reduced ($Fe_2O_3/FeO < 0.85$), albeit I-type, intrusions in the southwestern USA.

Ishihara (1981) pointed out that magnetite-series magmas tend to be enriched in S and Cl, components which, like the ore metals (Sillitoe, 1972, 1987), may have been derived, at least in part, from dehydration of subducted lithospheric slabs. This mechanism, involving incorporation of seawater sulphate (and chloride), accounts for both the positive $\delta^{34}S$ values and relatively high oxygen fugacities of magnetite-series magmas. It also explains the localization of most major intrusion-related gold deposits at sites of ancient or modern subduction.

As far as is known, all the gold deposits summarized in Tables 6.1 to 6.7, except perhaps for Muruntau and Navachab, are related to magnetite-series, I-type intrusions. The presence of magnetite as the principal opaque mineral (>0.1 vol.%), the general occurrence of magmatic hornblende and the absence of magmatic muscovite, garnet and cordierite all support this assumption (cf. Ishihara, 1981). However, it should be mentioned that Keith and Swan (1987b) have proposed a class of peraluminous gold deposits related genetically to ilmenite-series, S-type intrusions, although many workers do not accept connections between these intrusions and the nearby gold mineralization (e.g. at Mesquite, California; Willis *et al.*, 1989). This caveat notwithstanding, Keith and Swan (1987b) and Bloomstein (1987) claimed that the huge Muruntau gold deposit (Table 6.5) is assignable to the peraluminous category, as indeed may be several rhyolite-related gold occurrences in the Iditarod district of western Alaska (Buntzen *et al.*, 1987) and the aluminous gneiss-hosted American Girl and Padre y Madre gold deposits in south-eastern California (Tosdal and Smith, 1987).

Magnetite-series, I-type intrusions (and volcanics) may be subdivided into calcic, calc–alkaline, alkali-calcic and alkaline series on the basis of their alkalinity expressed by $K_{57.5}$ indices. Keith (1983) concluded that major gold deposits are linked to calcic, calc–alkaline and alkaline rocks but are generally absent in association with alkali-calcic rocks possessing $K_{57.5}$ indices between 2.45 and 3.8. The alkali-calcic series may, however, give rise to major precious-metal deposits with high Ag/Au ratios (say >100). Insufficient analytical data are available to the writer for the progenitor intrusions summarized in Tables 6.1 to 6.7 to test Keith's proposal properly. However, it is clear (Figure 6.9) that few analyses fall in the alkali-calcic field, and most of the analysed intrusions, with the possible exception of those at Ok Tedi and Star Pointer, are calcic, calc–alkaline or alkaline. It should be emphasized further that, in addition to the deposit-related intrusions plotted as alkaline in Figure

6.9, the progenitor intrusions at Zortman-Landusky, Gilt Edge (Table 6.2), Kendall, Foley Ridge (Table 6.4), Quesnel River (Table 6.5), Golden Sunlight and Ortiz (Table 6.6) are also considered to belong to the alkaline series (e.g. Mutschler *et al.*, 1985).

Leveille *et al.* (1988) attempted to reconcile previous hypotheses relating gold endowment to selected magmatic parameters by plotting alkalinity (expressed as $K_2O + Na_2O - 0.57SiO_2$) against oxidation state for 200 unaltered intrusions. Their results suggested that gold deposits accompany the more reduced rather than more oxidized intrusions across the full alkalinity range.

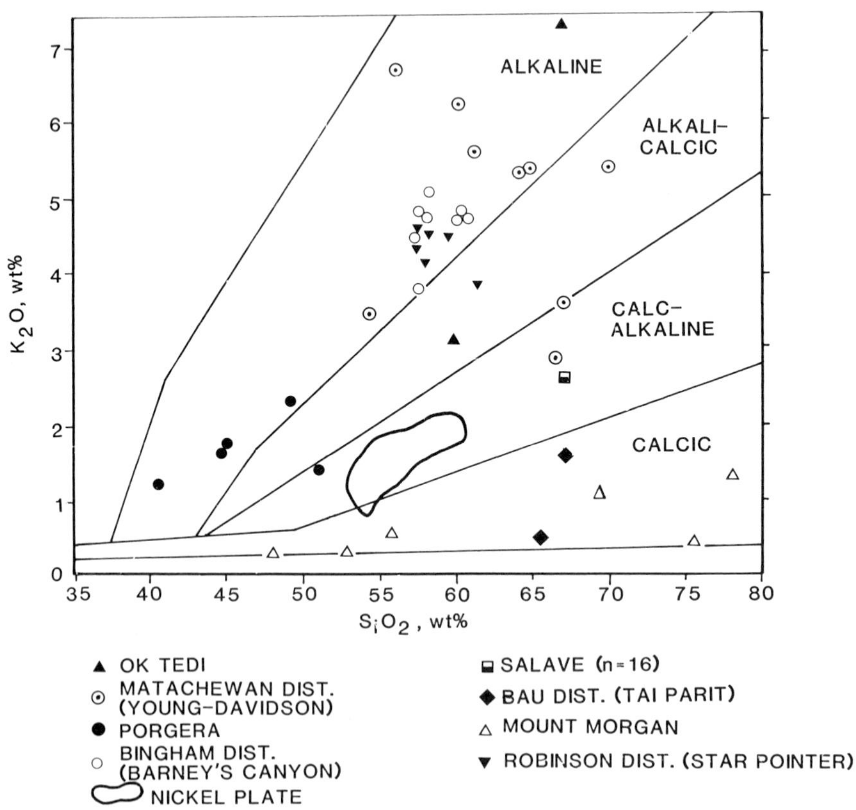

Figure 6.9 K_2O/SiO_2 diagram for unaltered intrusive rocks related genetically to nine of the gold deposits reviewed above. The quality of samples and analyses is largely unknown. Data sources: Ok Tedi – Mason and McDonald (1978); Young-Davidson – Sinclair (1982); Porgera – Fleming *et al.* (1986); Bingham district – Moore (1973); Nickel Plate – Ray *et al.* (1988); Salave – Harris (1980b); Tai Parit – Wolfenden (1965); Mount Morgan – Cornelius (1969); and Star Pointer – Bauer *et al.* (1966).

6.9.2 *Ore formation*

Many investigators (e.g. Henley and McNabb, 1978; Eastoe, 1982) identify three partly interdependent fluid regimes in intrusion-centred systems of porphyry type: relatively dense magmatic-hydrothermal brines that tend to reflux within the cooling intrusions; ascendant plumes of lower-density, sulphur-rich magmatic volatiles that condense upwards and mix with meteoric fluids; and convectively circulating meteoric cells marginal to intrusions that tend to 'collapse' progressively on to the intrusions as magmatic fluid generation wanes (Figure 6.10). Recognition of the former presence of one or more of these different fluids in intrusion-centred gold systems depends largely on alteration assemblages and interpretation of the results of fluid-inclusion and oxygen and hydrogen isotopic studies.

It is well documented that the magmatic-hydrothermal brines in association with magnetite-series, I-type (and other) magmas are capable of transporting the metals, including gold, and sulphur required for ore formation (Burnham and Ohmoto, 1980). The release of the brines by retrograde boiling is capable of dissipating sufficient mechanical energy to induce rock fragmentation, including the development of porphyry-type stockworks (Burnham, 1979) and breccia pipe formation (Burnham, 1985; the magmatic-hydrothermal breccia of Sillitoe, 1985).

Porphyry-type mineralization characterized by K-silicate alteration and variable quantities of Au, Cu and/or Mo is the hallmark of magmatic-hydrothermal brine activity, and has been documented in association with all the intrusion-related gold categories summarized in Tables 6.1 and 6.3–6.7. As shown in Figure 6.10, these magmatic-hydrothermal fluids possessed high temperatures (400°C to > 700°C) and salinities (30 to > 75 wt.% alkali chlorides) and, at least in the Bingham (Bowman *et al.*, 1987) and Robinson (Sheppard *et al.*, 1971) districts and at Red Dome (Ewers and Sun, 1988), exhibited $\delta^{18}O$ and δD values appropriate for a magmatic source. The brines coexisted with a lower-density volatile phase. The intrusion-related gold deposits that lack any recorded porphyry-type mineralization (Tables 6.2 to 6.7) are inferred to have either evolved smaller volumes of magmatic-hydrothermal brine less violently or, alternatively, have their porphyry-type stockworks concealed (e.g. at depth in some of the breccia-hosted deposits; Table 6.6).

Given the broadly coeval development of K-silicate alteration and prograde skarn assemblages, the latter are also likely to be the products of magmatic-hydrothermal brines (e.g. Red Dome; Ewers and Sun, 1988), even where skarn generation took place at distal sites, as in the Copper Canyon district (Theodore *et al.*, 1986).

Magmatic-hydrothermal brines might also be expected to have been responsible for generation of intrusion-hosted gold veins (Table 6.7), and indeed a fluid of dominantly magmatic origin is proposed at Zhao-Ye (summarized by Sang and Ho, 1987). However, at Charters Towers, Peters and Golding (1987) were unable to assign vein formation to any specific intrusive phase which, in conjunction with their oxygen isotopic results, led Morrison (1988) to propose a deep-seated source for the ore fluids that was not related directly to consolidation of the host intrusion (see below).

The low-pH fluids resulting from shallow condensation and groundwater-dissolution of acid magmatic volatiles are thought to give rise to the extensive zones of generally volcanic-hosted advanced argillic alteration that overlie porphyry-type stockworks (e.g. Meyer and Hemley, 1967; Sillitoe, 1983b; Heald *et al.*, 1987). Some of the metals, including Au and Cu, in these fluids, as well as the sulphur, may have

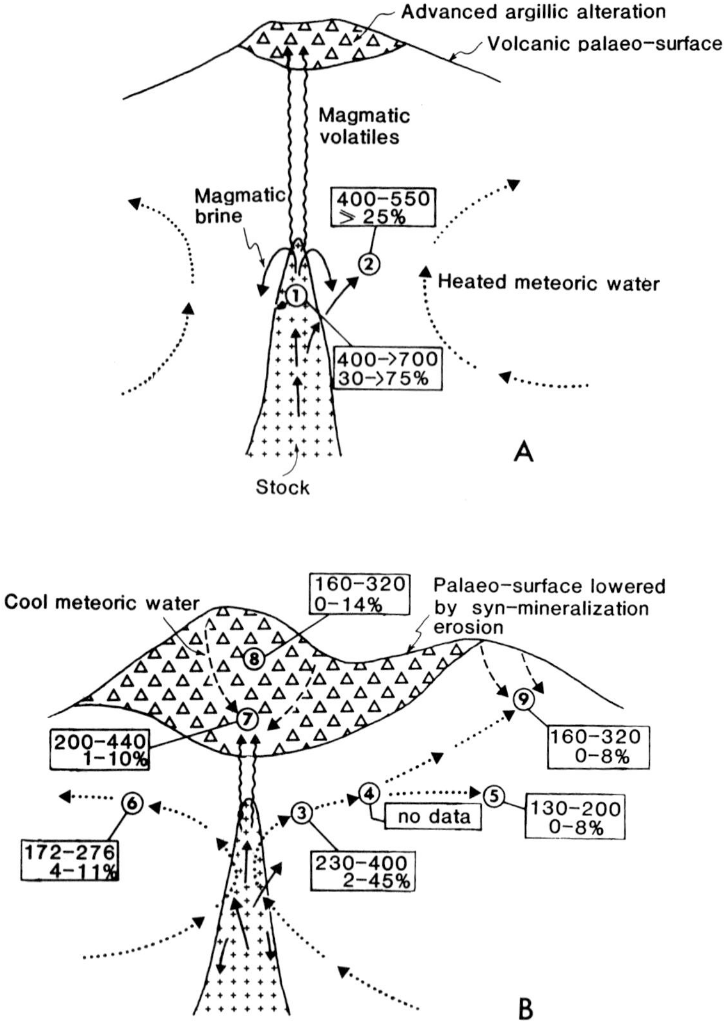

Figure 6.10 Genetic model for an intrusion-centred gold district, including the acid–sulphate- and adularia–sericite-type epithermal environments: (A) early stage dominated by magmatic-hydrothermal fluids; (B) late stage with 'collapse' of meteoric system. Arrows represent inferred flow-lines for different types of fluids. Circled numbers and attached boxes denote gold deposit types and the homogenization temperatures (°C) and salinities (wt.% NaCl equivalent) of representative ore fluids as determined for the intrusion-related deposits reviewed above and epithermal deposits in general: ① porphyry type (Alvarez and Noble, 1988; Baker and Andrew, 1988; Bowman *et al.*, 1987; Ewers and Sun, 1988; Morrison *et al.*, 1988; Takenouchi, 1981; Theodore and Blake, 1978; Wojdak and Sinclair, 1984); ② prograde skarn (Ewers and Sun, 1988; Theodore *et al.*, 1986; Torrey *et al.* 1986); ③ retrograde skarn (Ewers and Sun, 1988; Theodore *et al.*, 1986; Torrey *et al.*, 1986); ④ carbonate-replacement Au–base metals; ⑤ carbonate-replacement Au–As–Sb (Paterson *et al.*, 1987); ⑥ non-carbonate disseminated, stockwork and replacement (Eadington *et al.*, 1974; Handley and Bradshaw, 1986); ⑦ deep acid-sulphate (Wojdak and Sinclair, 1984); ⑧ acid–sulphate epithermal (summary by Sillitoe, 1989a); and ⑨ adularia–sericite epithermal (summary by Sillitoe, 1989a). Based on relationships detailed in Figure 6.8 and Sillitoe (1989a).

been contributed directly from the subjacent porphyry environment as volatile complexes (Sillitoe, 1983b; Symonds *et al.*, 1987). The acid-sulphate-type enargite–gold deposit at Lepanto (see above) forms part of such a shallow advanced argillic environment, whereas the intrusion-hosted enargite-bearing veins at Yauricocha (Alvarez and Noble, 1988; Table 6.4) and the Equity Silver gold deposit (Table 6.5) were emplaced at deeper levels, possibly in the roots of now-eroded advanced argillic zones (Figure 6.8; Sillitoe, 1988). At Equity Silver, the presence of both andalusite and pyrophyllite rather than alunite in the advanced argillic assemblages, and fluid-inclusion homogenization temperatures as high as 400°C (Figure 6.10; Wojdak and Sinclair, 1984) support this interpretation. However, as in most acid-sulphate-type epithermal gold deposits (e.g. Stoffregen, 1987), the fluids responsible for metallization at Equity Silver were relatively dilute (< 8 wt.% NaCl equivalent) and in all likelihood dominated by meteoric water (Wojdak and Sinclair, 1984).

The inferred positions of convectively circulating, meteoric-hydrothermal cells alongside intrusions (Figure 6.10) encompass the sites occupied by wallrock-hosted gold deposits (Tables 6.4, 6.5 and 6.7). The principal alteration characteristics of these deposits – decalcification and/or silicification of carbonate host rocks and feldspar-destructive alteration of silicate host rocks – are indeed compatible with magmatic-hydrothermal fluids that underwent cooling and dilution as a result of admixture with variable quantities of meteoric water during focused outward and upward flow through stratigraphic and/or structural conduits. Few wallrock-hosted gold deposits have been subjected to fluid-inclusion and isotopic studies, and the data available (Figure 6.10) are only of a reconnaissance nature. However, the results show that despite the contributions of meteoric-hydrothermal fluid, a magmatic signature apparently may be still decipherable at Foley Ridge (Paterson *et al.*, 1987), Porgera (Handley and Bradshaw, 1986), Mount Morgan (Eadington *et al.*, 1974) and probably Beal (Hastings and Harrold, 1988).

The inward and downward collapse of meteoric-hydrothermal cells during waning magmatic-hydrothermal activity gives rise to numerous important consequences in intrusion-centred gold systems. Firstly, prograde garnet-pyroxene skarn is partly retrograded to gold- and sulphide-rich, hydrous assemblages (Table 6.3; see above), although Theodore *et al.* (1986) and Ewers and Sun (1988) demonstrated that magmatic-hydrothermal brines (Figure 6.10) also participated in retrograde skarn stages in the Copper Canyon district and at Red Dome, respectively. Secondly, the sericitic alteration and epithermal characteristics of several non-porphyry-type intrusion-hosted gold deposits (Table 6.2) may be attributed to collapse of meteoric-hydrothermal cells on to the intrusions themselves, as supported by the results of fluid-inclusions and light stable isotopic studies at Zortman-Landusky (Wilson and Kyser, 1988). However, some samples from Gilt Edge (Table 6.2) also contain high-salinity fluids which, with homogenization temperatures as high as 700°C, are apparently magmatic-hydrothermal in origin (Paterson *et al.*, 1987, 1988). In contrast, comparable meteoric- hydrothermal overprinting of porphyry-type gold deposits, as at Marte and Dizon (Table 6.1; see above), does not appear to have introduced an appreciable amount of additional gold, although it did produce feldspar-destructive alteration.

As noted above, gold-bearing magmatic-hydrothermal and phreatomagmatic breccias post-dated the development of porphyry-type protore (Table 6.6). Whereas magmatic-hydrothermal breccias were emplaced largely without the involvement of meteoric-hydrothermal fluids, ingress of the latter to high-level magma chambers is proposed as a mechanism for the explosive emplacement of phreatomagmatic diatremes (Sillitoe, 1985) such as that which hosts the Montana Tunnels gold deposit (Sillitoe *et al.*, 1985). Light stable isotopic studies conducted on hydrothermal material from Golden Sunlight (Porter and Ripley, 1985), Mount Leyshon (Morrison *et al.*, 1988) and Kidston (Baker and Andrew, 1988) have all demonstrated the importance of magmatic-hydrothermal fluids. Moreover, at Kidston, the deuterium-depleted δD values for the main gold stage are interpreted by E.M. Baker and A.S. Andrew (in prep.) in terms of condensation of late magmatic volatiles rather than meteoric-hydrothermal dilution (cf. Taylor, 1988).

Given the oxidized, saline and high-temperature nature of the magmatic-hydrothermal fluids in the porphyry-type gold environment, gold – like copper – must have been introduced as chloride complexes, and underwent precipitation with the K-silicate assemblage as a consequence of cooling and/or pH increase (Seward, 1984). In the wallrock-, breccia- and non-porphyry-type intrusion-hosted gold environments, dilution and additional cooling of magmatic fluids by those of meteoric (or, locally, connate) origin would act as an effective mechanism for continued or renewed gold deposition (Seward, 1984). The tellurium contents of many wallrock-hosted gold deposits (Tables 6.3–6.7) may further support such a direct magmatic source (Afifi *et al.*, 1988).

The cooler, more dilute and more reduced character of meteoric water-dominated fluids may permit the outward transport of gold as bisulphide complexes to the fringes of intrusion-centred systems (cf. Seward, 1984). There they could precipitate gold in distal skarn or carbonate-replacement deposits. Such fluids could also mobilize As and Sb but, because of low chloride contents, would be rather ineffective transporters of base metals (Henley, this volume). If such fluids followed highly permeable stratigraphic and/or structural conduits characterized by decreasing pressure gradients (Henley *et al.*, 1986) and attained the distal parts of intrusion-centred systems, they could have been stripped efficiently of their base-metal contents, and perhaps were able to generate carbonate-replacement Au–As–Sb deposits (Table 6.4) upon encountering appropriate reactive rocks (Sillitoe and Bonham, 1990).

As an alternative or an addition to the direct input of gold-bearing magmatic fluids to meteoric-hydrothermal systems, the meteoric fluids themselves may scavenge, in bisulphide form, some of the gold that was precipitated previously by destabilization of gold chloride complexes (Sillitoe, 1989a). For example, early-deposited gold in porphyry-type ore or protore could undergo partial remobilization followed by its reprecipitation during later-stage distal and/or retrograde skarn development. Such hydrothermal upgrading of gold pre-concentrations may play an important role in the generation of multistage mega-deposits in intrusion-centred gold systems (e.g. Porgera; Fleming *et al.*, 1986; Sillitoe, 1989a).

6.10 Possible relationships with other gold deposit types

6.10.1 *Epithermal deposits*

In a number of geological models of hydrothermal systems, most notably those presented by Bonham (1986, 1988), Mutschler *et al.* (1985), Panteleyev (1986) and

Sillitoe (1983b, 1988, 1989a), some epithermal gold deposits are positioned at shallow levels above or marginal to intrusion-related porphyry-type mineralization (e.g. Figure 6.10). However, as noted above, telescoping, which is commonly caused by rapid erosion during the 2- to 3-Ma lifespans (Silberman, 1985) of major hydrothermal systems (Sillitoe, 1989a), may result in juxtaposition of epithermal and intrusion-related mineralization types.

Typically, acid-sulphate-type epithermal gold deposits are located above porphyry-type mineralization (Figure 6.10), albeit offset laterally because of structural factors in some districts (e.g. Lepanto; see above). In contrast, the adularia–sericite-type gold deposits tend to be situated in the shallow, marginal parts of intrusion-centred systems (Figure 6.10), although telescoping can result in their presence alongside (e.g. Masara) or even superimposed on (e.g. Porgera) intrusions. The giant Ladolam deposit in Lihir Island, Papua New Guinea (Davies and Ballantyne, 1987; Henley, this volume) seems to be the product of extreme telescoping of a gold-rich porphyry deposit, like those discussed above, and a major epithermal system.

6.10.2 *Sediment-hosted deposits*

Sediment-hosted or Carlin-type gold deposits are best known from Nevada and contiguous parts of the western USA. Typically they are hosted by thinly bedded, silty carbonate rocks, localized by high-angle faults, associated with decalcification, silicification (including jasperoid) and argillization, and carry Au with As, Sb, Hg and Tl (e.g. Bagby and Berger, 1985; Berger and Bagby, this volume). These deposits possess marked geological similarities to the carbonate-replacement gold deposits at Barney's Canyon, Tai Parit, Purísima Concepción, Kendall and Foley Ridge (Table 6.4), which are considered to be distal manifestations of intrusion-centred systems.

Furthermore, most sediment-hosted gold deposits contain intrusive rocks, commonly in the form of sills or dykes (Bagby and Berger, 1985; Berger and Bagby, this volume) and, as emphasized by Sillitoe and Bonham (in press), several major deposits in Nevada (e.g. Rain, Goldstrike, Deep Post, Gold Acres, Getchell, Chimney Creek) are near to, or even contain, intrusion-related base- and/or precious-metal mineralization. A genetic relationship between sediment-hosted gold mineralization and that of obvious intrusive affiliation is distinctly possible.

Sillitoe and Bonham (1990) suggest that many sediment-hosted gold deposits are the distal manifestations of intrusion-centred base- and precious-metal districts, and that commonly the progenitor intrusions and proximal ore deposit types remain unexposed. The gold, therefore, is assigned a direct or indirect magmatic origin rather than invoking leaching of the metal from sedimentary sequences by geothermal fluids in high heat-flow extensional settings (e.g. Radtke and Dickson, 1974; Nesbitt, 1988).

6.10.3 *Mother Lode-type deposits*

Mother Lode-type or mesothermal gold deposits occur as thick, massive to ribboned quartz–carbonate veins in association with major, steep, regional fault zones. They are hosted by a variety of wallrocks, are associated with sericite, carbonate and albite as alteration minerals and carry a generally base-metal-poor Au–As–Sb–Hg–W–B–Te suite of elements (see Nesbitt, this volume). Deposits of this type in the Grass Valley part of the Mother Lode belt in California, USA, are hosted exclusively by granitoid rocks (Johnston, 1940), and have been compared by Morrison (1988) with the Charters Towers gold veins (Table 6.7), although the latter do seem to carry a greater abundance of base metals. However, Morrison's (1988) analogy draws attention to the convergence in geological characteristics between the Mother Lode and pluton-hosted vein types of gold deposit – a topic beyond the scope of this review but requiring further attention.

6.11 Concluding remarks

Several distinct kinds of gold deposits in volcano-plutonic arcs are components of intrusion-centred base- and precious-metal systems, many of which possess porphyry-type Cu, Mo and/or Au mineralization as a central focus. The intrusion-centred systems may also encompass many epithermal and sediment-hosted (Carlin-type) gold deposits. There is a strong suggestion that systems possessing structurally prepared host rocks, especially if calcareous, tend to contain predominantly wallrock-hosted gold deposits, whereas 'tight' systems are more likely to be characterized by intrusion-hosted gold. However, some giant systems, like Bingham, exhibit metallization from their cores to their distal extremities.

An appreciation of the variety of gold deposit types in intrusion-centred systems, the geological parameters that controlled them, their mutual interrelationships and the resultant metal zoning patterns provides a cogent framework for gold exploration in volcano-plutonic arcs. When one reflects on several recent discoveries of intrusion-related gold deposits in long-active mining districts, such as those at Barney's Canyon, Cove, Fortitude, Lepanto and Purísima Concepción, it is obvious that numerous intrusion-centred base- and precious-metal districts world wide possess untested gold potential. No less than 25 of the 33 newly discovered (post-1979) deposits considered here (Tables 6.1–6.7) are located in old mining districts. Furthermore, given that districts possess radii as great as 8 km, individual exploration targets can be large.

At the practical level, however, accurate and expeditious geological diagnosis permits recognition of the style(s) of gold mineralization revealed during geological, geochemical and/or geophysical prospecting, and thereby helps to determine the most suitable exploration methodology to be employed. Geological analogy (e.g. Tables 6.1–6.7) often assists with identification of the style of newly discovered gold mineralization and of its overall economic potential.

Acknowledgements

Selection of gold deposits for this chapter was influenced heavily by personal familiarity. During work on or visits to the deposits, geologists too numerous to mention individually provided invaluable assistance, and their contributions are acknowledged with gratitude. The manuscript benefited from reviews by H.F. Bonham, Jr., R.W. Henley, A. Panteleyev and J.F.H. Thompson.

References

Abbott, G. (1986) Epigenetic mineral deposits of the Ketza-Seagull district, Yukon. *Yukon Geol.* **1**, 56–66.

Afifi, A.M., Kelly, W.C. and Essene, E.J. (1988) Phase relations among tellurides, sulfides, and oxides: I. Applications to telluride-bearing ore deposits. *Econ. Geol.* **83**, 395–404.

Alvarez A, A. and Noble, D.C. (1988) Sedimentary rock-hosted disseminated precious metal mineralization at Purísima Concepción, Yauricocha district, central Peru. *Econ. Geol.* **83**, 1368-1378.

Anzoleaga V, O. (1988) The La Joya copper–gold–silver district, Bolivia. In *Yacimientos Epitermales en Ambientes de Volcanismo Reciente*. Seminario-Taller, 4-16 September 1988, IUGS-UNESCO, La Paz, Bolivia, Preprint. 2pp.

Arnold, G.O. and Sillitoe, R.H. (1989) Mount Morgan gold–copper deposit, Queensland, Australia: Evidence for an intrusion-related replacement origin. *Econ. Geol.* **84** (in press).

Bagby, W.C. and Berger, B.R. (1985) Geologic characteristics of sediment-hosted, disseminated precious-metal deposits in the western United States. In *Geology and Geochemistry of Epithermal Systems*, eds. Berger, B.R. and Bethke, P.M. *Rev. Econ. Geol.* **2**, 169-202.

Baker, E.M. (1987) Brecciation, mineralization and alteration of the Kidston gold deposit. In *Proc. Pacific Rim Congress 87, Gold Coast, Queensland.* Australasian Inst. of Mining and Metallurgy, Parkville, Victoria, 29-33.

Baker, E.M. and Andrew, A.S. (1988) Processes associated with gold mineralization within the Kidston breccia pipe, north Queensland. In *Bicentennial Gold 88*, Melbourne, *Geol. Soc. Aust. Abstr.* **22**, 102-109.

Bamford, R.W. (1972) The Mount Fubilan (Ok Tedi) porphyry copper deposit, Territory of Papua and New Guinea. *Econ. Geol.* **67**, 1019-1033.

Barnes, H.L. (1975) Zoning of ore deposits: Types and causes. *Trans. Roy. Soc. Edin.* **69**, 295-311.

Bauer, H.L. Jr., Breitrick, R.A., Cooper, J.J. and Anderson, J.A. (1966) Porphyry copper deposits in the Robinson mining district, Nevada. In *Geology of the Porphyry Copper Deposits, Southwestern North America*, eds. Titley, S.R. and Hicks, C.L. Univ. of Arizona Press, Tucson, Arizona, 233-244.

Blake, D.W., Wotruba, P.R. and Theodore, T.G. (1984) Zonation in the skarn environment at the Tomboy-Minnie gold deposits, Lander County, Nevada. *Arizona Geol. Soc. Digest* **15**, 67- 72.

Blatchford, A. (1953) Charters Towers goldfield. In *Geology of Australian Ore Deposits*, ed. Edwards, A.B. 5th Empire Mining & Metallurgical Congress, Melbourne, 796-806.

Bloomstein, E.I. (1987) Geology and geochemistry of the Muruntau deposit, the biggest gold mine in the USSR. Society of Mining Engineers 116th Annual Meeting, Denver. Pocket Program, 83.

Bonham, H.F. Jr. (1986) Models for volcanic-hosted epithermal precious metal deposits; a review. In *Proc. Symposium 5: Volcanism, Hydrothermal Systems and Related Mineralization.* International Volcanological Congress, Auckland, 13-17.

Bonham, H.F. Jr. (1988) Models for volcanic-hosted precious metal deposits: A review. In *Bulk Mineable Precious Metal Deposits of the Western United States*, eds. Schafer, R.W., Cooper, J.J. and Vikre, P.G. Geol. Soc. of Nevada, Reno, 259-271.

Bowman, J.R., Parry, W.T., Kropp, W.P. and Kruer, S.A. (1987) Chemical and isotopic evolution of hydrothermal solutions at Bingham, Utah. *Econ. Geol.* **82**, 395-428.

Buntzen, T.K., Cox, B.C. and Veach, N.C. (1987) Heavy mineral provenance studies in the Iditarod and Innoko districts, western Alaska. In *Process Mineralogy VII*, eds. Vassiliou, A.H., Hausen, D.M. and Carson, D.J.T. The Metallurgical Society, Warrendale, Pennsylvania, 221-245.

Burnham, C.W. (1979) Magmas and hydrothermal fluids. In *Geochemistry of Hydrothermal Ore Deposits* (2nd edn.), ed. Barnes, H.L. Wiley, New York, 71-136.

Burnham, C.W. (1985) Energy release in subvolcanic environments: Implications for breccia formation. *Econ. Geol.* **80**, 1515-1522.

Burnham, C.W. and Ohmoto, H. (1980) Late-stage processes of felsic magmatism. In *Granitic Magmatism and Related Mineralization*, eds. Ishihara, S. and Takenouchi, S. *Mining Geol.* (Japan), Spec. Issue, **8**, 1-11.

Callow, K.J. (1967) The geology of the Thanksgiving mine, Baguio district, Mountain Province, Philippines. *Econ. Geol.* **62**, 472-481.

Cameron, D.E. and Garmoe, W.J. (1987) Geology of skarn and high-grade gold in the Carr Fork mine, Utah. *Econ. Geol.* **82**, 1319-1333.

Cameron, E.M. and Hattori, K. (1987) Archean gold mineralization and oxidized hydrothermal fluids. *Econ. Geol.* **82**, 1177-1191.

Chappell, B.W. and White, A.J.R. (1974) Two contrasting granite types. *Pacific Geol.* **8**, 173-174.

Cornelius, K.D. (1969) The Mount Morgan mine, Queensland – A massive gold–copper pyritic replacement deposit. *Econ. Geol.* **64**, 885-902.

Cyr, J.B., Pease, R.P. and Schroeter, T.G. (1984) Geology and mineralization at Equity Silver mine. *Econ. Geol.* **79**, 947-968.

Davies, R.M. and Ballantyne, G.H. (1987) Geology of the Ladolam gold deposit, Lihir Island, Papua New Guinea. In *Proc. Pacific Rim Conference 87, Gold Coast, Queensland.* Australasian Inst. of Mining and Metallurgy, Parkville, Victoria, 943-949.

DeWitt, E., Redden, J.A., Burack Wilson, A. and Buscher, D. (1986) Mineral resource potential and geology of the Black Hills National Forest, South Dakota and Wyoming. US Geol. Surv., Bull. 1580, 135pp.

Durgin, D.C. (1989) Gold in the Robinson mining district. Soc. Mining Engnrs Annual Meeting, Las Vegas. Preprint 89-42. 7pp.

Eadington, P.J., Smith, J.W. and Wilkins, R.W.T. (1974) Fluid inclusion and stable isotope research, Mount Morgan, Queensland. In *Australasian Institute of Mining & Metallurgy, Southern and Central Queensland Conf.*, Brisbane. Papers, 441-444.

Eastoe, C.J. (1982) Physics and chemistry of the hydrothermal system at the Panguna porphyry copper deposit, Bougainville, Papua New Guinea. *Econ. Geol.* **77**, 127-153.

Einaudi, M.T. (1982) Description of skarns associated with porphyry copper plutons: southwestern North America. In *Advances in Geology of the Porphyry Copper Deposits, Southwestern North America*, ed. Titley, S.R. Univ. of Arizona Press, Tucson, 139-183.

Einaudi, M.T., Meinert, L.D. and Newberry, R.J. (1981) Skarn deposits. *Econ. Geol.*, 75th Anniv. Vol., 317-391.

Emmons, D.L. and Coyle, R.D. (1988) Echo Bay details exploration activities at its Cove gold deposit in Nevada. *Mining Engng* **40**, 791-794.

Enders, M.S. and Rogers, L.M. (1983) Design of the geologic program at the Zortman and Landusky mines, Little Rocky Mountains, Montana. Soc. Mining Engnrs AIME Fall Meeting, Salt Lake City. Preprint 83-373, 9pp.

Ewers, G.R. and Sun, S.-S. (1988) Genesis of the Red Dome deposit, northeast Queensland. In *Bicentennial Gold 88*, Melbourne. *Geol. Soc. Aust. Abstr.* **22**, 110-115.

Fleming, A.W., Handley, G.A., Williams, K.L., Hills, A.L. and Corbett, G.J. (1986) The Porgera gold deposit, Papua New Guinea. *Econ. Geol.* **81**, 660-680.

Fox, P.E., Cameron, R.S. and Hoffman, S.J. (1987) Geology and soil geochemistry of the Quesnel River gold deposit, British Columbia. In *GEOEXPO/86: Exploration in the North American Cordillera*, eds. Elliot, L.L. and Smee, B.W. Assoc. of Exploration Geochemists, Rexdale, Ontario, 61-71.

Franklin, J.M. and Thorpe, R.L. (1982) Comparative metallogeny of the Superior, Slave and Churchill provinces. In *Precambrian Sulphide Deposits*, eds. Hutchinson, R.W., Spence, C.D. and Franklin, J.M. Geol. Assoc. Can., Spec. Paper 25, 3-90.

Frost, J.E. (1959) Notes on the genesis of the ore-bearing structures of the Paracale district, Camarines Norte, Philippines. *Philippine Geol.* **13**, 31-43.

Geikie, J.S. (1906) The occurrence of gold in Upper Sarawak. *Trans. Inst. Mining Metall.* **15**, 63-86.

Giles, D.L. (1983) Gold mineralization in the laccolithic complexes of central Montana. In *The Genesis of Rocky Mountain Ore Deposits: Changes with Time and Tectonics, Proc. Denver Regional Exploration Geologists Society Symp., Denver 1982*, 157-162.

Gonzalez, A. (1956) Geology of the Lepanto copper mine, Mankayan, Mountain Province. In *Copper Deposits of the Philippines*, eds. Kinkel, A.R. Jr. *et al.* Philippine Bur. Mines, Spec. Projects Ser. Pub. 16, 17-50.

Handley, G.A. and Bradshaw, P.M.D. (1986) The Porgera gold deposit, Papua New Guinea. In *Proc. Gold '86 Symposium*, ed. MacDonald, A.J. Toronto 1986, 416-424.

Harris, M. (1980a) Gold mineralization at the Salave gold prospect, northwest Spain. *Trans. Inst. Mining Metall.* **B89**, B1-B4.

Harris, M. (1980b) Hydrothermal alteration at Salave gold prospect, northwest Spain. *Trans. Inst. Mining Metall.* **B89**, B5-B15.

Hastings, J.S. (1987) Ore deposits of the Little Rocky Mountains, Montana. Soc. of Mining Engineers 116th Annual Meeting, Denver. Pocket Program, 74.

Hastings, J.S. (1988) Gold deposits of Zortman-Landusky, Little Rocky Mountains, Montana. In *Bulk Mineable Precious Metal Deposits of the Western United States*, eds. Schafer, R.W., Cooper, J.J. and Vikre, P.G. Geol. Soc. of Nevada, Reno, 187- 205.

Hastings, J.S. and Harrold, J.L. (1988) Geology of the Beal gold deposit, German Gulch, Montana. In *Bulk Mineable Precious Metal Deposits of the Western United States*, eds. Schafer, R.W., Cooper, J.J. and Vikre, P.G. Geol. Soc. Nevada, Reno, 207- 220.

Heald, P., Foley, N.K. and Hayba, D.O. (1987) Comparative anatomy of volcanic-hosted epithermal deposits: Acid-sulfate and adularia-sericite types. *Econ. Geol.* **82**, 1-26.

Hemley, J.J., Cygan, G.L. and d'Angelo, W.M. (1986) Effect of pressure on ore mineral solubilities under hydrothermal conditions. *Geology* **14**, 377-379.

Hemley, J.J., Cygan, G.C., Woo, C.L. and d'Angelo, W.M. (1987) Exploratory studies on metal sulfide solubilities in buffered systems: Implications to zoning in ore deposits. *Geol. Soc. Amer. Abstr. Programs* **19**, 699.

Henley, R.W. (1985) The geothermal framework for epithermal deposits. In *Geology and Geochemistry of Epithermal Systems*, eds. Berger, B.R. and Bethke, P.M. *Rev. Econ. Geol.* **2**, 1-24.

Henley, R.W. and McNabb, A. (1978) Magmatic vapor plumes and ground water interaction in porphyry copper emplacement. *Econ. Geol.* **73**, 1-20.

Henry, D.D. (1988) Porgera – the first fifty years. *Bicentennial Gold 88*, Melbourne. *Geol. Soc. Aust. Abstr.* **22**, 144-149.

Hewitt, W.V., Leckie, J.F., Lorraway, R., Moule, K., Rush, P.M., Seegers, H.-J. and Torr, G.L. (1980) Ok Tedi project – Geology and principal ore deposits. *BHP Tech. Bull.* **24** (1), 3-10.

Howell, W.J.S., Fardon, R.S.H., Carter, R.J. and Bumstead, E.D. (1978) History of the Ok Tedi porphyry copper prospect, Papua New Guinea. II. The years 1975 to 1978. *Econ. Geol.* **73**, 802-809.

Ishihara, S. (1981) The granitoid series and mineralization. *Econ. Geol.*, 75th Anniv. Vol., 458-484.

James, L.P. (1976) Zoned alteration in limestone at porphyry copper deposits, Ely, Nevada. *Econ. Geol.* **71**, 488-512.

Johnston, W.D., Jr. (1940) The gold–quartz veins of Grass Valley, California. US Geol. Surv., Prof. Paper 194. 101pp.

Jouravsky, G. (1952) Or. In *Géologie des Gîtes Mineraux Marocains.* Serv. Géol. Maroc, Notes et Mem. 87, 223-232.

Keith, S.B. (1983) Distribution of fossil metallogenic systems and magma geochemical belts within the Great Basin and vicinity from 145 million years ago to present (Abstr.). Geothermal Resources Council, Spec. Rep. 13, 285-286.

Keith, S.B. and Swan, M.M. (1987a) Oxidation state of magma series in the southwestern U.S.: Implications for geographic distribution of base, precious, and lithophile metal metallogeny. *Geol. Soc. Amer. Abstr. Programs* **19**, 723-724.

Keith, S.B. and Swan, M.M. (1987b) Peraluminous gold deposits. Soc. of Mining Engineers 116th Annual Meeting, Denver. Pocket Program, 82.

Krason, J. (1984) Muruntau: The world's largest gold producing mine complex. *Mining Engng* **36**, 1549-1550.

Kuyper, B.A. (1988) Geology of the McCoy gold deposit, Lander County, Nevada. In *Bulk Mineable Precious Metal Deposits of the Western United States*, eds. Schafer, R.W., Cooper, J.J. and Vikre, P.G. Geol. Soc. of Nevada, Reno, 173-185.

Lawrence, L.J. (1978) Porphyry type gold mineralization in shoshonite at Vunda, Fiji. *Proc. Aust. Inst. Mining Metall.* **256**, 1-14.

Leveille, R.A., Newberry, R.J. and Bull, K.F. (1988) An oxidation state-alkalinity diagram for discriminating some gold-favorable plutons: An empirical and phenomenological approach. *Geol. Soc. Amer. Abstr. Programs* **20**, A142.

Lindquist, W.F. (1980) The exploration of the Ortiz gold deposit, New Mexico. Geology and exploration. *Colorado Mining Assoc., 1980 Mining Yearbook*, 106-112.

Lindsey, D.A. (1985) A gold-mineralized breccia zone at Kendall, North Moccasin Mountains, Fergus County, Montana. US Geol. Surv., Prof. Paper 1301-C, 43-56.

Lindsey, D. A. and Fisher, F.S. (1985) Mineralized breccias in intrusive complexes of late Cretaceous and Paleocene age, north-central Montana. US Geol. Surv., Prof. Paper 1301-A, 1-34.

Lindsey, D.A. and Naeser, C.W. (1985) Relation between igneous intrusion and gold mineralization in the North Moccasin Mountains, Fergus County, Montana. US Geol. Surv., Prof. Paper 1301-B, 35-41.

Llaumett P., C. (1980) Antecedentes sobre producción, recursos y expectativas de la minería del oro en Chile. *Minerales* (Santiago de Chile) **35** (150), 9-20.

MacLeod, R.J. (1987) The geology of the Gilt Edge gold deposit, northern Black Hills, South Dakota. In *Third Western Regional Conf. on Precious Metals, Coal, and Environment*, eds. Han, K.N. and Kliche, C.A. Rapid City 1987, Black Hills Section AIME, Rapid City, South Dakota, 118-122.

Malihan, T.D. (1987) The gold-rich Dizon porphyry copper mine in the western central Luzon island, Philippines: Its geology and tectonic setting. In *Proc. Pacific Rim Congress 87, Gold Coast, Queensland.* Australasian Inst. of Mining and Metallurgy, Parkville, Victoria, 303-307.

Mason, D.R. and McDonald, J.A. (1978) Intrusive rocks and porphyry copper occurrences of the Papua New Guinea-Solomon Islands region: A reconnaissance study. *Econ. Geol.* **73**, 857-877.

McAllister, J.F., Flores W., H. and Ruiz F., C. (1950) Quicksilver deposits of Chile. US Geol. Surv., Bull. 964-E, 361- 400.

McMillan, W.J., Panteleyev, A. and Höy, T. (1987) Mineral deposits in British Columbia: A review of their tectonic settings. In *GEOEXPO/86: Exploration in the North American Cordillera*, eds. Elliot, L.L. and Smee, B.W. Assoc. of Exploration Geochemists, Rexdale, Ontario, 1-18.

Meinert, L.D. (1987) Skarn zonation and fluid evolution in the Groundhog mine, Central mining district, New Mexico. *Econ. Geol.* **82**, 523-545.

Meinert, L.D. (1988) Gold in skarn deposits – a preliminary overview. In *Proc. 7th Quadrennial IAGOD Symposium, Lulea, Sweden 1986*, ed. Zachrisson, E. E. Schweizerbart'sche, Stuttgart, 363-374.

Melling, D.R. and Watkinson, D.H. (1988) Alteration of fragmental basaltic rocks: The Quesnel River gold deposit, central British Columbia. British Columbia Ministry Energy Mines Petrol. Res., Geol. Fieldwork 1987, Paper 1988-1, 335-347.

Mercado, A.C., Estoque, J., Lodrigueza, L. and Rebillon, F. (1987) Geology and ore deposits of Masara mine, Davao del Norte, Philippines. In *Proc. Pacific Rim Congress 87, Gold Coast, Queensland.* Australasian Inst. of Mining and Metallurgy, Parkville, Victoria, 329-332.

Meyer, C. and Hemley, J.J. (1967) Wall rock alteration. In *Geochemistry of Hydrothermal Ore Deposits*, ed. Barnes, H.L. Holt, Rinehart, and Winston, New York, 166-235.

Miller, R.McG. (1983) The Pan African Damara orogen of South West Africa/Namibia. *Spec. Pub. Geol. Soc. S. Afr.* **11**, 431-515.

Moore, W.J. (1973) Igneous rocks in the Bingham mining district, Utah. US Geol. Surv. Prof. Paper 629-B, B1-B42.

Morrison, G.W. (1988) Paleozoic gold deposits of northeast Queensland. In *Bicentennial Gold 88*, Melbourne. *Geol. Soc. Aust. Abstr.* **22**, 91-101.

Morrison, G.W., Andrew, A.S. and Teale, G. (1988) Origin of gold mineralisation in the Mount Leyshon diatreme. In *Bicentennial Gold 88*, Melbourne. *Geol. Soc. Aust. Abstr.* **22**, 116-120.

Morrison, G.W., Teale, G.S. and Hodkinson, I. (1987) Geology and gold mineralisation at Mount Leyshon, north Queensland. In *Proc. Pacific Rim Congress 87, Gold Coast, Queensland.* Australasian Inst. of Mining and Metallurgy, Parkville, Victoria, 777-780.

Mustard, H. (1986) Geology and genesis of the Kidston gold deposit, Australia. In *Proc. Gold '86 Symposium*, ed. MacDonald, A.J. Toronto 1986, 404-415.

Mutschler, F.E., Griffen, M.E., Scott Stevens, D. and Shannon, S.S. Jr. (1985) Precious metal deposits related to alkaline rocks in the North American Cordillera – an interpretative review. *Trans. Geol. Soc. S. Afr.* **88**, 355-377.

Myers, G.L. and Meinert, L.D. (1989) Alteration zonation of the Fortitude gold skarn deposit, Lander County, NV. Soc. Mining Engnrs Annual Meeting, Las Vegas, Preprint 89-199, 11pp.

Nesbitt, B.E. (1988) Gold deposit continuum: A genetic model for lode Au mineralization in the continental crust. *Geology* **16**, 1044-1048.

Ney, C.N., Anderson, J.M. and Panteleyev, A. (1972) Discovery, geologic setting and style of mineralization, Sam Goosly deposit, B.C. *CIM Bull.* **65** (723), 53-64.

Orris, G.J., Bliss, J.D., Hammarstrom, J.M. and Theodore, T.G. (1987) Description and grades and tonnages of gold-bearing skarns. US Geol. Surv., Open-File Rep. 87-273, 50pp.

Panteleyev, A. (1986) Ore deposits #10. A Canadian Cordilleran model for epithermal gold–silver deposits. *Geosci. Can.* **13**, 101-111.

Paterson, C.J., Uzunlar, N. and Longstaffe, F.J. (1987) Epithermal Au–Ag deposits in the northern Black Hills: A variety of ore-forming fluids. In *Third Western Regional Conf. on Precious Metals, Coal, and Environment*, eds. Han, K.N. and Kliche, C.A. Rapid City 1987, Black Hills Section AIME, Rapid City, South Dakota, 102-108.

Paterson, C.J., Lisenbee, A.L. and Redden, J.A. (1988) Gold deposits in the Black Hills, South Dakota. In *Wyoming Geol. Assoc. 39th Field Conference Guidebook, Eastern Powder River Basin-Black Hills*, eds. Dietrich, R.P., Dyka, M.A.K. and Miller, W.R. 295-304.

Peters, S.G. and Golding, S.D. (1987) Relationship of gold quartz mineralisation to granodioritic phases and mylonites at Charters Towers goldfield, northeastern Queensland. In *Proc. Pacific Rim Congress 87, Gold Coast, Queensland.* Australasian Inst. of Mining and Metallurgy, Parkville, Victoria, 363-367.

Porter, E.W. and Ripley, E. (1985) Petrologic and stable isotope study of the gold-bearing breccia pipe at the Golden Sunlight deposit, Montana. *Econ. Geol.* **80**, 1689-1706.

Radtke, A.S. and Dickson, F.W. (1974) Genesis and vertical position of fine-grained disseminated replacement-type gold deposits in Nevada and Utah, U.S.A. In *Problems of Ore Deposition* 1, *Volcanic Ore Deposits, Proc. 4th IAGOD Symp., Varna, Romania 1974*, 71-78.

Ray, G.E., Dawson, G.L. and Simpson, R. (1988) Geology, geochemistry and metallogenic zoning in the Hedley gold-skarn camp. British Columbia Ministry Energy Mines Petrol. Res., Geol. Fieldwork 1987, Paper 1988-1, 59-80.

Redwood, S.D. (1987) The Soledad caldera, Bolivia: A Miocene caldera with associated epithermal Au–Ag–Cu–Pb–Zn mineralization. *Geol. Soc. Amer. Bull.* **99**, 395-404.

Roberts, R.J. (1966) Metallogenic provinces and mineral belts in Nevada. In *Papers Presented at the AIME Pacific Southwest Mineral Industry Conference*, Sparks, Nevada 1965, Nevada Bur. of Mines and Geol. Rep. 13, 47-72.

Rodriguez, C, C.J. and Pernett, M, A. (1983) Recursos minerales de Antioquia. *Bol. Geol.* (Ingeominas, Bogotá, Colombia) **26**(3), 116pp.

Rogers, L.M. and Enders, M.S. (1982) Syenite-hosted gold deposits of the Little Rocky Mountains, Montana. 88th Annual Northwest Mining Association Convention, Spokane, Washington, 10 December 1982. Preprint. 34pp.

Sang, J. and Ho, S.C. (1987) A review of gold deposits in China. Geol. Dept. & Univ. Extension, Univ. of W. Australia, Pub. 11, 307-320.

Schmidt, R.G. (1985) High-alumina hydrothermal systems in volcanic rocks and their significance to mineral prospecting in the Carolina Slate Belt. US Geol. Surv., Bull. 1562, 59pp.

Serafica, V.S. and Baluda, R.P. (1977) Geology of the Philex Santo Tomas II ore body. *5th Symposium on Mineral Resources Development*, Baguio City, Philippines, November 1977, Sec. 1, Paper 3.

Seward, T.M. (1984) The transport and deposition of gold in hydrothermal systems. In *Gold '82: The Geology, Geochemistry and Genesis of Gold Deposits*, ed. Foster, R.P. Proc. Symposium, Harare, Zimbabwe 1982, Balkema, Rotterdam, 165-181.

Sharp, J.E. (1984) A gold mineralized breccia pipe complex in the Clark Mountains, San Bernadino County, California. *Arizona Geol. Soc. Digest* **15**, 119-139.

Sheppard, S.M.F., Nielsen, R.L. and Taylor, H.P. Jr. (1971) Hydrogen and oxygen isotope ratios in minerals from porphyry copper deposits. *Econ. Geol.* **66**, 515-542.

Silberman, M.L. (1985) Geochronology of hydrothermal alteration and mineralization: Tertiary hydrothermal precious-metal deposits in the Great Basin. In *Geologic Characteristics of Sediment- and Volcanic-hosted Disseminated Gold Deposits – Search for an Occurrence Model*, ed. Tooker, E.W. US Geol. Surv., Bull. 1646, 55-70.

Sillitoe, R.H. (1972) A plate tectonic model for the origin of porphyry copper deposits. *Econ. Geol.* **67**, 184-197.

Sillitoe, R.H. (1979) Some thoughts on gold-rich porphyry copper deposits. *Mineral. Deposita* **14**, 161-174.

Sillitoe, R.H. (1983a) Unconventional metals in porphyry systems. In *Cameron Volume on Unconventional Mineral Deposits*, ed. Shanks, W.C. III. Soc. of Mining Engineers AIME, New York, 207- 221.

Sillitoe, R.H. (1983b) Enargite-bearing massive sulfide deposits high in porphyry copper systems. *Econ. Geol.* **78**, 348-352.

Sillitoe, R.H. (1983c) Styles of low-grade gold mineralization in volcano-plutonic arcs. In *Papers given at the Precious-Metals Symposium*, Sparks, Nevada 1980, Nevada Bur. of Mines and Geol. Rep. 36, 52-68.

Sillitoe, R.H. (1985) Ore-related breccias in volcanoplutonic arcs. *Econ. Geol.* **80**, 1467-1514.

Sillitoe, R.H. (1987) Copper, gold and subduction: A trans-Pacific perspective. In *Proc. Pacific Rim Congress 87, Gold Coast, Queensland.* Australasian Inst. of Mining and Metallurgy, Parkville, Victoria, 399-403.

Sillitoe, R.H. (1988) Gold and silver deposits in porphyry systems. In *Bulk Mineable Precious Metal Deposits of the Western United States*, eds. Schafer, R.W., Cooper, J.J. and Vikre, P.G. Geol. Soc. of Nevada, Reno, 233-257.

Sillitoe, R.H. (1989a) Gold deposits in western Pacific island arcs: The magmatic connection. In *The Geology of Gold Deposits: The Perspective in 1988*, eds. Keays, R.R., Ramsay, W.R.H. and Groves, D.I. *Econ. Geol. Monog.* **6**, 274-291.

Sillitoe, R.H. (1989b) Geotectonic setting of western Pacific gold deposits. In *Characterization and Comparison of Ancient (Precambrian-Mesozoic) Continental Margins* eds. Bartholomew, M.J., Hyndman, D.W., Mogk, D.W. and Mason, R., D. Reidel Publishing Co., Dordrecht, Holland (in press).

Sillitoe, R.H. and Bonham, H.F., Jr. (1990) Sediment-hosted gold deposits: Distal products of magmatic-hydrothermal systems. *Geology* **18**, 157-161.

Sillitoe, R.H. and Gappe, I.M. Jr. (1984) Philippine porphyry copper deposits: Geologic setting and characteristics. United Nations ESCAP, CCOP Tech Pub. 14. 89pp.

Sillitoe, R.H., Grauberger, G.L. and Elliott, J.E. (1985) A diatreme-hosted gold deposit at Montana Tunnels, Montana. *Econ. Geol.* **80**, 1707-1721.

Sinclair, W.D. (1982) Gold deposits of the Matachewan area, Ontario. In *Geology of Canadian Gold Deposits*, eds. Hodder, R.W. and Petruk, W. Can. Inst. of Mining and Metallurgy, Special Vol. 24, 83-93.

Smith, M.R., Wilson, W.R., Benham, J.A., Pescio, C.A. and Valenti, P. (1988) The Star Pointer gold deposit, Robinson mining district, White Pine County, Nevada. In *Bulk Mineable Precious Metal Deposits of the Western United States*, eds. Schafer, R.W., Cooper, J.J. and Vikre, P.G. Geol. Soc. Nevada, Reno, 221-231.

Stoffregen, R. (1987) Genesis of acid-sulfate alteration and Au–Cu–Ag mineralization at Summitville, Colorado. *Econ. Geol.* **82**, 1575-1591.

Strishkov, V.V. (1986) The Muruntau gold complex. *Mining Mag.* **155**, 207-209.

Symonds, R.B., Rose, W.I., Reed, M.H., Lichte, M.E. and Finnegan, D.L. (1987) Volatilization, transport and sublimation of metallic and non-metallic elements in high temperature gases at Merapi volcano, Indonesia. *Geochim. Cosmochim. Acta* **51**, 2083-2101.

Symons, P.M., Anderson, G., Beard, T.J., Hamilton, L.M., Reynolds, G.D., Robinson, J.M. and Staley, R.W. (1988) The Boddington gold deposit. In *Bicentennial Gold 88*, Melbourne, *Geol. Soc. Aust. Abstr.* **22**, 56-61.

Takenouchi, S. (1981) Fluid inclusion studies of several Philippine porphyry copper deposits. In *Metallogeny of Asia, 1980*, eds. Ishihara, S. and Sasaki, A. Geol. Surv. of Japan, Rep. 261, 149-167.

Tangonan, M. (1987) Geology and mineralization of Thanksgiving mine of Benguet Exploration, Inc. at Camp 6, Tuba, Benguet. In *Proc. Gold '87 in the Philippine Setting, Manila*, Vol. 2. Philippine Inst. of Mining, Metallurgy and Geology, 1/9- 12/9.

Taube, A. (1986) The Mount Morgan gold-copper mine and environment, Queensland: A volcanogenic massive sulfide deposit associated with penecontemporaneous faulting. *Econ. Geol.* **81**, 1322-1340.

Taylor, B.E. (1988) Degassing of rhyolitic magmas: Hydrogen isotope evidence and implications for magmatic-hydrothermal ore deposits. In *Recent Advances in the Geology of Granite-related Mineral Deposits*, eds. Taylor, R.P. and Strong, D.F. Can. Inst. of Mining and Metallurgy, Special Vol. 39, 33-49.

Taylor, J.D. and Bolin, D.S. (1988) Geology and gold mineralization at the Foley Ridge deposit, Bald Mountain mining district, Lawrence County, South Dakota (Abstr.). In *Bulk Mineable Precious Metal Deposits of the Western United States*, eds. Schafer, R.W., Cooper, J.J., and Vikre, P.G. Geol. Soc. of Nevada, Reno, 740.

Theodore, T.G. and Blake, D.W. (1978) Geology and geochemistry of the West ore body and associated skarns, Copper Canyon, Lander County, Nevada, with a section on microprobe analysis by N.J. Banks. US Geol. Surv., Prof. Paper 798-C. 85pp.

Theodore, T.G., Howe, S.S., Blake, D.W. and Wotruba, P.R. (1986) Geochemical and fluid zoning in the skarn environment at the Tomboy-Minnie gold deposits, Lander County, Nevada. *J. Geochem. Explor.* **25**, 99-128.

Titley, S.R. and Megaw, P.K.M. (1985) Carbonate-hosted ores of the western Cordillera – an overview. Soc. Mining Engnrs AIME Annual Meeting, New York, Preprint 85-115. 17pp.

Torrey, C.E., Karjalainen, H., Joyce, P.J., Erceg, M. and Stevens, M. (1986) Geology and mineralization of the Red Dome (Mungana) gold skarn deposit, north Queensland, Australia. In *Proc. Gold '86 Symposium*, ed. MacDonald, A.J. Toronto 1986, 504-517.

Tosdal, R.M. and Smith, D.B. (1987) Gneiss-hosted kyanite gold and gneiss-hosted epithermal gold: A supplement to U.S. Geological Survey Bulletin 1693. US Geol. Surv., Open-File Rep. 87-272b. 8pp.

Walker, S.D. and Cregheur, P. (1982) The Chadbourne mine, Noranda, Quebec: a gold-bearing breccia. In *Geology of Canadian Gold Deposits*, eds. Hodder, R.W. and Petruk, W. Can. Inst. of Mining and Metallurgy, Special Vol. 24, 58-66.

Watanabe, T. (1943) Geology and mineralization of the Suian district, Yuosen (Korea). *J. Fac. Sci. Hokkaido Imperial Univ., Ser. IV, Geol. & Mineral.* **6**(3–4), 205-303.

Willis, G.F., Tosdal, R.M. and Manske, S.L. (1989) Structural control on epithermal gold veins and breccias in the Mesquite district, southeastern California. In *USGS Research on Mineral Resources – 1989 Program and Abstracts*, ed. Schindler, K.S. US Geol. Surv., Circ. 1035, 78-79.

Wilson, G.I., Lewis, R.W., Gallo, J.B. and Tullemans, F.J. (1986) The geology of the Kidston gold mine. In *Publications of the 13th Congress of the Council of Mining & Metallurgical Institutions & Australasian Institute of Mining & Metallurgy*, ed. Berkman, D.A. Vol. 2. Geology & Exploration, Singapore 1986, 235-242.

Wilson, M.R. and Kyser, T.K. (1988) Geochemistry of porphyry-hosted Au-Ag deposits in the Little Rocky Mountains, Montana. *Econ. Geol.* **83**, 1329-1346.

Wolfenden, E.B. (1965) Bau mining district, West Sarawak, Malaysia. Part I. Bau. Geol. Surv. Borneo Region, Malaysia, Bull. 7. 147pp.

Wojdak, P.J. and Sinclair, A.J. (1984) Equity Silver silver-copper-gold deposit: Alteration and fluid inclusion studies. *Econ. Geol.* **79**, 969-990.

Wotruba, P.R., Benson, R.G. and Schmidt, K.W. (1988) Geology of the Fortitude gold-silver skarn deposit, Copper Canyon, Lander County, Nevada. In *Bulk Mineable Precious Metal Deposits of the Western United States*, eds. Schafer, R.W., Cooper, J.J. and Vikre, P.G. Geol. Soc. of Nevada, Reno, 159-171.

Wright, A. (1983) The Ortiz gold deposit (Cunningham Hill) – Geology and exploration. In *Papers given at the Precious-Metals Symposium*, Sparks, Nevada 1980. Nevada Bur. of Mines, and Geol. Rep. **36**, 42-51.

Wyman, D. and Kerrich, R. (1988) Alkaline magmatism, major structures, and gold deposits: Implications for greenstone belt gold metallogeny. *Econ. Geol.* **83**, 454-461.

Zhu, F. (1988) Study on the geochemistry, lithology and metallogeny of Zhao-Ye migmatitic hydrothermal gold deposits belt. In *Bicentennial Gold 88*, Melbourne, *Geol. Soc. Aust. Abstr.* **23**, Vol. 1, 336-338.

7 The geology and origin of Carlin-type gold deposits

B.R. BERGER and W.C. BAGBY

7.1 Introduction

The static price of gold coupled with increased costs for underground mining in the post-World War II era in the United States led companies such as the Newmont Mining Company to seek precious-metal properties amenable to low-cost mining (Ramsey, 1973). Following concepts proposed on the relationship of gold and silver mineralization to windows through the Palaeozoic Roberts Mountains thrust fault in north-central Nevada (Roberts, 1960), Newmont conducted an exploration programme in the Tuscarora Mountains north of Carlin, Nevada, that resulted in the discovery of the Carlin gold deposit in 1962. Because of both the exploration and research attention paid to the Carlin Mine in the years following its discovery, deposits with similar mineralogy, host rocks, and trace elements to Carlin have become known as 'Carlin-type deposits' (Figure 7.1).

Carbonate-replacement gold deposits associated with alkaline igneous intrusions similar in many, but not all, respects to the Carlin-type are here termed 'Carlin-like deposits'. Data on these deposits are included to aid the reader in discriminating them from the Carlin-type.

In this chapter, we first discuss the classification of sedimentary rock-hosted gold and silver deposits as a general class in order to better define the nature of Carlin-type deposits. This is followed by a discussion of the regional geological and tectonic settings of Carlin-type deposits in both North America and China. The general characteristics of these deposits and those of selected, specific deposits are discussed in the third section. The final sections of this chapter deal with deposit models, speculations regarding the genesis of Carlin-type deposits, and some exploration guidelines. A complete review of all of the individual deposits is beyond the scope of this chapter and the reader is referred to the Reference section and bibliographies contained in those papers for further information. Recent general reviews of Carlin-type deposits include Bagby and Berger (1985) and Percival *et al.* (1988).

7.2 Classification of Carlin-type deposits

7.2.1 *A historical perspective*

What are referred to in this chapter as Carlin deposits were classified by Lindgren (1913) in a subcategory of mesothermal deposits in which gold-bearing jasperoids replace limestone (e.g. Black Hills, South Dakota; Mercur, Utah). Lindgren noted that few examples of this deposit type were known outside the western US Cordillera, and he did not discriminate between Carlin-type (Mercur) and Carlin-like (Black Hills). Joralemon (1951) published a description of the Carlin-type Getchell Mine in north-central Nevada, and noted open-space filling textures and the similarity of the geochemistry to that found in active hot springs and, from this period on, the common usage has been to include Carlin deposits in the epithermal rather than in the mesothermal category.

Weed and Pirsson (1898) described the Judith Mountains district, Montana, where limestone intruded by alkalic igneous rocks is replaced by silica with associated gold and fluorite. Lindgren (1913) observed that the Black Hills silicification replaces dolomite 'with great fidelity', a common textural feature in jasperoids of Carlin-type deposits. Giles (1983) mentioned the similarity of the Montana alkalic-rock-related replacement gold deposits to Carlin-type deposits, including low base-metal concentrations and regional anomalies in arsenic, antimony, and mercury (e.g. Kendall Mine, Montana).

By analogy with modern hot springs, Joralemon (1951) interpreted the Getchell, Nevada, deposit to have formed close to the surface based on the occurrence of vugs and breccia textures and the presence of gold in association with realgar, orpiment, stibnite, and cinnabar. Hausen (1967) interpreted the silicification at the Carlin gold mine, Nevada, as being similar to surficial siliceous sinter found around hot-spring vents, and the trace-element associations as similar to those in the Steamboat hot springs, Nevada.

Bagby and Berger (1985) accepted the epithermal classification for Carlin-type deposits, but included them along with a wider range of gold and silver replacement deposits in a single class they termed 'sediment-hosted, disseminated precious-metal deposits'. Bagby and Berger (1985) further subdivided this class into subsets based upon the metal suites and host-rock alteration characteristics of individual deposits. By contrast, Sawkins (1984) likened Carlin-type deposits to 'distal, low-temperature equivalents of gold-bearing skarns'. He based this interpretation on their association with igneous rocks, with high-angle faults that controlled the flow of hydrothermal fluids, and with the presence of reactive carbonate-bearing rocks.

7.2.2 *Current perspectives*

Recent research at Carlin and at other Carlin-type deposits has provided new insights into the genesis of these deposits. It is now clear that, as Lindgren (1913) suspected, many of the deposits (e.g. Carlin, Getchell, Jerritt Canyon) were formed at considerable depth relative to the depths of formation for most volcanic-rock-related epithermal deposits (e.g. Bodie and McLaughlin, California; Comstock Lode and Tonopah, Nevada). In addition, there are other attributes that indicate an environment of deposition dissimilar to most epithermal-type deposits including the considerable vertical continuity of mineral assemblages without apparent zonation and, in several districts, the spatial and temporal association of gold mineralization with granitic intrusive complexes that form tungsten- and molybdenum-bearing skarns (e.g. Getchell, Gold Acres, White Caps, Northumberland). There may be a continuum of depths over which the ore-forming processes may take place in Carlin-type deposits, but more data on the palaeosurfaces are needed. These facts tend to support

Lindgren's (1913) original designation of these deposits as having formed in geological environments different from the classic epithermal veins related to subaerial volcanic rocks.

The close spatial and temporal association of Carlin-like deposits with the intrusion of alkalic igneous complexes and geochemical studies of stable isotopes have led workers to relate them genetically to the igneous rocks (e.g. Giebink and Paterson, 1986), but the deposits are normally classified with volcanic-rock-related epithermal deposits.

The tectonic settings and local thermal histories within the western Cordilleras of North America may have played key roles in determining the nature of Carlin deposit-forming fluids, and may have been important controls on whether a deposit is Carlin-type or Carlin-like. If this is true, then the grouping of both Carlin-type and Carlin-like carbonate-replacement deposits as subtypes in the same class is appropriate, but this class is distinct enough from epithermal deposits *as defined by* Lindgren (1913, 1933) to warrant a separate classification.

7.2.3 *Relationship of Carlin-type deposits to polymetallic replacements*

Carbonate-replacement deposits which contain relatively high amounts of silver and associated base metals as well as typical vein structures, like at Candelaria and Taylor, Nevada, show some affinities to polymetallic replacement deposits such as those at Eureka, Nevada, and Tintic, Utah. Nevertheless, because these polymetallic deposits have produced only small quantities of base metals, common usage amongst exploration geologists has tended to combine these deposits with the Carlin deposits in a single class separate from the polymetallic replacement deposits. Bagby and Berger (1985) used such a classification scheme calling them 'sediment-hosted, disseminated precious-metal deposits', but separated them into different end-member types based in part upon their contained metal values. We believe that there has been insufficient research on the base-metal-deficient polymetallic carbonate-replacement deposits to clearly understand their relationship, if any, to the Carlin-type deposits. Where appropriate, in this chapter we present data from polymetallic carbonate replacement deposits in order to illustrate their differences with respect to Carlin-type deposits, but the main focus of this chapter will be on the Carlin-type deposits *per se*.

7.3 Regional geological and tectonic setting

The Carlin-type deposits of western North America and southern China all occur predominantly in sequences of marine sedimentary rocks within or adjacent to overthrust terranes related to continental margin tectonics. The majority of the known deposits in western North America occur in Nevada, although there are deposits from Montana to northern Mexico. The Chinese deposits are only now being studied, but it appears that they are widespread within a terrane analogous to parts of western North America. Locations of the known deposits on both continents are shown on Figures 7.1 and 7.2.

7.3.1 *Regional geological and tectonic setting in North America*

The North American Carlin-type deposits occur predominantly in Palaeozoic and early Mesozoic marine sedimentary and volcanic rocks deposited along a complexly

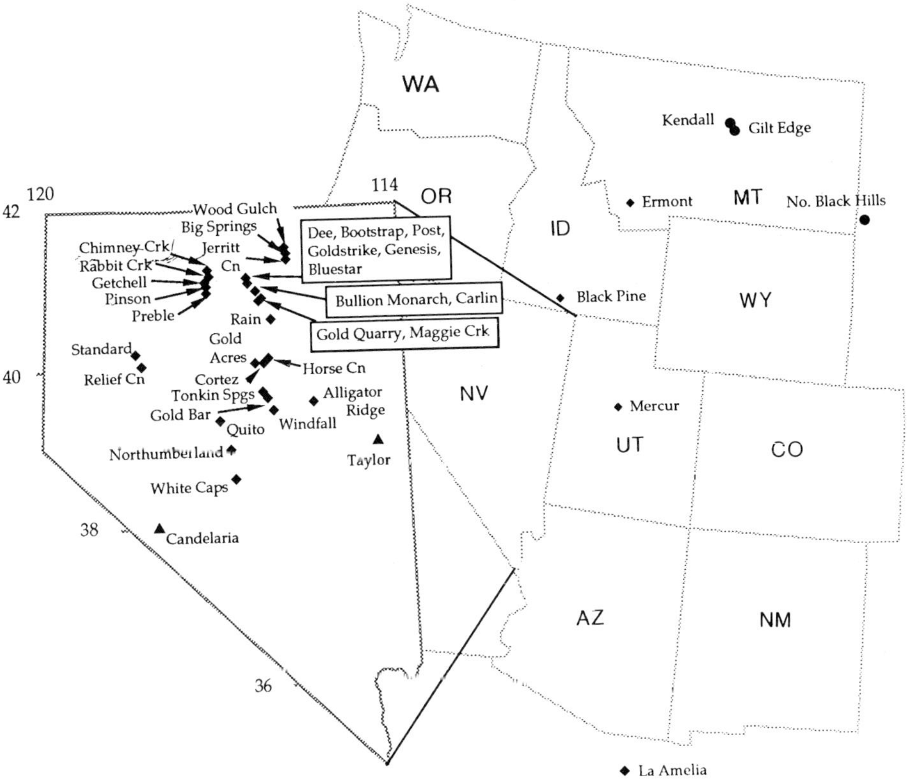

Figure 7.1 Location of Carlin-type (◆) and selected Carlin-like (●) and polymetallic replacement (▲) deposits in the western Cordillera of North America.

evolving continental margin. The highest grade deposits are in carbonaceous, finely laminated silty carbonate, commonly dolomitic, and in carbonate-bearing siltstones, although substantial tonnages of ore also occur in other lithologies. Table 7.1 lists the major deposits in the western United States, the age of the host rocks, and the generalized nature of the predominant ore-bearing lithology.

The tectonic history is an important aspect of the genesis of Carlin-type deposits (Figure 7.3). Three major orogenic events affected the nature of pre-Tertiary sedimentation in western North America. The Antler event, which occurred in the Late Devonian to Early Mississippian, thrust eugeoclinal volcanic and siliciclastic rocks eastward in an imbricate manner over transitional and miogeoclinal shelf-slope carbonate and siliciclastic facies rocks along the Roberts Mountains thrust fault

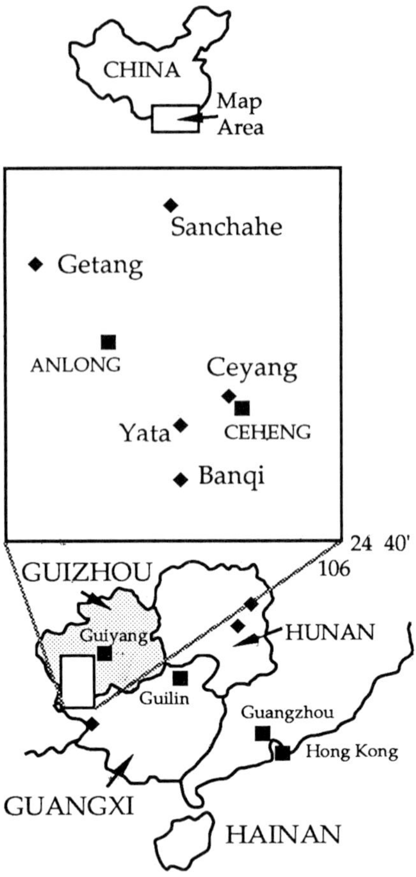

Figure 7.2 Location of Carlin-type deposits (◆) in south-eastern China. (From Cunningham *et al.*, 1988.)

(Merriam and Anderson, 1942; Roberts, 1951). Magmatism did not accompany this orogenic activity. The Permian Sonoma event thrust post-Antler siliciclastic and carbonate rocks eastwards over the Roberts Mountains allochthon. In the Late Jurassic to Early Cretaceous, there was a period of imbricate thrusting behind the island arc along a series of thrusts (Speed, 1983), and in the Late Cretaceous thrust faults developed near the old Palaeozoic geoclinal shelf-slope hinge line during the Sevier orogeny. Unlike the Palaeozoic orogenies in the region, the Mesozoic orogenies were accompanied by both magmatism and metamorphism.

The repeated orogenic events of the Palaeozoic and Mesozoic resulted in crustal thickening. Although the actual thickness of the accumulated thrust sheets is largely unknown, our conservative estimate totals 10 to 20 km (cf. Langenheim and Larson, 1973).

Tectonic elements important to ore genesis in the western Cordilleras are more apparent in Mesozoic and younger rocks than in older sequences. For example, in the

Table 7.1 Host lithologies of selected Carlin-type and polymetallic replacement deposits in the western United States and south-eastern China

Deposit	Location	Host formation	Original character of major mineralized horizons						Reference
			Dolo ls	Calc sltst	Ls	Silty ls/do	Carbon-aceous	Other	
CARLIN-TYPE DEPOSITS									
Montana									
Ermont Mine	Pioneer Mts.	Dev. Jefferson Dolo	✓				✓		Geach (1972)
Gilt Edge Mine	Judith Mts.	Miss. Mission Cn. Ls			✓	✓	✓		Weed and Pirsson (1898)
South Dakota									
Ross-Hannibal Mine	N. Black Hills	Cam. Deadwood Fm.	✓	✓		✓	✓		Irving (1904)
Idaho									
Tolman Mine	Black Pine R.	Penn. Oquirrh Fm.		✓		✓	✓		Brady (1984)
Nevada									
Getchell Region									
Chimney Mine	Dry Hills	Penn. Etchart Ls.			✓	✓	✓		After Osterberg (1988)
		Ord. Valmy Fm.(?)						basalt	
Getchell Mine	Osgood Mts.	Cam. Preble Fm.	✓	✓		✓	✓		Berger (1985)
Pinson Mine	Osgood Mts.	Ord. Comus Fm.	✓	✓			✓		Powers (1978)
		Cam. Preble Fm.		✓		✓	✓		
Preble Mine	Osgood Mts.	Cam. Preble Fm.		✓		✓	✓	sh	Madrid and Bagby (1988)
Rabbit Crk deposit	Dry Hills	Penn. Etchart Ls.(?)			✓	✓	✓		E. Bloomstein (writ. comm., 1988)
		Ord. Valmy Fm.(?)						basalt	
Carlin Region									
Bluestar Mine	Tuscarora Mts.	Sil. Roberts Mts. Fm.		✓		✓	✓		Evans (1980)
		Ord. Vinini Fm.		✓			✓	qtzte, cherty sltst	
Bootstrap Mine	Tuscarora Mts.	Sil. Roberts Mts. Fm.		✓		✓	✓		MacLeod (1987)
		Ord. Vinini Fm.		✓			✓	chert, sh	
Carlin Mine	Tuscarora Mts.	Sil. Roberts Mts. Fm.		✓		✓	✓		Baker (1987)

Table 7.1 *Continued*

Deposit	Location	Host formation	Dolo ls	Calc sltst	Ls	Silty ls/do	Carbon-aceous	Other	Reference
		Original character of major mineralized horizons							
Dee Mine	Tuscarora Mts.	Dev. Bootstrap Fm.		✓	✓		✓		Ellis (1987)
Genesis deposit	Tuscorora Mts.	Sil. Roberts Mts. Fm.		✓		✓	✓		Christensen *et al*. (1987)
Gold Quarry Mine	Tuscarora Mts.	Ord. Vinini Fm.				✓	✓	sltst, chert, sh	Rota (1987)
Goldstrike Mine	Tuscarora Mts.	Sil. Roberts Mts. Fm.		✓	✓		✓		Christensen *et al*. (1987)
		Ord. Vinini Fm.	✓				✓	chert, sh	
Maggie Crk Mine	Tuscarora Mts.	Sil. Roberts Mts. Fm.				✓	✓		McFarlane (1987)
Post Mine	Tuscarora Mts.	Ord. Vinini Fm.		✓		✓	✓		Christensen *et al*. (1987)
Rain deposit	Carlin-Pinon Range	Miss. Webb Fm.		✓			✓		Thoreson (1987)
Independence range									
Jerritt Canyon	Independence Mts.	Sil. Roberts Mts. Fm.		✓		✓	✓		Birak and Hawkins (1985)
		Ord. Sil. Hanson Crk Fm.			✓		✓		
Cortez area									
Cortez Mine	Cortez Mts.	Sil. Roberts Mts. Fm.		✓		✓	✓		Wells *et al*. (1971)
Gold Acres Mine	Shoshone Range	Sil. Roberts Mts. Fm.		✓		✓	✓		Wrucke and Armbrustmacher (1975)
		Ord. Valmy Fm.		✓			✓		
Horse Canyon Mine	Cortez Mts.	Dev. Wenban Ls.			✓		✓		Foo and Hebert (1987)
		Ord. Vinini Fm.		✓			✓	cherty sltst	
Central Nevada region									
Alligator Ridge Mine	Alligator Ridge	Dev./Miss. Pilot Shale		✓			✓		Ilchik *et al*. (1986)
Gold Bar Mine	Roberts Mts.	Dev. Nevada Fm.		✓		✓	✓		Broili *et al*. (1988)
Northumberland Mine	Toquima Range	Sil. Roberts Mts. Fm.				✓	✓		Motter and Chapman (1984)
		Ord. Gatecliff Fm.				✓	✓		
Quito deposit	Toiyabe Range	Ord. Antelope Valley Ls.		✓	✓		✓		Droste *et al*. (1988)
Tonkin Springs Mine	Simpson Park Range	Ord. Vinini Fm.				✓	✓		Gesick (1987)
White Caps Mine	Toquima Range	Cam. Gold Hill Fm.					✓		Ferguson (1924)
Windfall Mine	Fish Crk. Range	Cam. Hamburg Dolo.				✓	✓		Wilson and Wilson (1986)

Lovelock Area								
Relief Canyon Mine	W. Humboldt Range	Trias. Cane Spring Fm.		✓		✓		Wallace (1989)
Standard Mine	Humboldt Range	Trias. Cane Spring Fm.		✓		✓		A.R. Wallace (pers. comm., 1989)
Utah								
Mercur Mine	Oquirrh Mits.	Miss. Great Blue Ls.	✓	✓	✓	✓	ss	Gilluly (1932)
POLYMETALLIC REPLACEMENT DEPOSITS								
Candelaria Mine, NV	Candelaria Hills	Miss.-Tr. Pickhandle Gulch		✓			qtzte, sh	Page (1959)
		Trias. Candelaria Fm.			✓		blk sh	
Taylor Mine, NV	Schell Crk Range	Dev. Guilmette Ls.		✓		✓		Havenstrite (1984)
CARLIN-TYPE DEPOSITS IN SOUTH-EASTERN CHINA								
Xiang-Gian zone	Guizhou Province	Cam. ls-dolo			✓	✓		Cunningham *et al.* (1988)
Yata deposit	Guizhou Province	Trias. Xinyuan Fm.			✓	✓	ss	Cunningham *et al.* (1988)
Getang deposit	Guizhou Province	Per. Longtan Fm.			✓	✓		Cunningham *et al.* (1988)
Sanchahe deposit	Guizhou Province	Per. Longtan Fm	✓			✓		Cunningham *et al.* (1988)
		Per. Changxing Fm.			✓	✓		
		Per. Dalong Fm.			✓	✓		
Ceyang deposit	Guizhou Province	Trias. Xinyuan Fm.				✓		Cunningham *et al.* (1988)
Banqi deposit	Guizhou Province	Trias. Xinyuan Fm.		✓		(?)		Cunningham *et al.* (1988)

Osgood Mountains and vicinity, central Nevada, north-south faulting of at least Jurassic age and cross-cut by northeast-trending faults is of considerable importance in controlling mineralization along the Getchell trend. This north-south faulting is truncated to the north and east by a series of subparallel north-western tectonic trends that are important in the control of a large number of mineral deposits including those of the Carlin trend *per se*. In general, the Mesozoic tectonic fabric in the northern Cordillera is predominantly one of north-west- and north-east-trending high-angle faults.

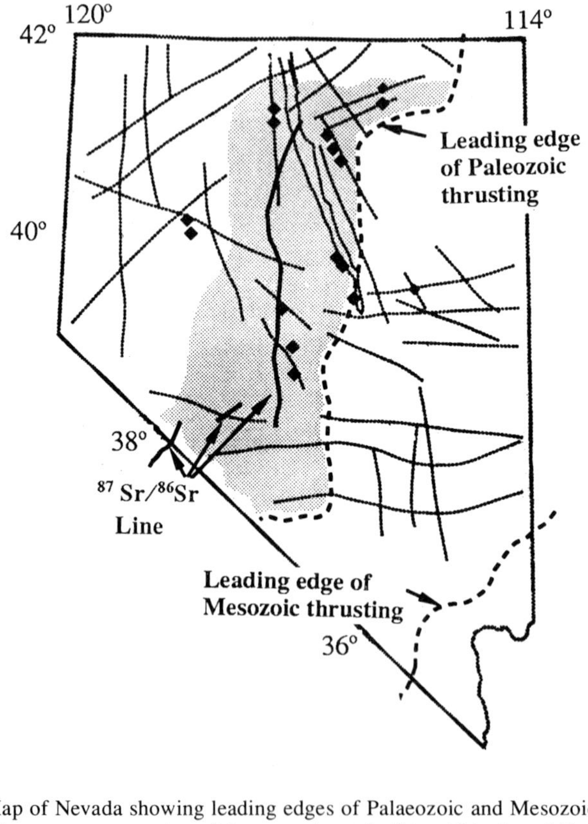

Figure 7.3 Map of Nevada showing leading edges of Palaeozoic and Mesozoic thrust faulting, aeromagnetic lineaments suggestive of deep-continental rifting [thin dotted lines] (cf. Blakely, 1988), the western boundary of continental crust as inferred from $^{87}Sr/^{86}Sr \geq 0.706$ (Stewart, 1988), and the location of selected Carlin-type deposits (◆). The stipple pattern shows the distribution of Palaeozoic allochthonous rocks (Stewart, 1980).

Widespread extensional faulting became important in the western United States commencing in the Eocene, and by Middle Oligocene normal faulting was widespread throughout the Cordillera. The greatest period of rifting in the Basin and Range Province appears to have taken place from the Middle Miocene onwards. For reasons discussed in a later section, we believe that all of the Carlin-type deposits are no younger than earliest Miocene and, in some cases, could be considerably older.

7.3.2 *Magmatism in western North America*

Pre-Early Triassic igneous activity in the overthrust zones of western North America is primarily confined to the oceanic volcanic terranes along the continental margin. In the Mesozoic, plutonic and volcanic activity increased throughout the Cordillera climaxing in the Cretaceous (Stewart, 1980). A number of the Carlin-type deposits are spatially associated with Cretaceous plutons, a relationship that is discussed further in other sections of this chapter.

Initially, early to mid-Tertiary magmatism was widespread throughout most of the Cordillera. However, by middle to late Tertiary, activity became confined to the western margin of the Basin and Range Province and in the Rio Grande Rift of central New Mexico and central Colorado.

7.3.3 *Regional geological, tectonic, and magmatic settings in south-eastern China*

Within the past decade, several Carlin-type deposits have been recognized in south-eastern China (Cunningham *et al.*, 1988). The deposits occur in south-western and eastern Guizhou, eastern Guangxi, and eastern Hunan Provinces (Figure 7.2), and are located near the southern edge of the Yangtze Craton adjoining an accretionary fold belt (Zhang *et al.*, 1984). The Yangtze Craton consists of Proterozoic crystalline rocks overlain by Palaeozoic to Middle Triassic marine sedimentary rocks with a transition to non-marine deposition evident in the Upper Triassic (Cunningham *et al.*, 1988). The Phanerozoic marine carbonate rocks were laid down on a broad cratonic platform in environments similar to those that existed during the same time-period in western North America. Turbidites interbedded with argillaceous rocks were deposited on the south-eastern China continental shelf and slope (Cunningham *et al.*, 1988).

The Carlin-type deposits in eastern Guizhou Province (Table 7.1) and those in Hunan Province occur in a Cambrian stratigraphic succession consisting of sandstones, commonly carbonaceous, and carbonaceous dolomite and limestone with intercalated black shale (I-Ming Chou, US Geological Survey, written communication, 1986; Hua, 1988). The south-western Guizhou Province deposits are in a Triassic sequence of carbonaceous, thinly laminated, argillaceous limestone with interbedded shale and sandstone. Rocks along the south-western margin of the Yangtze Craton were folded and faulted in the Jurassic and Cretaceous, developing gentle folds in the vicinity of the Guizhou deposits (Cunningham *et al.*, 1988). Granodioritic intrusions accompanied the Mesozoic orogenic activity in the eastern Guangxi to western Hunan region, which is the location of a major, deep-penetrating fracture zone. In addition, granites with associated tin deposits at Dachang are known in the vicinity of Carlin-type gold deposits (H.F. Bonham, Nevada Bureau Mines, personal communication, 1989).

7.4 Characteristics of the deposits

Features common amongst the Carlin-type deposits were recognized after the discovery of the Carlin deposit. These features have been summarized in some detail by a number of workers including Radtke and Dickson (1976), Bagby and Berger

(1985), and Percival *et al.* (1988) for the Carlin-type deposits, and are evident in the publications by Weed and Pirsson (1898) and Irving (1904) for Carlin-like deposits. Consequently, these features will only be discussed briefly in this chapter.

7.4.1 *Nature of the host rocks*

Carlin-type deposits occur in a number of different lithologies (Table 7.1). The most favourable rocks are very finely laminated, carbonaceous silty carbonates and carbonate-bearing siltstones and shales. Host rock ages vary from Cambrian to Triassic, but the most favourable hosts in the western United States are Cambrian to Mississippian.

In Nevada, the Lower Silurian to Lower Devonian Roberts Mountains Formation is a host to ores in several Carlin-type deposits and serves as a useful example of important host-rock characteristics. In the Carlin Mine area it is a dark-grey, fine-grained, laminated dolomitic limestone which contains some silt (Evans, 1980). The grain size averages 0.05 mm and ranges from a calcite:dolomite ratio of 3:1 to 1:9. Detrital quartz makes up 5–40% of the rock with lesser amounts of chlorite, illite, kaolinite, collophane, epidote, hornblende, plagioclase, K-feldspar, and tourmaline. The total organic carbon content varies from about 0.4 to 1.6%, occurring as lenticular seams along bedding planes in dolomitic siltstones, as patches filling open spaces in limestone, and as disseminated material in siltstones, cherts, and dolomitic limestones (Hausen and Park, 1986). The finer grained, more detritus-rich parts of the formation are most susceptible to replacement in the Carlin-type deposits.

7.4.2 *Structural setting of the deposits*

In north-central Nevada, the doming of autochthonous rocks is an important control on the locus of mineralization (Roberts, 1960). Normal faults related to the doming served to channel hydrothermal fluids to favourable horizons. Organic geochemical studies at several deposits lend some support to the doming concept, as there is a suggestion of increased content of carbonaceous material in the vicinity of some orebodies (Hausen and Park, 1986; Leventhal *et al.*, 1987). Hausen and Park (1986) interpreted the Carlin Mine area as a 'failed' petroleum reservoir.

High-angle, normal faults are important elements in each of the Carlin-type deposits, but details of the structural setting differ from district to district. The faulting increased host-rock permeability, creating breccias that focused ore deposition in several deposits. Replacement orebodies do not occur at great distances from mappable fractures.

All of the Carlin-type deposits in the western United States occur within the overthrust belt, and the same tectonic setting appears to be present in south-eastern China.

7.4.3 *Associated igneous rocks*

Igneous rocks occur within, or in the vicinity of, nearly all of the Carlin-type deposits (Table 7.2). Granodioritic stocks as well as granodiorite to granite dykes are found in the Getchell, Gold Acres, Goldstrike, Northumberland, and White Caps deposits, and magnetic anomalies suggest the presence of large granitic bodies beneath other deposits such as Carlin (Grauch, 1988). The Alligator Ridge deposit is unique in that igneous intrusions have not been exposed in the mine workings. The Carlin-like deposits in central Montana and the Black Hills are closely associated with alkaline igneous rocks including syenite and alkaline granite. In most districts, the igneous rocks are intensely altered by the hydrothermal fluids, and ore commonly occurs in fractures parallel to the intrusions.

7.4.4 *Geochronology of the deposits*

The age of formation of the Carlin-type deposits is difficult to determine. The host rocks intrinsically contain illite and(or) chlorite, the argon systematics of which may be variably changed during hydrothermal activity, commonly resulting in hybrid ages. Also, alteration in dyke rocks is typically extremely fine grained, and the material is commonly unsuitable for dating. In many deposits, supergene effects have altered the exposed portions of many orebodies and rendered difficult the interpretation of radiometric ages. Nevertheless, there have been a number of attempts to date the different deposits and associated igneous rocks, the results of which are summarized in Table 7.2. The limited geochronological studies indicate that there is no single time of formation for all of the Carlin-type deposits in the western United States. Geological and geochronological evidence from Getchell, Pinson, Northumberland, and Gold Acres suggest a Cretaceous age for mineralization, while similar evidence from Cortez indicates an Oligocene mineralization age.

The most comprehensive chronology studies to date have been in the Carlin and Getchell mine areas, but the published dates are controversial. In the Carlin Mine region, we believe that either the Mesozoic or Oligocene ages are reasonable for the age of mineralization, with the lower limit being an age of ~ 28 Ma based on supergene alunite from Gold Quarry (W.C. Bagby, unpublished data). At the Getchell Mine, Silberman *et al.* (1974) interpreted the gold mineralization to be related to the thermal event accompanying the emplacement of the Osgood Mountains stock and related tungsten-bearing skarns. The same samples of altered granodiorite were later re-dated using the $^{40}Ar/^{39}Ar$ method in a six-step experiment with plateau ages obtained (B.R. Berger, unpublished data) corroborating the earlier K-Ar dates, and indicating that the altered minerals have not been heated to more than about 250°C since the Cretaceous (unpublished fluid-inclusion data indicate that late-stage mineralization occurred at 250°C [W.C. Bagby, unpublished data, 1989]). It is most likely that the Getchell deposits are Cretaceous.

7.4.5 *Alteration and metallization*

Carlin-type deposits are characterized by the dissolution of carbonate minerals and the precipitation of silica. In some deposits, pre-gold-ore stage jasperoid replaced carbonate along and adjacent to fault structures. In many instances, these jasperoids display a delicate silica-replacement of the carbonate host that preserves primary features such as fine laminations, bedding, and many fossils. During the gold-deposition stage, replacement silicification continued; quartz veins and silicified breccias formed that cross-cut the earlier, pre-ore silicification, and additional silicification of the host rocks continued elsewhere in the evolving hydrothermal system. Some clay alteration of detrital material took place in the host rocks during the

Table 7.2 Composition of igneous rocks and geochronology of Carlin-type and polymetallic replacement deposits in the western United States

Deposit	Igneous rock type	Igneous rock age (Ma)	Alteration age (Ma)	Mineral dated	Comments	Reference
Carlin region						
Goldstrike deposit	granodiorite	121.0 ± 5		biotite	'Fresh' sample from altered area	Hausen and Kerr (1968)
Goldstrike deposit	diorite		78.4 ± 3.9	biotite	Biotite bronzy with low K_2O; ore	Morton *et al.* (1977)
Goldstrike deposit	felsic dyke		36.4 ± 1.8	serite	Alteration in dyke	Bagby and Pickthorn (unpublished data)
Carlin Mine	felsic dyke	149.0 ± 6		biotite	'Fresh', but with low K_2O; ore zone	Radtke (1985)
Carlin Mine	felsic dyke	128.0 ± 4		biotite	'Fresh', but partially altered	Radtke (1985)
Carlin Mine	felsic dyke		131.0 ± 4	biotite	Partially sericitized, in ore zone	Morton *et al.* (1977)
Carlin Mine	felsic dyke		57.6 ± 2.5	sericite	Along dyke boundary, suspect	Radtke (1985)
So. of Carlin Mine	granite	108.0 ± 3		biotite	Alteration in wallrocks	Morton *et al.* (1977)
So. of Carlin Mine	diorite	38.5 ± 1.2		biotite	Evans (1980) noted slight alteration	Morton *et al.* (1977)
So. of Carlin Mine	qtz latite	36.6 ± 0.7		biotite	Fresh	Evans (1980)
Gold Quarry Mine			≈ 27.9	alunite	From ore zone	Bagby and Pickthorn (unpublished data)
Carlin region	rhyolite	14.3.0 ± 0.3		sanidine	Unaltered rock on jasperoid	Radtke (1985)
Getchell Trend region						
Nr Alpine WO_3 Mine	granodiorite	89.9 ± 1.8		biotite	Fresh	Silberman *et al.* (1974)
Nr Valley View WO_3 Mine	granodiorite	92.2 ± 1.8		biotite	Fresh	Silberman *et al.* (1974)
		89.5 ± 1.8		hornblende		
Kirby WO_3 Mine			87.6 ± 3.4	sericite	WO_3 skarn	Silberman *et al.* (1974)
Marcus WO_3 Mine			88.4 ± 3.3	sericite	WO_3 skarn	Silberman *et al.* (1974)
Mt. King WO_3 Mine			92.6 ± 2.8	sericite	WO_3 skarn	Silberman *et al.* (1974)
Getchell Mine	dacite	89.9 ± 1.8		biotite	Fresh	Silberman *et al.* (1974)
Getchell Mine	granodiorite		92.2 ± 2.8	sericite	Coarse grained, Sec. 4 Pit	Silberman *et al.* (1974)
Getchell Mine	granodiorite		80.9 ± 3.3	sericite	Coarse grained, So. Pit	Silberman *et al.* (1974)
Getchell Mine	granodiorite		74.7 ± 2.2	sericite	Fine grained, So. Pit	Silberman *et al.* (1974)
Getchell Mine	granodiorite		67.0 ± 3.3	sericite	Fine grained, So. Pit	Silberman *et al.* (1974)
Preble Mine	granodiorite		100.4 ± 1.6[1]	sericite	Dyke in ore zone	Bagby and Berger (1985)
Cortez–Gold Acres region						
Gold Acres Mine	felsic sill		94.3 ± 1.9	sericite	From ore zone	Silberman and McKee (1971)
Gold Acres Mine	qtz monzonite	98.8 ± 8.2		biotite	Partially alt. to chl + ser	Silberman and McKee (1971)
Gold Acres Mine	qtz monzonite		92.8 ± 1.9	sericite	170 m below pit bottom	Silberman and McKee (1971)
Cortez Mine	rhyolite	35.0 ± 1.1		whole rk	Glassy rind; ore zone	Morton *et al.* (1977)

No. of Cortez Mine	rhyolite	34.4 ± 1.1		biotite	Caetano tuff	Wells *et al.* (1971)
No. of Cortez Mine	qtz porphyry	35.0 ± 1.0		whole rk		Wells *et al.* (1971)
Central Nevada region						
No. Toquima Range	qtz monzonite	151.0 ± 3		biotite	Clipper Gap stock	Silberman and McKee (1971)
Northumberland Mine	granodiorite	154.0 ± 3		biotite	Northumberland pluton	Silberman and McKee (1971)
Northumberland Mine	qtz porphyry		84.6 ± 1.7	whole rk?	Alt dyke in ore zone	Motter and Chapman (1984)
Windfall Mine	dacite	36.4 ± 1.3		biotite	Similar to alt dykes in mine	McKee *et al.* (1971)
Western Nevada region						
Sterling Mine			21.6 ± 1.0	jarosite	Vein material	M.L. Silberman (unpublished data)
Montana						
Judith Mts., MT	granite	62.0 ± 1		Rb-Sr age	Alk gr of Judith Peak	Marvin *et al.* (1980)
Spotted Horse Mine, Judith Mts, MT			58.8 ± 2.1	roscoelite	Qtz-fluorite-calcite vein	Marvin *et al.* (1980)
Kendall Mine, Judith Mts., MT	syenite	59.8 ± 3.1		zircon[2]	Argillically altered	Lindsey and Naeser (1985)
Utah						
Mercur Mine	rhyolite	31.6 ± 0.9		biotite	Eagle Hill rhyolite plug	Moore (1973)
Mercur Mine	granodiorite	36.7 ± 0.5		biotite	Granodiorite porphyry	Moore and McKee (1983)
Polymetallic replacement deposits						
Candelaria Mine, Nevada	qtz monzonite	126.0 ± 4		sericite	Mt. Diablo Mine area	Silberman *et al.* (1975)
Taylor Mine, Nevada	rhyolite dyke		≈ 35[3]	whole rk?	Argillically altered	Havenstrite (1984)
Taylor Mine area Schell Crk. Range	qtz latite	38.0 ± 4		biotite	Unaltered vitrophyre	McKee *et al.* (1976)

1. Age determination done by L.W. Snee of the US Geological Survey using an $^{40}Ar/^{39}Ar$ technique.
2. Age determination done using a fission-track technique.
3. Specifics of the analytical method, analytical results, and mineral analysed were not given in the cited publication.

Table 7.3 Paragenesis of silicification and post-silicification veining and the textures of the silicification at the Mercur, Pinson, Jerritt Canyon, and Carlin deposits

Paragenesis	Deposit							
	Mercur[1]		Pinson[2]		Jerritt Canyon[3]		Carlin[4] +	
	Replacement and veining	Textures of silica	Replacement and veining	Texture of silica	Replacement and veining	Texture of silica	Replacement and veining	Texture of silica
Pre-gold-ore stage	*Silver chert mineralization*							
	Jasperoid; silica-replacement of silty ls and cal siltst preserving some sedimentary textures	Xenomorphic and jigsaw-puzzle texture	Jasperoid; silica-replacement of silty dolomite preserving some sedimentary textures	Xenomorphic and jigsaw-puzzle texture	Jasperoid; silica replacement of silty ls and chert preserving some sedimentary textures; quartz veins	Xenomorphic in jasperoid; jigsaw-puzzle in chert; grandular quartz in veins	Jasperoid; silica replacement of silty ls; abundant sericite and carbonaceous material; quartz microveinlets	Jigsaw-puzzle texture
	Hydrothermal solution brecciation at base of jasperoid		Radial masses of silica in jasperoid	Jigsaw-puzzle texture				
	Cementation of jasperoid and jasperoid breccia with quartz	Xenomorphic and reticulated textures						
	Silica veinlets[5]	Chalcedonic						
	Quartz in vugs and interstices	Euhedral to subhedral coxcomb overgrowths						

	Gold mineralization							
	Jasperoid; silica replacement of silty ls and cal siltstone preserving sedementary textures	Jigsaw-puzzle texture						
Gold-ore stage	Disseminated silica-replacement of carbonate minerals	Euhedral to subhedral qtz	Multiple episodes of quartz veins	Reticulate texture	Multiple episodes of quartz veins	Xenomorphic to reticulate texture	Silica-replacement of silty ls; quartz veinlets	Silicification is jigsaw-puzzle texture; veinlets xenomorphic to reticulate texture
	Silica veinlets	Subhedral quartz			Quartz in open spaces	Granular to euhedral		
Post-main stage	Calcite veins often with barite, orpiment, or realgar		Calcite veins		Calcite veins often with realgar, orpiment or barite; stibnite + barite veins		Chalcedony veins; quartz in vugs	Quarts in vugs is xenomorphic

1. Data from Jewel (1984) and Tafuri (1987)
2. Data from Powers (1978)
3. Data from Birak and Hawkins (1985) and Hofstra *et al*. (1988)
4. Unpublished data from T.G. Lovering (US Geological Survey)
5. Silver mineralization took place some time after this episode of silicification in interstices and along fractures in the jasperoid breccia and jasperoid

ore-stage silicification and veining. Post-ore calcite veins are common and may contain associated barite, orpiment, realgar, stibnite, cinnabar, and (or) complex thallium minerals (e.g. Radtke, 1985; Tafuri, 1987). The variable textures of silica deposited during the different stages of alteration are indicative of pre-ore, ore-stage, and post-ore events during the life of the hydrothermal system, and are summarized for the Mercur, Pinson, Jerritt Canyon, and Carlin deposits in Table 7.3.

The gold is native and extremely fine-grained. Native silver and electrum have also been reported. In the alkaline-rock-related deposits, tellurides are common, and tellurides are reported from Mercur (Tafuri, 1987). Pyrite is the most abundant ore-stage sulphide, and is a host for gold at many deposits. Marcasite is also common and arsenopyrite is generally present, but is uncommon. Realgar is a typical post-main-stage sulphide, often acompanied by lesser amounts of orpiment, cinnabar, and stibnite. Barite is commonly present in considerable amounts in the post-main-stage veining, and fluorite is found locally (fluorite is ubiquitous in 'Carlin-like' ores). Barite is also typical as a constituent of pre-ore-stage silicification in the Nevada deposits. With the exception of Mercur, where base metals and silver are early in the paragenesis, base-metal sulphides generally occur only in small quantities as late-stage vein constituents, normally associated with barite.

Table 7.4 gives the alteration paragenesis in the Carlin Mine and elucidates features found in many deposits. Calcite + quartz and calcite + hydrocarbon veins (Kuehn and Bodnar, 1984), generally referred to as 'metamorphic' in character, are early and are found at many deposits in Nevada (e.g. Carlin, Pinson, Alligator Ridge, Jerritt Canyon; Holland *et al.*, 1988; Hofstra *et al.*, 1988). Bakken and Einaudi (1986) noted that these early veins at Carlin are fractured and that ore-stage mineralization commonly fills the fractures.

At Carlin, gold is spatially related to the development of silicified zones and no correlation exists between the density of veinlets and gold tenor (Bakken and Einaudi, 1986). Kuehn and Rose (1987) found the alteration to be zoned away from the most intensely silicified rock as follows: (1) quartz + dickite/kaolinite; (2) quartz + dickite/kaolinite ± K-mica/illite; (3) quartz + dolomite + illite/K-mica; (4) quartz + dolomite + illite + calcite; and (5) through quartz + calcite + dolomite + illite($2M_1$) + K-feldspar, progressively into 'unaltered' host rocks (Kuehn and Rose, 1987). The highest grades of gold are not in the most intensely silicified or jasperoidal lenses, but in zone (3) where both silicification and potassium metasomatism are co-extensive (Kuehn and Rose, 1987). Radtke (1985) indicated that some deposition of gold in fact pre-dated some of the more intense silicification, because gold grains are locally encapsulated in quartz. Kuehn and Rose (1987) found that post-ore-stage calcite veins contain barite, realgar, and local cinnabar, tetrahedrite, and other Sb–As–Tl phases. Radtke (1985) noted that some scheelite occurs in the upper parts of the Carlin deposit.

Hofstra *et al.* (1988) studied the alteration across a typical orebody in the Jerritt Canyon deposit and found a zonation from the ore outwards of: (a) quartz; (b) quartz + dolomite; (c) dolomite + calcite; and (d) calcite. The ore is generally within the intermediate zones of carbonate dissolution and silicification (Hofstra *et al.*, 1988), similar to that recognized at the Carlin deposit by Kuehn and Rose (1987).

The Mercur district is somewhat unusual amongst the Carlin-type deposits in having a paragenetically early silver-base metal mineralized area referred to as the Silver Chert (Table 7.3) where high-grade silver ore was produced from oxidized rock.

Table 7.4 Published paragenetic models of the Carlin, Nevada, deposit

Paragenesis	Carlin deposit paragenetic models		
	Radtke *et al.* (1980)	Bakken and Einaudi (1986)	Kuehn and Rose (1986)
Pre-mineralization stage		Cal veins; (Type A) cal ± qtz ± hydrocarbons veinlets (Type B)	Dark cal stylolites cut by qtz ± cal ± pyrobitumen veins; emplaced at 750–1600 bars pressure; barite veins with minor gal ± sph; late pyrobitumen veins; Cretaceous igneous dykes cross-cut hydrocarbon veins
Pre-ore stage	Calcite veins	Cal ± dolo veins (Type C); cal-dolo ± bar ± qtz (Type E); grey, turbid qtz veins (Type F)	Calcite veinlets due to progressive silicification
Main ore stage	'*Main stage*': qtz-py-K-clays-K-mica + Au-Hg-As-Sb-Tl; hydrocarbon-mobilization; early jasperoid; minor bar veinlets '*Late main stage*': jasperoid and qtz veins; As-Sb-Hg-Tl-S minerals; Pb-Zn-Cu-S minerals; cal veins	Grey, translucent qtz veins (Type G) cross-cut Type E; jasperoid; silica-breccia veinlets (Type H) related to Type G	Jasperoid breccias and stockworks cut by qtz veins with CO_2-rich inclusions emplaced at 500-800 bars pressure and cross-cut by argillically altered dykes
Post-main ore stage	'*Acid-leaching oxidation*': jasperoid and qtz veins; Pb-Zn-Cu-S veins; cal veins above leached zone; anhydrite and kaol veins	Euhedral qtz and bar in vugs; bar ± gal ± sph	Cal ± Fe-dolo veinlets; cal ± real veins; Orp-cal-bar-real-stib; massive bar with minor stib; bar, cal, qtz crystals in vugs in jasperoid; cal veinlets
Weathering	Calcite veins	Clear, euhedral calcite veinlets (Type D)	'Peach-coloured' calcite veins; vug- and fracture-filling calcite with iron oxide; iron oxide and kaolinite on fractures

These silver ores differ from silver–lead ores in other districts near Mercur in that they lack lead. The gold ores at Mercur are very similar to those of other Carlin-type deposits, although Hills (1894) noted the presence of local rhodochrosite in late-stage carbonate veins in the gold ores.

Occurrence of gold. Most of the gold in Carlin-type deposits is very irregularly distributed. At Carlin, Radtke (1985) found the gold to occur as thin coatings or films on pyrite, and locally on the surfaces of amorphous carbon, encapsulated in silica, and dispersed in auriferous pyrite within grains of realgar. The majority of the gold at Carlin is within or on pyrite. Wells and Mullens (1973) examined the occurrence of gold at the Carlin and Cortez Mines and found it mainly within arsenian pyrite and in arsenian rims around pyrite grains. Wells and Mullens (1973) detected little or no gold in carbonaceous material at either deposit.

Tafuri (1987) observed gold in the Mercur deposit in two forms – as free gold surrounded by quartz and calcite in disseminated ores, and as inclusions in marcasite. Tafuri (1987) also speculated that arsenian pyrite contains gold, because separates of this pyrite always contain assayable amounts of gold.

Joralemon (1949) observed gold at Getchell associated with pyrite, marcasite, quartz, and realgar. He recognized both native gold and electrum, and reported laboratory tests that indicate that most of the gold occurs as inclusions in pyrite. The more porous grains are the best hosts, with the gold occurring near grain margins and rarely in the cores. Joralemon (1949) observed gold rimming quartz grains, both as individual, large grains and as clusters of very tiny grains. The clusters are elongate parallel to the bedding. The gold with the calcite–realgar veins is paragenetically late and followed the main period of gold deposition.

7.4.6 *Geochemistry of the deposits*

Trace-element geochemistry. A characteristic sulphide-mineral assemblage (pyrite, realgar, orpiment, stibnite, cinnabar) and barite accompany the gold in Carlin-type deposits. At the Getchell Mine, Erickson *et al.* (1964) identified a suite of anomalous As, Sb, Hg, and W; concentrations of base metals were consistently low. Erickson *et al.* (1966) found this same trace-element suite at Cortez, Bootstrap, Carlin, and Gold Acres. Analytical data that we have acquired from selected Carlin gold deposits are shown in Table 7.5.

The Carlin-type deposits in south-eastern China also contain anomalous Hg, As, and Sb and therefore show a similar trace-element suite to that found in the North American deposits. In addition, the Chinese deposits are in a region that has a variety of other types of gold deposits with the most common being those containing W–Sb–Au, W–Au, and Sb–Au (Hua, 1988).

Organic geochemistry. The nature and role of organic matter in the mineralizing processes responsible for the genesis of Carlin-type deposits has been studied by a number of workers. At Carlin, Radtke (1985) found that mineralized rocks contain significantly more organic carbon (0.5–0.6 wt.%) than unmineralized rock (0.2–0.4 wt.%). A similar study at Jerritt Canyon found < 0.1wt.% organic carbon in unmineralized rock and 0.3–5 wt.% in mineralized rock (Leventhal *et al.*, 1987). Hausen and Park (1986) concluded from a study of the Carlin deposit and from comparisons with other deposits that the domed and faulted areas associated with some Carlin-type deposits had been petroleum reservoirs.

A number of studies have been conducted to characterize the nature of the carbonaceous material. Hausen and Park (1986) concluded that the ores at Carlin, Getchell, and Alligator Ridge contain bituminous matter that is mostly petroleum residues emplaced in the host rocks prior to gold mineralization. Kuehn and Bodnar (1984) found pyrobituminous material within fluid inclusions in calcite ± quartz ± hydrocarbon veins at Carlin. Gize (1986) studied the thermal history, oxidation, and re-working of organic matter at Carlin, and concluded that there was evidence for multiple generations of bitumens indicative of a pulsing of incoming brines during original migration of the organic matter. During hydrothermal activity, the carbonaceous material was heated, partially volatilized, and mobilized locally by the hydrothermal solutions (Hausen and Park, 1986).

The role of carbonaceous material in the transport and precipitation of gold has been somewhat controversial. Radtke and Scheiner (1970) determined empirically that, when the organic content of the ore exceeded 0.3 wt.%, there was a significant percentage of the gold with the organic material, and concluded that organic carbon played an important role in the deposition of gold at Carlin. Wells and Mullens (1973) studied the distribution of gold in the Carlin and Cortez mines and found that most of the gold was within or on pyrite, and little, if any, gold was contained in carbonaceous material. Hausen and Park (1986) also examined the carbonaceous ore at Carlin and found that most of the gold was on and within pyrite and to a much lesser degree in the carbonaceous matter. Using a high-resolution synchrotron X-ray fluorescence technique, Chao *et al.* (1986) also failed to discover any gold in areas of carbonaceous material within a gold-rich sample from the Horse Canyon deposit (Figure 7.1). We believe that the carbonaceous material in Carlin-type deposits may have been important in the chemistry of the ore-carrying solutions, but played no significant role in metal transport.

The hydrogen index of maturity of organic matter in the vicinity of the Alligator Ridge deposit increases near the orebodies (Ilchik *et al.*, 1986). Kettler *et al.* (1986) studied the distribution of n-alkane and total aliphatic hydrocarbons within and around the same orebodies and found a systematic variation related to distance from the Vantage orebodies. Within 0.5 and 1.9 km of these orebodies, there is a unimodal n-alkane distribution with a maximum at nC_{15}. Close to the ore, but not within it, Kettler *et al.* (1986) found a depletion of short-chain n-alkanes, but an increase in the abundance of n-alkanes and total aliphatic hydrocarbons relative to those further out. Within the ore, there is a decrease in the total concentrations of both types of hydrocarbons. Kettler *et al.* (1986) interpreted the zonation as indicative of hydrothermal washing or oxidation of the hydrocarbons. Their data also suggest a relatively higher concentration of organic material in the ore zone at the time of mineralization relative to the organic content of the host rocks away from mineralization. Gerdenich *et al.* (1986) studied the amounts of C_1 through C_4 hydrocarbon gases within fluid inclusions in jasperoids in Carlin-type deposits. They were unable to detect any C_2H_6, but all samples contained C_2H_4 and CH_4 . They thus concluded that meteoric water was predominant in the hydrothermal recharge zones, and that the alkene/alkane of hydrocarbons may be useful in delineating recharging and mixing zones within deposits.

Table 7.5 Trace-element geochemical data from selected Carlin and polymetallic replacement deposits in the western United States. (Analyses are in parts per million unless otherwise noted. For the sample listed, n.d. indicates that no analysis was done for that particular element.)

Deposit	Au[1]	Ag	As	Sb	Hg	W	Tl	Ba	Mo	Zn	Pb	Cu	Comment
CARLIN-TYPE DEPOSITS													
Nevada													
Carlin region deposits													
Bluestar Mine	<10	3	1500	200	n.d.	≤50	n.d.	1500	10	1500	20	7000	Limy sltst
Bullion Monarch	<10	1.5	1500	200	n.d.	<50	n.d.	700	<5	700	70	100	Arg-FeOx vein
Carlin Mine	13	0.5	1200	30	n.d.	<50	n.d.	>5000	≤5	1100	100	20	Silicified ls
Carlin Mine	<10	2	100	10	n.d.	<50	n.d.	150	5	200	100	100	Carbonaceous ls
Carlin Mine	<10	1.5	200	44	n.d.	<50	n.d.	>5000	≤5	700	1000	20	Cal-FeOx vein
Dee Mine[2]	3	7	130	66	0.34	64	6.6	1000	10	230	<10	70	Silicified sltst
Gold Quarry Mine	<10	100	70	700	n.d.	<50	n.d.	>5000	5	10	1500	20	Jasperoid brec
Maggie Crk Mine	3.2	5	1200	5000	n.d.	200	n.d.	1000	7	500	70	40	Argillized ls
Maggie Crk Mine	10	7	1500	1500	n.d.	150	n.d.	100	≤5	3000	100	100	Silicified sltst
Rain deposit	<10	<0.5	1500	12	n.d.	<50	n.d.	>5000	<5	20	20	10	Bar-jasper brec
Independence range													
Jerritt Canyon[2]	0.2	0.5	10	6	0.8	0.5	0.4	>5000	<5	<5	<10	<5	Blk jasperoid
Cortez region deposits													
Cortez Mine	<10	10	200	85	n.d.	<50	n.d.	700	5	120	30	30	Calcite veins
Gold Acres Mine	<0.05	10	55	15	n.d.	<50	n.d.	150	<5	60	100	15	Calcareous sltst
Horse Canyon Mine	<10	1.5	1000	150	n.d.	<50	n.d.	500	10	200	50	70	Silicified sltst
Horse Canyon Mine[2]	10	<0.5	6500	140	1.3	73	1.4	200	15	240	10	50	Oxide ore
Getchell region deposits													
Getchell Mine	4.7	5	>1%	3000	n.d.	350	n.d.	100	10	50	15	50	Carbonaceous ls
Getchell Mine[2]	0.6	20	4200	50	4.4	25	3.3	1000	<5	35	10	100	Qtz vein
Iron Point prospect	<10	5	1000	500	n.d.	100	n.d.	>5000	20	<200	50	500	Silicified sh
Pinson Mine	0.10	<0.5	440	380	n.d.	100	n.d.	2000	50	500	10	50	Jasperoid
Pinson Mine[2]	4.8	1.5	1100	52	30.0	250	7.4	1000	15	75	<10	50	Jasperoid
Preble Mine	1.5	1	1500	150	n.d.	≤50	n.d.	3000	20	60	50	150	Carbonaceous sh
Preble Mine[2]	2.3	2.0	400	10	34.0	9	3.6	2000	<5	<5	50	30	Carbonaceous sh
Buffalo Mt.–Battle Mt. region													
Marigold Mine	9.5	3	780	150	n.d.	<50	n.d.	300	5	190	15	20	Altered ss
Central Nevada region deposits													
Alligator Ridge Mine	1.4	10	1500	700	n.d.	<50	n.d.	300	15	300	20	50	Silicified sh
Alligator Ridge Mine[2]	1.0	7	1900	100	0.54	30	8.0	5000	7	160	<10	30	Carb sandy sh
Northumberland Mine	0.05	2	1500	100	n.d.	<50	n.d.	>5000	70	800	30	100	Carbonaceous sh

Shale Pit (Round Mt. District)	0.20	3	780	18	n.d.	<50	n.d.	1 000	7	120	100	70	Qtz–cal–clay vein
Tonkin Springs Mine[2]	5.7	50	2100	100	4.3	3.0	3.2	500	<5	5	70	20	Silicified breccia
White Caps Mine	27	2	>1%	3000	n.d.	<50	n.d.	20	15	45	50	20	Carbonaceous ls
White Caps Mine[2]	8.7	0.5	3000	2000	1000	7	20	300	<5	50	30	20	Calcite vein
Windall Mine	0.25	1	>1%	220	n.d.	<50	n.d.	150	20	1000	150	10	Silicified dolo
Windfall Mine[2]	1.2	2	890	120	2.3	3.0	0.2	50	10	500	700	10	Hm sandy dolo
Lovelock region													
Relief Canyon Mine	1.7	20	400	46	n.d.	<50	n.d.	150	10	30	100	7	Silicified ls
Standard Mine	0.90	30	170	2000	n.d.	<50	n.d.	300	30	60	100	100	Silicified ls-sh
South-Central region													
Sterling Mine	10	5	>1%	1500	n.d.	<50	n.d.	150	200	700	2000	50	Jasperoid
Sterling Mine[2]	17.0	3.0	2500	24	2.4	4.5	0.6	200	<5	10	50	30	Silicified dolo
POLYMETALLIC REPLACEMENT DEPOSITS													
Antimony Pit prospect[2] (Taylor District)	0.10	10.0	200	1.2%	3.8	<0.5	2.4	>5000	<5	190	300	50	Limestone
Argus Pit area (Taylor District)	≤0.5	1500	1500	1%	n.d.	<50	n.d.	100	≤5	1%	2%	1%	Calcite vein
Candelaria Mine	0.60	1000	2000	1%	n.d.	<50	n.d.	1000	10	>1%	2%	1500	Manganese ox
Candelaria Mine[2]	3.2	2000	2700	>1%	3.7	<0.5	0.2	300	<5	2100	2%	1000	Siliceous ore
Enterprise Mine (Taylor District)	≤.05	1000	1500	>1%	n.d.	<50	n.d.	1000	5	880	1%	1500	Silicified ls
Idaho													
Tolman Mine	28	1	5500	10	550	6	n.d.	>1%	<2	<200	15	20	Carb sltst
Utah													
Mercur[3]	3.2	n.d.	4600	3	11	n.d.	270	n.d.	n.d.	n.d.	n.d.	n.d.	Sulphide ore
Sonora, Mexico													
La Amelia	9.5	7	800	190	0.48	<50	n.d.	300	5	200	70	20	Qtz-hm breccia
CARLIN–TYPE DEPOSITS IN SOUTH–WESTERN GUIZHOU PROVINCE, CHINA[4]													
Getang deposit	30	<2	520	260	17	n.d.	6.4	25	7	25	5	10	Silicified ls
Sanchahe deposit	10	<2	1600	31	30	n.d.	2.7	380	2	130	33	150	Calcareous sh
Yata deposit	12	4	5500	81	20	n.d.	2.7	250	<2	120	26	77	Carbonaceous sh

1. Gold analyses listed with a detection level of 10 ppm were analysed using a six-step emission spectrographic technique. Analyses with lower levels of detection were analysed using an atomic absorption spectrophotometric technique. 2. Data from Hill *et al.* (1986) 3. Data from Tafuri (1987) 4. Data from Dean *et al.* (1988)

Isotope geochemistry. Detailed stable-isotopic studies have been undertaken at Jerritt Canyon (Northrop *et al.*, 1987) and Carlin (Radtke *et al.*, 1980). The paragenetically earliest stage of these deposits was characterized by an isotopically heavy fluid of basinal or 'metamorphic' origin involved in early, pre-ore jasperoid formation. This was followed by the progressive mixing of the isotopically heavy fluid with a lighter, overlying, oxygenated fluid during the gold-depositing stage, and finally the swamping of the system totally by meteoric waters during the last stages when cross-cutting calcite veins were deposited. A less comprehensive data set was published by Holland *et al.* (1988), but their data are consistent with the above interpretation.

In the Enfield Bell Mine in the Jerritt Canyon district, Northrop *et al.* (1987) found a range of $\delta^{18}O = +25$ to $+28‰$ in pre-ore jasperoid, and associated quartz veins exhibited a variation with time from $\delta^{18}O$ of about +24 to +26‰ to +17.2‰. The calculated $\delta^{18}O_{H_2O}$ of the fluid that precipitated the early pre-ore silica varies from about $+18$ to $\leq +7‰$. Methane-rich inclusions in this early phase of silicification contain $\delta D = -118‰$. Barite was formed late in the initial silicification and records the lightest fluid at $\delta^{18}O = +12.0‰$ (calculated $\delta^{18}O_{H_2O} \approx +5‰$). The ore-stage jasperoid varies from $\delta^{18}O$ of = 15.2 to 1.1‰, and the grade of gold in the jasperoid is directly proportional to the oxygen isotopic composition (Northrop *et al.*, 1987; Hofstra *et al.*, 1988) (Figure 7.4). Calculated $\delta^{18}O_{H_2O}$ for these gold ore-stage fluids is about +5 to −5‰ (Hofstra *et al.*, 1988). The latest stage of silica veining shows a $\delta^{18}O$ isotopic variation from +9.2 to −1.0‰, and a corresponding calculated $\delta^{18}O_{H_2O}$ variation of −3 to −15‰ (Northrop *et al.*, 1987). These data reflect a hydrothermal system evolving through time, a system which Hofstra *et al.* (1988) interpreted to indicate the mixing of deeper fluids of original meteoric origin with shallow, more oxidizing fluids, for which mixing was the predominant mechanism of gold deposition.

Fluids in the jasperoids of the Carlin deposit are isotopically heavy, varying from about $\delta^{18}O$ of +13.0 to +18.1‰, and remain so in jasperoid with detectable gold. Quartz veins, interpreted by Radtke *et al.* (1980) to be related to the time of jasperoid

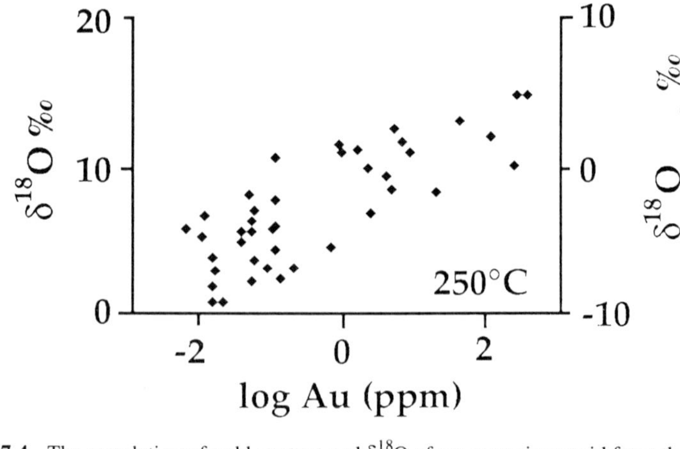

Figure 7.4 The correlation of gold content and $\delta^{18}O$ of ore-stage jasperoid from the Enfield Bell Mine in the Jerritt Canyon mining district, Nevada. (Data from Hofstra *et al.*, 1988.)

formation, range in $\delta^{18}O$ from +18.6 to +21.8%, whereas quartz veins containing gold in the main stage of ore deposition vary from +18.1 to +19.1‰. A jasperoid sample associated with late-stage stibnite mineralization is considerably lighter at +9.3‰. There is some suggestion in the data that the Carlin ore fluids were isotopically lighter during late stages of the mineralization.

Typical sulphur-isotope values for sulphides at the Pinson, Preble, Getchell, White Caps, Mercur, Jerritt Canyon, Cortez, and Carlin deposits are heavy, varying from $\delta^{34}S$ of +4.2 to +16.1‰, with the exception of Getchell which contains some $\delta^{34}S$ values near 0‰ in late-stage realgar (Rye, 1985). Rye found barite to range from +27.5 to +37.7‰ at each of these deposits. Northrop *et al.* (1987) analysed the sulphur isotopic composition of minerals in each alteration stage in jasperoid at Jerritt Canyon and discovered the early pre-ore barite to have $\delta^{34}S = +29.7$‰, averaging +29.3‰ during the ore-stage event, and averaging +19.5‰ in the post-ore-stage veining.

Fluid-inclusion geochemistry. At Carlin, Kuehn and Rose (1986, 1987) and Rose and Kuehn (1987) found pre-ore veins to contain methane-bearing and high-CO_2 inclusions indicative of pressures of formation on the order of 750–1600 bar (≈ 2–4 km). In the ore-forming paragenesis, they found two broad generations of fluids, one a low-salinity, high-CO_2, high-H_2S fluid associated with the gold-deposition stage (depth of formation of 500–800 bar [≈ 2–3 km]) and the other a low-salinity, low-CO_2 fluid in late-stage and peripheral minerals (Rose and Kuehn, 1987). The latter authors found no evidence for widespread boiling of either the ore-depositing fluids or the late-stage fluids and concluded that the ore was deposited through cooling, most likely during the mixing of two distinct fluids.

Jewell (1984) measured fluid-inclusion filling temperatures at Mercur, Utah, and found that some quartz in the silver- mineralized Silver Chert jasperoid formed from solutions with a mean of about 7 wt % NaCl equiv. and an average temperature range of 237 to 270°C. Gold-bearing rock contains fluid inclusions that yield a mean salinity of about 6.6 wt.% NaCl equiv. and a temperature variation of 222 to 256°C. Post-silicification calcite veins (Table 7.3) give a range in mean salinity from 5 to 7.2 wt.% NaCl equiv. and a mean temperature range of 152 to 192°C.

Cunningham *et al.* (1988) measured fluid-inclusion filling temperatures for several deposits in south-western Guizhou Province, China. At the Yata deposit, the temperatures vary from 150 to 240°C with salinities up to 5 wt.% NaCl equiv., although no data are given for CO_2 concentrations in the inclusions. Lu (1988) studied fluid inclusions from some unspecified Carlin-type deposits in south-eastern China and reported a variation in homogenization temperatures from about 75°C to about 200°C in what we infer to be a single deposit, and from about 100°C to over 300°C in another. Lu (1988) also noted extremely high salinities at 30–35 wt.% NaCl equiv.

7.4.7 *Geophysical studies*

Little has been published regarding the geophysical characteristics of Carlin-type deposits. Heran and Smith (1984) conducted induced polarization, electromagnetic, gravity, and magnetic geophysical surveys across the Getchell and Preble deposits. The electrical surveys, particularly those using the Very Low Frequency (VLF) method, were useful for identifying fault structures and present a view from the near-surface zone to 1–2 km depth. The gravity data largely outline intrusive rocks in the region as do the aeromagnetic and ground magnetic data. Because of the presence of clay along the Getchell fault, the induced polarization method primarily outlined faults which contain argillic material.

Grauch *et al.* (1987) and Grauch (1988) examined the relationships among steep magnetization boundaries, gravity, and geological map data, and used these data to interpret the locations of the margins of plutons. They found that many Carlin gold deposits are located along the boundaries of such features.

7.4.8 *Sizes, shapes, and grades of deposits*

Carlin-type deposits vary considerably in size and ore grade (Figure 7.5). The tonnages and mineability of several deposits depend to a great extent on the current price of gold. However, in higher grade deposits, the considerable vertical extent that some orebodies can attain has only recently been recognized, thereby greatly enlarging actual and potential mineable tonnages in some deposits.

The distribution of ore has not been published for most deposits. However, published reports commonly refer to abrupt assay boundaries on ore grades and the intimate control on ore of faults and favourable horizons in the host rocks immediately adjacent to the faults. Berger (1985) described the distribution of gold grades in a portion of the South Pit at the Getchell Mine (Figure 7.6), where the orebodies tend to be irregularly shaped pods elongated along favourable beds adjacent to and within the faults that served as the main fluid conduits. Berger (1985) also showed from drillhole data that the mineralization at Getchell persists to a depth exceeding 1 km down dip from the present outcrop, and that the character of the mineralization is the same at the deep intersection as it is in outcrop. Recent exploration north-east of Getchell has found deep ores in the Chimney and Rabbit Creek deposits, and along the Carlin trend deep ores have been discovered in the Post deposit. This persistence of gold ore to great depths *without apparent zonation* in the ore mineralogy is not typical of volcanic-rock-related epithermal deposits, and is additional evidence for a different classification of Carlin-type deposits.

7.5 Ore deposit models

7.5.1 *Published models*

The earliest narrative models of Carlin deposits were the detailed descriptions and theories of origin presented by Spurr (1896) for the Mercur, Utah, Carlin-type deposits and by Weed and Pirsson (1898) for the Judith Mountains, Montana, Carlin-like deposits. The main period of model development was much later, and followed the discovery of the Carlin deposit in the 1960s. Radtke and Dickson (1976) proposed that igneous activity provided heat to convecting meteoric waters, which leached metals from sedimentary rocks and transported them to favourable horizons in the near-surface zone, depositing quartz, pyrite, gold, and associated elements. An integral part of their model was the development of a late-stage, hypogene, acid-leaching phase at higher than ore-stage temperatures due to boiling that produced much of the oxide ores at Carlin. Cox and Singer (1986) presented a descriptive model for Carlin-type deposits that summarizes all of the geological and mineralogical

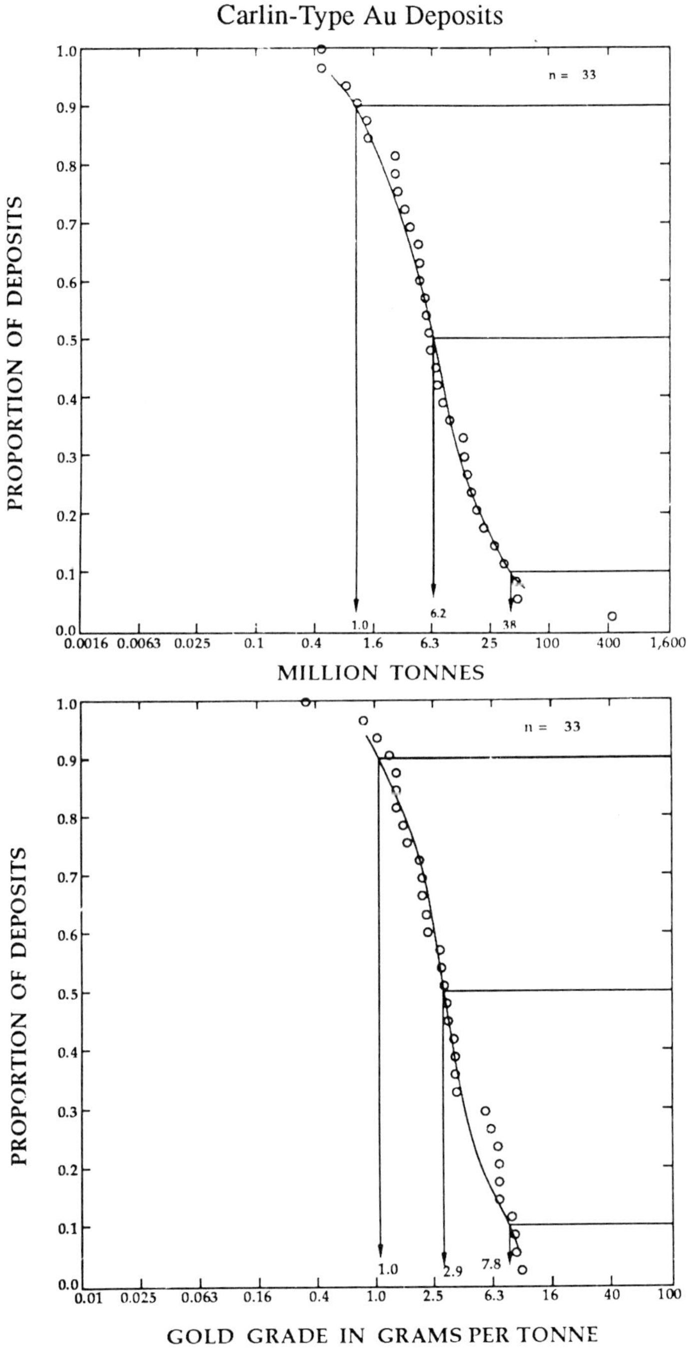

Figure 7.5 Grade–tonnage distribution for Carlin-type deposits in the western United States. (Data from Cox and Singer, 1986.)

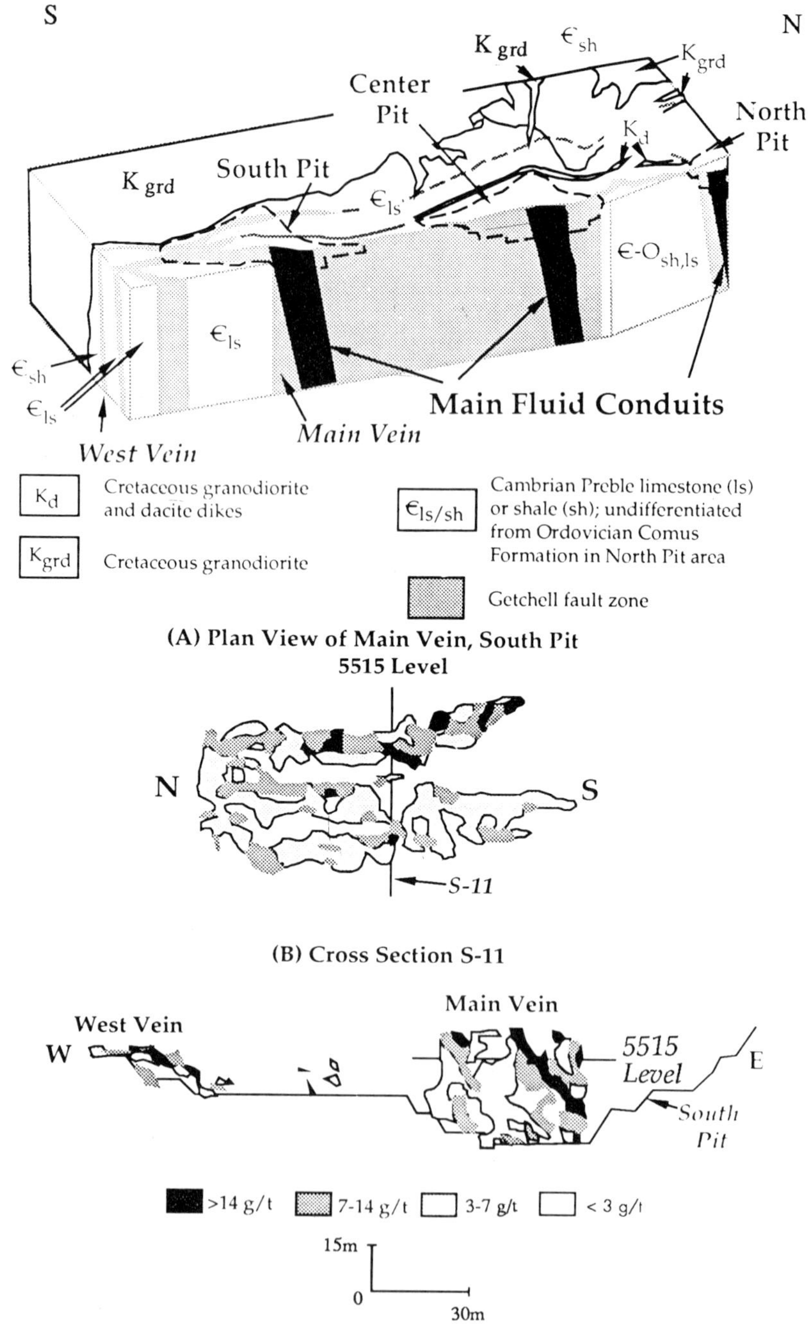

Figure 7.6 A plan and cross-section of gold grades in the South Pit of the Getchell mine at the end of mining activity in the 1960s. A similar distribution pattern occurred in the central and northern pits suggesting that the ore was fed from three discrete fluid channels as shown schematically in the top part of the figure (after Joralemon, 1951).

characteristics and grade–tonnage relationships of this deposit type, but the model contains no genetic concepts. Rose and Kuehn (1987) tied the formation of Carlin-type deposits to intrusions that de-gas CO_2 and mix it with CO_2 generated from the contact metasomatism of adjacent carbonate rocks. This low-salinity mixed fluid then rises until it encounters and mixes with a lower temperature, low-CO_2 fluid where gold is precipitated. The low- salinity, high-CO_2, and high-H_2S contents would enhance gold transport and inhibit the transport of Ag, Zn, and Pb. Rose and Kuehn (1987) did not consider boiling to be a factor in their genetic model for the Carlin deposit.

7.5.2 *A speculative model*

The restriction of Carlin-type deposits to the western Cordilleras and a similar regional setting in China implies to us that some relationship exists between the genesis of the deposits and the magmatic and tectonic processes that occur in active continental margins. The following model reflects our contention that there is a strong link between magmatism as a probable heat and metal source and tectonism in the formation of Carlin-type deposits.

As with all hydrothermal ore deposits, the essential components to form Carlin-type deposits are sources of heat, fluids, and metals, and sufficient permeability to permit large-scale fluid convection and metal deposition. The following discussion is organized accordingly.

Constraints on the nature of the heat source. High- enthalpy, high-gas, metal-bearing geothermal systems are large-scale features with steep lateral and vertical thermal gradients (Elder, 1981). The high-temperature upflow regime may have a cross-section of 3–5 km^2 (Henley and Hoffman, 1987) and extend from depths of 8 km (Henley and McNabb, 1978) with outflow rates of 100–400 kg/s at the surface (Henley and Hoffman, 1987). Assuming the fluid contains on the order of 10 μg/kg Au in solution, then, at 250°C, approximately 1.3×10^{20} J of heat energy are required to form a typical 30×10^6 gold deposit; additional heat of about 0.6×10^{20} J is required to bring the upflow region to the ambient temperature of the convecting fluids (Henley and Hoffman, 1987). Carlin-type deposits are not related to periods of metamorphism; therefore, the heat from about 100 km^3 of cooling basaltic magma or approximately 200 km^3 of granitic magma (a large stock or batholithic complex) is required (Henley and Hoffman, 1987). High concentrations of Te in some deposits (e.g. Mercur, Getchell), and mass balance constraints on H_2S and possibly CO_2 in all deposits are suggestive of a magmatic heat source. Rye's (1985) report of $\delta^{34}S$ near zero per mil for realgar at Getchell is also supportive of this hypothesis.

The apparent contemporaneity of magmatic activity and mineralization at Getchell, Preble, Gold Acres, Cortez, and at the Carlin-like deposits in central Montana (Table 7.2) is compelling evidence for a magmatic heat input into the formation of many, if not all, Carlin-type deposits.

Alternatively, the conductive heating of fluids in an elevated, but non-magma related, continental geothermal gradient can form convective systems that may transport and deposit Carlin-type ores. Above-average continental heat flows are known in such areas as the Great Basin of the western United States (e.g. Lachenbruch and Sass, 1977), but recent deep-crustal seismic surveys indicate that the lower crust in this region is being both underplated and intruded by magmas (Potter *et al.*, 1987). Therefore, the high heat flow in the Great Basin is, in fact, related to magmatism. In other tectonic regimes, such as the Tibetan plateau, convective geothermal systems are related to rapid uplift, where large volumes of fluid are rapidly expelled upward, the rate of expulsion being evidenced by the lack of retrograde metamorphism (Fyfe *et al.*, 1978). The tectonic pumping mechanism of Sibson *et al.* (1975) is a possible way to induce fluid flow from deep crustal levels to shallower ones during the rapid uplift. However, in both of the above tectonic settings, geothermal exploration has not found fluids of comparable temperature or chemistry to those in metal-bearing hydrothermal systems more directly related to magmatism. For extensional terranes such as the Great Basin, the main source for the high heat flow is probably basaltic magmas, and this same heat source may have predominated during the formation of many of the Carlin-type deposits.

Constraints on the origin of the hydrothermal fluids. The light, stable-isotopic data presented earlier indicate that fluids from multiple sources formed Carlin-type deposits. Entrapped, slightly acidic and oxidizing sedimentary basin waters are indicated by heavy ^{18}O in early jasperoids at a number of deposits including Pinson, Alligator Ridge, Carlin, and Jerritt Canyon (Northrop *et al.*, 1987; Holland *et al.*, 1988). The data of Radtke *et al.* (1980) and Northrop *et al.* (1987) suggest the presence of an upwelling fluid, with δD indicative of original meteoric derivation, mixing with a second dilute, meteoric water that became predominant in the waning stages of mineralization.

Between 200 and 300°C, gold is transported as a bisulphide complex (Henley, 1985); consequently, the solubility of $Au(HS)_2^-$ is a function of the concentration of dissolved H_2S. Fluid-mineral equilibria in geothermal systems couple H_2S and CO_2 (Henley *et al.*, 1984) and both species are required in any model. Both may be derived from magma sor, at elevated temperatures, from sedimentary rocks. However, the high $^3He/^4He$ in ores at Zacatecas (Simmons *et al.*, 1988) and in volcanic emanations and active geothermal systems in the Taupo Volcanic Zone, New Zealand (Giggenbach, 1986) suggest that metalliferous geothermal fluids obtain some components such as H_2S and CO_2 from magmatic sources ultimately with a mantle origin. In the Taupo Volcanic Zone, relative to the regional groundwater, the thermal waters show a small δD shift consistent with the input of magmatic fluids (Hedenquist, 1986).

Constraints on the sources of ore metals. There is no direct way to determine the source of gold in Carlin-type ore deposits. Indirect evidence is provided by the clustering of several large deposits, implying closely spaced, individual hydrothermal systems in a number of districts including Carlin, Jerritt Canyon, Cortez, and Getchell. This is analogous to the clustering of geothermal systems in the Taupo Volcanic Zone, New Zealand, and in Yellowstone National Park in the United States. As stated previously, gold-depositing geothermal systems transport on the order of 5–10 μg/kg gold. If the gold is derived from leaching of metal from crustal rocks, then from a typical rock averaging 3 mg/t gold, there must be 100% efficiency of extraction from a volume of about 130 km^3 to supply the observed metal to the geothermal system. It is more likely that extraction processes are less efficient and so an even larger volume of rock is required for each individual system. This implies that in specific areas with multiple deposits of a similar age, each separated by 10–20 km (e.g. Carlin, Jerritt Canyon, Getchell), the possible leach areas would significantly overlap. This renders lateral leaching as an unlikely source for all or most of the gold. A more probable, yet speculative, source is the derivation of gold from basaltic magmas emplaced into the lower or middle crust.

Constraints on the composition of genetically related igneous rocks. Reliable ages of intrusion and hydrothermal alteration are known only from a few deposits (e.g. Getchell, Preble, Gold Acres, Cortez, Northumberland; Table 7.2). In these deposits, the associated igneous rocks have granodioritic (dacitic) to granitic (rhyolitic) compositions and textures. These same compositions are found for intrusive bodies in most Carlin-type deposits, whereas in the Carlin-like deposits in the Black Hills, South Dakota, and central Montana the intrusions are syenites to alkaline granites. For the south-eastern China deposits, Cunningham *et al.* (1988) reported no igneous rocks in the immediate vicinity of these deposits. Additional discussion of the composition of associated igneous rocks follows below.

A model: Carlin deposits as distal, magma-related, replacement deposits. Given the above constraints, several aspects of Carlin-type deposits lead us to consider them as spatially and temporally related to thermal events associated with relatively reduced, primarily granodioritic or alkaline magmas that may also form tungsten-bearing skarn deposits and (or) molybdenum-bearing stockwork deposits.

A tungsten–molybdenum connection? Tungsten-bearing skarns are present within and in the vicinity of several Carlin deposits (e.g. Getchell, Gold Acres, Carlin, Goldstrike, Ermont, and Standard). In addition, scheelite occurs in the later stages of ore deposition, (e.g. Carlin, Getchell, and Gold Acres) and tungsten is a trace constituent of many of the deposits (Table 7.3). We infer from this that there may be a genetic link between tungsten skarn-producing magmas and Carlin-type hydrothermal systems. The Carlin-type systems may be distal replacement-type deposits of such tungsten-deposit-producing granodioritic or granitic intrusions. In the Getchell and Ermont regions, the intrusions also produced stockwork molybdenum deposits, and skarns in the vicinity of the Getchell, Chimney Creek, and Rabbit Creek deposits contain some molybdenite. The parent magmas associated with tungsten deposits have relatively low Fe/Al, low water contents, and were emplaced with relatively high pressures of crystallization. These factors result in relatively uniform, non-porphyritic granitic textures, common late-stage aplites and pegmatites, and the retention of tungsten in the residual melt (Newberry and Swanson, 1986). The tectonic thickening of the western Cordillera of North America was possibly instrumental in the creation of an environment amenable to tungsten-deposit-producing intrusions – the same environment ostensibly necessary for the formation of Carlin-type deposits.

Evolution of a deposit: the tectonic setting. Mixed carbonate, siliciclastic, and carbonaceous material accumulated as finely laminated sediments along a continental margin. When lithified, these rock are highly favourable host rocks. Where present, overthrusting in the western Cordillera provided: (a) an over-thickened wedge of favourable host rocks; (b) entrapped aquifers of meteoric water within the thrust slices; (c) a structural setting for later migration of bituminous material and fluids into favourable structural traps; and (d) a conduit for later, near-surface, oxygenated meteoric waters to migrate into deeper parts of the sedimentary section. When deep-seated, high-angle normal faults, or rifts, developed or were reactivated in the back-arc region and were superimposed on earlier thrust structures, fluid-flow channels were provided for deep circulation as well as for the entrapped basin brines to flow along in a thermal gradient. All of the known deposits occur in structural settings where there are shallow structures coincident with a regional-scale deep-seated fracture.

Evolution of a deposit: ore-forming processes. We infer from the available data that ore formation took place in stages analogous to those described below and presented in schematic form in Figure 7.7.

(i) Magmatic, metalliferous and gas-bearing fluids emanate into convecting meteoric water at relatively high pressure. Evidence for this is derived from the studies of the Osgood Mountains, Nevada, where Taylor and O'Neil (1977) found that magmatic fluids pervaded the contact zone with the limestones, but that emanations along the near-vertical margins of the stock immediately mixed and re-equilibrated with waters in the sedimentary section. $\delta^{18}O$-rich CO_2 was produced as a consequence of reactions involving wollastonite thereby enriching the fluids (initially with $\delta^{18}O_{calc} = +9.4‰$; $\delta D_{calc} = -30$ to $-45‰$) in ^{18}O. Minerals in the margin of the stock and in the skarn became markedly enriched in ^{18}O (e.g. quartz $\delta^{18}O = +11.1‰$ in core of stock to $+13.4‰$ several metres outside of the pluton). The ^{18}O-enriched fluids flowed along the contact and permeated the contact zone of the stock. Thus, the magmatic fluids interacting with the sedimentary section produced an isotopically heavy, CO_2-enriched, hydrothermal solution which was 20–50% meteoric water (Taylor and O'Neil, 1977) during late-stage, post-skarn formation of the quartz vein.

(ii) Initially, entrapped chloride-bearing, ^{18}O-laden waters within the sedimentary rocks begin to convect and heat up the country rock to the ambient temperature of the fluid. Fluid-country rock disequilibrium causes exchange, and, in the H_2S field, sulphur will react with iron-bearing minerals such as magnetite or chlorite (e.g. $Fe_3O_4 + 6H_2S_{aq} = 3FeS_2 + 4H_2O + 2H_2$) and the solution will not carry sufficient quantities of H_2S to complex gold because the reaction to stabilize the $Au(HS)_2^-$ complex is of the form

$$Au + 2H_2S = Au(HS)_2^- + H^+ + 0.5H_2$$

Therefore, the initial fluids reflect the isotopic and chemical composition of the sedimentary brines, are barren of gold, and lack sufficient aqueous sulphide to precipitate any base metals carried as chloride complexes in this early brine. These initial fluids react with carbonate rocks, replacing selected sedimentary beds with silica to form the early, pre-ore jasperoids seen at most Carlin-type deposits.

(iii) As a steady state is reached, a magmatic input continues, but the basin brines that were the predominant fluid source begin to be replaced by deeply circulating meteoric waters. In addition, the system reaches thermal

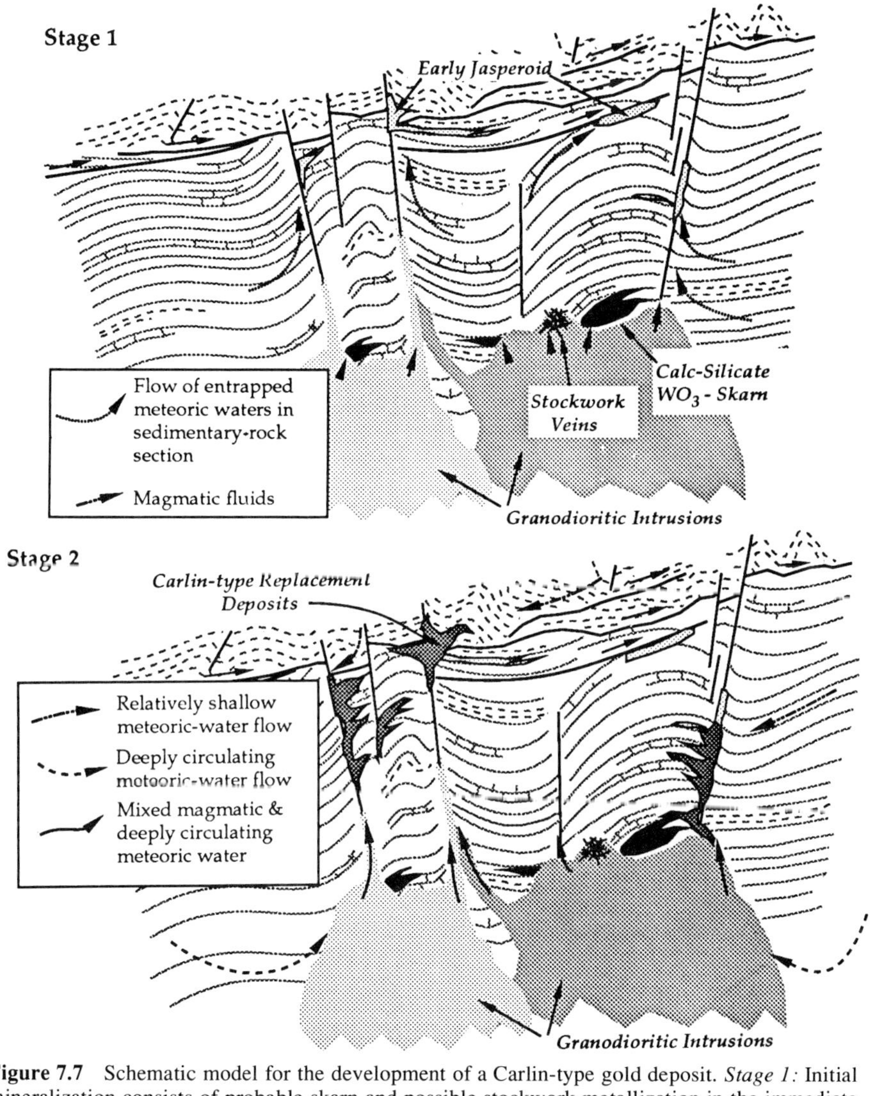

Figure 7.7 Schematic model for the development of a Carlin-type gold deposit. *Stage 1:* Initial mineralization consists of probable skarn and possible stockwork metallization in the immediate vicinity of the intrusive complex and concurrent formation of early jasperoid distal to the complex. *Stage 2:* Schematic model for the development of a Carlin gold deposit. Gold mineralization may occur either close in, or distal to, the intrusive complex. Gold deposition typically coincides with early, pre-ore jasperoid. The early jasperoid may be above, within, and/or beneath the gold mineralization.

equilibrium with the country rock within the upwelling plume, aqueous sulphide becomes available for metal complexing, and significant gold transport takes place.

(iv) Fluid-inclusion studies by Radtke *et al.* (1980) at Carlin and Hofstra *et al.* (1988) at Jerritt Canyon indicate that boiling was not the predominant mechanism in the gold-depositing process. Ore deposition was likely through the mixing of two chemically distinct fluids based on: (a) the substantial vertical extent of ores and the lack of clear, vertical zoning in some Carlin-type deposits (indicative of metal deposition concurrently under essentially identical conditions); (b) the unlikely possibility that the temperatures were fluctuating rapidly over a considerable range at the depths of ore deposition indicated for Carlin, Jerritt Canyon, and Getchell; and (c) oxygen-isotope data that show a wide range of $\delta^{18}O$ values in ore (Rye, 1985; Northrop *et al.*, 1987) and an apparent relationship to gold tenor at Jerritt Canyon (Hofstra *et al.*, 1988).

(v) Paragenetically early marcasite (e.g. Getchell, Mercur, China) implies that the hydrothermal conditions were initially quite acid at the site of gold deposition, although the replacement of this marcasite by pyrite at Getchell and the association of pyrite and gold in most deposits indicate a change in chemical conditions during gold deposition. At Carlin, Rose and Kuehn (1987) suggested that the most favourable sites for gold deposition were in the transition from calcite-stable to dolomite-stable alteration, further implying a chemical control on the actual gold deposition somewhat independent of the degree of mixing of fluids.

(vi) With time, the input of magmatic constituents decreases and dilute meteoric waters dominate the chemistry of the system. Some gold is still being transported and deposited, but the amount is considerably less than earlier. The ubiquitous occurrence of marcasite and barite in the late paragenesis at most deposits implies a lower pH in the waning stages of the hydrothermal system.

Deposits may be very close to their igneous source as at Getchell or more distal as at Jerritt Canyon. The development of regionally extensive jasperoids such as at Jerritt Canyon and Mercur may reflect a more distal heat source. These jasperoids, therefore, better preserve the geochemical record of the flushing of sedimentary basin waters through the site of the deposit during the early stages of the system.

7.6 Exploration guidelines

The most important guides to Carlin deposits are: (a) the occurrence of silty to argillaceous, thinly laminated, marine, carbonate rocks or carbonate-bearing siliciclastic rocks; (b) an area that has undergone extensional tectonics; and (c) concurrent mid- to upper-crustal plutonism. If deep-seated extensional fault zones that controlled the emplacement of plutons can be identified, such as the north–south trend mentioned previously in the Osgood Mountains region, Nevada, or the Northern Nevada Rift or Carlin rift zones, then the area is promising providing that favourable host rocks are present. Highly prospective structural zones related to early to late Palaeozoic continental-margin rifting probably have yet to be identified, but could be located through careful geological mapping and geophysical surveys. Those areas that contain favourable host rocks and tectonic features, but no geological or geophysical indication of igneous intrusive activity, are generally unfavourable, such as the 'quiet zone' of south-central Nevada (Blakely, 1988).

The evidence that Jerritt Canyon, Carlin, and Getchell may have been 'failed' petroleum reservoirs suggests that the crests and flanks of folds in favourable host lithologies, where they straddle deep-penetrating extensional fault zones, may provide the best sites for ore formation. This implies that detailed structural mapping of Palaeozoic–Mesozoic stratigraphy in north-central Nevada or analogous regions elsewhere in the world could help target specific areas for exploration.

The clustering of deposits in several districts suggests that magmas of appropriate composition were abundant in the area and that tectonism was permissive for widespread ore-forming processes. Such is the case for the Getchell and Carlin trends.

The occurrence of tungsten and molybdenum deposits appears to us to be a good exploration guide for Carlin-type deposits, although the incompleteness of mineral-occurrence records for most regions is a limiting factor. Where deposits do not appear to be clustered, such as in the vicinity of Alligator Ridge, or where additional deposits are only now being discovered, such as the Getchell region, there is a real potential for the discovery of additional deposits.

Jasperoid has been used as an exploration tool, and the paragenesis outlined in Table 7.3 serves as a visual guide to jasperoid evaluation and sampling for geochemical analysis (see Holland *et al.*, 1988). The geochemical distinction between jasperoid formed from early, pre-ore stages of the system and japeroid developed during ore deposition is quite important in exploration. Careful petrographic study of the jasperoid is needed to indentify possible ore-related stages, and is essential before applying analyses of light-stable isotopes or fluid inclusions in exploration.

As has been the case in exploration for many years, the characteristic trace-element suite of Au, As, Sb, Hg, and Tl (with generally low concentrations of base metals) is still a good indicator that Carlin-type hydrothermal processes may have taken place.

7.7 Summary

We suggest that Carlin-type deposits formed above active, deep-penetrating extensional faults where plutons were emplaced into the middle and upper crust in regions of thick marine carbonate–siliciclastic sequences commonly thickened by overthrusting. The ores were deposited from slightly to highly acidic, low-salinity, high-CO_2, and high-H_2S fluids which commonly form tungsten–skarn deposits and skarn- or porphyry-type molybdenum deposits. Carlin-type ores are characterized by high gold to silver ratios, a paucity of base metals, consistently associated arsenic, antimony, and mercury, and geochemically anomalous tungsten in many deposits. The Carlin-type deposits differ from base-metal-bearing, polymetallic-replacement deposits in that the latter are related to intrusions with relatively higher Fe/Al ratios which form base-metal skarn and porphyry-copper-type deposits.

Carlin-type deposits may form at depths of up to 3 km as evidenced by high-density, CO_2-bearing fluid inclusions that commonly occur in alteration zones and ore minerals. The physical and geochemical data suggest that gold precipitation was related to a mixing process, and, for the most part, was not due to boiling.

References

Bagby, W.C. and Berger, B.R. (1985) Geologic characteristics of sediment-hosted, disseminated precious-metal deposits in the western United States. In *Geology and Geochemistry of Epithermal Systems*; eds. Berger, B.R. and Bethke, P.M. Reviews in Economic Geology, Soc. of Econ. Geol., Vol. 2, 169-202.

Bakken, B.M. and Einaudi, M.T. (1986) Spatial and temporal relations between wall rock alteration and gold mineralization, Main Pit, Carlin gold mine, Nevada, U.S.A. In *Gold '86*, ed. MacDonald, A.J. Proceedings Volume, Gold '86, Toronto, 388- 403.

Baker, E.D. (1987) Carlin mine geology. In Symp. *Bulk Mineable, Guidebook for Field Trips*, ed. Johnson, J.L. Geol. Soc. of Nevada, Reno, 276-280.

Berger, B.R. (1985) Geological and geochemical relationships at the Getchell mine and vicinity, Humboldt County, Nevada. In *Discoveries of Epithermal Precious Metal Deposits*; ed. Hollister, V.F. Soc. of Mining Engineers, New York, 51–59.

Birak, D.J., and Hawkins, R.B., 1985, The geology of the Enfield Bell mine and Jerritt Canyon district, Elko County, Nevada. In *Geologic Characteristics of Sediment- and Volcanic-Hosted Disseminated Gold Deposits – Search for an Occurrence Model*, ed. Tooker, E.W. US Geol. Surv., Bull. 1646, 95-105.

Blakely, R.J. (1988) Curie temperature isotherm analysis and tectonic implications of aeromagnetic data from Nevada. *J. Geophys. Res.* **93**, 11,817-11,832.

Brady, B.T. (1984) Gold in the Black Pine mining district, southeast Cassia County, Idaho. US Geol. Surv. Bull. 1382-E. 15pp.

Broili, C., French, G.M., Shaddrick, D.R. and Weaver, R.R. (1988) Geology and gold mineralization of the Gold Bar deposit, Eureka County, Nevada. In *Proc. Symp. Bulk Mineable Precious Metal Deposits of the Western United States*, eds. Schafer, R.W., Cooper, J.J. and Vikre, P.G. Geol. Soc. of Nevada, Reno, 57-72.

Chao, E.C.T., Minkin, J.A., Back, J.M., Chen, J.R., Bagby, W.C., Rivers, M.L., Sutton, S.R., Hanson, A.L., Gordon, B.M. and Hones, K.W. (1986) Occurrence of gold in an unoxidized Carlin-type ore sample – A preliminary synchrotron and micro-optical restudy (abs.). *Geol. Soc. Amer. Abstr. Programs* **18**(6), 562.

Christensen, O.D., Knutsen, G.C. and Ekburg, C.E. (1987) Disseminated gold deposits of the Carlin trend, Eureka and Elko counties, Nevada. Soc. of Mining Engineers, Preprint 87-84. 7pp.

Cox, D.P. and Singer, D.A. (eds.) (1986) Mineral deposit models. US Geol. Surv., Bull. 1693. 379pp.

Cunningham, C.G., Ashley, R.P., Chou, I-M., Huang, Z., Wan, C. and Li, W. (1988) Newly discovered sedimentary-rock hosted disseminated gold deposits in the People's Republic of China. *Econ. Geol.* **83**, 1462-1467.

Dean, W.E., Bostick, N.H., Bartel, A.J., Brandt, E.L., Daws, T.A., Doughten, M., Gent, C.A., Jaunarajs, S.R., Libby, B., Malcolm, M.J., Robb, E.C., Taggart, J.E., Threlkeld, C.N., Vuletich, A.K., Cunningham, C.G., Ashley, R.P. and Chou, I. (1988) Data on the geochemistry and thermal maturation of sedimentary-rock hosted, disseminated gold deposits and associated rocks, southwestern Guizhou Province, People's Republic of China. US Geol. Surv., Open-File Report 88-271. 22pp.

Droste, K.J., Bruff, S.R., Johnson, E.A., Brooks, W.E. and Johnson, R.O. (1988) The Austin gold venture. In *Proc. Symp. Bulk Mineable Precious Metal Deposits of the Western United States*, eds. Schafer, R.W., Cooper, J.J., and Vikre, P.G. Geol. Soc. of Nevada, Reno, 73-99.

Elder, J. (1981) *Geothermal Systems*. Academic Press, New York. 508pp.

Ellis, R. (1987) The Dee gold mine. In *Symp. Bulk Mineable, Guidebook for Field Trips*; ed. Johnson, J.L. Geol. Soc. of Nevada, Reno, 285-288.

Erickson, R.L., Marranzino, A.P., Oda, U. and Janes, W.W. (1964) Geochemical exploration near the Getchell mine, Humboldt County, Nevada. US Geol. Surv., Bull. 1198-A. 26pp.

Erickson, R.L., Van Sickle, G.H., Nakagawa, H.M., McCarthy, J.H., Jr. and Leong, K.W. (1966) Gold geochemical anomaly in the Cortez district, Nevada. US Geol. Surv., Circ. 534. 9pp.

Evans, J.G. (1980) Geology of the Rodeo Creek NE and Welches Canyon quadrangles, Eureka County, Nevada. US Geol. Surv., Bull. 1473. 81pp.

Ferguson, H.G. (1924) Geology and ore deposits of the Manhattan district, Nevada. US Geol. Surv., Bull. 723. 163pp.

Foo, S.T. and Hebert, J.P. (1987) Geology of the Horse Canyon deposit, Eureka County, Nevada. In *Symp. Bulk Mineable, Guidebook for Field Trips*, ed. Johnson, J.L. Geol. Soc. of Nevada, Reno, 326-332.

Fyfe, W.S., Price, N.J. and Thompson, A.B. (1978) *Fluids in the Earth's Crust.* Elsevier, New York. 383pp.

Geach, R.D. (1972) Mines and mineral deposits, Beaverhead County, Montana. Montana Bur. of Mines and Geol., Bull. 85. 194pp.

Gerdenich, M.J., Hanel, I.V. and Kesler, S.E. (1986) Alkene/alkane hydrocarbon gas ratios in fluid inclusions from Archean and Tertiary gold deposits (abs.). *Geol. Soc. Amer. Abstr. Programs*, **18**(6), 611.

Gesick, T.E. (1987) Tonkin Springs gold deposits. In *Symp. Bulk Mineable, Guidebook for Field Trips*, ed. Johnson, J.L. Geol. Soc. of Nevada, Reno, 305-315.

Giebink, R.G. and Paterson, C.J. (1986) Stratigraphic controls on sediment-hosted epithermal gold mineralization: Evidence from the Annie Creek mine in the Cambrian Deadwood Formation, Black Hills, South Dakota. *Geol. Soc. Amer. Abstr. Programs* **18**(6), 613.

Giggenbach, W.F. (1986) The use of gas chemistry in delineating the origin of fluids discharged over the Taupo Volcanic Zone. *Proc. Symp. V, Internat. Volcanological Congress, Auckland, New Zealand*, 47-50.

Giles, D.L. (1983) Gold mineralization in the laccolithic complexes of central Montana. In *The Genesis of Rocky Mountain Ore Deposits: Changes with Time and Tectonics*: Denver Region Exploration Geol. Soc., 157-162.

Gilluly, J. (1932) Geology and ore deposits of the Stockton and Fairfield quadrangles, Utah. US Geol. Surv., Prof. Paper 173. 171pp.

Gize, A.P. (1986) The development of a thermal mesophase in bitumens from high temperature ore deposits. In *Proc. Symp. Organics and Ore Deposits*; ed. Dean, W.E. Denver Region Exploration Geol. Soc. 137-150.

Grauch, V.J.S. (1988) Geophysical tools for defining covered geologic features: Significance for disseminated gold deposits in Nevada, U.S.A. *Bicentennial Gold 88*, Extended Abstracts, Poster Programme, eds. Goode, A.D.T., Smyth, E.L., Birch, W.D., and Bosma, L.I. *Geol. Soc. Aust. Abstr.* **23**(2), 527- 529.

Grauch, V.J.S., Blank, H.R., Blakely, R.J., Oliver, H.W., Plouff, D. and Ponce, D. (1987) Geophysical delineation of granitic plutons in Nevada and their relationship to gold deposits. *Geol. Soc. Amer. Abstr. Programs*, **19**(7), 682.

Hausen, D.M. (1967) Fine gold occurrence at Carlin, Nevada. Unpublished Ph.D. Dissertation, Columbia Univ. 166pp.

Hausen, D.M. and Kerr, P.F. (1968) Fine gold occurrence at Carlin, Nevada. In *Ore Deposits of the United States, 1933- 1967, The Graton-Sales Volume*, ed. Ridge, J.D. Amer. Inst. of Mining, Metallurgical, and Petroleum Engineers, 908-940.

Hausen, D.M. and Park, W.C. (1986) Observations on the association of gold mineralization with organic matter in Carlin- type ores. In *Proc. Symp. Organics and Ore Deposits*, ed. Dean, W.E. Denver Region Exploration Geol. Soc., 119-136.

Havenstrite, S.R. (1984) Geology and ore deposits of the Taylor silver district. In *Exploration for Ore Deposits of the North American Cordillera, Field Trip Guidebook*; ed. Johnson, J.L. Assoc. Exploration Geochemists, 1984 Regional Symp. Reno, FT2, 37-45.

Hedenquist, J.W. (1986) Geothermal systems in the Taupo Volcanic Zone: Their characteristics and relation to volcanism and mineralization. In *Late Cenozoic Volcanism in New Zealand*; ed. Smith, I.E.M. Royal Soc. of New Zealand, Bull. 23, 134-168.

Henley, R.W. (1985) The geothermal framework for epithermal deposits. In *Geology and Geochemistry of Epithermal Systems*, eds. Berger, B.R. and Bethke, P.M. *Reviews in Econ. Geol.* 2, 1-24.

Henley, R.W. and Hoffman, C.F. (1987) Gold: Sources to resources. *Proc. Conf. Australasian Institute of Mining and Metallurgy, Pacific Rim Congress 87*, 155-168.

Henley, R.W. and McNabb, A. (1978) Magmatic vapor plumes and ground-water interaction in porphyry copper emplacement. *Econ. Geol.* **73**, 1-20.

Henley, R.W., Truesdell, A.H. and Barton, P.B., Jr. (1984) *Fluid-Mineral Equilibria in Hydrothermal Systems.* Reviews in Economic Geology, Soc. of Econ. Geol., Vol. 1. 267pp.

Heran, W.D. and Smith, B.D. (1984) Geophysical surveys at the Getchell and Preble disseminated gold deposits, Humboldt County, Nevada. US Geol. Surv., Open-File Report 84-795. 78pp.

Hill, R.H., Adrian, B.M., Bagby, W.C., Bailey, E.A., Goldfarb, R.J. and Pickthorn, W.J. (1986) Geochemical data for rock samples collected from selected sediment-hosted disseminated precious- metal deposits in Nevada. US Geol. Surv., Open-File Report 86- 107. 30pp.

Hills, R.C. (1894) Ore deposits of Camp Floyd district, Tooele County, Utah. *Proc. Colorado Scientific Soc.*, Vol. 5, 54- 65.

Hofstra, A.H., Northrop, H.R., Rye, R.O., Landis, G.P. and Birak, D.J. (1988) Origin of sediment-hosted disseminated gold deposits by fluid mixing – Evidence from jasperoids in the Jerritt Canyon gold district, Nevada, U.S.A. In *Bicentennial Gold 88*, Extended Abstracts, Poster Programme, eds. Goode, A.D.T., Smyth, E.L., Birch, W.D. and Bosma, L.I. *Geol. Soc. Aust. Abstr.* **22**, 284-289.

Holland, P.T., Beaty, D.W. and Snow, G.G. (1988) Comparative elemental and oxygen isotope geochemistry of jasperoid in the northern Great Basin: Evidence for distinctive fluid evolution in gold-producing hydrothermal systems. *Econ. Geol.* **83**, 1401-1423.

Hua, Y. (1988) Preliminary study on the origin and mineralization feature of mercury and gold deposits in Guizhou Province east part. In *Bicentennial Gold '88*, Extended Abstracts Poster Programme, eds. Goode, A.D.T., Smyth, E.L., Birch, W.D. and Bosma, L.I. *Geol. Soc. Aust. Abstr.* **23**(1), 219-223.

Ilchik, R.P., Brimhall, G.H. and Schull, H.W. (1986) Hydrothermal maturation of indigenous organic matter at the Alligator Ridge gold deposits, Nevada. *Econ. Geol.* **81**, 113-130.

Irving, J.D. (1904) Economic resources of the northern Black Hills. US Geol. Surv., Prof. Paper 26. 222pp.

Jewell, P.W. (1984) Chemical and thermal evolution of hydrothermal fluids, Mercur gold district, Tooele County, Utah. Unpublished M.S. thesis, Univ. of Utah. 77pp.

Joralemon, P. (1949) The occurrence of gold at the Getchell mine, Nevada. Unpublished Ph.D. Dissertation, Harvard Univ. 175pp.

Joralemon, P. (1951) The occurrence of gold at the Getchell mine, Nevada. *Econ. Geol.* **46**, 267-310.

Kettler, R.M., Meyers, P.A., Kesler, S.E., Steininger, R.C., and Glaser, A. (1986) Aliphatic hydrocarbons in Pilot Shale adjacent to the Alligator Ridge gold deposit, Nevada. Indicators of ore- forming processes (abs.). *Geol. Soc. Amer. Abstr. Programs.* **18**(6), 656.

Kuehn, C.A. and Bodnar, R.J. (1984) P-T-X characteristics of fluids associated with the Carlin sediment-hosted gold deposit (abs.). *Geol. Soc. Amer. Abstr. Programs* **16**(6), 566.

Kuehn, C.A. and Rose, A.W. (1986) Temporal framework for the evolution of fluids at the Carlin gold mine (abs.). *Geol. Soc. Amer. Abstr. Programs* **18**(6), 663.

Kuehn, C.A. and Rose, A.W. (1987) Chemical and mineralogical zonation associated with low pH hydrothermal alteration at the Carlin gold deposit, NV (abs.). *Geol. Soc. Amer. Abstr. Programs* **19**(7), 734.

Lachenbruch, A.H. and Sass, J.H. (1977) Heat flow in the United States and the thermal regime of the crust. In *The Earth's Crust: Its Nature and Physical Properties*; ed. Heacock, J.G. Amer. Geophys. Union, Geophys. Monograph 20, 626-675.

Langenheim, R.L., Jr. and Larson, E.R. (1973) Correlation of Great Basin stratigraphic units. Nevada Bur. of Mines and Geol., Bull. 72. 36pp.

Leventhal, J.S., Hofstra, A.H., Vuletich, A.K. and Mancuso, T.B. (1987) Sediment-hosted disseminated gold mineralization at Jerritt Canyon, Nevada. III – Role of organic carbon (abs.). *Geol. Soc. Amer. Abstr. Programs* **19**(7), 745.

Lindgren, W. (1913) *Mineral Deposits.* McGraw-Hill New York. 883pp.

Lindgren, W. (1933) *Mineral Deposits* (4th edn.). McGraw- Hill, New York. 930pp.

Lindsey, D.A. and Naeser, C.W. (1985) Studies of mineralized intrusive complexes in north-central Montana. US Geol. Surv., Prof. Paper 1301-B, 35-41.

Lu, H. (1988) Fluid inclusion studies on different types of Chinese gold deposits. In *Bicentennial Gold 88*, Extended Abstracts Poster Programme, eds. Goode, A.D.T., Smyth, E.L., Birch, W.D. and Bosma, L.I. *Geol. Soc. Aust. Abstr.* **23**(2), 442-444.

MacLeod, R. (1987) Geology of the Bootstrap mine. In *Symp. Bulk Mineable, Guidebook for Field Trips*; ed. Johnson, J.L. Geol. Soc. of Nevada, Reno, 283-284.

Madrid, R.J. and Bagby, W.C. (1988) Gold occurrence and its relation to vein and mineral paragenesis in selected sedimentary-rock hosted, Carlin-type deposits in Nevada. In *Bicentennial Gold '88*, Extended Abstracts Oral Programme, eds. Goode, A.D.T., Smyth, L.E., Birch, W.D. and Bosma, L.I. *Geol. Soc. Aust. Abstr.* **22**, 161-166.

Marvin, R.F., Hearn, B.C., Jr., Mehnert, H.H., Naeser, C.W., Zartman, R.E. and Lindsey, D.A. (1980) Late Cretaceous-Paleocene-Eocene igneous activity in north-central Montana. *Isochron/West* **29**, 5-25.

McFarlane, D.N. (1987) Maggie Creek deposit, In *Symp. Bulk Mineable, Guidebook for Field Trips*; ed. Johnson, J.L. Geol. Soc. of Nevada, Reno, 274-275.

McKee, E.H., Silberman, M.L., Marvin, R.E. and Obradovich, J.D. (1971) A summary of radiometric ages of Tertiary volcanic rocks in Nevada and eastern California. Part I: Central Nevada. *Isochron/West* **2**, 21-42.

McKee, E.H., Tarshis, A.L. and Marvin, R.F. (1976) Summary of radiometric ages of Tertiary volcanic and selected plutonic rocks in Nevada. Part V: Northeastern Nevada. *Isochron/West* **16**, 15-27.

Merriam, C.W. and Anderson, C.A. (1942) Reconnaissance survey of the Roberts Mountains, Nevada. Geol. Soc. of Amer., Bull. 53, 1675-1726.

Moore, W.J. (1973) A summary of radiometric ages of igneous rocks in the Oquirrh Mountains, north-central Utah: *Econ. Geol.* **68**, 97-101.

Moore, W.J. and McKee, E.H. (1983) Phanerozoic magmatism and mineralization in the Tooele $1^\circ \times 2^\circ$ quadrangle, Utah. In *Tectonic and Stratigraphic Studies in the Eastern Great Basin*; eds. Miller, D.M., Todd, V.R. and Howard, K.A. Geol. Soc. of Amer., Memoir 157, 183-190.

Morton, J.L., Silberman, M.L., Bonham, H.F., Garside, L.J. and Noble, D.C. (1977) K-Ar ages of volcanic rocks, plutonic rocks, and ore deposits in Nevada and eastern California – Determinations run under the USGS–NBMG cooperative program. *Isochron/West* **20**, 19-29.

Motter, J.W. and Chapman, P.E. (1984) Northumberland gold deposit, Nye County, Nevada. In *Western Geological Excursions*, Volume 3; ed. Lintz, J. Jr. Univ. of Nevada, Mackey School of Mines, 181-199.

Newberry, R.J. and Swanson, S.E. (1986) Scheelite skarn granitoids: An evaluation of the roles of magmatic source and process. *Ore Geol. Rev.* **1**, 57-81.

Northrop, H.R., Rye, R.O., Landis, G.P., Lustwerk, R., Jones, M.B. and Daly, W.E. (1987) Sediment-hosted disseminated gold mineralization at Jerritt Canyon, Nevada. V. – Stable isotope geochemistry and a model of ore deposition (abs.). *Geol. Soc. Amer. Abstr. Programs* **19**(7), 791.

Osterberg, M. (1988) Geology of the Chimney Creek sediment-hosted gold deposit, Osgood Mountains, Nevada (abs.). Geol. Soc. Nevada, October meeting, 4-6.

Page, B.M. (1959) Geology of the Candelaria mining district, Mineral County, Nevada. Nevada Bur. of Mines, Bull. 56. 67pp.

Percival, T.J., Bagby, W.C. and Radtke, A.S. (1988) Physical and chemical features of precious metal deposits hosted by sedimentary rocks in the western United States. In *Proc. Symp. Bulk Mineable Precious Metal Deposits of the Western United States*, eds. Schafer, R.W., Cooper, J.J., and Vikre, P.G. Geol. Soc. of Nevada, Reno, 11-34.

Potter, C.J., Liu, C., Huang, J., Zheng, L., Hauge, T.A., Hauser, E.C., Allmendinger, R.W., Oliver, J.E., Kaufman, S. and Brown, L. (1987) Crustal structure of north-central Nevada: Results from COCORP deep seismic profiling. Geol. Soc. of Amer., Bull. 98, 330-337.

Powers, S.L. (1978) Jasperoid and disseminated gold at the Ogee- Pinson mine, Humboldt County, Nevada. Unpublished M.S. thesis, Univ. of Nevada, Reno. 112pp.

Radtke, A.S. (1985) Geology of the Carlin gold deposit. US Geol. Surv., Prof. Paper 1267. 124pp.

Radtke, A.S. and Dickson, F.W. (1976) Genesis and vertical position of fine-grained disseminated replacement-type gold deposits in Nevada and Utah, U.S.A. In *Problems of Ore Deposition, Proc. Fourth IAGOD Symp., Varna, Bulgaria, 1974.* Sofia, Bulgarian Academy of Sciences, Vol. 1, 71-78.

Radtke, A.S. and Scheiner, B.J. (1970) Studies of hydrothermal gold deposition (I). Carlin gold deposit, Nevada: The role of carbonaceous materials in gold deposition. *Econ. Geol.* **65**, 87-102.

Radtke, A.S., Rye, R.O. and Dickson, F.W. (1980) Geology and stable isotope studies of the Carlin gold deposit, Nevada. *Econ. Geol.* **75**, 641-672.

Ramsey, R.H. (1973) *Men and Mines of Newmont–A Fifty Year History.* New Octagon Books, New York. 344pp.

Roberts, R.J. (1951) Geology of the Antler Peak quadrangle, Nevada. US Geol. Surv. Geologic Quadrangle Map GQ-10, scale 1 : 62 500.

Roberts, R.J. (1960) Alinement of mining districts in north- central Nevada. US Geol. Surv., Prof. Paper 400-B, B-17-B19.

Rose, A.W. and Kuehn, C.A. (1987) Ore deposition from highly acidic CO_2-enriched solutions at the Carlin gold deposit, Eureka County, Nevada (abs.). *Geol. Soc. Amer. Abstr. Programs* **19**(7), 824.

Rota, J.C., 1987, The Gold Quarry deposit. In *Symp. Bulk Mineable, Guidebook for Field Trips*, ed. Johnson, J.L. Geol. Soc. of Nevada, Reno, 271-273.

Rye, R.O. (1985) A model for the formation of carbonate-hosted disseminated gold deposits based on geologic, fluid-inclusion, geochemical, and stable-isotope studies of the Carlin and Cortez deposits, Nevada. In *Geologic Characteristics of Sediment- and Volcanic-Hosted Disseminated Gold Deposits–Search for an Occurrence Model*; ed. Tooker, E.W. US Geol. Surv., Bull. 1646, 35-46.

Sawkins, F.J. (1984) *Metal Deposits in Relation to Plate Tectonics*. Springer-Verlag, New York. 325pp.

Sibson, R.H., McMoore, J. and Rankin, R.H. (1975) Seismic pumping – a hydrothermal fluid transport mechanism. *J. Geol. Soc. London* **131**, 653-659.

Silberman, M.L. and McKee, E.H. (1971) K–Ar ages of granitic plutons in north-central Nevada. *Isochron/West*, **1**, 15-32.

Silberman, M.L., Berger, B.R. and Koski, R.A. (1974) K–Ar age relations of granodiorite emplacement and tungsten and gold mineralization near the Getchell mine, Humboldt County, Nevada. *Econ. Geol.* **69**, 646-656.

Silberman, M.L., Bonham, H.F., Jr. and Osborne, D.H. (1975) New K–Ar ages of volcanic and plutonic rocks and ore deposits in western Nevada. *Isochron/West* **13**, 13-21.

Simmons, S.F., Gemmell, J.B. and Sawkins, F.J. (1988) The Santo Niño silver–lead–zinc vein, Fresnillo district, Zacatecas: Part II. Physical and chemical nature of ore forming solutions. *Econ. Geol.* **83**, 1619-1641.

Speed, R.C. (1983) Pre-Cenozoic tectonic evolution of northeastern Nevada. In *The Role of Heat in the Development of Energy and Mineral Resources in the Northern Basin and Range Province*. Geothermal Resources Council, Special Report 13, 11- 24.

Spurr, J.E. (1896) Economic geology of the Mercur mining district, Utah. US Geol. Surv. 16th Annual Report, Pt. 2, 343- 455.

Stewart, J.H. (1980) Geology of Nevada. Nevada Bur. of Mines and Geol., Special Pub. 4. 136pp.

Stewart, J.H. (1988) Tectonics of the Walker Lane belt, western Great Basin: Mesozoic and Cenozoic deformation in a zone of shear. In *Metamorphism and Crustal Evolution of the Western United States*, ed. Ernst, W.G. Prentice-Hall, New Jersey, 681- 713.

Tafuri, W.J. (1987) Geology and geochemistry of the Mercur mining district, Tooele County, Utah. Unpublished Ph.D. Dissertation, Univ. of Utah. 180pp.

Taylor, B.E. and O'Neil, J.R. (1977) Stable isotope studies of metasomatic Ca–Fe–Al–Si skarns and associated metamorphic and igneous rocks, Osgood Mountains, Nevada. *Contrib. Mineral. Petrol.* **63**, 1-49.

Thoreson, R.F. (1987) The Rain gold deposit. In *Symp. Bulk Mineable, Guidebook for Field Trips*; ed. Johnson, J.L. Geol. Soc. of Nevada, Reno, 269-270.

Wallace, A.R. (1989) The Relief Canyon gold deposit, southern Humboldt Range, Nevada: A mineralized solution breccia. *Econ. Geol.* **84** (in press).

Weed, W.H. and Pirsson, L.V. (1898) Geology and mineral resources of the Judith Mountains of Montana. US Geol. Surv., 18th Annual Report, Pt. 3, 437-616.

Wells, J.D., Elliott, J.E. and Obradovich, J.D. (1971) Age of the igneous rocks associated with ore deposits, Cortez-Buckhorn area, Nevada. US Geol. Surv., Prof. Paper 750-C, C127-C135.

Wells, J.D. and Mullens, T.E. (1973) Gold-bearing arsenian pyrite determined by microprobe analysis, Cortez and Carlin gold mines, Nevada. *Econ. Geol.* **68**, 187-201.

Wilson, W.L. and Wilson, W.B. (1986) Geology of the Eureka- Windfall and Rustler gold deposits, Eureka County, Nevada. In *Sediment-Hosted Precious-Metal Deposits of Northern Nevada*, eds. Tingley, J.V. and Bonham, H.F., Jr. Nevada Bur. of Mines and Geol., Report 40, 81-84.

Wrucke, C.T. and Armbrustmacher, T.J. (1975) Geochemical and geologic relations of gold and other elements at the Gold Acres open-pit mine, Lander County, Nevada. US Geol. Surv., Prof. Paper 860. 27pp.

Zhang, Z.M., Liou, J.G. and Coleman, R.G. (1984) An outline of the plate tectonics of China. *Geol. Soc. of Amer., Bull.* 95, 295- 312.

8 Auriferous hydrothermal precipitates on the modern seafloor

M.D. HANNINGTON, P.M. HERZIG and S.D. SCOTT

8.1 Introduction

Submarine hot springs were probable sources for gold enrichment in a variety of rock types which host mineable gold deposits. These include iron formations, mixed chemical and clastic sediments, tuffaceous exhalites, and disseminated or massive sulphides in both volcanic- and sediment-dominated sequences. Gold-bearing iron formations and interflow metalliferous sediments associated with seafloor hydrothermal activity also have been implicated as potential source rocks for some non-stratabound gold deposits in ancient greenstone belts (Foster and Wilson, 1984; Keays, 1984). Recent studies of gold in volcanogenic massive sulphides indicate a strong genetic relationship between gold and sulphide mineralization in seafloor hydrothermal systems (Hannington and Scott, 1989; Large *et al.*, 1989; Huston and Large, 1989). The total past production and current reserves of gold in massive sulphides world wide amount to nearly 2900 t Au and indicate that modified seawater is capable of transporting and depositing significant amounts of gold. In addition, the discovery of gold-rich sulphides actively forming at hydrothermal vents on the modern seafloor has confirmed the existence of gold-bearing fluids in submarine hot springs and supports a seafloor hydrothermal origin for gold in many preserved deposits now on land. The documentation of fluid chemistry at active vents also has served to constrain the conditions of gold mineralization on the present-day seafloor. In this chapter, we describe the occurrence and distribution of gold in modern hot spring deposits and discuss aspects of gold transport and deposition in seafloor hydrothermal systems with reference to possible implications for the origin of gold deposits in auriferous chemical sediments.

8.2 Gold in seafloor polymetallic sulphide deposits

Since the first discovery of sulphide deposits on the East Pacific Rise (CYAMEX, 1979), hydrothermal precipitates associated with active hot springs have been found in a variety of geologic settings on the modern seafloor (Figure 8.1), including mid-ocean ridges, axis and off-axis seamounts, back-arc spreading centres, and sedimented intercratonic rifts. Rona (1988) provided a thorough compilation of the occurrence of recent hydrothermal deposits on the seafloor. Active vents are found along rift zones, punctuated by recent volcanism and frequent tectonic activity, and

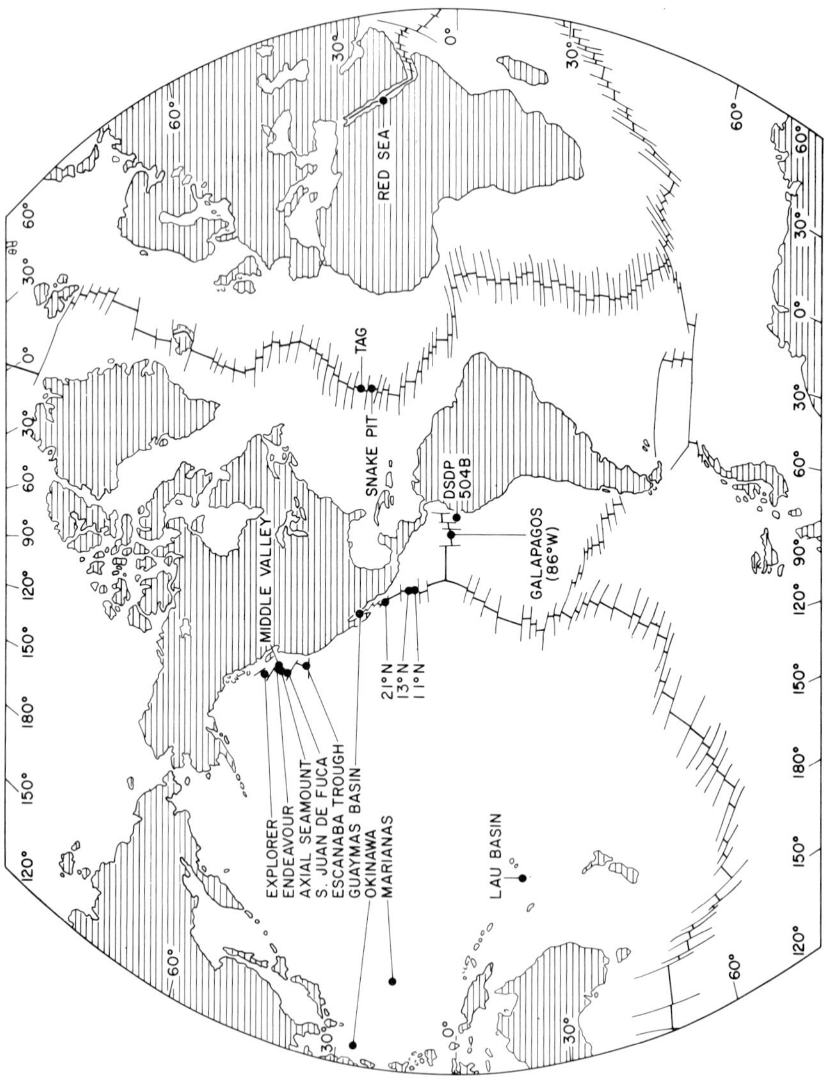

Figure 8.1 Location of hydrothermal deposits on the seafloor considered in this study.

where heat for the deep convection of seawater is provided by the intrusion of magma to within a few kilometres of the seafloor. Sustained water–rock interactions at temperatures up to 400°C leach base and precious metals from the permeable volcanic rocks in quantities large enough to produce ore-forming fluids (Rosenbauer and Bischoff, 1983; Mottl, 1983; Seyfried and Janecky, 1985; Seyfried *et al.*, 1988). High concentrations of dissolved H_2S are derived from a combination of reduced seawater-sulphate and basaltic sulphur leached from the volcanics along with the metals. Sulphides, sulphates, and silica are then precipitated from ascending hydrothermal fluids in response to decreasing temperature and increasing pH following conductive cooling and mixing with ambient seawater at or near the seafloor.

The hydrothermal precipitates accumulate locally as mineral-lined chimneys surrounding high-temperature vents and as large hydrothermal mounds. These typically consist of pyrrhotite–chalcopyrite–isocubanite assemblages formed at temperatures near 350°C (black smokers), pyrite–chalcopyrite assemblages at 300-350°C, and sulphur-rich, pyrite–marcasite–sphalerite assemblages at temperatures less than about 300°C (white smokers). Anhydrite is an important constituent of the high-temperature black smokers, and amorphous silica and barite are common in the lower-temperature assemblages.

Most of the known hydrothermal deposits are small, but a few exceed several million tonnes (Scott, 1987; Rona, 1988). Their gold and base metal contents vary widely (Table 8.1) but are similar to grades encountered in ores from ancient sulphide deposits now found on land. Average gold contents range from 0.07 to 4.9 ppm Au, but significant gold enrichment is found in examples from each of the geologic settings in which active hot springs occur.

8.2.1 *Mid-ocean ridges*

Basaltic mid-ocean ridges are currently the best known and most completely studied sites of hot spring activity on the seafloor. Gold-bearing deposits are known on the Mid-Atlantic Ridge, the Southern Explorer Ridge in the north-east Pacific, near 13°N on the East Pacific Rise, and in the Galapagos Rift (Figure 8.1). Two deposits (the TAG hydrothermal field and Explorer Ridge) are estimated to contain 1 to 5 million tonnes of massive sulphide each (Scott *et al.*, 1989; Rona *et al.*, 1986) and, if analysed samples are representative, could contain several thousand kilograms of gold. However, other deposits in the eastern Pacific are uniformly gold-poor.

The TAG hydrothermal field (3620–3700 m depth) occurs in the central rift valley of the Mid-Atlantic Ridge and consists of high-temperature, black smoker chimneys and pervasive, lower- temperature venting from a large hydrothermal mound (Rona *et al.*, 1986). The vent field covers an area approximately 580 m in diameter, with the central mound measuring 250 m in diameter and 50 m in height. The large size of the deposit suggests that hydrothermal activity has occurred over a long period of time with extensive hydrothermal reworking and subsurface recrystallization (Thompson *et al.*, 1988). Measured temperatures for the high-temperature chimneys range from 290°C to 321°C and associated mineral assemblages consist of pyrite, chalcopyrite, and anhydrite. The lower-temperature assemblages contain pyrite, marcasite, sphalerite, chalcopyrite, minor bornite, amorphous silica, and carbonate (Thompson *et al.*, 1988). Gold contents in the pyrite–chalcopyrite chimneys range from 0.2 to 1 ppm

Table 8.1 Average composition of samples from volcanic-hosted deposits on the seafloor

	Mid-Atlantic Ridge		Southern[3] Explorer Ridge	East Pacific[4] Rise 13°N	Galapagos[5] Rift	East Pacific[6] Rise 11°N	East Pacific[7] Rise 21°N	Southern[8] Juan de Fuca Ridge	Axial[9] Seamont	Marianas[10] Back-arc	Okinawa[11] Trough
	TAG Field[1]	Snakepit[2]									
	(18)*	(12)	(48)	(33)	(28)	(11)	(5)	(3)	(14)	(11)	(5)
Au (ppm)	2.1	1.5	0.63	0.42	0.35	0.15	0.15	0.11	4.9	0.80	5.1
Ag	72	50	97	49	46	38	98	178	189	184	772
As	73	304	544	154	139	399	296	359	569	126	31 000
Sb	17	23	27	–	11	30	23	18	349	190	–
Co	113	56	513	1 002	288	16	14	11	3	2	–
Se	62	101	100	157	65	27	40	13	2	10	–
Cd	301	160	199	233	109	886	376	519	522	465	620
Mo	77	25	173	132	129	30	22	13	38	5	–
Hg	15.9	2.7	10.8	–	4.9	2.5	2.0	2.0	20.2	22.0	–
Tl	27	25	43	–	12	19	13	23	1.1	18	–
Cu (wt.%)	9.2	2.0	3.2	7.8	4.5	1.9	0.6	0.2	0.4	1.2	1.4
Fe	24.4	34.0	25.9	26.0	32.6	22.4	12.4	19.7	5.6	2.4	8.8
Zn	7.6	4.8	5.3	8.2	4.0	28.0	19.8	36.7	22.2	10.0	23.4
Pb	0.05	0.03	0.11	0.05	0.04	0.07	0.21	0.26	0.35	7.4	15.9
S	36.0	36.9	28.3	35.1	32.4	35.7	31.4	39.3	18.8	17.8	–
SiO_2	6.0	1.8	9.1	9.2	21.5	1.2	19.0	5.1	28.1	1.2	10.2
Ba	<0.01	<0.01	7.4	0.4	0.04	0.06	0.15	0.06	9.6	33.3	3.4
Ca	5.0	1.5	1.2	4.5	0.3	0.02	11.2	0.05	0.21	3.7	0.06
CO_2	3.4	<0.1	0.2	–	0.3	–	–	–	0.36	–	–
Au/Ag	0.029	0.030	0.006	0.009	0.008	0.004	0.002	<0.001	0.026	0.004	0.007

*Figures in parentheses show the total number of samples analysed for different elements from each site.

Data are from: (1) Herzig *et al.* (1988a), Hannington *et al.* (1988b), Thompson *et al.* (1988), Schroeder *et al.* (1986), Hannington *et al.* (1989); (2) Herzig *et al.* (1988a), Hannington *et al.*(1989); (3) Hannington *et al.* (1986), Scott *et al.* (1989), M.D. Hannington (unpublished); (4) Fouquet *et al.* (1988); (5) Garimas II Cruise Report (1987), Garimas I Cruise Report (1986), Tufar *et al.* (1986); (6) McConachy (1988), M.D. Hannington (unpublished); (7) Zierenberg *et al.* (1984), Bischoff *et al.* (1983); (8) USGS Study Group (1986), Koski *et al.* (1984), Bischoff *et al.* (1983); (9) Hannington and Scott (1988), Hannington *et al.* (1986), M.D. Hannington (unpublished); (10) Kastner *et al.* (1987), M.D. Hannington (upublished); (11) Halback *et al.* (1989).

Au, but the lower-temperature pyrite-marcasite-sphalerite assemblages contain up to 4.9 ppm Au and 335 ppm Ag (Table 8.2 and Figure 8.2).

Table 8.2 Analysis of representative samples from the TAG hydrothermal field, Mid-Atlantic Ridge

	Active vents		Inactive deposits			
	Black smoker 1675–1	White smoker 1676–6–15b	Cu–Fe-rich 1–44	Zn–Fe-rich 1–8	Zn–Fe-rich 1–45	Fe-rich 1–10
Au (ppm)	0.30	4.9	0.83	3.6	3.0	1.6
Ag	12	184	14	95	37	17
As	10	24	28	272	49	50
Sb	1	50	5	30	15	4
Co	247	107	32	3	3	2
Se	474	–	13	4	1	2
Cd	<1	1234	20	490	200	50
Mo	180	21	197	67	45	75
Hg	–	–	<0.1	19.7	19.1	9.7
Tl	–	–	0.2	23.9	41.2	51.2
Cu (wt.%)	31.2	8.6	25.0	0.2	0.2	0.1
Fe	26.1	6.4	29.0	29.3	18.8	41.6
Zn	0.1	10.6	0.5	17.4	7.2	2.1
Pb	<0.01	0.03	<0.01	0.13	0.06	0.10
S	–	–	35.8	40.4	29.1	52.1
SiO_2	–	–	2.6	7.0	0.2	1.0
Ba	<0.01	<0.01	<0.01	<0.01	<0.01	<0.01
Ca	–	–	0.01	0.35	0.02	0.02
CO_2	–	–	<0.01	1.01	<0.01	<0.01
Au/Ag	0.025	0.027	0.059	0.038	0.081	0.094

Data are from Hannington *et al.* (1988b), Herzig *et al.* (1988a), Thompson *et al.* (1988), Schroeder *et al.* (1986), and Hannington *et al.* (1989).

The Snakepit vent field (3500 m depth) is situated about 300 km south of TAG and consists of coalesced sulphide chimneys and mounds together with 350°C black smokers and lower-temperature (226–350°C) sulphide-sulphate chimneys (Thompson *et al.*, 1988; Campbell *et al.*, 1988c). Unlike high-temperature sulphides in the TAG field, black smokers at the Snakepit site contain both pyrite and pyrrhotite but have similar low gold contents (Table 8.3). Precipitates at lower-temperature, white smoker vents consist of pyrite, marcasite, and sphalerite (Thompson *et al.*, 1988) with gold contents up to 3.9 ppm Au and silver locally exceeding 200 ppm Ag (Figure 8.2).

More than 60 sulphide-rich mounds occur along an 8 km segment of Southern Explorer Ridge (1800 m depth) in the NE Pacific (Scott *et al.*, 1989). One large deposit, 200–250 m in diameter, currently hosts active vents at temperatures up to 306°C (Tunnicliffe *et al.*, 1986). High-temperature sulphides at this site consist of pyrite, chalcopyrite, and minor pyrrhotite and contain 0.1–0.5 ppm Au. Lower-temperature, pyrite–marcasite–sphalerite assemblages with abundant amorphous silica and barite contain 0.5–3.3 ppm Au (Table 8.4). Gold-rich, Zn–Fe–Ba–SiO_2 precipitates are characterized by the presence of Fe-poor sphalerite, galena, argentiferous Pb–Sb–As sulphosalts, and minor tetrahedrite (Hannington *et al.*, 1986).

In the Galapagos Rift (2600 m depth), sulphide- and silica- rich mounds occur along steep faults parallel to a neovolcanic zone within the rift valley. Active vents in the area have measured temperatures of only 30°C (Corliss *et al.*, 1979; Edmond *et al.*,

1979a,b), but the presence of extensive sulphide mineralization indicates the former existence of high-temperature fluids. Sulphides from the inactive mounds consist mainly of pyrite, marcasite, and chalcopyrite with minor sphalerite and rare pyrrhotite (Tufar *et al.*, 1986; Embley *et al.*, 1988). Gold contents are typically less than 0.5 ppm Au (Figure 8.2) but reach 1.4 ppm Au with up to 220 ppm Ag in some sphalerite-rich samples.

Near 13°N on the East Pacific Rise (2600 m depth), sulphide deposits are located along the margins and within the central graben of the axial rift (Hekinian and

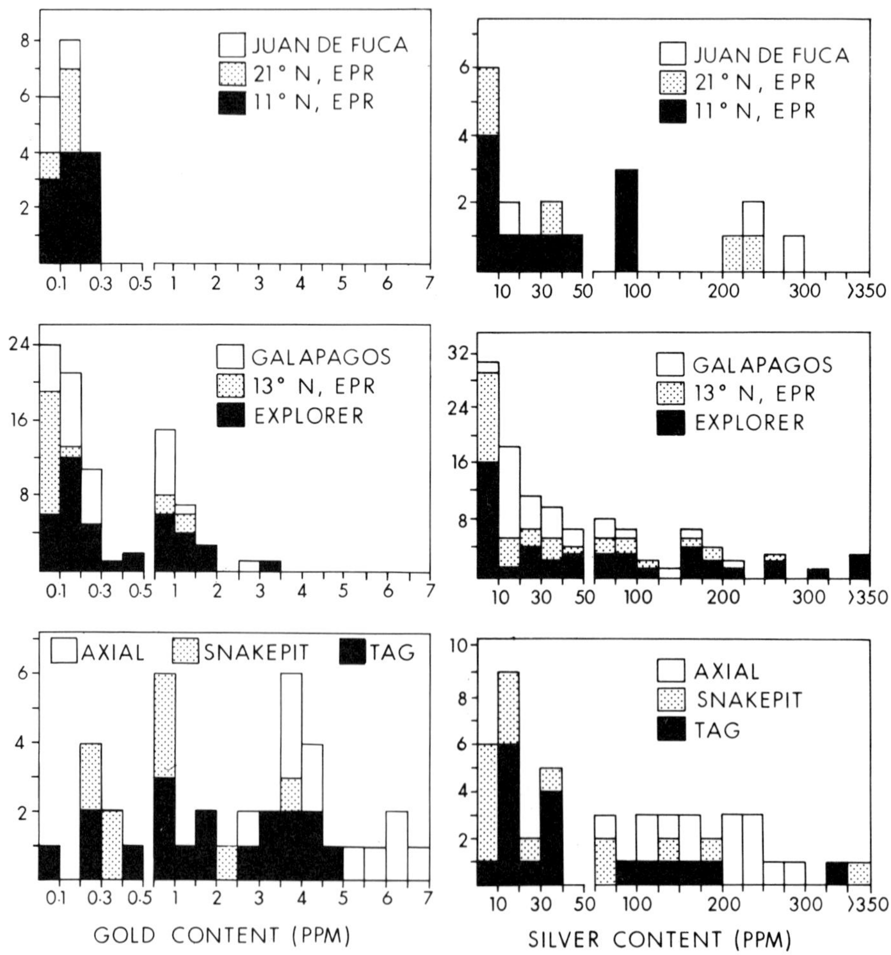

Figure 8.2 Distribution of gold and silver contents in samples of primary sulphides from seafloor polymetallic deposits on mid-ocean ridges (n=143). (See Table 8.1 for sources).

Table 8.3 Analyses of representative samples from the Snakepit vent field, Mid-Atlantic Ridge

	Active vents				Inactive deposits	
	Black smoker 1683–1-1b	335°C Vent 1683–4–1a	227°C Mound 1683–6–1	227°C Mound 1683–8–1a	Zn–Fe-rich 1683–5–1	Fe-rich 1683–3–1
Au (ppm)	0.35	2.3	0.88	2.6	3.9	0.94
Ag	30	69	13	140	57	5
As	46	307	340	197	1346	473
Sb	5	29	8	58	40	5
Co	129	8	73	7	2	36
Se	363	115	12	26	<1	120
Cd	22	216	15	419	152	<10
Mo	9	23	36	24	20	52
Hg	–	1.5	1.5	–	7.0	1.3
Tl	–	17.7	16.2	–	38.5	14.1
Cu (wt.%)	5.8	1.7	0.1	0.8	2.1	0.8
Fe	34.8	34.7	43.4	18.7	40.2	44.4
Zn	0.7	6.3	0.5	13.2	4.4	0.1
Pb	<0.01	0.05	0.02	–	0.11	0.02
S	29.4	27.4	49.2	26.9	42.2	49.6
SiO_2	0.9	0.2	0.4	13.6	0.7	0.4
Ba	<0.01	<0.01	<0.01	<0.01	<0.01	<0.01
Ca	5.6	0.01	0.01	0.03	0.04	0.02
CO_2	–	0.06	0.07	0.06	0.04	0.06
Au/Ag	0.012	0.033	0.068	0.019	0.068	0.190

Data are from Herzig *et al.* (1988a) and Hannington *et al.* (1989)

Fouquet, 1985). Active, black smoker chimneys at this site have measured temperatures up to 354°C and are precipitating pyrite, chalcopyrite, and pyrrhotite with < 0.1 ppm Au. Lower-temperature (245–283°C) pyrite–sphalerite chimneys contain up to 1.3 ppm Au and 180 ppm Ag (Fouquet *et al.*, 1988).

Hydrothermal vents at 21°N on the East Pacific Rise (2600 m depth) are among the best described in the literature (Haymon and Kastner, 1981; Styrt *et al.*, 1981; Goldfarb *et al.*, 1983; Zierenberg *et al.*, 1984). High-temperature (350°C) black smoker chimneys at this site consist of pyrite, chalcopyrite, and sphalerite, with minor pyrrhotite, isocubanite, and marcasite. As in other deposits, gold contents in the high-temperature sulphides from 21°N EPR are uniformly low, but lower-temperature sphalerite-rich samples are also gold-poor (Bischoff *et al.*, 1983; Zierenberg *et al.*, 1984). Similar deposits are forming at 347°C vents near 11°N on the East Pacific Rise (2500 m depth) (McConachy *et al.*, 1986) and also contain <0.2 ppm Au (Table 8.1). Despite the low gold contents of these sulphides, some of the samples from 21°N and 11°N locally contain as much as 250 ppm Ag (Fig. 8.2).

Hydrothermal vents on the Southern Juan de Fuca Ridge (2200 m depth) are precipitating pyrite–marcasite–sphalerite assemblages at temperatures below 284°C (Koski *et al.*, 1984; USGS Study Group, 1986). Samples recovered from this site resemble the gold-rich, Zn–Fe–Ba–SiO_2 precipitates from the Southern Explorer Ridge, but contain virtually no barite and < 0.1 ppm Au (Table 8.1).

8.2.2 *Seamounts*

Basaltic volcanoes on or adjacent to mid-ocean ridges are recognized as preferred sites of hot spring activity on the modern seafloor (Scott, 1985, 1987; Rona, 1988).

Table 8.4 Analyses of representative samples from the inactive deposit at Southern Explorer Ridge

	Cu-Fe-rich	Fe-rich	Zn-Fe-rich	Zn–Fe–Ba–SiO_2		
	1505–3	1504–3	1505–7	1505–10	1504–4	1505–5
Au (ppm)	0.18	2.0	1.0	0.63	1.0	0.67
Ag	7	43	102	69	41	37
As	280	1 400	540	430	660	240
Sb	3	44	40	38	100	26
Co	3 100	48	–	2	3	1
Se	130	35	21	8	3	7
Cd	1	34	1 500	880	4	500
Mo	350	160	65	140	92	32
Hg	0.2	21.2	6.5	5.1	13.6	0.1
Tl	32.5	70.7	33.1	47.4	13.3	16.7
Cu (wt.%)	15.6	1.7	5.1	3.5	<0.1	1.4
Fe	38.9	40.4	20.1	8.3	26.3	4.5
Zn	<0.1	2.7	34.3	19.5	1.5	10.5
Pb	0.01	0.04	0.04	0.03	0.03	0.02
S	25.9	25.2	32.1	21.5	23.6	17.6
SiO_2	1.0	2.6	1.3	25.0	7.6	28.5
Ba	<0.01	0.08	5.8	7.9	13.0	22.1
Ca	<0.01	0.03	0.65	0.40	0.05	0.16
CO_2	<0.01	<0.01	0.30	0.20	<0.01	<0.01
Au/Ag	0.026	0.047	0.010	0.009	0.024	0.018

Data are from Hannington *et al.* (1986), Scott *et al.* (1989), and M.D. Hannington (unpublished).

Low-temperature hydrothermal venting is also known on mid-plate seamounts (e.g. Karl *et al.*, 1988), but gold contents in hydrothermal precipitates at these sites have not been determined.

Axial Seamount (1490 m depth) is a large volcano situated on the central Juan de Fuca Ridge in the NE Pacific. Widespread hydrothermal activity occurs along the walls of a 21 km^2 caldera at the summit of the volcano, with low-temperature (<250°C) hydrothermal deposits at the north end of the caldera (CASM II, 1985) and relatively high-temperature (300 – 330°C) active vents in the south (ASHES Expedition, 1986). The low-temperature deposits consist of abundant amorphous silica and barite as well as Fe-poor sphalerite, wurtzite, marcasite, and minor chalcopyrite. As in the deposits on Southern Explorer Ridge, gold and silver contents in the Zn–Fe–Ba–SiO_2 precipitates are consistently high, ranging from 2.9 to 6.7 ppm Au and up to 260 ppm Ag (Table 8.5 and Figure 8.2). Gold enrichment is closely associated with the presence of galena, argentiferous Pb–As–Sb sulphosalts, tetrahedrite, and local native sulphur (Hannington *et al.*, 1986). Fluid inclusions in the sphalerite indicate a temperature of ≤ 235°C for gold–sulphide mineralization with formation temperatures of ≤ 185°C for associated barite and amorphous silica (Hannington and Scott, 1988). Gold-rich precipitates are also forming in the vicinity of high-temperature hot springs in the southern vent field, but the highest gold contents (up to 5 ppm Au) are consistently associated with late-stage, lower-temperature phases (Harvey-Kelly *et al.*, 1988).

Pyritic sulphides recovered from a large deposit (covering an area of 800 x 200 m) on the flank of an off-axis seamount near 13°N EPR contain < 0.1 ppm Au (Fouquet *et al.*, 1985). However, pyrite from a similar deposit in the caldera of a small seamount

Table 8.5 Analyses of representative material from a 160 kg sample from Axial Seamount

	$Zn–Fe–SiO_2$ precipitates			$Ba–SiO_2$ precipitates		
	1324–2(2)	1324–2 –17	1324–2–41	1324–2(3)	1324–2–53	1324–2–15
Au (ppm)	6.7	6.4	3.7	2.9	3.6	6.1
Ag	229	260	152	122	117	210
As	600	590	420	620	560	740
Sb	500	380	130	160	150	380
Co	<1	3	5	<1	6	1
Se	3	2	5	2	1	2
Cd	560	540	1 200	13	150	60
Mo	37	32	32	38	49	43
Hg	14.7	27.5	16.1	0.1	26.8	45.4
Tl	97.9	94.8	64.6	112	95.9	160
Cu (wt.%)	0.5	0.4	0.8	0.1	0.1	0.1
Fe	6.7	6.5	5.6	2.8	5.1	6.6
Zn	30.8	24.7	34.7	3.2	8.6	4.3
Pb	0.59	0.44	0.10	0.44	0.19	0.55
S	20.3	19.8	20.8	12.7	11.3	16.0
SiO_2	29.0	32.7	24.6	22.7	36.5	32.0
Ba	2.0	6.0	0.05	34.2	23.5	18.3
Ca	0.30	0.21	0.05	0.36	0.34	0.31
CO_2	–	0.40	0.50	–	0.05	–
Au/Ag	0.029	0.025	0.024	0.024	0.031	0.029

Data are from Hannington and Scott (1988b), Hannington *et al.* (1986), and M.D. Hannington (unpublished).

near 21°N EPR locally contains 1.3 ppm Au (Alt, 1988). Low-temperature (100–170°C) opal-rich precipitates from this site contain up to 0.65 ppm Au.

8.2.3 *Island-arc settings*

Modern examples of island-arc settings are common in the south-west Pacific (e.g. Mariana back-arc, Okinawa Trough, Lau Basin). Unlike spreading centres on the mid-ocean ridges, the rifts associated with active magmatic arcs typically erupt bimodal volcanics. Felsic lavas and mafic volcanics with arc-like components have been reported from the Mariana back-arc (Lonsdale and Hawkins, 1985; Macdougall *et al.*, 1987) and in the Okinawa Trough (Kimura *et al.*, 1988). Hydrothermal deposits have been located in these and several other back-arc basins throughout the south-west Pacific (Rona, 1988). Gold-bearing hydrothermal precipitates associated with back-arc rifting are broadly similar to the gold-rich, $Zn–Fe–Ba–SiO_2$ assemblages from the mid-ocean ridges (e.g. Axial Seamount and Explorer Ridge).

In the Mariana back-arc (3600–3700 m depth), hydrothermal venting at temperatures up to 287°C has precipitated numerous sulphide–sulphate chimneys and mounds consisting of barite, opaline silica, sphalerite, galena, pyrite, and rare chalcopyrite (Campbell *et al.*, 1987; Craig *et al.*, 1987; Kastner *et al.*, 1987). Gold contents in these samples range from 0.1 to 1.7 ppm Au and average 0.8 ppm Au, with silver contents close to 200 ppm Ag (Table 8.1).

Similar hydrothermal precipitates have been reported from a caldera in the Okinawa Trough (1480 m depth) (Halbach *et al.*, 1989). Recovered samples consist of barite, amorphous silica, sphalerite, pyrite, galena, and chalcopyrite, together with tetrahedrite, tennantite, enargite, bornite, covellite, native sulphur, and As-sulphides.

Individual samples have reported gold contents of up to 9.8 ppm Au and silver contents up to 0.1 wt.% Ag, with averages of 5.1 ppm Au and 772 ppm Ag in the massive sulphides (Table 8.1). The highest concentrations of gold occur with late-stage barite, orpiment, and realgar (Halbach *et al.*, 1989).

8.2.4 *Sedimented-rift environments*

Hydrothermal deposits are also forming in response to seafloor spreading and basaltic volcanism in areas of high sedimentation near continental margins. The variable mineralogy and geochemistry of these deposits reflects the influence of sediments on the chemistry of the hydrothermal fluids.

Active rifting of the Guaymas Basin (2000 m depth), in the central Gulf of California, has resulted in dyke and sill intrusion into unconsolidated terrigenous and biogenic muds which cover the basin floor to a thickness of 500 m (Lonsdale and Becker, 1985). Hydrothermal fluids driven by heat from large deep-seated intrusions are venting through the sediments at temperatures up to 315°C (Lonsdale *et al.*, 1980; Koski *et al.*, 1985). Sulphides and sulphates from these fluids have precipitated both on the seafloor and within the sediments. Amorphous silica, barite, anhydrite, and calcite occur with pyrrhotite, isocubanite, pyrite, marcasite, chalcopyrite, Fe-rich sphalerite, and galena in the hydrothermal chimneys and mounds which have formed on the seafloor (Koski *et al.*, 1985; Peter and Scott, 1988). Gold contents in these deposits are typically < 0.2 ppm Au (Table 8.6), but silver contents may reach 350 ppm Ag (Hannington *et al.*, 1986; Peter and Scott, 1988).

Middle Valley (2500 m depth) is a sedimented rift on the northern Juan de Fuca Ridge with up to 1500 m of turbidites and hemipelagic sediment (Davis *et al.*, 1987; Goodfellow and Blaise, 1988). Heat-flow measurements indicate that temperatures at the base of the sediments may exceed 300°C (Davis *et al.*, 1987). Hydrothermal, mound-like structures up to 60 m in height and several hundred metres across, occupy the sedimented floor of the rift and locally host active hydrothermal vents and sulphide chimneys. Sulphides also occur as alternating beds of clastic material in the altered sediments (Goodfellow and Blaise, 1988). Most of the recovered samples consist of pyrrhotite, pyrite, marcasite, Fe-rich sphalerite, chalcopyrite, and galena together with talc, barite, amorphous silica, Mg-rich clays, and Fe-oxides. As in the Guaymas Basin, gold contents in the sulphides are uniformly low (Table 8.6).

Sulphide deposits in the Escanaba Trough (3250–3300 m depth), on the southern Gorda Ridge, are associated with volcanic edifices that penetrate up to 500 m of turbiditic sediments (Morton *et al.*, 1987; Koski *et al.*, 1988). Venting of hydrothermal fluids includes low-temperature seeps at 18–100°C and at least one vent with a maximum temperature of 217°C (Campbell *et al.*, 1988b). The hydrothermal mounds consist dominantly of massive pyrrhotite with talc and barite, but local polymetallic sulphides also contain sphalerite, marcasite, and arsenopyrite, together with loellingite, galena, boulangerite (Pb–Sb sulphosalts), native bismuth, tetrahedrite, and Ag-sulphides (Koski *et al.*, 1988). Gold contents in pyrrhotite-rich samples are typically < 0.2 ppm Au, but barite-rich crusts on the polymetallic sulphides contain up to 2 ppm Au (Table 8.6).

The existence of metal-rich brines associated with seafloor spreading in the Red Sea has been known since 1965, when density-stratified brine pools were discovered at the bottom of several deep, anoxic basins (Degens and Ross, 1969). In the Atlantis II

Table 8.6 Average chemical composition of hydrothermal precipitates in sedimented rifts

	Guaymas Basin[1]			Middle Valley[2]	Escanaba Trough[3]		
	273–312°C Vents (3)*	Inactive vents (5)	Inactive mound (6)	Sediment core (43)	Massive pyrrhotite (4)	Poly-metallic sulphides (2)	Barite-rich crust (1)
Au (ppm)	0.15	0.12	0.15	0.14	<0.20	<0.20	2.0
Ag	49	114	63	9	14	433	387
As	139	164	57	227	3 600	22 000	7 200
Sb	173	171	140	33	109	980	2 700
Co	–	4	6	19	470	–	–
Se	58	100	465	88	285	860	<10
Cd	57	26	18	40	16	1 240	830
Mo	2	4	17	104	<40	–	–
Hg	0.8	7.2	4.4	–	–	–	–
Tl	2.5	9.7	12.7	–	–	–	–
Cu (wt.%)	0.4	0.2	0.1	0.4	0.9	1.4	0.5
Fe	4.1	10.1	3.2	30.3	43.6	25.0	6.4
Zn	1.8	1.2	0.3	2.5	0.8	31.8	17.3
Pb	0.23	0.69	0.16	0.04	0.18	4.4	4.4
S	5.3	6.5	3.4	26.0	41.3	32.5	21.1
SiO_2	11.1	26.2	39.0	17.4	0.3	–	–
Ba	0.9	18.9	18.6	2.0	0.04	1.3	23.0
Ca	20.6	6.4	0.13	0.17	0.03	–	–
CO_2	18.6	7.6	0.65	–	–	–	–
Au/Ag	0.003	0.001	0.002	0.016	<0.014	<0.001	0.005

*Figures in parenthesis show the total number of samples analysed for different elements.

Data are from: (1) Peter and Scott (1988), Hannington *et al.* (1986), M.D. Hannington (unpublished); (2) Davis *et al.* (1987), Goodfellow and Blaise (1988) (Au, Sb, Co, Ni, Mo, Mn, and Ca based on 10 analyses); (3) Koski *et al.* (1988).

Deep (2000–2100 m depth), the standing lower brine has a stable temperature of about 60°C, but was probably derived from hydrothermal fluids injected into the basin at temperatures up to 420°C (Shanks and Bischoff, 1977, 1980; Oudin *et al.*, 1984; Ramboz *et al.*, 1988). Precipitation of laminated Fe-silicates, Fe-oxides, and sulphides (dominantly pyrite, chalcopyrite, and sphalerite) from the brine pool has resulted in a large, stratiform accumulation of metalliferous mud interbedded with normal marine sediments to a thickness of 10–30 m and covering an area of approximately 40 km^2 (Hackett and Bischoff, 1973; Backer, 1976; Pottorf and Barnes, 1983; Zierenberg and Shanks, 1983; Oudin, 1987). Facies transitions from dominantly biogenic carbonate to dominantly sulphide-rich horizons are related to the rate of influx of the hydrothermal brines and the size of the brine pool (e.g. Shanks and Bischoff, 1980). The conduits for the injection of high-temperature fluids consist of veins which cross-cut the sediments and contain anhydrite, pyrrhotite, isocubanite, and Fe-rich sphalerite (Pottorf and Barnes, 1983; Zierenberg and Shanks, 1983, 1988; Oudin *et al.*, 1984; Ramboz *et al.*, 1988). Approximately 227 million tonnes of metalliferous sediment occur in the Atlantis II Deep, including proven reserves of 90 million tonnes (dry weight, salt free) averaging 2% Zn, 0.5% Cu, and 39 ppm Ag (Mustafa *et al.*, 1984; Nawab, 1984). Gold contents in the sulphide-rich horizons range from < 0.5 up to 4.6 ppm Au and average nearly 2 ppm Au (Table 8.7 and Figure 8.3). Based on a conservative gold grade of 0.5 ppm Au, these deposits could contain at least 45 000 kg

Table 8.7 Analyses of representative samples from the metalliferous sediments of the Red Sea

	Metalliferous sediments[1]				Sulphide-rich horizons2 Atlantis II Deep			
	(Dry weight, including NaCl)			Average (93)	(Dry weight, NaCl-free)			Average (53)
Au (ppm)	0.6	1.1	5.6	0.58	1.4	2.1	4.2	2.0
Ag	10	41	109	28	135	105	214	85
As	100	400	900	130	190	140	1 070	345
Sb	20	20	90	36	50	20	70	32
Co	35	105	198	48	73	125	153	137
Cd	4	70	313	42	168	320	480	194
Mo	81	166	1 290	100	–	–	–	–
Hg	–	–	–	–	–	4.0	5.8	5.9
Tl	–	20	40	6	–	–	–	–
Mn	42 100	20 100	140	11 558	–	–	–	–
Sr	180	3 750	140	406	–	–	–	–
Cu (wt.%)	0.3	0.6	1.0	0.2	0.7	1.0	1.7	1.0
Fe	33.1	11.9	7.9	14.6	28.5	27.6	10.3	21.1
Zn	0.4	1.4	10.1	1.0	2.8	12.3	23.3	5.8
Pb	0.02	0.11	0.05	0.02	0.12	0.12	0.33	0.10
S	0.5	4.2	18.3	2.1	9.5	8.5	20.4	10.3
SiO_2	3.7	7.9	9.5	6.9	13.8	20.3	27.2	20.7
Ba	0.31	0.35	0.03	0.19	–	–	–	–
Ca	0.96	1.75	8.9	6.4	–	–	–	–
Na	3.8	9.4	7.1	13.7	–	–	–	–
CO_2	–	6.7	–	8.9	–	–	–	–
Au/Ag	0.060	0.027	0.051	0.021	0.010	0.020	0.020	0.024

Data are from: (1) Hendricks *et al.* (1969), including combined results for silicate, oxide, and sulphide components of metalliferous sediments from the Atlantis II, Chain, and Discovery Deeps; (2) Oudin (1987), results for sulphide-rich horizons in the Atlantis II Deep (excluding sulphide veins from the boiling zone).

of gold (Mustafa *et al.*, 1984). Local boiling is believed to be responsible for the precipitation of minor electrum and argentite in the feeder-veins, but is not the cause of widespread gold enrichment in the metalliferous sediments (Oudin, 1987).

8.3 Mineralogy and geochemistry of gold in seafloor hydrothermal systems

In volcanic-associated hot springs, gold enrichment occurs primarily in the late stages of mineralization and almost exclusively in pyritic assemblages formed at temperatures below 300°C. Zn–Fe–Ba–SiO_2 precipitates are commonly the most gold-rich (Tables 8.2 to 8.5), although several deposits with high Zn contents do not contain significant gold (21°N EPR, 11°N EPR, Southern Juan de Fuca Ridge). No free gold has been reported in any of the gold-rich sulphides at 1250 × magnification of polished specimens, suggesting that individual gold-bearing phases are present as grains smaller than about 1 μm (Hannington *et al.*, 1986; Herzig *et al.*, 1988a; Fouquet *et al.*, 1988). Indirect geochemical evidence indicates that submicroscopic gold occurs as inclusions in sphalerite, pyrite, or marcasite, and may be intergrown with fine-grained sulphosalts commonly found in the lowest- temperature assemblages of the gold-rich deposits. A similar situation exists in the hot spring deposits of geothermal systems in New Zealand, where gold occurs at concentrations of up to 50 ppm Au but does not appear to have crystallized as a discrete mineral phase (Browne and Ellis, 1970; Weissberg *et al.*, 1979; Krupp and Seward, 1987). Seward (1984), and more recently Renders and Seward (1989), have indicated that gold precipitated at the

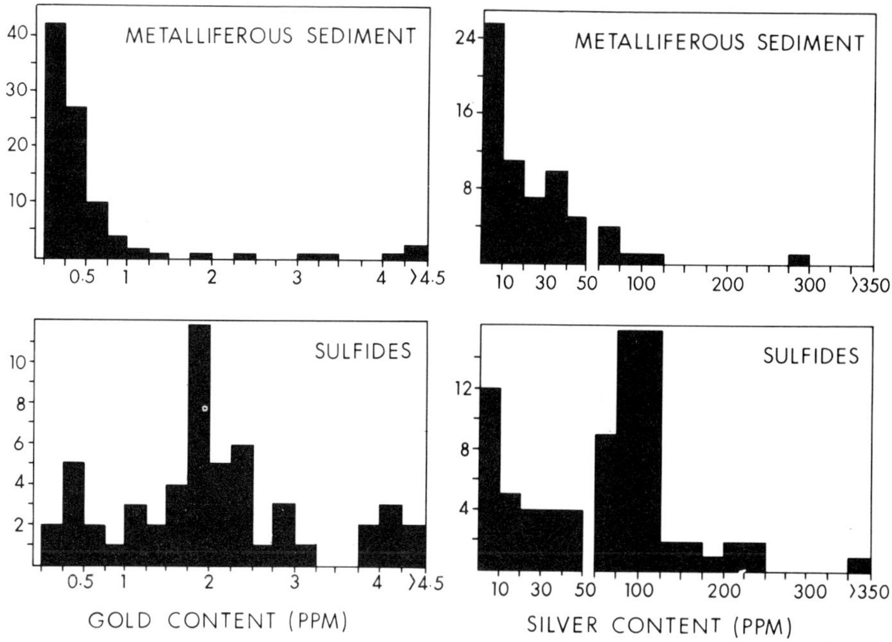

Figure 8.3 Distribution of gold and silver contents in samples from the Red Sea metalliferous sediments (data from Hendricks *et al.*, 1969) and sulphide-rich horizons from the Atlantis II Deep. (Data from Oudin, 1987.)

margins of the geothermal pools may be concentrated by adsorption on to amorphous antimony and arsenic sulphides and could account for the absence of identifiable gold minerals. Hannington *et al.* (1986) noted that a similar process may be responsible for gold enrichment associated with sulphosalts in the low-temperature assemblages from Axial Seamount and Southern Explorer Ridge.

Gold-rich precipitates in modern seafloor deposits typically contain high concentrations of Ag, As, Sb, Hg, Tl, and Pb. Samples containing >1 ppm Au from Axial Seamount, Southern Explorer Ridge, the TAG hydrothermal field, and the Snakepit vent site locally contain 280–1100 ppm Ag, 270–1400 ppm As, 70-660 ppm Sb, 9-45 ppm Hg, 39–160 ppm Tl, and 1300–9700 ppm Pb. The correlations with silver and arsenic are imperfect, as these elements may also occur at high concentrations, up to 290 ppm Ag and 770 ppm As, in relatively gold-poor assemblages (Table 8.1 and Figure 8.2). Au/Ag ratios in the volcanic-hosted deposits range from 0.003 to 0.03, but tend to be higher in the low-temperature assemblages and reflect the late-stage enrichment of gold relative to silver. Gold-poor sulphides within individual deposits commonly have high Cu and Fe contents and may be enriched in Co, Se, and Mo (Tables 8.2 to 8.4).

Despite their low gold contents, the sediment-hosted deposits have high concentrations of Ag, As, Sb, and Pb (Table 8.6). Interaction of hydrothermal seawater with greywacke or organic-rich marine sediment may release large quantities of these elements compared to water–rock reactions involving mid-ocean ridge basalt (e.g. Bischoff *et al.*, 1981; Thornton and Seyfried, 1987). The unusual geochemistry of polymetallic sulphides in the Escanaba Trough (av. 2.2% As and 0.1% Sb, as well as 70 ppm Bi and 700 ppm Sn) is apparently a consequence of extensive reaction between normal ridge-crest hydrothermal fluids and turbidites derived from the nearby continental margin (Koski *et al.*, 1988). In the sediments of the Atlantis II Deep, gold concentrations correlate with total sulphur and reflect the deposition of gold in sulphidic rather than oxide- or silicate-rich layers (Oudin, 1987).

Table 8.8 Gold and associated elements in hydrothermal plumes and near-field metalliferous sediment

	Particulates in hydrothermal plumes			Plume fall-out	Near-field metalliferous sediment		
	12°N EPR	11°N EPR	Explorer Ridge	21°N EPR	21°N EPR		Explorer Ridge
	(1)*	(7)	(1)	(1)	(3)	(5)	(1)
T (°C)	350	347	306	NGS site	OBS site	Oasis site	
Au (ppb)	200	200	150	247	66	207	28
Ag (ppm)	47	37	11	38	20	42	1
As	29	216	53	61	60	210	140
Sb	4	17	7	10	5	13	4.6
Co	151	322	12	213	228	186	–
Se	21	66	<20	38	498	285	50
Cd	207	348	33	320	113	447	1
Mo	19	87	59	23	31	80	180
Mn	901	2 500	–	420	157	1 130	18 700
Cu (wt.%)	0.7	3.2	0.1	0.8	22.8	5.8	3.2
Fe	30.4	23.9	1.7	36.0	25.5	30.4	–
Zn	7.2	9.3	1.0	12.0	4.0	13.8	0.03
Pb	0.05	0.14	0.04	0.05	0.02	0.04	0.01
S	31.3	37.3	–	–	25.0	–	5.2
SiO_2	8.6	6.9	–	2.6	7.7	7.0	–
Ba	0.52	0.11	1.9	0.04	0.02	0.05	0.46
Ca	0.29	2.3	–	0.31	0.24	0.28	–
Au/Ag	0.004	0.005	0.013	0.007	0.003	0.005	0.028

Data are from McConachy (1988), Mottl and McConachy (in press), Hall *et al.* (1988), and M.D. Hannington (unpublished).
*Figures in parenthesis show the total number of samples analysed for different elements.

8.4 Gold in sub-seafloor stockwork mineralization

A few examples of sub-seafloor stockwork mineralization have been exposed by faulting along mid-ocean ridges or by scientific drilling through recent ocean crust. Stockwork-like veins in 6-Ma-old pillow basalts were intersected by DSDP Hole 504B during deep-sea drilling near the Costa Rica Rift in 1983. Although these veins are not connected to sulphide deposits, they are interpreted to be the product of axial hydrothermal fluids upwelling through the dyke section of the crust and depositing pyrite, sphalerite, chalcopyrite, and quartz at 200–340°C in response to mixing with seawater within the overlying pillow lavas (Honnorez *et al.*, 1985; Alt *et al.*, 1986; Kawahata and Shikazono, 1988). A significant amount of gold appears to have been removed from the dyke section in 504B during alteration (Nesbitt *et al.*, 1987), but vein sulphides from the stockwork contain < 0.1 ppm Au (Hannington *et al.*, 1988a).

Shallow stringer-sulphides in altered pillow lavas also have been recovered from a fault-exposed stockwork immediately beneath massive sulphides in the Galapagos Rift (Embley *et al.*, 1988; Jonasson *et al.*, 1988). Although massive sulphides at this site contain up to 1.4 ppm Au, sulphides from the underlying stockwork contain < 0.2 ppm Au (Jonasson *et al.*, 1988; Tufar *et al.*, 1986). These observations suggest either that ascending hydrothermal fluids did not precipitate significant gold before reaching the seafloor or that gold was remobilized by subsequent generations of fluid traversing the stockworks.

8.5 Gold in hydrothermal plumes and associated metalliferous sediments

Sudden mixing of high-temperature fluids with ambient seawater at the seafloor commonly results in the precipitation of dissolved metals and sulphur as a suspension of fine-grained sulphides which are carried into the water column as a buoyant hydrothermal plume. This process gives rise to the familiar black smokers which characterize high-temperature vents on sediment-starved mid-ocean ridges (e.g. RISE Project, 1980). Particulates in the smoke consist mainly of pyrrhotite or pyrite, with lesser amounts of sphalerite, chalcopyrite, anhydrite, and amorphous silica (Haymon and Kastner, 1981; McConachy, 1988). Samples of black smoke from vents at 21°N (Mottl and McConachy, in press) and 11°N (McConachy, 1988) on the East Pacific Rise contain only about 200 ppb Au (Table 8.8). Despite these low concentrations, hydrothermal plumes may account for up to 90% of the total mass flux of metals (Edmond *et al.*, 1982; Converse *et al.*, 1984) and probably contain most of the gold arriving at the seafloor in high-temperature vents. At 21°N, where venting is dominated by high-temperature chimneys (Haymon and Kastner, 1981; Styrt *et al.*, 1981; Goldfarb *et al.*, 1983), a significant amount of gold may be carried out of the system and lost to a diffuse hydrothermal plume. Sub-seafloor mixing and cooling of the vent fluids will significantly reduce the fraction of the total mass flux lost to the hydrothermal plume (Converse *et al.*, 1984) and will allow the base metals and gold to accumulate in lower-temperature hydrothermal precipitates forming at or near the seafloor. This is evident in gold-rich deposits on Explorer Ridge which have been dominated recently by pervasive, low-temperature venting through large hydrothermal mounds (Tunnicliffe *et al.*, 1986; Scott *et al.*, 1989). In the Guaymas Basin and Escanaba Trough, sub-seafloor precipitation of metals in the sediments has severely depleted the vent fluids such that black smokers and related hydrothermal plumes are less common (e.g. Lonsdale and Becker, 1985; Campbell *et al.*, 1988b).

Hydrothermal plumes are the major dispersal mechanism for chemical species introduced into the oceans by submarine hot springs (Edmond *et al.*, 1982; Mottl and McConachy, in press). Except in rare circumstances, where the density of the hydrothermal fluids may be greater than that of seawater (e.g. Atlantis II Deep, Red Sea), hydrothermal plumes will rise to several hundred metres above the seafloor and drift laterally with prevailing near-bottom currents (Lupton *et al.*, 1985; Baker and Massoth, 1986, 1987; McConachy and Scott, 1987). Although plume fall-out has contributed significantly to the metalliferous component of sediments within a few kilometres of some active vents (e.g. Nelsen *et al.*, 1986), it is unlikely in normal circumstances that settling of particles from hydrothermal plumes will form sizeable deposits on modern mid-ocean ridges (Cathles, 1983; Campbell *et al.*, 1984; Converse *et al.*, 1984). Most of the near-field metalliferous sediment associated with sulphide deposits on the East Pacific Rise is a product of the mass-wasting of chimneys and mounds (McConachy, 1988). The gold and base metal contents of this debris are similar to those of the parent sulphide chimneys (Table 8.8). In contrast, distal marine sediments on the crests and flanks of mid-ocean ridges probably derive a significant fraction of their gold from the direct hydrothermal input of high- temperature plumes. Normal marine sediments on or adjacent to the East Pacific Rise locally contain up to 28 ppb Au on a carbonate-free basis (Piper and Graef, 1974). However, the large dilution of the hydrothermal component by detrital and biogenic material (e.g. Walther and Stoffers, 1985) results in a bulk gold content of only a few ppb Au in most sediments (Piper and Graef, 1974). Unlike the buoyant hydrothermal plumes on mid-ocean ridges, bottom-seeking brines in the Atlantis II Deep are confined to a restricted anoxic basin, and gold and base metals are deposited efficiently and uniformly in the associated metalliferous sediments.

8.6 Transport and deposition of gold in seafloor hydrothermal systems

Recent studies by Cathles (1986), Hannington and Scott (1989), Large *et al.* (1989), and Huston and Large (1989) have emphasized that any process responsible for gold mineralization in seafloor hydrothermal systems first requires a suitable fluid chemistry and a favourable precipitation mechanism.

8.6.1 *The chemistry of seafloor hydrothermal fluids in volcanic environments*

The compositions of high-temperature, hydrothermal fluids from basaltic mid-ocean ridges are remarkably similar at all of the vent sites. Most sampled fluids from active vents are mixtures of these high-temperature fluids and ambient seawater, and the concentrations of dissolved constituents in the undiluted end-members can be calculated by assuming linear mixing with seawater (e.g. Von Damm *et al.*, 1985a). Results for a number of vent sites are given in Table 8.9. The similarities in the chemistry of end-member fluids from different hydrothermal systems as well as the long-term stability of the solution compositions are due to equilibrium with a common mineral assemblage (i.e. a greenschist facies alteration assemblage) at about the same temperature and effective water-rock ratio (Bowers *et al.*, 1988; Campbell *et al.*, 1988a; Von Damm, 1988). Experimental seawater–basalt reactions indicate that significant mobilization of base metals from the crust requires temperatures in excess of 350°C and solutions which remain consistently acid at high temperatures (Seyfried and Janecky, 1985; Seyfried, 1987; Seyfried *et al.*, 1988). These results are consistent with the end-member fluids at depth being somewhat hotter (i.e. 385–400°C at 300–400 bar) than sampled hot springs at the seafloor. Depletion of gold in the sheeted dykes from DSDP Hole 504B confirms its mobility during hydrothermal alteration of the crust to greenschist facies (Nesbitt *et al.*, 1987).

Temperatures in most host springs on the mid-ocean ridges are less than 350°C, with pH's greater than 3.5 and salinities close to that of contemporaneous seawater (3.5–6.4 wt.% NaCl equiv.). The reduced sulphur contents of these fluids are high and commonly exceed the concentrations of dissolved metals by a factor of 2 to 3 (Table 8.9). At 21°N EPR, end-member concentrations of H_2S range from 6.6 to 8.4

Table 8.9 Composition of hydrothermal fluids at modern seafloor vents

	21°N EPR[1]	13°N EPR[2]	11°N EPR[3]	Galapagos Rift[4]	Juan de Fuca Ridge[5]	Explorer Ridge[6]	Snakepit vents,[7] MAR
T (°C)	350	354	347	30	284	276–306	350
pH (25°C)	3.5	3.1	3.1	6.7	3.2	4.8	3.9
NaCl (eq. wt.%)	3.5	4.7	3.7	3.5	6.4	3.7	3.7
H_2S (mg/kg)	253	279	273	11.2	124	–	200
CO_2	250	475–7 353	–	510	–	–	–
SO_4	0	0	0	2 500	0	0	0
SiO_2	1 052	1 166	1 130	74.4	1 378	900	1 094
Ca	649	2 153	902	436	3 454	1823	397
Na	10 488	12 673	10 856	–	17 181	–	11 730
K	939	1 075	1 251	–	1 753	1 290	923
Mg	0	0	0	1.2	0	0	0
Mn	48.6	161.0	42.1	3.1	195.4	16.1	27.0
Ba	>1.4	4.8	–	0.3	–	1.7	–
Sr	7.1	15.9	7.0	7.5	23.6	12.9	4.4
Rb	2.5	1.6	2.1	0.2	2.8	4.9	0.9
Li	7.2	4.1	6.1	0.6	8.1	2.9	5.9
NH_4	<0.18	–	–	–	–	0.05	–
Cu	1.4	–	–	–	<0.1	<0.1	1.1
Fe	79.7	527.8	294.4	0.01	870.0	1.8	121.6
Zn	5.5	0.3	6.9	–	40.0	1.0	3.3
	TAG Field[7] MAR	Axial Seamount[9]		Guaymas Basin[9]	Atlantis II Deep[10] Lower Brine	Seawater[11]	
T (°C)	290–320	29	350	315	56	–	
pH (25°C)	–	6.2	–	5.9	5.6	7.7	
NaCl (eq. wt.%)	4.3	3.5	3.5	3.5	8.2	3.5	
H_2S (mg/kg)	–	11.2	224	145	–	0	
CO_2	–	1 760	–	satd.9	100	120	
SO_4	–	2 490	0	0	840	2 672	
SiO_2	1 322	66	1 139	654	59	9.6	
Ca	1 043	443	1 146	1 116	5 150	406	
Na	13 432	–	–	9 962	92 600	10 672	
K	665	–	–	1 485	1 870	0.38	
Mg	0	1200	0	0	764	1 265	
Mn	54.9	1.5	30.4	7.9	82.0	<0.001	
Ba	–	0.2	3.2	>2.2	0.9	0.02	
Sr	8.7	8.1	14.7	15.8	48.0	7.7	
Rb	0.9	0.4	5.4	5.5	–	0.11	
Li	2.9	0.4	4.4	5.5	4.4	0.19	
NH_4	–	–	–	218.0	–	–	
Cu	–	–	–	up to 0.07	0.3	<0.001	
Fe	91.5	–	–	3.3	81.0	<0.001	
Zn	–	–	–	0.7	5.4	<0.001	

Based on average calculated end-member concentrations. Results for low-temperature fluids from the Galapagos Rift, Axial Seamount (29°C), and the Atlantis II Deep are direct measurements. Data are from: (1) Von Damm *et al.* (1985a); (2) Bowers *et al.* (1988), Merlivat *et al.* (1987), Grimaud *et al.* (1984), Michard *et al.* (1984); (3) Bowers *et al.* (1988); (4) Edmond *et al.* (1979a, b); (5) Von Damm and Bischoff (1987); (6) Tunnicliffe *et al.* (1986); (7) Campbell *et al.* (1988b); (8) McDuff *et al.* (1983), CASM II (1985), Hannington and Scott (1988); (9) Von Damm *et al.* (1985b) (the fluid is saturated with respect to calcium carbonate); (10) Brewer and Spencer (1969) as quoted in Shanks and Bischoff (1977); (11) Von Damm (1983);

mmole/kg (225–285 ppm H_2S), while dissolved metal concentrations are only 0.9 to 2.6 mmole/kg (about 50–150 ppm) (Von Damm *et al.*, 1985a). Figure 8.4 shows the H_2S contents of the hydrothermal fluids from 120 active vents and indicates that H_2S concentrations remain high even at temperatures as low as 200°C. Preliminary results for vent fluids from the Mariana back-arc indicate a depleted H_2S concentration (≤ 2.6 mmol/kg H_2S) likely caused by sub-seafloor precipitation (Campbell *et al.*, 1987). However, the fluid chemistry of hydrothermal systems in island-arc settings has yet to be fully documented. The concentrations of H_2S measured in seafloor vents are comparable to those of deep geothermal fluids in New Zealand (e.g. 6.8 mmol/kg H_2S at Rotokawa: Krupp and Seward, 1987) and may reflect similar controls on dissolved sulphur in both subaerial and submarine hydrothermal systems (e.g. Giggenbach, 1980, 1981). The oxidation state of end-member fluids at 21°N is also similar to that of geothermal systems on land, with a hydrogen partial pressure of 0.13 bar (Welhan and Craig, 1983).

The chemistry of seafloor hot springs at different temperatures is a consequence of the mixing–cooling history of the end-member fluids. This is illustrated at a number of vent sites where analyses of vent fluids, fluid inclusions, and mineral chemistry have constrained the temperature, pH, and oxidation state at the time of mineralization

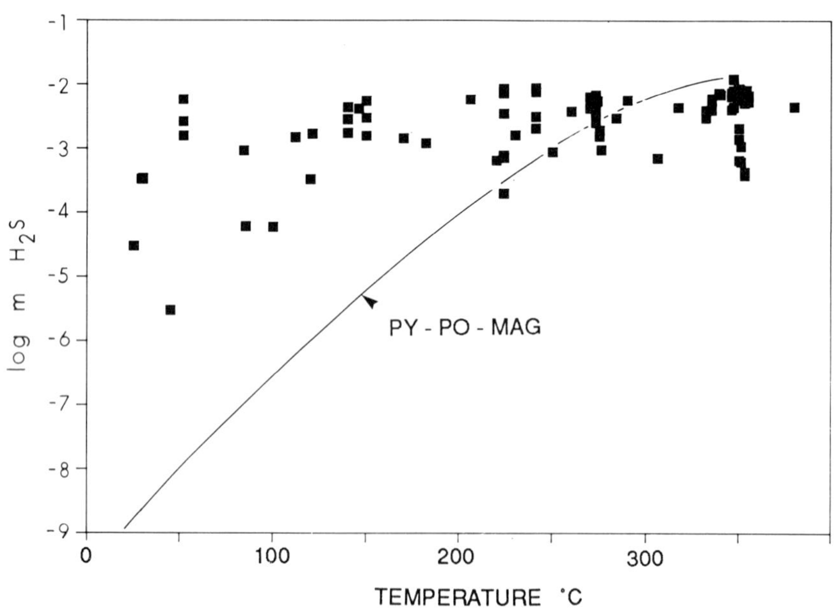

Figure 8.4 Total dissolved H_2S concentrations in 120 mid-ocean ridge hydrothermal vents from 350°C to 25°C: data are from 21°N (*n*=95: Von Damm, 1983), Explorer Ridge (*n*=5: Tunnicliffe *et al.*, 1986), Axial Seamount (*n*=1: McDuff *et al.*, 1983), 13°N (*n*=4: Michard *et al.*, 1984; Grimaud *et al.*, 1984; Bowers *et al.*, 1988), 11°N (*n*=3: Bowers *et al.*, 1988), the Snakepit vents (*n*=2: Campbell *et al.*, 1988c), the Southern Juan de Fuca Ridge (*n*=10: USGS Study Group, 1986; Von Damm and Bischoff, 1987), and the Galapagos Rift (*n*=1: Edmond *et al.*, 1979). H_2S concentrations for the buffer assemblage pyrite–pyrrhotite–magnetite are shown by the solid line.

(Figure 8.5). These data indicate that the composition of hydrothermal fluids at 21°N EPR, Southern Explorer Ridge, and Axial Seamount lie along a T–pH–aO_2 path (A to D in Figure 8.5) which is similar to that predicted for mixing between a 21°N-type end-member fluid and cold seawater (Janecky and Seyfried, 1984; Bowers *et al.*, 1985).

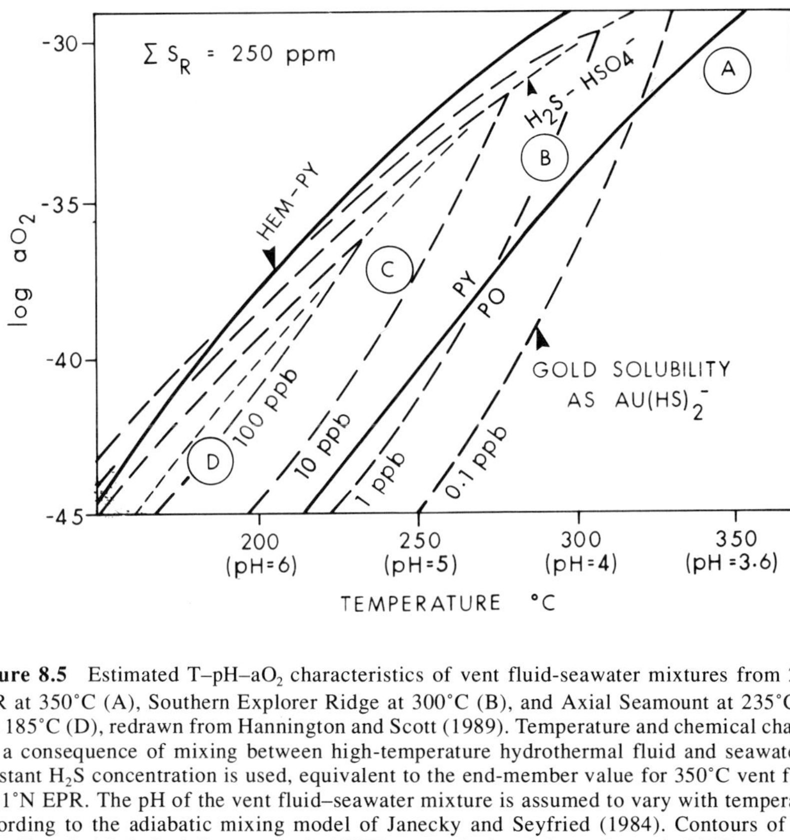

Figure 8.5 Estimated T–pH–aO_2 characteristics of vent fluid-seawater mixtures from 21°N EPR at 350°C (A), Southern Explorer Ridge at 300°C (B), and Axial Seamount at 235°C (C) and 185°C (D), redrawn from Hannington and Scott (1989). Temperature and chemical changes are a consequence of mixing between high-temperature hydrothermal fluid and seawater. A constant H_2S concentration is used, equivalent to the end-member value for 350°C vent fluids at 21°N EPR. The pH of the vent fluid–seawater mixture is assumed to vary with temperature according to the adiabatic mixing model of Janecky and Seyfried (1984). Contours of gold solubility (bold dashed lines: ppb $Au(HS)_2^-$) are calculated from Seward (1973) assuming activity coefficients of one. High-temperature, low-pH fluids from 21°N (A), precipitate sulphides with an average gold content of 0.2 ppm. Lower-temperature, less acid fluids from Explorer Ridge (B) precipitate pyritic sulphides averaging 0.7 ppm Au. Fluids from Axial Seamount precipitate sulphides at 235°C (C) and associated barite and silica at 185°C (D) with average gold contents of 4.9 ppm Au. The fields of the principal aqueous sulphur species (H_2S and HSO_4^-) and the stability fields for pyrite (PY), pyrrhotite (PO), and hematite (HEM) are shown.

8.6.2 *The chemistry of seafloor hydrothermal fluids in sedimentary environments*

The composition of hydrothermal fluids in settings like the Guaymas Basin are the product of a 21°N-type end-member fluid reacting with sediments which overlie the volcanic basement (Von Damm *et al.*, 1985b; Bowers *et al.*, 1985). The resulting vent fluids differ from those at sediment-starved, mid-ocean ridges because of buffering by the sediments to a higher pH and alkalinity and a lower aO_2 (Von Damm *et al.*, 1985b). The high pH of the fluids, as well as sub-seafloor precipitation of sulphides within the sediments, lead to significantly lower base metal contents in the calculated end-members (Table 8.9). Dissolved sulphur concentrations are also about 40% lower than those of vent sites on sediment-starved ridges. Preliminary results for vent fluids in the Escanaba Trough are similar to those from the Guaymas Basin (Campbell *et al.*, 1988b). The abundance of pyrrhotite in these deposits is a likely consequence of the fluids being somewhat more reducing following interaction with organic-rich sediments (Peter and Scott, 1988; Koski *et al.*, 1988).

Hydrothermal seawater beneath the Red Sea has acquired high concentrations of dissolved salts by circulating through Miocene evaporites in sediments adjacent to the rift zone (Shanks and Bischoff, 1977, 1980; Pottorf and Barnes, 1983). These sub-seafloor brines have salinities of 15–32 wt.% NaCl equiv. at 420°C (Oudin *et al.*, 1984; Ramboz *et al.*, 1988). A component of the high salinity has also been attributed to boiling and phase separation (Ramboz *et al.*, 1988). Despite the influence of evaporitic sediments on the chemistry of the brines, isotopic evidence also indicates an important contribution by basalt-seawater interaction typical of mid-ocean ridges (Zierenberg and Shanks, 1986, 1988). The composition of the standing lower brine at the bottom of the Atlantis II Deep is given in Table 8.9.

8.6.3 *The solubility of gold in seafloor hydrothermal fluids*

End-member fluids which circulate at depth along the mid-ocean ridges (1–2 km sub-seafloor) may transport gold as a chloride complex [$AuCl_2^-$]. Using thermo-dynamic data from Helgeson (1969), the calculated solubility of gold as $AuCl_2^-$ in the end-member fluid from 21°N is at least 1 ppb, but decreases to < 0.1 ppb as the fluid cools below 350°C. Seward (1984) has suggested that stable sulphur complexes (e.g. $HAu(HS)_2^0$) are also potentially important at high temperatures and low pH and could account for the transport of gold in sulphur-rich, end-member fluids. Below 350°C, $Au(HS)_2^-$ probably accounts for the bulk of gold in solution (Seward, 1973), and at temperatures below 300°C the calculated solubility of gold as $Au(HS)_2^-$ in typical vent fluids exceeds that of $AuCl_2^-$ by several orders of magnitude. On this basis, Hannington *et al.* (1986) and Hannington and Scott (1989) concluded that the transport of gold in vent fluids near the seafloor is mainly due to $Au(HS)_2^-$.

Given the similarities in the temperature and bulk composition of the solutions and their uniform source rocks, the concentrations of gold in end-member fluids at different vent sites on the mid-ocean ridges are expected to be very similar. A consequence of this observation is that differences in the gold content of hydrothermal precipitates on the seafloor may be simply a result of the different mixing-cooling histories of otherwise similar end-member fluids and the extent to which the vent fluid-seawater mixtures may become saturated with respect to gold as $Au(HS)_2^-$. The solubility of gold as $Au(HS)_2^-$ is a strong function of the temperature, pH, and oxidation state of the fluid and is given by the reaction:

$$Au + 2H_2S + \tfrac{1}{4}O_2 = Au(HS)_2^- + H^+ + \tfrac{1}{2}H_2O$$

Figure 8.5 compares isopleths of this reaction at a constant H_2S concentration with the T-pH-aO_2 characteristics of vent fluid–seawater mixtures from 21°N, Explorer Ridge,

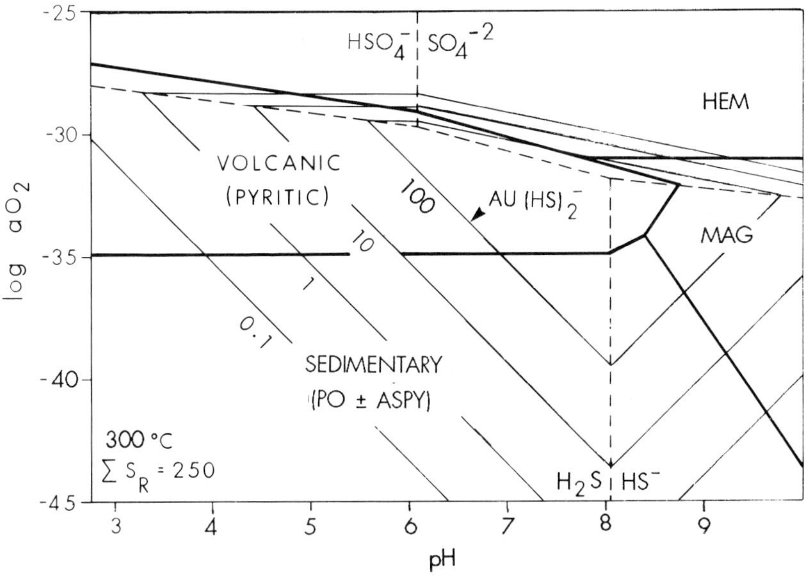

Figure 8.6 Approximate pH–aO_2 characteristics of vent fluids from volcanic-dominated and sediment-dominated hydrothermal systems on the modern seafloor. The principal aqueous sulphur species, stability fields for the iron oxides and sulphides, and solubility contours for gold are calculated as for Figure 8.5 (see Hannington and Scott, 1989). Arsenopyrite is stable at low aO_2 with pyrrhotite in the sediment-hosted deposits. (Heinrich and Eadington, 1986.)

and Axial Seamount. The solubility of gold as $Au(HS)_2^-$ increases in the early stages of mixing in response to the higher pH of the vent fluid-seawater mixture and the increasing dissociation of H_2S at temperatures below 350°C. Due to the high reduced sulphur content of the vent fluids, gold will remain in solution down to at least 200°C. Recent experimental work by Shenberger and Barnes (1989) defined a solubility maximum which occurs between 150°C and 250°C, depending on the equilibria which buffer the pH and oxidation state of the fluids. Gold may be precipitated by a large decrease in temperature below about 150°C or by the oxidation which follows sustained mixing with seawater. This latter process is illustrated by the dramatic decrease in the stability of $Au(HS)_2^-$ across the H_2S–HSO_4^- boundary (Figures 8.5 and 8.6). The common occurrence of barite in gold-rich polymetallic sulphides suggests that late-stage mixing plays an important role in precipitation of gold. Most high-temperature fluids will vent on to the seafloor before significant mixing occurs and, in the absence of an effective precipitation mechanism, will lose their dissolved gold to a diffuse hydrothermal plume.

Depending on the temperature and relative solubilities of other minerals in the same solutions, ascending hydrothermal fluids will likely deposit gold-poor, Cu-Fe-sulphides first, followed by relatively gold-rich, Zn-Fe-Ba-SiO_2 assemblages. In a complex hydrothermal mound, a vent fluid–seawater mixture may dissolve gold at low temperatures from pre-existing sulphides in its path and reprecipitate the gold in response to mixing at the top of the mound. This is consistent with the zone-refining model proposed by Hannington *et al.* (1986). In some deposits the absence of gold in low-temperature, Zn-rich sulphides may be a consequence of low-pH vent fluids in which the solubility of gold as $Au(HS)_2^-$ is uniformly low (e.g. pH = 3.2 at the Southern Juan de Fuca site: USGS Study Group, 1986; Von Damm and Bischoff, 1987). Elsewhere, Zn-rich sulphides are deposited at relatively high temperatures (e.g. up to 300°C at 21°N and 11°N on the East Pacific Rise) or close to the field of pyrrhotite stability where significant gold enrichment from $Au(HS)_2^-$ is also unlikely (Hannington and Scott, 1989). Silver may be transported in solution as chloride complexes (Seward, 1976) or sulphur complexes (Sugaki *et al.*, 1987; Gammons and Barnes, 1989) under conditions similar to those for the transport of gold. However, the low Au/Ag ratios in many high-temperature assemblages apparently reflect the more effective transport and deposition of silver as a chloride complex.

The elevated chloride content of the Red Sea brines may sustain a high initial concentration of gold as $AuCl_2^-$ and probably accounts for the high gold contents of associated sulphides in the Atlantis II Deep. Vent fluids in other sediment-hosted deposits, such as the Guaymas Basin, are unlikely to carry significant amounts of gold as either $Au(HS)_2^-$ or $AuCl_2^-$. By virtue of their having reacted extensively with organic-rich sediments, the strongly reducing vent fluids at these sites lie well within the field of pyrrhotite stability at nearly all temperatures. Under these conditions, significant transport of gold as $Au(HS)_2^-$ is not generally possible except at high pH (Figure 8.6).

8.6.4 *The flux of gold in seafloor hydrothermal fluids*

High-temperature, end-member fluids at basaltic mid-ocean ridges contain about 300–500 mg/kg of dissolved metals and sulphur which may readily precipitate as smoke at the vent fluid-seawater interface. Gold contents of 200 ppb Au in this smoke (Table 8.8) would have required a minimum concentration of about 0.06–0.1 ppb Au in the end-member fluid, assuming complete precipitation of the gold (Hannington and Scott, 1989). This estimate is similar to the 0.2 ppb Au reported from preliminary analyses of end-member fluids at 21°N EPR (Campbell *et al.*, 1987). The active vents at 21°N EPR have a combined mass flux of about 150 kg/s (Converse *et al.*, 1984) and therefore may transport as much as 500–1000 g Au/yr in solution. For comparison, the geothermal fluids at Broadlands, New Zealand, transport about 4700 g Au/yr at a concentration of 1.5 ppb Au and an upflow rate of 100 kg/s (Brown, 1986). These calculations suggest that seafloor hydrothermal systems may have been responsible for the removal of a large amount of gold from the ancient oceanic crust and could have carried enough gold to account for the observed gold grades in volcanogenic massive sulphides and related auriferous hydrothermal precipitates now found on land. The total mass flux of undiluted fluids in ancient seafloor hydrothermal systems is unknown but could have exceeded that of the 21°N vent field by at least an order of magnitude (e.g. Cathles, 1981, 1983).

8.7 Secondary enrichment of gold in supergene sulphides and gossans

The oxidized ores of exposed sulphide deposits on land are commonly enriched in gold in a zone of secondary Cu-sulphides or Fe-oxide gossans. This enrichment is sometimes spectacular, as in the Mt. Morgan massive sulphide deposit in Queensland, Australia, where 2.4 million tonnes of oxidized ore were mined at an average grade of 31 ppm Au (Taube, 1986). Supergene enrichment of gold in this environment is commonly attributed to the exposure of sulphides to oxidation by near-surface ground water following uplift and erosion. However, high concentrations of secondary gold have also been found in oxidized sulphide deposits on the modern seafloor (Table 8.10). In the TAG hydrothermal field, coarse-grained, native gold (up to 21 ppm Au) and native copper in supergene sulphides and Fe-oxide gossans are believed to have formed by the chemical transport and redeposition of these elements during submarine weathering (Hannington *et al.*, 1988b).

These observations refute the conventional wisdom that supergene enrichment of gold is exclusively a subaerial process and suggest that high concentrations of secondary gold may occur in ancient deposits that have never been exposed to near-surface weathering on land. This has been confirmed recently by the discovery of high concentrations of gold in 80 Ma-old seafloor gossans, known as ochres, associated with Cretaceous massive sulphide deposits in Cyprus (Table 8.10). The

Table 8.10 Gold and associated elements in smaples of supergene sulphides and Fe-oxides

	TAG hydrothermal field		13°N EPR	Cyprus	
	Secondary Cu-sulphides (16)*	Fe- oxides (12)	Fe-oxides (2)	Ochres (6)	Devil's Mud (2)
Au (ppm)	9.1	12.4	3.5	7.0	14.2
Ag	62	38	<1	<5	86
As	108	90	263	276	56
Sb	16	20	–	18	12
Co	3	3	130	27	31
Se	<2	<2	4	107	1 605
Sd	10	<10	7	<5	<5
Mo	121	213	–	60	12
Cu (wt.%)	24.6	>40	–	0.1	–
Fe	23.9	20.7	27.3	46.7	15.2
Zn	1.2	0.2	–	0.1	<0.1
Pb	0.02	<0.01	<0.01	0.02	–
S	24.5	1.2	–	–	–
SiO_2	6.6	1.2	–	10.3	57.4
Au/Ag	0.15	0.33	>3.5	>1.4	0.17

*Figures in parenthesis show the total number of samples analysed for different elements.

High Cu contents in the Fe-oxides samples from the TAG hydrothermal field are due to abundant Cu-chloride (i.e. atacamite). Data are from Hannington *et al.* (1988b), Herzig *et al.* (1988b), Herzig *et al.* (1989), Thompson *et al.* (1988), Schroeder *et al.* (1986), Fouquet *et al.* (1988), and Constantinou (1972).

ochres are commonly found at stratigraphic horizons overlying massive ore and buried by nearly contemporaneous marine sediments or pillow lavas. Previous workers (e.g. Constantinou and Govett, 1972; Constantinou, 1973; Robertson, 1976) recognized that the ochres are products of ancient submarine weathering, but the secondary enrichment of gold in these deposits has been documented only recently

(Herzig *et al.*, 1989). Gold-rich ochres from the Skouriotissa deposit contain coarse-grained native gold in a matrix of Fe-oxides and jarosite and are interpreted to be the fossil analogues of gold-rich seafloor gossans at the TAG hydrothermal field. Intermittent, siliceous horizons known as 'devil's mud', also occur within the Cyprus deposits and may contain up to 14 ppm Au and 100 ppm Ag (Bear, 1963). The high concentrations of precious metals in the devil's mud from Skouriotissa (Table 8.10) have previously been attributed to a residual enrichment caused by differential leaching of the sulphides in acid-sulphate groundwaters (Constantinou, 1973; Bischoff *et al.*, 1983).

8.8 Gold in ancient seafloor hydrothermal systems

Auriferous hydrothermal precipitates associated with active volcanism on the seafloor are found throughout the geologic record, and many of these ancient deposits have similarities to gold-bearing hydrothermal precipitates on the modern seafloor.

8.8.1 *Volcanogenic massive sulphides*

Hydrothermal deposits that have formed in island-arcs are well-preserved in the geologic record because of failed rifting or accretion on to stable cratons (Cathles, 1983; Scott, 1987). Major concentrations of massive sulphide such as the Kuroko deposits in Japan, the Bathurst district deposits in Canada, and the Rio Tinto deposits in Spain are believed to have formed in island-arc settings which experienced hydrothermal processes similar to those observed on the modern seafloor (Franklin *et al.*, 1981; Sawkins, 1984; Scott, 1985, 1987). Modern examples of sulphide mineralization in the back-arc basins of the western Pacific are considered to be close analogues of the Kuroko massive sulphides (e.g. Halbach *et al.*, 1989). Sulphide deposits currently forming on mid-ocean ridges more closely resemble ophiolite-hosted, pyrite deposits in Cyprus, Oman, and Newfoundland (e.g. Scott *et al.*, 1989), although these deposits also may have formed in back-arc basins (Miyashiro, 1973; Moores *et al.*, 1984).

Many ancient volcanic-associated sulphide deposits exhibit patterns of gold enrichment similar to those observed in modern seafloor sulphides, suggesting similar controls on the occurrence and distribution of gold (Large *et al.*, 1989; Huston and Large, 1989; Hannington and Scott, 1989). About one in twenty massive sulphide deposits world wide contains more than 30 000 kg Au (1 million ounces) (Figure 8.7). Most of the gold-producing deposits occur in mixed mafic and felsic volcanic sequences and are closely associated with felsic volcanic centres. Fewer than 20% of massive sulphides occur in mafic-dominated ophiolitic sequences, although this may reflect an exploration bias rather than a real difference in the proportion of deposits which occur exclusively in mafic volcanics. Several gold-rich massive sulphides are found in the pillow lavas of explored ophiolite terranes (2.2 ppm Au at Kinousa, Cyprus; 3.8 ppm Au at Turner-Albright, Oregon). Deposits in mixed volcanic and sedimentary rocks are also few in number, but they tend to be large and therefore may contain significant amounts of gold even at low grades (125 million tonnes at 0.6 ppm Au in San Guillermo, Spain; 45 million tonnes at 1.7 ppm Au in Anaconda-Caribou, Bathurst). Large stratiform deposits of this kind were formed in sediment-dominated sequences along rifted continental margins (e.g. Sawkins, 1984). Although most of these deposits have not been significant gold producers, the potential exists for the discovery of a Red Sea-type deposit on land which could contain more than 50 000 kg of recoverable gold.

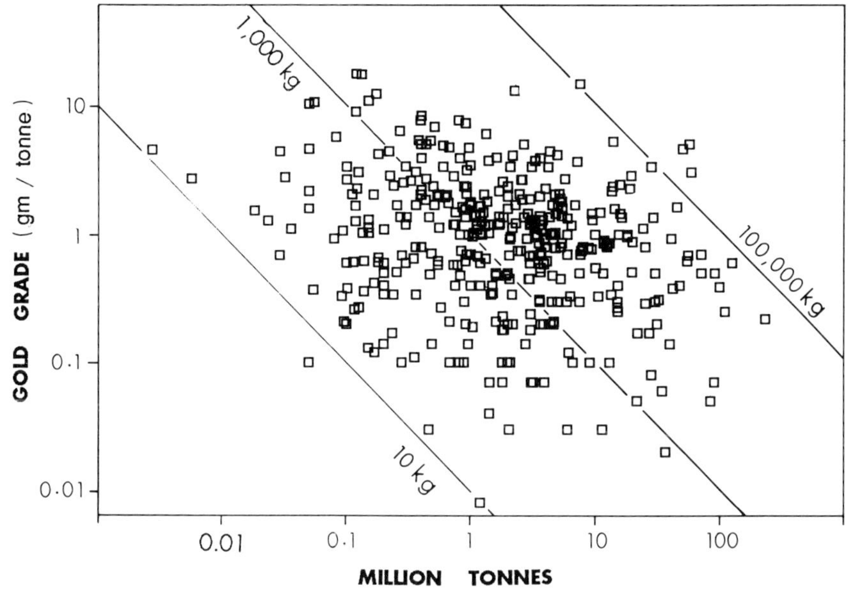

Figure 8.7 Gold grade and tonnage of volcanic-associated massive sulphide deposits world-wide, including current reserves and past production (*n*=372: data from Mosier *et al.*, 1983; Canadian Mines Handbook, 1988; Divi *et al.*, 1980).

8.8.2 *Metalliferous sediments*

Fe-hydroxide and silica deposits, together with microbial mats and non-tronitic sediments, surround many of the low-temperature seafloor vents (Corliss *et al.*, 1979; Alt *et al.*, 1987; Alt, 1988; Karl *et al.*, 1988; Kimura *et al.*, 1988). In addition, some recent volcanogenic sediments have been described as possible modern analogues for ancient silicate-oxide iron formation (e.g. Quinby- Hunt *et al.*, 1986). Except for opal-rich precipitates from the small seamount near 21°N, which contain 0.65 ppm Au, few samples of this material have been analysed for gold. Many of the deposits resemble the late-stage, cherty and Fe-rich chemical sediments which overlie or grade laterally into volcanogenic massive sulphide ores (Sangster and Scott, 1976; Franklin *et al.*, 1981). In Japan, ferruginous chert beds in tuffaceous sediments (tetsusekiei) overlying the Kuroko deposits contain up to 1.5 ppm Au (Kalogeropoulos and Scott, 1983). Opal-barite deposits locally associated with the tetsusekiei are interpreted to be the product of low-temperature venting similar to that observed on the modern seafloor (Kalogeropoulos and Scott, 1986). Pyritic cherts associated with the sulphide deposits at Noranda, Quebec, are only slightly enriched in gold, averaging 0.04 ppm Au (Main Contact Tuff: Kalogeropoulos and Scott, 1983, 1989).

On a regional scale, outside the confines of massive sulphide deposits, some ancient chemical sediments of apparent volcanogenic origin and in particular sulphide facies iron formation are commonly enriched in gold (Ridler and Shilts, 1974; Fripp, 1976; Boyle, 1979; Saager *et al.*, 1982). If recirculated seawater was the source for gold in these ancient chemical sediments, an explanation is required for their low base-metal contents. This has led some authors to suggest their formation from low-salinity, metamorphic secretions (e.g. Fyfe and Kerrich, 1984; Cathles, 1986). Despite the strong mobility of base metals in seafloor hydrothermal fluids, their concentrations in end-member solutions are subject to the prevailing temperature, pressure, and pH during alteration of the crust, as well as the crystallinity of the source rocks and the influence of secondary mineral assemblages. Complex variations in these parameters during experimental seawater–basalt reactions indicate that the mobility of Fe, Cu, and Zn in the reaction zone is not uniform, even at high temperatures (Seyfried and Janecky, 1985; Seyfried, 1987; Seyfried *et al.*, 1988). In contrast, reduced seawater, which has reacted with basalt at temperatures as low as 300°C, may contain as much as 100 ppm H_2S (Mottl *et al.*, 1979; Seyfried and Janecky, 1985) and could carry a large amount of gold as $Au(HS)_2^-$ in the absence of significant base metals. Base-metal mineralization is a product of only the most intense hydrothermal activity which is focused along the axial zones of active rifts (e.g. Rona, 1988). However, thermal modelling of convective heat flow in the modern oceans also requires extensive low-intensity, off-axis hydrothermal circulation (Morton and Sleep, 1985) which may be effective in transporting gold but not base metals. Future off-axis exploration of the seafloor will test this hypothesis.

8.8.3 *Gold deposits in auriferous chemical and clastic sediments*

Large-tonnage, low-grade stratabound and stratiform gold deposits in ancient greenstone belts commonly exhibit a close association with sulphide-bearing chemical and clastic sediments. Examples of these deposits occur in sulphide-carbonate facies iron formation, sulphide-silicate-oxide iron formation, mixed chemical and clastic sediments (greywackes, turbidites), interflow metalliferous sediments within mafic volcanic sequences, and disseminated or massive sulphides in pyritic quartz–sericite schists. A detailed assessment of evidence for and against a seafloor hydrothermal origin for these deposits is beyond the scope of this chapter – the origins of many are disputed and may be unrelated to seafloor hydrothermal processes (e.g., iron formation replacement deposits: Phillips *et al.*, 1984). However, a number of features in many of the deposits associated with auriferous chemical sediments are common to gold-bearing hydrothermal precipitates on the modern seafloor. Sulphide facies iron formation or disseminated and massive sulphides are commonly the most important ore host. In many cases, facies transitions involving sulphides occur within the mine sequence, and these have been widely interpreted in terms of proximity to a proposed hydrothermal vent. Deposits in pyritic interflow sediments (argillites and mudstones) are distinctly similar to metalliferous muds on the modern seafloor, and locally contain up to several weight per cent of combined Cu, Pb, and Zn in addition to gold. Deposition of these sediments from a buoyant hydrothermal plume would have required multiple vent sources with a high total mass flux in order to compete with dispersion and dilution from normal marine sediments.

This suggests that special circumstances akin to the Atlantis II Deep (i.e. restricted anoxic basins) may have been required for the effective accumulation of gold-rich metalliferous sediment in the ancient oceans.

Gold deposits associated with horizons of massive pyrrhotite or arsenopyrite in turbidites and metagreywackes (e.g. Homestake, Lupin) resemble the sulphide deposits actively forming within turbiditic and hemipelagic sediments in the Escanaba Trough. Hydrothermal seawater in this environment may carry significant gold as $Au(HS)_2^-$, with relatively low base-metal concentrations, if the fluids are buffered by the sediments to a high pH (e.g. Figure 8.6). Gold mineralization in turbidite-hosted deposits commonly occurs in horizons of massive pyrrhotite and arsenopyrite. In contrast, the mineralogy of gold deposits in disseminated or massive pyritic sulphides in volcanic- dominated sequences is consistent with a more acidic, sulphur-rich hydrothermal fluid, closely resembling that of sulphide deposits forming on sediment-starved mid-ocean ridges.

At present, the solubility of gold and its availability in the oceanic crust is not sufficiently understood for an appropriate range of conditions to accurately predict its behaviour relative to base metals in the source fluids. Nevertheless, the importance of hydrothermal seawater in the formation of auriferous chemical sediments in ancient oceans cannot be easily dismissed. The prospects of finding a modern analogue for large-tonnage, low-grade gold deposits may improve as exploration of the seafloor proceeds to island-arc settings and active marginal basins.

Acknowledgements

We would like to thank Energy, Mines and Resources Canada and the Natural Sciences and Engineering Research Council of Canada for their financial support of studies on gold in modern seafloor deposits by the Marine Geology Research Group, University of Toronto. Some of the samples for this study were provided by G. Thompson (WHOI), P.A. Rona (NOAA), M. Kastner (Scripps), and J. Lange (Preussag).

References

Alt, J.C. (1988) The chemistry and sulfur isotope composition of massive sulfide and associated deposits on Green Seamount, Eastern Pacific. *Econ. Geol.* **83**, 1026–1033.

Alt, J.C., Lonsdale, P., Haymon, R. and Muehlenbachs, K. (1987) Hydrothermal sulfide and oxide deposits on seamounts near 21°N, East Pacific Rise. *Geol. Soc. Amer. Bull.* **98**, 157–168.

Alt, J.C., Muehlenbachs, K. and Honnorez, J. (1986) An oxygen isotopic profile through the upper kilometer of the oceanic crust, DSDP Hole 504B. *Earth Planet. Sci. Letters* **80**, 217-229.

ASHES Expedition (1986) Pisces submersible exploration of a high- temperature vent field in the caldera of Axial Seamount [abs.]. *Amer. Geophys. Union Trans.* **67**, 1027.

Backer, H. (1976) Fazies und chemische zusammensetzung rezenter ausfallungen aus mineralquellen in Roten Meer. *Geol. Jahrb.* **17**, 151-172.

Baker, E.T., and Massoth, G.J. (1986) Hydrothermal plume measurements: a regional perspective. *Science* **234**, 980-982.

Baker, E.T. and Massoth, G.J. (1987) Characteristics of hydrothermal plumes from two vent fields on the Juan de Fuca Ridge, northeast Pacific Ocean. *Earth Planet. Sci. Lett.* **85**, 59-73.

Bear, L.M. (1963) The mineral resources and mining industry of Cyprus. Nicosia, Cyprus, Ministry Commerce Industry, Geol. Surv. Dept., Bull. 1.

Bischoff, J.L., Radtke, A.S. and Rosenbauer, R.J. (1981) Hydrothermal alteration of graywacke by brine and seawater : Roles of alteration and chloride complexing on metal solubilization at 200°C and 350°C. *Econ. Geol.* **76**, 659-676.

Bischoff, J.L., Rosenbauer, R.J., Aruscavage, P.J., Baedecker, P.A. and Crock, J.G. (1983) Sea-floor massive sulfide deposits from 21°N, East Pacific Rise; Juan de Fuca Ridge; and Galapagos Rift: Bulk chemical composition and economic implications. *Econ. Geol.* **78**, 1711-1720.

Bowers, T.S., Campbell, A.C., Measures, C.I., Spivack, A.J. and Edmond, J.M. (1988) Chemical controls on the composition of vent fluids at 13°N–11°N and 21°N, East Pacific Rise. *J. Geophys. Res.* **93**, 4522–4536.

Bowers, T.S., Von Damm, K.L. and Edmond, J.M. (1985) Chemical evolution of mid-ocean ridge hot springs. *Geochim. Cosmochim. Acta* **49**, 2239-2252.

Boyle, R.W. (1979) The geochemistry of gold and its deposits. Geol. Surv. of Can., Bull. 280.

Brown, K.L. (1986) Gold deposition from geothermal discharges in New Zealand. *Econ. Geol.* **81**, 979-983.

Browne, P.R.L. and Ellis, A.J. (1970) The Ohaki-Broadlands hydrothermal area, New Zealand: Mineralogy and related chemistry. *Amer. J. Sci.* **296**, 97-131.

Campbell, A.C., Bowers, T.S., and Edmond, J.M. (1988a) A times-series of vent fluid compositions from 21°N, EPR (1979, 1981, 1985) and the Guaymas Basin, Gulf of California (1982, 1985). *J. Geophys. Res.* **93**, 4537-4549.

Campbell, A.C., Edmond, J.M., Colodner, D., Palmer, M.R. and Falkner, K.K. (1987) Chemistry of hydrothermal fluids from the Mariana Trough back arc basin in comparison to mid-ocean ridge fluids [abs.]. *Amer. Geophys. Union Trans.* **68**, 1531.

Campbell, A.C., German, C., Palmer, M.R. and Edmond, J.M. (1988b) Preliminary report on the chemistry of hydrothermal fluids from the Escanaba Trough [abs.]. *Amer. Geophys. Union Trans.* **69**, 1271.

Campbell, A.C., Palmer, M.R., Klinkhammer, G.P., Bowers, T.S., Edmond, J.M., Lawrence, J.R., Casey, J.F., Thompson, G., Humphris, S., Rona, P. and Karson, J.A. (1988c) Chemistry of hot springs on the Mid-Atlantic Ridge. *Nature* **335**, 514-519.

Campbell, I.H., McDougall, T.J. and Turner, J.S. (1984) A note on fluid dynamic processes which can influence the deposition of massive sulfides. *Econ. Geol.* **79**, 1905-1913.

Canadian Mines Handbook (1987–1988), ed. Gardiner, C.D. Northern Miner Press Ltd., Toronto.

CASM II (1985) Hydrothermal vents on an axis seamount of the Juan de Fuca Ridge. *Nature* **313**, 212-214.

Cathles, L.M. (1981) Fluid flow and genesis of hydrothermal ore deposits. *Econ. Geol.*, 75th Anniv. Vol., 424-457.

Cathles, L.M. (1983) An analysis of the hydrothermal system responsible for massive sulfide deposition in the Hokuroku Basin of Japan. *Econ. Geol. Mon.* **5**, 439-487.

Cathles, L.M. (1986) The geologic solubility of gold from 200-350°C and its implications for gold-base metal ratios in vein and stratiform deposits. Can. Inst. of Mining and Metallurgy, Special Vol. 38, 187-211.

Constantinou, G. (1973) Geology, geochemistry, and genesis of Cyprus sulphide deposits. *Econ. Geol.* **68**, 843-858.

Constantinou, G., and Govett, G.J.S. (1972) Genesis of sulphide deposits, ochre and umber of Cyprus. *Trans. Inst. Min. Mettall.* (Sect. B, Appl. Earth Sci.) **81**, 34-36.

Converse, D.R., Holland, H.D. and Edmond, J.M. (1984) Flow rates in the axial hot springs of the East Pacific Rise (21°N): Implications for the heat budget and the formation of massive sulfide deposits. *Earth Planet. Sci. Letters* **69**, 159-175.

Corliss, J.B., Dymond, J., Gordon, L.I., Edmond, J.M., von Herzen, R.P. and van Andel, T.H. (1979) Submarine thermal springs on the Galapagos Rift. *Science* **203**, 1073-1083.

Craig, H., Horibe, Y., Farley, K.A., Welhan, J.A., Kim, K.-R. and Hey, R.N. (1987) Hydrothermal vents in the Mariana Trough: Results of the first Alvin dives [abs.]. *Amer. Geophys. Union Trans.* 68, 1531.

CYAMEX (1979) Massive deep-sea sulphide ore deposits discovered on the East Pacific Rise. *Nature* **277**, 523-528.

Davis, E.E., Goodfellow, W.D., Bornhold, B.D., Adshead, J., Blaise, B., Villinger, H. and Le Cheminant, G.M. (1987) Massive sulfides in a sedimented rift valley, northern Juan de Fuca Ridge. *Earth Planet. Sci. Letters* **82**, 49-61.

Degens, E.T. and Ross, D.A., eds. (1969) *Hot Brines and Recent Heavy Metal Deposits in the Red Sea*. Springer-Verlag, New York.

Divi, S.R., Thorpe, R.I. and Franklin, J.M. (1980) Use of discriminant analysis to evaluate compositional controls of stratiform massive sulfide deposits in Canada. Can. Geol. Surv., Paper 79-20.

Edmond, J.M., Measures, C.I., McDuff, R.E., Chan, L.H., Collier, R., Grant, B., Gordon, L.I. and Corliss, J.B. (1979a) Ridge crest hydrothermal activity and the balance of the major and minor elements in the ocean: The Galapagos data. *Earth Planet. Sci. Letters* **46**, 1-18.

Edmond, J.M., Measures, C., Mangum, B., Grant, B., Sclater, F.R., Collier, R., Hudson, A., Gordon, L.I. and Corliss, J.B. (1979b) On the formation of metal-rich deposits at ridge crests. *Earth Planet. Sci. Letters* **46**, 19-30.

Edmond, J.M., Von Damm, K.L., McDuff, R.E. and Measures, C.I. (1982) Chemistry of hot springs on the East Pacific Rise and their effluent dispersal. *Nature* **297**, 187-191.

Eldridge, C.S., Barton, P.B., Jr. and Ohmoto, H. (1983) Mineral textures and their bearing on formation of the Kuroko orebodies. *Econ. Geol. Mon.* **5**, 241-281.

Embley, R.W., Jonasson, I.R., Perfit, M.R., Franklin, J.M., Tivey, M.A., Malahoff, A., Smith, M.F. and Francis, T.J.G. (1988) Submersible investigation of an extinct hydrothermal system on the Galpagos Ridge: Sulfide mounds, stockwork zone, and differentiated lavas. *Can. Mineral.* **26**, 517-539.

Foster, R.P. and Wilson, J.F. (1984) Geological setting of Archaean gold deposits in Zimbabwe. In *The Geology, Geochemistry, and Genesis of Gold Deposits, Proc. of Gold '82, Harare, Zimbabwe*, ed. Foster, R.P. Balkema, Rotterdam, 521- 551.

Fouquet, Y., Auclair, G., Cambon, P. and Etoubleau, J. (1988) Geological setting, mineralogical and geochemical investigations on sulfide deposits near 13° North, on the East Pacific Rise. *Marine Geol.* **84**, 145-178.

Franklin, J.M. (1986) Volcanic associated massive sulfide deposits – an update. Irish Assoc. of Econ. Geol., Special Pub. 4, 49-69.

Franklin, J.M., Lydon, J.W. and Sangster, D.F. (1981) Volcanic-associated massive sulfide deposits. *Econ. Geol.*, 75th Anniv. Vol., 458-627.

Fripp, R.E.P. (1976) Stratabound gold deposits in Archean banded iron formation, Rhodesia. *Econ. Geol.* **71**, 58-75.

Fyfe, W.S. and Kerrich, R. (1984) Gold: Natural concentration processes. In *The Geology, Geochemistry, and Genesis of Gold Deposits, Proc. of Gold '82, Harare, Zimbabwe*, ed. Foster, R.P. Balkema, Rotterdam, 99-126.

Gammons, C.H. and Barnes, H.L. (1989) The solubility of Ag_2S in near-neutral aqueous sulfide solutions at 25 to 300°C. *Geochim. Cosmochim. Acta* **53**, 279-290.

GARIMAS I Cruise Report (1986) Galapagos Rift massive sulfides (SO-32). Preussag Marine Technology, Hannover.

GARIMAS II Cruise Report (1987) Galapagos Rift massive sulfides (SO-39). Preussag Marine Technology, Hannover.

Giggenbach, W.F. (1980) Geothermal gas equilibria. *Geochim. Cosmochim. Acta* **44**, 2021-2032.

Giggenbach, W.F. (1981) Geothermal mineral equilibria. *Geochim. Cosmochim. Acta* **45**, 393-410.

Goldfarb, M.S., Converse, D.R., Holland, H.D. and Edmond, J.M. (1983) The genesis of hot spring deposits on the East Pacific Rise, 21°N. *Econ. Geol. Mon.* **5**, 184-197.

Goodfellow, W.D. and Blaise, B. (1988) Sulfide formation and hydrothermal alteration of hemipelagic sediment in Middle Valley, northern Juan de Fuca Ridge. *Can. Mineral.* **26**, 675-696.

Grimaud, D., Michard, A. and Michard, G. (1984) Composition chimique et composition isotopique du strontium dans les eaux hydrothermals sous-marines de la dorsale Est Pacifique à 13° Nord. *C.R. Acad. Sc. Paris* **299**, Serie II, no.13, 865-870.

Hackett, J.P. and Bischoff, J.L. (1973) New data on the stratigraphy, extent, and geologic history of the Red Sea geothermal deposits. *Econ. Geol.* **68**, 553-564.

Halbach, P., Nakamura, K., Wahsner, M., Lange, J., Sakai, H., Kaselitz, L., Hansen, R.-D., Yamano, M., Post, J., Prause, B., Seifert, R., Michaelis, W., Teichmann, F., Kinoshita, M., Marten, A., Ishibashi, J., Czerwinski, S. and Blum, N. (1989) Probable modern anlogue of Kuroko-type massive sulphide deposits in the Okinanwa Trough back-arc basin. *Nature* **338**, 496- 499.

Hall, G.E.M., Vaive, J., Hannington, M.D., McConachy, T.F. and Scott, S.D. (1988) Gold and associated trace elements in vent waters and suspended particulates from Southern Explorer Ridge [abs.]. *Amer. Geophys. Union Trans.* **69**, 1501.

Hannington, M.D. and Scott, S.D. (1988) Mineralogy and geochemistry of a hydrothermal silica-sulfide-sulfate spire in the caldera of Axial Seamount, Juan de Fuca Ridge. *Can. Mineral.* **26**, 603-625.

Hannington, M.D. and Scott, S.D. (1989) Gold mineralization in volcanogenic massive sulfides: Implications of data from active hydrothermal vents on the modern seafloor. *Econ. Geol. Mon.* **6**. *Proceedings of Gold '88, Melbourne, Australia* 491-507.

Hannington, M.D., Herzig, P.M. and Alt, J.C. (1988a) The distribution of gold in sub-seafloor stockwork mineralization from DSDP Hole 504B and the Agrokipia B deposit, Cyprus [abs.]. *Geol. Assoc. Can. Program Abstr.* **13**, A51-A52.

Hannington, M.D., Herzig, P.M., Scott, S.D., Thompson, G. and Rona, R.A. (1989) The mineralogy and geochemistry of gold-bearing polymetallic sulfides from seafloor spreading centers in the N.E. Pacific and Mid-Atlantic (submitted).

Hannington, M.D., Peter, J.M. and Scott, S.D. (1986) Gold in sea-floor polymetallic sulfide deposits. *Econ. Geol.* **81**, 1867-1883.

Hannington, M.D., Thompson, G., Rona, P.A. and Scott, S.D. (1988b) Gold and native copper in supergene sulfides from the Mid-Atlantic Ridge. *Nature* **333**, 64-66.

Harvey-Kelly, F.E.L., Jonasson, I.R., Franklin, J.M. and Embley, R.W. (1988) Sulfide deposits of Axial Seamount: Mineralogy and chemistry [abs.]. *Amer. Geophys. Union. Trans.* **69**, 1499-1500.

Haymon, R.M. and Kastner, M. (1981) Hot spring deposits on the East Pacific Rise at 21°N: Preliminary description of mineralogy and genesis. *Earth Planet. Sci. Letters* **53**, 363-381.

Heinrich, C.A. and Eadington, P.J. (1986) Thermodynamic predictions of the hydrothermal chemistry of arsenic, and their significance for the paragenetic sequence of some cassiterite–arsenopyrite–base metal sulfide deposits. *Econ. Geol.* **81**, 511-529.

Hekinian, R. and Fouquet, Y. (1985) Volcanism and metallogenesis of axial and off-axial structures on the East Pacific Rise near 13°N *Econ. Geol.* **80**, 221-249.

Helgeson, H.C. (1969) Thermodynamics of hydrothermal systems at elevated temperatures and pressures. *Amer. J. Sci.* **267**, 729-804.

Hendricks, R.L., Reisbeck, F.B., Mahaffey, E.J., Roberts, D.B. and Peterson, M.N.A. (1969) Chemical composition of sediments and interstitial brines from the Atlantis II, Discovery and Chain Deeps. In *Hot Brines and Recent Heavy Metal Deposits in the Red Sea*, eds. Degens, E.T. and Ross, D.A. Springer-Verlag, New York, 407-440.

Herzig,P.M., Hannington, M.D., Scott, S.D., Maliotis, G., Thompson, G. and Rona, R.A. (1989) Secondary gold in submarine gossans from the Troodos Ophiolite and the Mid-Atlantic Ridge (submitted).

Herzig, P.M., Hannington, M.D., Scott, S.D., Thompson, G. and Rona, P.A. (1988a) Comparison of sulfides from the TAG and Snakepit Hydrothermal Fields (Mid-Atlantic Ridge) and the Troodos Ophiolite (Cyprus) [abs.]. *Amer. Geophys. Union Trans.* **69**, 1488.

Herzig, P.M., Hannington, M.D., Scott, S.D., Thompson, G. and Rona, P.A. (1988b) The distribution of gold and associated trace elements in modern submarine gossans from the TAG Hydrothermal Field, Mid-Atlantic Ridge and in ancient ochres from Cyprus [abs.]. *Geol. Soc. Amer. Abstr. Programs* **20**, A240.

Honnorez, J., Alt, J.C., Honnorez-Guerstein, B.-M., Laverne, C., Muehlenbachs, K., Ruiz, J. and Saltzman, E. (1985) 10. Stockwork- like sulfide mineralization in young oceanic crust: Deep Sea Drilling Project Hole 504B, In *Initial Reports of the Deep Sea Drilling Project LXXXIII*, eds. Anderson, R.N. *et al.* US Govt. Printing Office, Washington, 263-282.

Huston, D.L. and Large, R.R. (1989) A chemical model for the concentration of gold in volcanogenic massive sulphide deposits. *Ore Geol. Rev.* **4**, 171-200

Janecky, D.R. and Seyfried, W.E. Jr. (1984) Formation of massive sulfide deposits on oceanic ridge crests: Incremental reaction models for mixing between hydrothermal solutions and seawater. *Geochim. Cosmochim. Acta* **48**, 2723-2738.

Jonasson, I.R., Perfit, M.R., Franklin, J.M., Embley, R.W. and Malahoff, A. (1988) Precious metal levels in sulfide and host basaltic lavas from the Galapagos ridge crest at 85°50'W [abs.]. *Bicentennial Gold '88. Geol. Soc. Aust. Abstr.* **23**, 596-601.

Kalogeropoulos, S.I. and Scott, S.D. (1983) Mineralogy and geochemistry of tuffaceous exhalites (tetsusekiei) of the Fukazawa mine, Hokuroku District, Japan. *Econ. Geol. Mon.* **5**, 412-432.

Kalogeropoulos, S.I. and Scott, S.D. (1986) On the genesis of barite associated with volcanogenic massive sulfides, Fukazawa mine, Hokuroku District, Japan. In *Geology and Metallogeny of Copper Deposits*, eds. Friedrich, G.H. *et al.* Springer- Verlag, Berlin, 370-388.

Kalogeropoulos, S.I. and Scott, S.D. (1989) Mineralogy and geochemistry of an Archean tuffaceous exhalite: The Main Contact Tuff, Millenbach Mine area, Noranda, Quebec. *Can. J. Earth Sci.* **26**, 88-105.

Karl, D.M., McMurtry, G.M., Malahoff, A. and Garcia, M.O. (1988) Loihi Seamount, Hawaii: A mid-plate volcano with a distinctive hydrothermal system. *Nature* **335**, 532-535.

Kastner, M., Craig, H. and Sturz, A. (1987) Hydrothermal deposition in the Mariana Trough: Preliminary mineralogical investigation [abs.]. *Amer. Geophys. Union Trans.* **68**, 1531.

Kawahata, H. and Shikazono, N. (1988) Sulfur isotope and total sulfur studies of basalts and greenstones from DSDP Hole 504B, Costa Rica Rift: Implications for hydrothermal alteration. *Can. Mineral.* **26**, 555-566.

Keays, R.R. (1984) Archean gold deposits and their source rocks: The upper mantle connection. In *The Geology, Geochemistry, and Genesis of Gold Deposits, Proc. of Gold '82, Harare, Zimbabwe*, ed. Foster, R.P. Balkema, Rotterdam, 17-51.

Keays, R.R. (1987) Principles of mobilization (dissolution) of metals in mafic and ultramafic rocks – The role of immiscible magmatic sulphides in the generation of hydrothermal gold and volcanogenic massive sulphide deposits. *Ore Geol. Rev.* **2**, 47-63.

Kimura, M., Uyeda, S., Kato, Y., Tanaka, T., Yamano, M., Gamo, T., Sakai, H., Kato, S., Izawa, E. and Oomori, T. (1988) Active hydrothermal mounds in the Okinawa Trough backarc basin, Japan. *Tectonophys.* **145**, 319-324.

Koski, R.A., Clague, D.A. and Oudin, E. (1984) Mineralogy and chemistry of massive sulfide deposits from the Juan de Fuca Ridge. *Geol. Soc. of Amer., Bull.* 95, 930-945.

Koski, R.A., Lonsdale, P.F., Shanks, W.C. III, Berndt, M.E. and Howe, S.S. (1985) Mineralogy and geochemistry of a sediment- hosted hydrothermal sulfide deposit from the Southern Trough of Guaymas Basin, Gulf of California. *J. Geophys. Res.* **90**, 6695-6707.

Koski, R.A., Shanks, W.C. III, Bohrson, W.A. and Oscarson, R.L. (1988) The composition of massive sulfide deposits from the sediment-covered floor of Escanaba Trough, Gorda Ridge: Implications for depositional processes. *Can. Mineral.* **26**, 655-673.

Krupp, R.E. and Seward, T.M. (1987) The Rotokawa geothermal systems, New Zealand: An active epithermal gold-depositing environment. *Econ. Geol.* **82**, 1109-1129.

Large, R.R., Huston, D.L., McGoldrick, P.J., Ruxton, P.A. and McArthur, G. (1989) Gold distribution and genesis in Australian volcanogenic massive sulfide deposits and significance for gold transport models. *Econ. Geol. Mon.* **6**, *Proc. of Gold '88, Melbourne, Australia*, 520-536.

Lonsdale, P. and Becker, K. (1985) Hydrothermal plumes, hot springs, and conductive heat flow in the Southern Trough of Guaymas Basin. *Earth Planet. Sci. Lett.* **73**, 211- 225.

Lonsdale, P. and Hawkins, J.W. (1985) Silicic volcanism at an off-axis geothermal field in the Mariana Trough back-arc basin. *Geol. Soc. Amer., Bull.* 96, 940-951.

Lonsdale, P., Bischoff, J.L., Burns, V.M., Kastner, M. and Sweeney, R.E. (1980) A high-temperature hydrothermal deposit on the seabed at a Gulf of California spreading center. *Earth Planet. Sci. Lett.* **49**, 8-20.

Lupton, P., Delaney, J.R., Johnson, H.P., and Tivey, M.K. (1985) Entrainment and vertical transport of deep-ocean water by buoyant hydrothermal plumes. *Nature* **316**, 621-623.

Macdougall, J.D., Volpe, A. and Hawkins, J.W. (1987) An arc-like component in Mariana Trough basalts [abs.]. *Amer. Geophys. Union Trans.* **68**, 1531.

McConachy, T.F. (1988) Hydrothermal plumes and related deposits over spreading ridges in the northeast Pacific Ocean: The East Pacific Rise near 11°N and 21°N, Explorer Ridge, and the J. Tuzo Wilson Seamounts. Unpub. Ph.D. thesis, Univ. of Toronto.

McConachy, T.F. and Scott, S.D. (1987) Real-time mapping of hydrothermal plumes above Southern Explorer Ridge, N.E. Pacific Ocean. *Marine Mining*, **6**, 181-204.

McConachy, T.F., Ballard, R.D., Mottl, M.J. and Von Herzen, R.P. (1986) Geologic form and setting of a hydrothermal vent field at lat 10°56'N, East Pacific Rise: A detailed study using Angus and Alvin. *Geology* **14**, 295-298.

McDuff, R.E., Edmond, J.M., Juniper, K., Lupton, J.E. and Scott, S.D. (1983) The chemistry of hydrothermal water, Axial Seamount, Juan de Fuca Ridge [abs.]. *Amer. Geophys. Union Trans.* **64**, 723.

Merlivat, L., Pineau, F. and Javoy, M., 1987, Hydrothermal vent waters at 13°N on the East Pacific Rise: Isotopic composition and gas concentration. *Earth Planet. Sci. Lett.* **84**, 100-108.

Michard, G., Albarede, F., Michard, A., Minster, J.F., Charlou, J.-L. and Tan, N. (1984) Chemistry of solutions from the 13°N East Pacific Rise hydrothermal site. *Earth Planet. Sci. Letters* **67**, 297-307.

Miyashiro, A. (1973) The troodos ophiolite complex was probably formed in an island arc. *Earth Planet. Sci. Letters* **19**, 218-224.

Moores, E.M., Robinson, P.T., Malpas, J. and Xenophanas, C. (1984) Model for the origin of the Troodos massif, Cyprus, and other mideast ophiolites. *Geology* **12**, 500-503.

Morton, J.L. and Sleep, N.H. (1985) A mid-ocean ridge thermal model: Constraints on the volume of axial heat flux. *Geophys. Res.* **90**, 11345-11353.

Morton, J.L., Holmes, M.L. and Koski, R.A. (1987) Volcanism and massive sulfide formation at a sedimented spreading center, Escanaba Trough, Gorda Ridge, northeast Pacific Ocean. *Geophys. Res. Letters* **14**, 769-772.

Mosier, D.L., Singer, D.A. and Salem, B.B. (1983) Geologic and grade-tonnage information on volcanic-hosted copper-zinc-lead massive sulfide deposits. *US Geol. Surv., Open-File Report* 83-89.

Mottl, M.J. (1983) Metabasalts, axial hot-springs, and the structure of hydrothermal systems at mid-ocean ridges. *Geol. Soc. of Amer., Bull.* 94, 161-180.

Mottl, M.J., Holland, H.D. and Corr, R.F. (1979) Chemical exchange during hydrothermal alteration of basalt - II. Experimental results for Fe, Mn, and sulfur species. *Geochim. Cosmochim. Acta* **43**, 869-884.

Mottl, M.J. and McConachy, T.F. (in press) Chemical processes in buoyant hydrothermal plumes on the East Pacific Rise near 21°N. *Geochim. Cosmochim. Acta* (submitted).

Mustafa, H.E.Z., Nawab, Z., Horn, R. and Le Lann, F. (1984) Economic interest of hydrothermal deposits: Atlantis II Project. In *Proc. 2nd Internat. Seminar on Offshore Mineral Resources.* Centre Oceanologique de Bretagne, Group d'Etude et de Recherche de Mineralisations au Large, Brest, France, 509- 539.

Nawab, Z.A. (1984) Red Sea mining: a new era. *Deep-Sea Res.* **31**, 813-822.

Nelsen, T.A., Metz, S., Trefry, J. and Pimmel, A. (1986) Sedimentology and composition of sediment near a black smoker field on the Mid-Atlantic Ridge [abs.]. *Amer. Geophys. Union Trans.* **67**, 1022.

Nesbitt, B.E., St. Louis, R.M. and Muehlenbachs, K. (1987) Distribution of gold in altered basalts of DSDP Hole 504B. *Can. J. Earth Sci.* **24**, 201-209.

Oudin, E. (1981) Etudes mineralogiques et geochemique des depots sulfures sous-marins actuel de la ride est-pacifique (21N). Documents du Bureau de Recerches Geologiques et Minieres 25.

Oudin, E. (1987) Geochemistry of submarine sulphides. In *Marine Minerals: Resource Assessment Strategies, Proc. NATO Advanced Research Workshop*, Series C 194, eds. Teleki, P.G., Dobson, M.R., Moore, J.R. and von Stackelberg, U. Reidel, Boston, 349-362.

Oudin, E., Thisse, Y. and Ramboz, C. (1984) Fluid inclusion and mineralogical evidence for high-temperature saline hydrothermal circulation in the Red Sea metalliferous sediments: Preliminary results. *Marine Mining* **5**, 3-31.

Peter, J.M. and Scott, S.D. (1988) Mineralogy, composition, and fluid inclusion microthermometry of seafloor hydrothermal deposits in the southern trough of Guaymas Basin, Gulf of California. *Can. Mineral.* **26**, 567-588.

Phillips, G.N., Groves, D.I. and Martyn, J.E. (1984) An epigenetic origin for banded iron formation-hosted gold deposits. *Econ. Geol.* **79**, 162-171.

Piper, D.Z. and Graef, P.A. (1974) Gold and rare-earth elements in sediments from the East Pacific Rise. *Marine Geol.* **17**, 287-297.

Pottorf, R.J. and Barnes, H.L. (1983) Mineralogy, geochemistry, and ore genesis of hydrothermal sediments from the Atlantis II Deep, Red Sea. *Econ. Geol. Mon.* **5**, 198-223.

Quinby-Hunt, M.S., Wilde, P., Corrigan, D. and Dengler, A.T. (1986) Very recent analogs of volcanogenic Archean sequences. *Geology* **14**, 48-51.

Ramboz, C., Oudin, E. and Thisse, Y. (1988) Geyser-type discharge in the Atlantis II Deep, Red Sea: Evidence from of boiling from fluid inclusions in epigenetic anhydrite. *Can. Mineral.* **26**, 765-786.

Renders, P.J. and Seward, T.M. (1989) The adsorption of thio gold(I) complexes by amorphous As_2S_3 and Sb_2S_3 at 25 and 90°C. *Geochim. Cosmochim. Acta* **53**, 255-267.

Ridler, R.H. and Shilts, W.W. (1974) Mineral potential of the Rankin Inlet, Ennadai Belt. *Can. Mining. J.* **95**, 32- 42.

RISE Project (1980) East Pacific Rise: Hot springs and geophysical experiments. *Science* **207**, 1421-1433.

Robertson, A.H.F. (1976) Origins of ochres and umbers: evidence from Skouriotissa, Troodos Massif, Cyprus. *Trans. Inst. Min. Metall.* (Section B, Applied Earth Science) **85**, 245-251.

Rona, P.A. (1988) Hydrothermal mineralization at oceanic ridges. *Can. Mineral.* **26**, 431-465.

Rona, P.A., Klinkhammer, G., Nelsen, T.A., Tefry, J.H. and Elderfield, H. (1986) Black smokers, massive sulfides, and vent biota at the Mid-Atlantic Ridge. *Nature* **321**, 33-37.

Rosenbauer, R.J. and Bischoff, J.L. (1983) Uptake and transport of heavy metals by heated seawater: A summary of the experimental results. In *Hydrothermal Processes at Seafloor Spreading Centers*, NATO Conference, Series IV, Marine Sciences 12, eds. Rona, P.A., Bostrom, K., Laubier, L. and Smith, K.L. Jr. Plenum, New York, 177-198.

Saager, R., Meyer, M. and Muff, R. (1982) Gold distribution in supracrustal rocks from Archean greenstone belts of Southern Africa and from Paleozoic ultramafic complexes of the European Alps: Metallogenic and geochemical implications. *Econ. Geol.* **77**, 1-24.

Sangster, D.F. and Scott, S.D. (1976) Precambrian strata-bound massive Cu–Zn–Pb sulfide ores of North America. In *Handbook of Strata-Bound and Stratiform Ore Deposits* 6, ed. Wolf, K.H. Elsevier, New York, 129-222.

Sawkins, F.J. (1984) *Metal Deposits in Relation to Plate Tectonics.* Springer-Verlag, New York.

Schroeder, B., Thompson, G., Humphris, S.E., Sulanowska, M. and Rona, P.A. (1986) Hydrothermal mineralization, TAG area, Mid- Atlantic Ridge 26°N [abs.]. *Amer. Geophys. Union Trans.* **67**, 1022.

Scott, S.D. (1985) Seafloor polymetallic sulfide deposits: Modern and ancient. *Marine Mining* **5**, 191-212.

Scott, S.D. (1987) Seafloor polymetallic sulfides: Scientific curiosities or mines of the future? In *Marine Minerals: Resource Assessment Strategies, Proc. NATO Advanced Research Workshop*, Series C 194, eds. Teleki, P.G., Dobson, M.R., Moore, J.R. and von Stackelberg, U. Reidel, Boston, 277- 300.

Scott, S.D., Chase, R.L., Hannington, M.D., Michael, P.J., McConachy, T.F. and Shea, G.T. (1989) Sulfide deposits, tectonics, and petrogenesis of Southern Explorer Ridge, northeast Pacific Ocean. *Proc. Troodos '87*, Ophiolite Symposium Volume (in press).

Seyfried, W.E. Jr. (1987) Experimental and theoretical constraints on hydrothermal alteration processes at mid-ocean ridges. *Ann. Rev. Earth Planet. Sci.* **15**, 317-335.

Seyfried, W.E. Jr. and Janecky, D.R. (1985) Heavy metal and sulfur transport during subcritical and supercritical hydrothermal alteration of basalt: Influence of fluid pressure and basalt composition and crystallinity. *Geochim. Cosmchim. Acta* **49**, 2545-2560.

Seyfried, W.E. Jr. Berndt, M.E. and Seewald, J.S. (1988) Hydrothermal alteration processes at mid-ocean ridges: Constraints from diabase alteration experiments, hot-spring fluids and composition of the oceanic crust. *Can. Mineral.* **26**, 787-804.

Seward, T.M. (1973) Thio complexes of gold and the transport of gold in hydrothermal ore solutions. *Geochim. Cosmochim. Acta* **37**, 379-399.

Seward, T.M. (1976) The stability of chloride complexes of silver in hydrothermal solutions up to 350°C. *Geochim. Cosmochim. Acta* **40**, 1329-1341.

Seward, T.M. (1984) The transport and deposition of gold in hydrothermal systems. In *The Geology, Geochemistry, and Genesis of Gold Deposits, Proc. of Gold '82, Harare, Zimbabwe*, ed. Foster, R.P. Balkema, Rotterdam, 165-181.

Shanks, W.C. III and Bischoff, J.L. (1977) Ore transport and deposition in the Red Sea geothermal systems: A geochemical model. *Geochim. Cosmochim. Acta* **41**, 1507-1519.

Shanks, W.C. III and Bischoff, J.L. (1980) Geochemistry, sulfur isotope composition, and accumulation rates of the Red Sea geothermal deposits. *Econ. Geol.* **75**, 445-459.

Shenberger, D.M. and Barnes, H.L. (1989) Gold solubility in aqueous sulfide solutions from 150 to 350°C. *Geochim. Cosmichim. Acta* **53**, 269-278.

Styrt, M.M., Brackmann, A.J., Holland, H.D., Clark, B.C., Pisutha-Arnond, V., Eldridge, E.S. and Ohmoto, H. (1981) The mineralogy and the isotopic composition of sulfur in hydrothermal sulfide/sulfate deposits on the East Pacific Rise, 21°N latitude. *Earth Planet. Sci. Letters* **53**, 382-390.

Sugaki, A., Scott, S.D., Hayashi, K. and Kitakaze, A. (1987) Ag_2S solubility in sulfide solutions up to 250°C. *Geochem. J.* **21**, 291-305.

Taube, A. (1986) The Mount Morgan gold-copper mine and environment, Queensland: A volcanogenic massive sulfide deposit associated with penecontemporaneous faulting. *Econ. Geol.* **81**, 1322-1340.

Thompson, G., Humphris, S.E., Schroeder, B., Sulanowska, M. and Rona, P.A. (1988) Hydrothermal mineralization on the Mid-Atlantic Ridge. *Can. Mineral.* **26**, 697-712.

Thornton, E.C. and Seyfried, W.E. Jr. (1987) Reactivity of organic-rich sediment in seawater at 350°C, 500 bars: Experimental and theoretical constraints and implications for the Guaymas Basin hydrothermal system. *Geochim. Cosmochim. Acta* **51**, 1997-2010.

Tufar, W., Tufar, E. and Lange, J. (1986) Ore paragenesis of recent hydrothermal deposits at the Cocos-Nazca plate boundary (Galapagos Rift) at 85° 51' and 85° 55'W: complex massive sulfide mineralizations, non-sulfidic mineralizations and mineralized basalts. *Geol. Rundschau* **75**, 829-861.

Tunnicliffe, V., Botros, M., de Burgh, M.E., Dinet, A., Johnson, H.P., Juniper, S.K. and McDuff, R.E. (1986) Hydrothermal vents of Explorer Ridge, northeast Pacific. *Deep Sea Res.* **33**, 401-412.

US Geological Survey Study Group (1986) Submarine fissure eruptions and hydrothermal vents on the southern Juan de Fuca Ridge: Preliminary observations from the submersible Alvin. *Geology* **14**, 823-827.

Von Damm, K.L. (1983) Chemistry of submarine hydrothermal solutions at 21°N, East Pacific Rise and Guaymas basin, Gulf of California. Unpub. Ph.D. thesis, Massachusetts Inst. Technology/Woods Hole Oceanograph. Inst.

Von Damm, K.L. (1988) Systematics of and postulated controls on submarine hydrothermal solution chemistry. *J. Geophys. Res.* **93**, 4551-4561.

Von Damm, K.L. and Bischoff, J.L. (1987) Chemistry of hydrothermal solutions from the southern Juan de Fuca Ridge. *J. Geophys. Res.* **92**, 11334-11346.

Von Damm, K.L., Edmond, J.M., Grant, B. and Measures, C.I. (1985a) Chemistry of submarine hydrothermal solutions at 21°N, East Pacific Rise. *Geochim. Cosmochim. Acta* **49**, 2197-2220.

Von Damm, K.L., Edmond, J.M., Measures, C.I. and Grant, B. (1985b) Chemistry of submarine hydrothermal solutions at Guaymas Basin, Gulf of California. *Geochim. Cosmochim. Acta* **49**, 2221-2237.

Walther, P. and Stoffers, P. (1985) Chemical characteristics of metalliferous sediments from eight areas on the Galapagos Rift and East Pacific Rise between 2°N and 24°N. *Marine Geol.* **65**, 271-287.

Weissberg, B.G., Browne, P.R.L. and Seward, T.M. (1979) Ore metals in active geothermal systems. In *Geochemistry of Hydrothermal Ore Deposits* (2nd edn.), ed. Barnes, H.L. Wiley, New York, 738-780.

Welhan, J.A. and Craig, H. (1983) Methane, hydrogen, and helium in hydrothermal fluids at 21°N on the East Pacific Rise. In *Hydrothermal Processes at Seafloor Spreading Centers, Proc. NATO Advanced Research Institute*, Series IV, 12, eds. Rona, P.A., Bostrom, K., Lanbier, L. and Smith, K.L.Jr. Plenum Press, New York, 391-409.

Zierenberg, R.A. and Shanks, W.C. III (1983) Mineralogy and geochemistry of epigenetic features in metalliferous sediment, Atlantis II Deep, Red Sea. *Econ. Geol.* **78**, 57-72.

Zierenberg, R.A. and Shanks, W.C. III (1986) Isotopic constraints on the origin of the Atlantis II, Suakin, and Valdivia brines, Red Sea. *Geochim. Cosmochim. Acta* **50**, 2205-2214.

Zierenberg, R.A. and Shanks, W.C. III (1988) Isotopic studies of epigenetic features in metalliferous sediment, Atlantis II Deep, Red Sea. *Can. Mineral.* **26**, 737-753.

Zierenberg, R.A., Shanks, W.C. and Bischoff, J.L. (1984) Massive sulfide deposits at 21°N, East Pacific Rise: Chemical composition, stable isotopes, and phase equilibria. *Geol. Soc. of Amer., Bull.* 95, 922-929.

9 Ancient placer gold deposits

W.E.L. MINTER

9.1 Introduction

In this chapter, ancient placers refer to deposits of Archaean and early Proterozoic age. Because each deposit may cover hundreds of square kilometres, they present economically attractive exploration targets. Individual ancient placer deposits like the Steyn/Basal placer in the Welkom goldfield of South Africa have produced in excess of 4500 t (metric tons) of gold from heavy mineral concentrates distributed over a palaeosurface area of 170 km^2 (Minter *et al.*, 1986). This amounts to about 26 t of gold/km^2. Another important deposit, the Vaal placer, which lies in the adjacent Klerksdorp goldfield, extracts about 32 t of gold/km^2. Production figures from the well-known Witwatersrand Supergroup placers illustrate their productivity relative to other types of gold deposits (Figure 9.1).

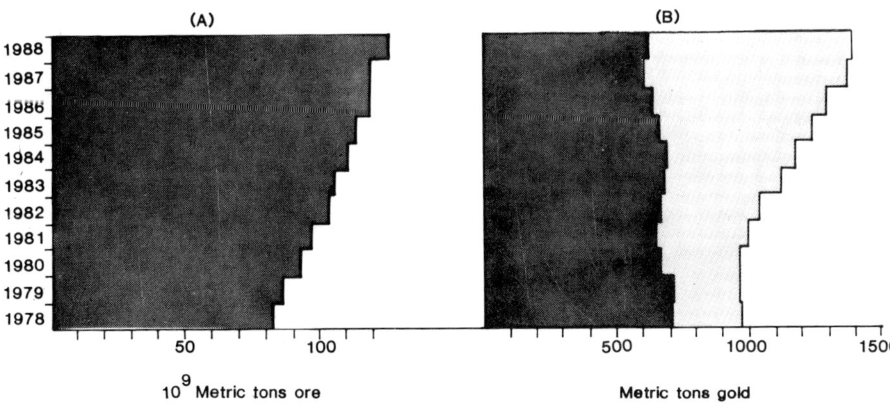

Figure 9.1 Histogram illustrating: (A) the annual amount of ancient placer ore mined from the Witwatersrand Supergroup during the past ten years in metric tons; (B) the number of metric tons of gold produced annually from this ore compared with total gold production in the western world.

Although the ancient Witwatersrand Supergroup placers are the best-known examples in this class of gold deposit, they are not unique. Elsewhere on the Kaapvaal Craton, the Pongola, Dominion, Ventersdorp and Transvaal Supergroups also contain placer deposits that have been identified and mined. They range in age from 3.0 to 2.3 Ga. Other areas in Gondwana that had cratonized prior to 2.0 Ga also contain ancient placers. For instance, the shield that extended from Central Africa across to Brazil, in the 150 Ma reconstruction of Gondwana (Figure 9.2), encompassed the São Francisco

Craton on which the Jacobina and Moeda placers, dated at about 2.6 Ga, are located. These, and the Tarkwa placers on the West African Craton in Ghana, dated at 2.0 Ga, have all had a long history of exploration.

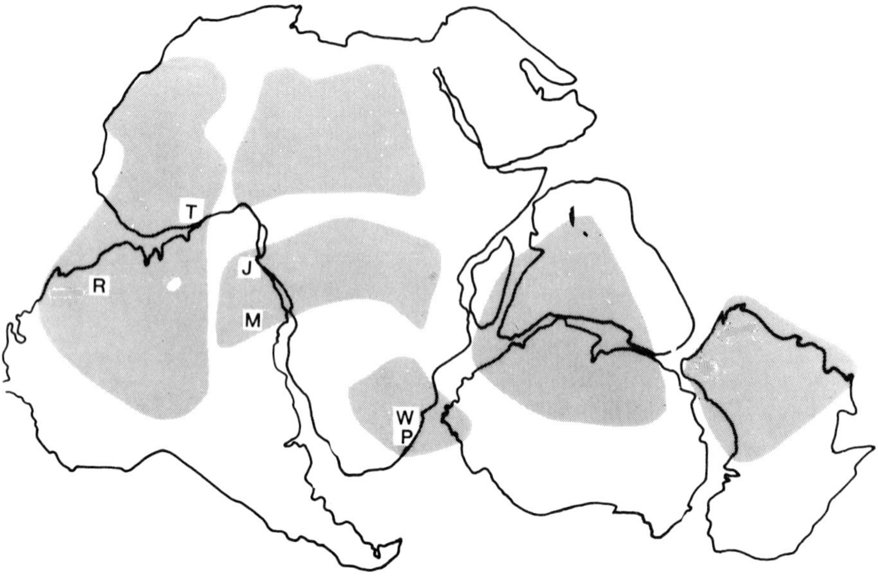

Figure 9.2 Distribution of some of Gondwana's crustal domains which had cratonized by about 2.0 Ga. The boundaries of these cratonic regions are schematic, in that they may have been subsequently tectonothermally overprinted. Other, smaller cratonic areas of this age are omitted. The distribution of these cratons in this Gondwana framework (at about 150 Ma) does not necessarily imply that this was their relative geographic distribution at 2.0 Ga. Placer localities: T – Tarkwa, R – Roraima, J – Jacobina, M – Moeda, W – Witwatersrand, P – Pongola. (After de Wit *et al.*, 1988.)

9.2 Geological setting

9.2.1 *Pongola*

The Pongola Supergroup, which outcrops south of Mbabane in Swaziland (Figure 9.3(a)) and to the north and south of Vryheid in Natal (South Africa), developed on a stabilized basement between 3.1 and 2.9 Ga. Evidence of palaeosols preserved on granitic bedrock (Button and Tyler, 1981) and a thick basal unit of fluvial sandstones and conglomerates indicate a continental environment. Altered Nsuze volcanics of uncertain magma type, but with tholeiitic affinities, dominate the lower part of the 5000-m sequence. They have a bimodal silica content, contain pillow lavas, and are associated with some of the oldest known shelf stromatolites. They are believed to have erupted in a volcanic arc setting (Tankard, 1982).

Pongola Supergroup placer deposits occur on an unconformity at the base of the Mozaan Group where they are associated with oligomict conglomerates in a fluvial sequence that is up to 90 m thick. Overlying this are intertidal and subtidal sandstones

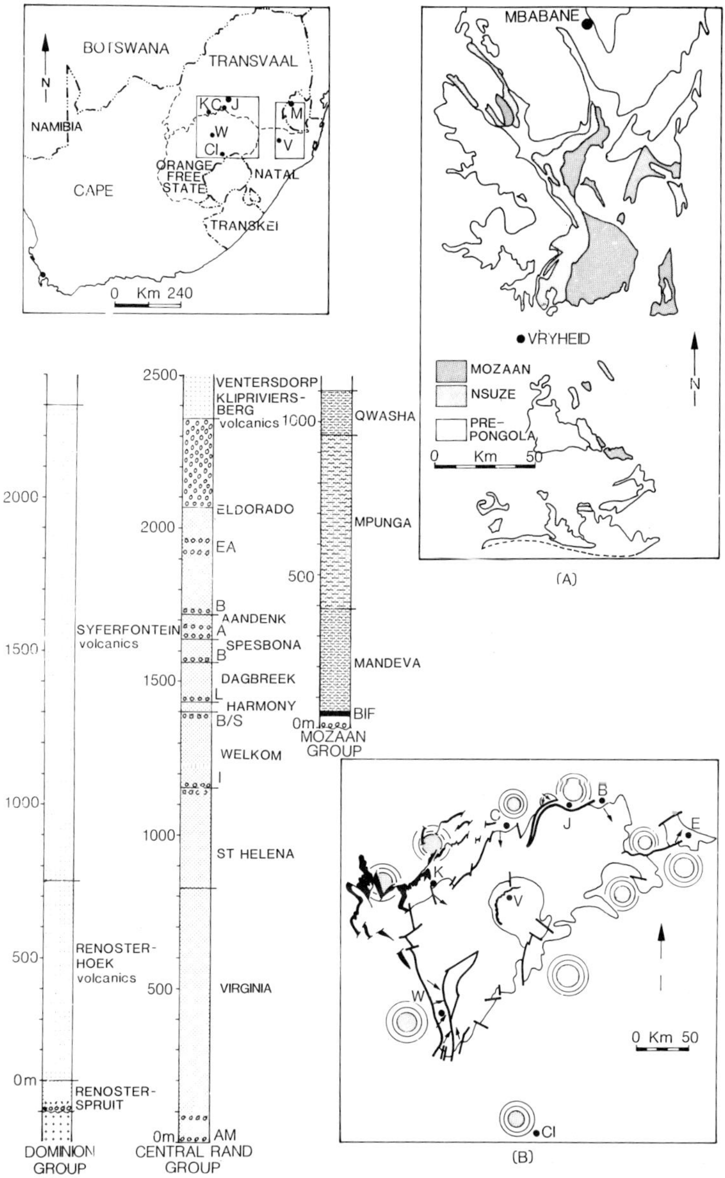

Figure 9.3 Distribution of the Pongola, Dominion, and Witwatersrand deposits on the Kaapvaal Craton and their stratigraphy. In (A), the Mozaan of the upper Pongola is distinguished from the lower Nsuzi and basement rocks. In (B), the Dominion Group is shown in black around the western margin of the Central Rand Group. (modified after Pretorius, 1986). Goldfields and circled granitic basement domes are shaded; cover rocks removed. Towns indicated by initial letters are: K – Klerksdorp, C – Carletonville, J – Johannesburg, B – Benoni, E – Evander, V – Vredefort, W – Welkom.

and thin iron-formation sediments. This sequence has been interpreted as having been deposited on braid plains flanked and transgressed by shallow marine conditions (Watchorn, 1980). Burke *et al.* (1985) have proposed that the Pongola represents the world's oldest intracratonic rift. Volcanism and sedimentation had terminated by about 2.8 Ga when the Usushwana Intrusive Suite was emplaced and was followed by granitoid magmatism between 2.7 and 2.5 Ga.

9.2.2 *Dominion*

The Dominion Group and Witwatersrand Supergroup deposits are superimposed over the centre of the Kaapvaal Craton. The Dominion Group outcrops around the northern and south-western flanks of the Witwatersrand Supergroup Basin and is exposed in the upturned western rim of the Vredefort Dome (Figure 9.3). The Dominion Group placer lies at the base of the Renosterspruit Quartzite Formation, 60 m thick, which rests non-conformably on Archaean granite that has been dated at 3.1 Ga (Walraven *et al.*, in press). This initial sedimentation took place under fluvial conditions, and palaeocurrent measurements indicate a south-westward drainage pattern. Subsequent instability and subsidence is reflected by 2650 m of basaltic andesite, tuffs, acid lavas, and quartz–feldspar porphyry lava in what is believed to be a proto-basinal phase of the greater Witwatersrand Supergroup Basin (Tankard *et al.*, 1982). Recent dating of zircons from the Dominion Group lavas gives a minimum age of 3060 Ma (Armstrong *et al.*, 1986).

9.2.3 *Witwatersrand*

The West Rand Group, comprising the lower part of the Witwatersrand Supergroup, was deposited in marine shelf and tidal environments with minor periods of fluvial accumulation, and conformably overlaps the Dominion Group. The thickness of the sequence averages 4650 m and its distribution may have extended over an area of 100 000 km^2 (Tankard *et al.*, 1982). Extensive macrotidal, ebb-dominated deposits indicate a wide continental shelf, open to the south-west. Small, braid-delta placer deposits occur west of the Klerksdorp goldfield in the Jeppestown Subgroup, near the top of the West Rand Group succession. The West Rand Group is folded along a north-east axis and encloses the Central Rand Group within which practically all the Witwatersrand Supergroup placers are confined.

This upper part of the Witwatersrand Supergroup is more arenaceous and is believed to have been deposited by numerous peripheral, fluvially dominated, braid deltas, during closure of the Witwatersrand Supergroup Basin. These sedimentary rocks are structurally preserved in downwarped or downfaulted parts of an extensive intracratonic basin, 9750 km^2 in area, that extends from Johannesburg for 290 km to the south-west (Figure 9.3). Only about 10% of the periphery of the Central Rand Group outcrops, the rest being concealed by younger Proterozoic and Phanerozoic cover rocks. Recent dating of zircons by ion-microprobe analysis (Armstrong *et al.*, 1986) has bracketed the age of the placers at between 3060 Ma for underlying Dominion lavas and 2708 Ma for overlying Ventersdorp lavas. It is clear then that these continental sediments were deposited during the Archaean.

9.2.4 *Ventersdorp*

Venterspost placers at the base of the Ventersdorp Supergroup are more restricted than the Central Rand Group placers in their distribution. They represent yoked deposits in so far as they occur along the north-western margin of the intracratonic Central Rand Group Basin, between Klerksdorp and north of Carletonville (Figure 9.3(b)). They are structurally controlled both by post-Witwatersrand Supergroup block faults that produced a number of half-grabens, and by unconformities that strike obliquely to those in the Central Rand Group. The placers appear to have incorporated new detritus with reworked Witwatersrand Supergroup placers.

9.2.5 *Transvaal*

Where Witwatersrand Supergroup and Ventersdorp Supergroup placers subcrop beneath the Transvaal Supergroup on the Kaapvaal Craton, the older lithified placers have been eroded by fluvial processes and the detritus has been deposited as immature sediment within entrenched channelways of the Black Reef Formation.

9.2.6 *Jacobina*

The São Francisco Craton in Brazil comprises the western end of an early Proterozoic shield that extended from Central Africa to Brazil until 150 Ma ago (Figure 9.2). Basement rocks are dated at 2.8 Ga and older.

Jacobina is located in Bahia at the northern end of the craton (Figure 9.4(a)) where an arenite sequence, 780 m thick, rests non-conformably on a granite–gneiss basement. The deposit, the Serra da Corrego, is a tabular structural block that strikes north for 60 km and dips steeply east at 70°. A major fault against overthrust metamorphosed basement limits its downdip continuation. The Serra da Corrego is entirely fluvial, was derived from granite–greenstone terrain in the east, and fines upward into finer grained intertidal and marine shelf deposits of the Rio do Ouro. The setting appears to represent a continental margin with braid deltas coalescing along an extended shoreline.

9.2.7 *Moeda*

At the southern end of the São Francisco Craton, immediately south of Belo Horizonte in Minas Gerais (Figure 9.4(b)), the Moeda Formation is preserved in a number of separated synclinal keels that have been folded to produce the Archaean Rio das Velhas granite–greenstone terrane. In the Gandarela Syncline, the Moeda is 120 m thick and comprises three units of sandstone with a placer deposit at the base, resting unconformably on deformed, lithified basement. These are the first cratonic sediments and fine upwards into marine shelf Batatal shale, Caue carbonates, and ferruginous shales, indicating deposition on a cratonic continental shelf. The source of the sandstones was from the north-west quadrant into what appears to have been individual basins, separated by resistant basement ridges of chert and iron formation (Minter *et al.*, 1989).

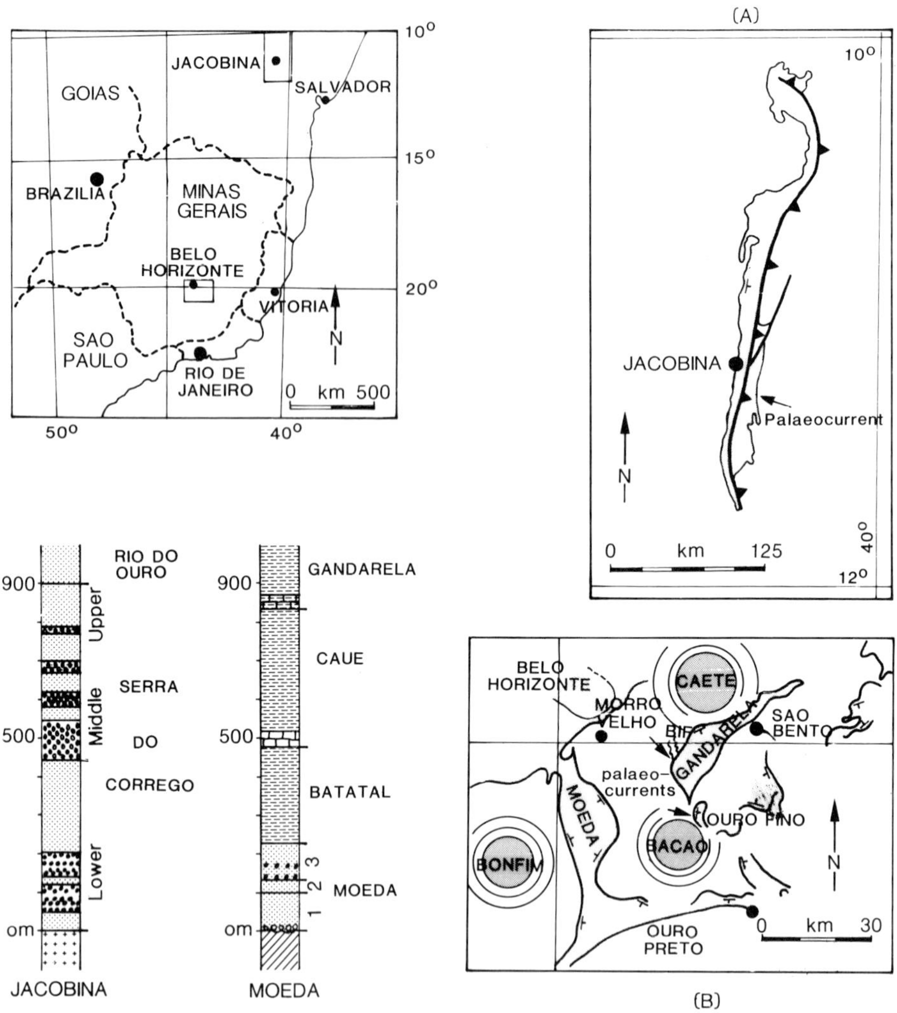

Figure 9.4 Distribution of the Jacobina and Moeda deposits on the São Francisco Craton and their stratigraphy. In (A) the Jacobina deposits shaded are in non-conformable contact with metamorphosed gneissic basement in the west whereas basement is thrust over the Jacobina deposits along most of its eastern strike length. Palaeocurrents from the east. In (B), the Moeda remnants are shaded. The overlying Batatal and iron formations have been omitted. Circled granitic domes and gold deposits in the basement at São Bento and Morro Velho are shown. Palaeocurrents from north-west arrowed.

9.2.8 *Tarkwa*

The Tarkwaian placers are located in Ghana near the edge of the large cratonized area that linked West Africa to Venezuela (Figure 9.1) in the 150 Ma reconstruction of Gondwana (de Wit *et al.*, 1988). Five belts of Birimian basement, folded within 2.1 Ga granitoids, form the easternmost domain of the Man Shield (Hirdes and Leube, 1986).

The belts comprise tholeiitic volcanics with bimodal silica contents and volcaniclastic, turbidite and chemical exhalative sediments. Each belt is believed to represent a separate intracratonic rift within which the Birimian sediments accumulated. The placer-bearing Tarkwaian succession, derived from Birimian and granitoid basement rocks, is believed to rest unconformably on the Birimian and to have been deposited as alluvial fans (Sestini, 1973) in an intermontane piedmont setting, within the separate rift grabens. The best development of Tarkwaian sediments occurs in a 250 km belt striking north-east between Tarkwa and Agogo (Figure 9.5). These mid-Proterozoic gold placers contain neither pyrite nor uraninite in the detrital mineral assemblage. They are characterized by containing detrital hematite.

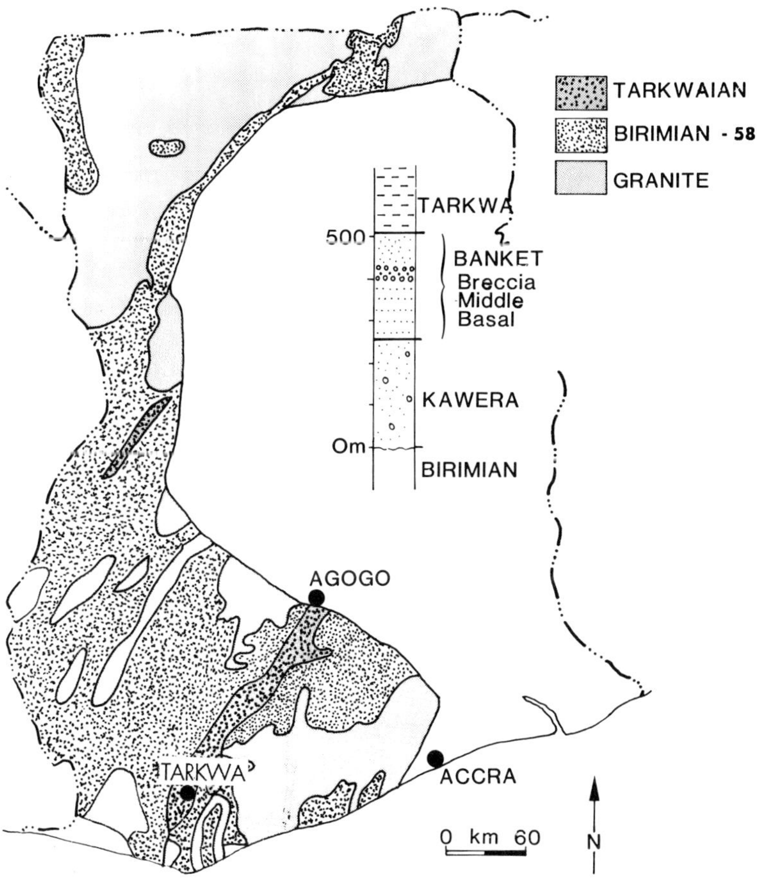

Figure 9.5 Distribution and stratigraphy of the Tarkwaian succession, Ghana, on the eastern margin of the West African Craton. The older Birimian and granitic basement are also distinguished.

9.3 The palaeosurfaces

9.3.1 *Areal dimensions*

The lateral extent of individual ancient placer deposits can be reasonably well estimated only where intensive mining has taken place. In less well-explored deposits, estimates are more speculative. The Witwatersrand Central Rand Group Basin, therefore, provides the best examples. The first point to note is that the scale of the placer palaeosurfaces is localized. For instance, the only two regional palaeosurfaces that can be identified with any degree of certainty across the Central Rand Group Basin (Figure 9.6) are stratigraphically located near the lithostratigraphic bases of the Turffontein and Johannesburg Subgroups.

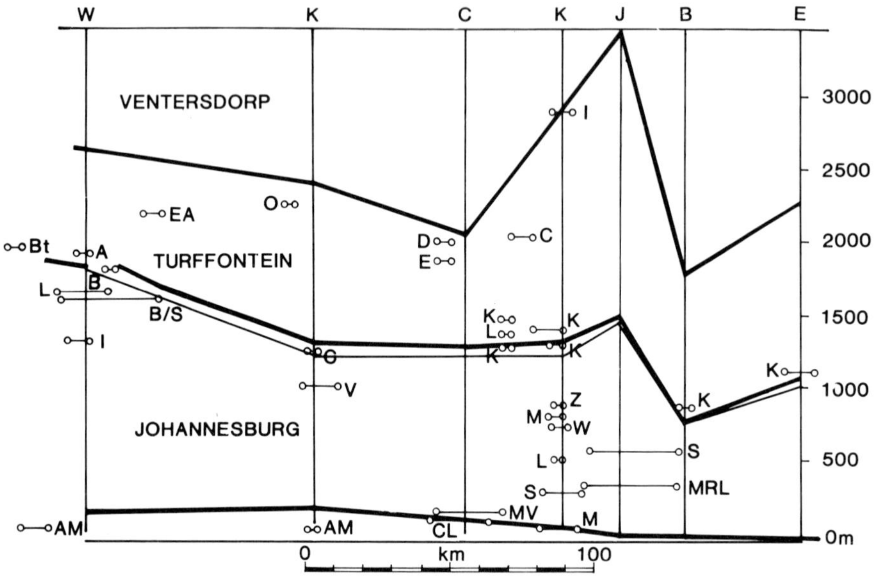

Figure 9.6 Stratigraphic profile around the western periphery of the Central Rand Group basin distinguishing extent of regional palaeosurfaces by thick lines and local placer palaeosurfaces by thin lines. Goldfield positions located in letters at top, viz. **W** – Welkom, **K** – Klerksdorp, **C** – Carletonville, **K** – Krugersdorp, **J** – Johannesburg, **B** – Benoni, **E** – Evander. Placers identified by letters within diagram in order of goldfields: EA – multiple Eldorado placers, Bt – Beatrix, A - A, B - B, L – Leader, B/S – Basal and Steyn, I – Intermediate, AM – Ada May, O – Orkney, C – Cristal Kop, V – Vaal, D – Deelkraal, E – Elsburg, MV – Middelvlei, CL – Carbon Leader, I – Individual, C - Composite, K - Kimberley, Z - Zone Two, M - Monarch, W – White, L – Livingstone, S – South, M – Main, MRL – Main Reef Leader.

These preserved surface areas may each encompass 9750 km^2 and are characterized by being associated with debris-flow diamictites. Their development may have been the result of particular regional tectonic and climatic conditions, but they are not associated with extensive placers.

The palaeosurfaces with which the important placers are associated are more numerous, but of local extent. They are the erosion surfaces that separate unconformity-bounded formations in the Witwatersrand stratigraphy, and their limits are defined by the lateral continuity of the placers (Figure 9.6). Using this criterion, they have been identified over areas of 150–400 km^2.

In many instances the placers are composed of two laterally coalescing deposits. For instance, the Vaal placer (Minter, 1976; Antrobus *et al.*, 1986), the Carbon Leader placer (Buck and Minter, 1985), the Basal/Steyn placers (Minter, 1980), and the Middelvlei placer (Els, 1987) all owe their palaeostrike extent of 20–30 km to this factor. The longest palaeodip distance explored is 25 km on the Steyn placer (Minter *et al.*, 1986). The Tarkwa placers also have palaeostrike continuities of 25 and 45 km (Sestini, 1973) but, like the Central Rand Group placers, their palaeosurfaces were developed on unlithified sandy strata by unconfined drainage systems.

Placer palaeosurfaces cut on lithified or crystalline bedrock, like the Moeda, Jacobina, Dominion, and Ventерspost examples, have been confined and controlled by the nature of the footwall. For instance, the lower Dominion and Jacobina placer palaeosurfaces are controlled by gently rolling granitic topography, and chert and iron-formation ridges appear to constrict the Moeda palaeosurface in the Gandarela Syncline (Minter *et al.*, 1990). The Venterspost palaeosurface is confined to a width of 5 km by major block faults in the Venterspost graben, while, in the East Driefontein graben, the surface is 20 km wide and has been mined down the palaeodip for 30 km. More deeply channelled placers on lithified bedrock, like the Black Reef placers, are laterally confined to less than 1 km, but may be followed for tens of kilometres down palaeoflow (Papenfus, 1957).

9.3.2 *Topographic relief*

The two regional-scale palaeosurfaces in the Central Rand Group are characterized by deeply incised channels more than 100 m deep and in places more than 1 km wide. The channel courses across gold mines to the east of Johannesburg are known to be sinuous and have been traced for tens of kilometres (Papenfus, 1957), whereas in the Carletonville, Klerksdorp and Welkom goldfields they appear to be linear. Load-casted pod-like remnants and slumped sheets of reworked placer deposits, eroded and recycled from underlying placers, occur associated with the diamictites in these erosion channels, and have in some instances been exploited successfully. (A new mine in the Welkom goldfield, named the Oryx, lies in such a channel and is set to exploit a placer estimated to contain 400 t of gold.)

The topographic relief of the erosion surfaces bounding the formations on which placers generally occur is much more subdued. For instance, although channelling of up to 6 m depth has been recorded in proximal reaches of the Steyn placer in the Welkom goldfield (Minter, 1985), the average relief over an area of 200 km^2 is 0.5 m. In the Klerksdorp goldfield, the Vaal placer covers an area of 260 km^2, and along a palaeostrike distance of 14 km the deepest channel encountered was 1 m deep and 1000 m wide (Minter, 1976). The Carbon Leader palaeorelief is similar to that of the Vaal placer over an area of 160 km^2 (Buck and Minter, 1985). In all of these more recently mined placers, systematic sedimentological studies have revealed that the drainage etches on the unconformities have a very low sinuosity. Such drainageways would have had the form of braid belts comprising shallow sinuous streams that transported and deposited the placer sediments. These aspects of areal extent, surface relief, and drainage patterns resemble those of braid plains, in which unconfined flow on sandy substrate promoted the lateral shift of channels.

In rare instances where placer channels have eroded into mudstones and siltstones, the cohesiveness of the substrate has resulted in discrete, steeply sided channelways with more sinuous distribution patterns, as in the B placer in the Welkom goldfield (Minter, 1978).

The topographic relief on crystalline basement is much greater and had the effect, in both the Jacobina and Dominion placers, of limiting their lateral continuity. Elevation differences of 20 m between valley and hill have resulted in the upper Dominion placer resting on bedrock in places and thereby being misidentified as the lower placer.

The Venterspost placer, in the East Driefontein graben, rests on a composite geomorphic palaeosurface that has planed down through 2500 m of underlying Witwatersrand Supergroup strata over an area of 300 km^2. The average angle of unconformity is 3–5°, but on a local scale it is evident that the palaeosurface is terraced. Risers documented by McWha (1988) indicate elevation differences of up to 4 m between erosional terraces cut in bedrock (Figure 9.7). They appear to be unpaired and are the result of fluvial downcutting with lateral shift of the channels, probably controlled by continuous tectonic rejuvenation. The terraces are from 20 to 100 m wide and, collectively, may account for hundreds of metres of topographic elevation. This factor, as with the Dominion and Jacobina placers, has a bearing on placer correlation and valuation.

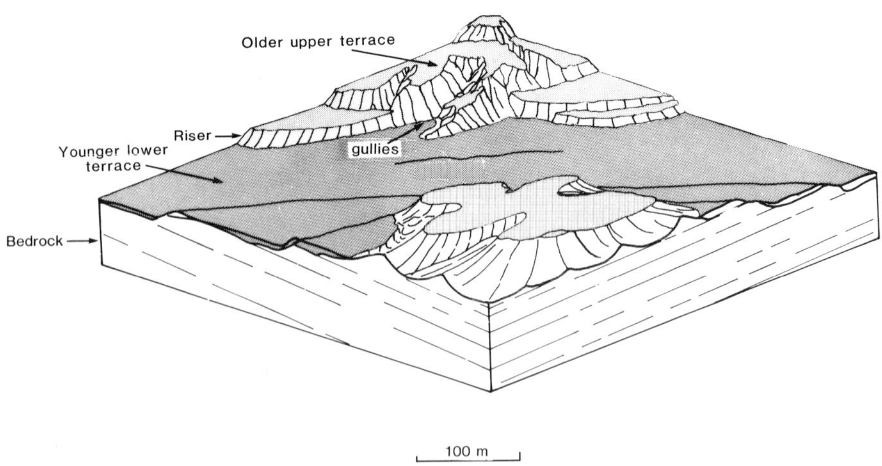

Figure 9.7 Sketch illustrating the terraced nature of the palaeosurface beneath the Venterspost placer at Western Deep Levels gold mine near Carltonville. Multiple unpaired terraces cut in bedrock are separated by risers sloping at 30° and have been gullied. (After McWha, 1988.)

9.3.3 *Stratigraphic position*

In the Dominion Group and Moeda Formation placers (Figures 9.3 and 9.4) the important palaeosurfaces represent basal contacts between cratonized basement rocks and the first continental sediments, which in both instances are only about 100 m thick. The Ventersposť palaeosurface is more complex, however, because each terrace could be regarded as a stratigraphic entity if sufficiently extensive. In this case, the placers are viewed as residual sediments on surfaces of degradation and as having been in a transport system when buried by lava flows. Evidence of this is demonstrated by soft sediment deformation of placer sediment into load-casted and pillowed lavas (Tankard *et al.*, 1982). A preserved sedimentary depository, accommodating the eroded Witwatersrand Supergroup strata, has not yet been identified. The upper reaches of the Moeda placer in the Gandarela Syncline is of a similar nature in so far as residual, fluvial, placer sediments are preserved beneath a transgressive marine sandstone. Towards the south-eastern margin of the syncline, however, this stratigraphic package of residual fluvial sediments quite abruptly wedges open into a fan 100 m thick, representing fan deposition at a site of accumulation. The Jacobina and Tarkwa placer palaeosurfaces have more in common with Witwatersrand Supergroup examples because they occur within thick sedimentary sequences.

In the Witwatersrand Supergroup deposits, the placer palaeosurfaces are located in discrete areas peripheral to the Central Rand Group Basin margin (Figures 9.3 and 9.6). They mark the unconformable boundaries between synthems composed essentially of sandstone, and are repeated stratigraphically, the placer sediments apparently being supplied repeatedly from geographically similar entry points. In the Welkom goldfield, it is clear from isopach analysis that structural deformation occurred synchronously with sedimentation (Figure 9.8) and that the stratigraphic stacking of placers represents intermittent tectonic rejuvenation of the basin margin (Minter *et al.*, 1986). The tectonic activity at each entry point, however, when measured by the number of placer events recorded in the sequence, is different in each goldfield, the number ranging from 1 to 8 or more. In detail, the position of major discharge points and the sites of better mineralization are related to the particular geometry and distribution of each placer deposit.

Insufficient information is available to comment further on the middle and upper placers at Jacobina, and the middle and Breccia placers at Tarkwa, other than to say that they too reflect intermittent tectonic rejuvenation of their basin margins.

9.3.4 *Correlation*

The number of formations at each entry point of the basin in the dominantly fluvial Central Rand Group are different because they are evidently related to different degrees of local tectonic activity. The sandstone packages comprising these formations have subtly different lithological characteristics, different geochemical signatures and different palaeocurrent distribution configurations. This all reflects the separateness of their local provenance areas. Therefore, it is unreasonable to expect to be able to correlate individual placer palaeosurfaces beyond the depositional environment of each placer.

9.4 The placer sediments

In the Witwatersrand Supergroup, the Central Rand Group sediments are up to 2880 m thick and mostly comprise coarse-grained subwackes. Interbedded with these are units

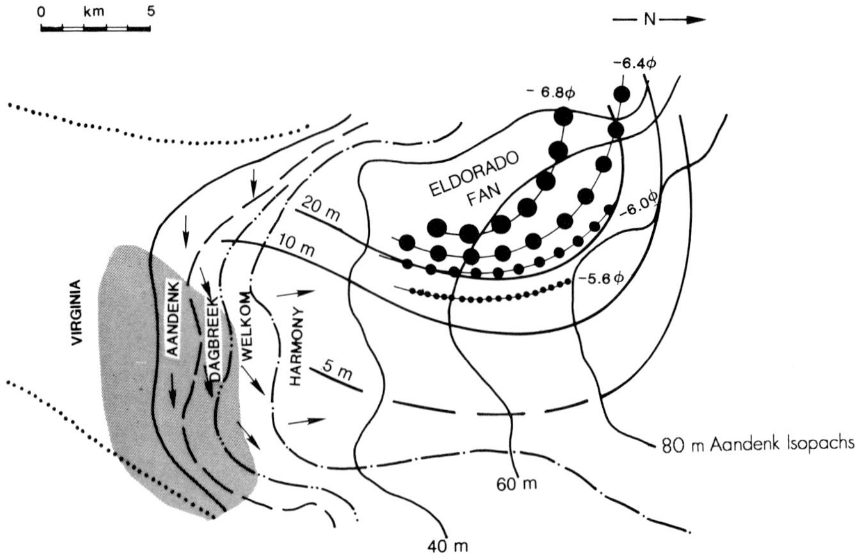

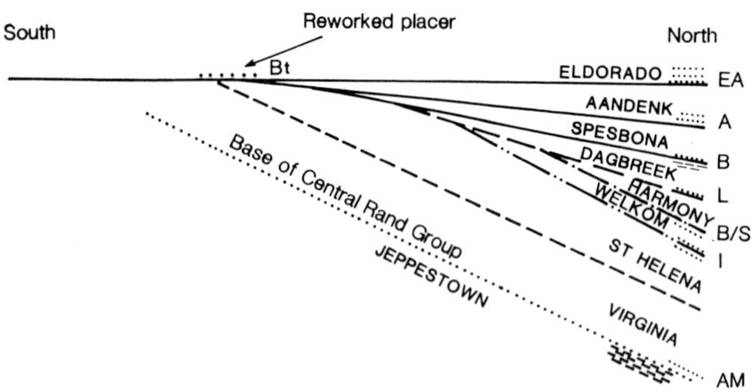

Figure 9.8 Plan and section illustrating subcrop limits of separate onlapping formations in the stratigraphic sequence of the Central Rand Group in the Welkom goldfield demonstrate local syndepositional tectonism. The placers are associated with unconformity boundaries (see Figure 9.6). A reworked placer (shaded) at the subcrop of the Aandenk Formation is preserved beneath the Eldorado fan. Palaeocurrents arrowed. Eldorado clast-size isopleths (circles proportional to sizes) and cumulative conglomerate-thickness isopachs shown overlapping the Aandenk Formation.

of conglomerate and quartz arenite that collectively total less than 10% of the entire sequence. It is with these minor fractions of the succession that the placers are associated. This lithological distinction applies to all the ancient placer deposits and denotes a degree of greater maturity than in the bounding lithologies, which is attributed to the nature of the hydraulic flow regime that produced them. It has been suggested that perennial flow produced the more mature sediments whereas more ephemeral flow conditions could account for the deposition of the subwackes (Smith and Minter, 1980).

9.4.1 *Gravel and sand fraction*

The gravel facies in ancient gold placers comprise beds of small-pebble to cobble-size clasts that, on average, are well rounded and spheroidally shaped with a small proportion displaying ventifacts (Minter, 1976). These features are believed to indicate abrasion by dominantly fluvial processes with an element of aeolian modification. The sorting, indicated by the standard deviation of apparent long-axis measurements and calculated by moments, ranges from well sorted (0.4 phi) in the small-pebble Vaal placer, which has a mean pebble size of -3.2 phi (Minter, 1976), to moderately sorted (0.7 phi) in coarse Ventersdpost placer that has a mean pebble size of up to -5.6 phi. This upper value translates to an equivalent weighed mean of -6.1 phi (70 mm). This is comparable to coarse gravels in the Tarkwa and Pongola placers, but boulder beds have been recorded from the northern end of the East Driefontein graben (Krapez, 1985) and also from the most proximal preserved limits of Moeda placers in the Ouro Fino and Gandarela synclines (Minter *et al.*, 1989).

The clasts are composed of an oligomict assemblage of vein quartz, chert, and quartz arenite, with minor silicified shale and quartz porphyry components. This is a very mature assemblage and is identified with intensive erosion and considerable transport from a distant source. Coarse facies of the Witwatersrand Steyn placer and of the Moeda placer contain up to 30% quartz–porphyry components. The chert component, in particular, generally comprises 10% of the suite, but in some cases, as in the Witwatersrand Evander goldfield (Tweedie, 1986), is up to 50% of the total pebble population.

The placer sand-size facies are pale grey coloured, coarse-grained quartz arenites. The cement is mostly silica with a very small proportion (10%) of phyllosilicates in the matrix. In the Witwatersrand Vaal placer, the micaceous minerals are 95% mica and 5% chlorite. The sand facies occur interstitially in clast supported conglomerates, in matrix-supported conglomerates, and also as beds interlayered with conglomerates. These sandstones are, on the basis of the low phyllosilicate content, more mature than non-placer sandstones either above or below the placers. Debris-flow placers with a dark muddy matrix do exist in the record but are rare. They occur, for instance, as slope deposits on risers between terraces of the Ventersdpost placer, in the Beatrix placer (Minter *et al.*, 1988) where debris-flow facies of a fan have entrained placer concentrates weathered out from older placers in the substrata, and as proto-ore beneath the Moeda on the Archaean palaeosurface (Minter *et al.*, 1990).

9.4.2 *Geometry*

The external geometry of the placers is generally tabular and sheet-like with a slightly irregular channelled base and a flat top. The very coarse Ventersdpost placer described by Krapez (1985) is associated with more rugged topography than is encountered in the Witwatersrand Supergroup, where proximal Steyn placer basal channels comprise an interconnecting system of broad, low-sinuosity, ribbon bodies up to 500 m wide and 4.5 km long (Minter, 1985). This pattern of drainage on the same scale is also evident in more distal deposits like the Carbon Leader placer (Buck and Minter, 1985) and the Vaal placer. In some instances, the placers are confined within channels and form shoestring orebodies.

Internally, the gravel facies occur as single pebble layers on scour surfaces or as sheets that have accumulated in increments of up to 20 cm at a time. Although placer bodies average 0.5 m in total thickness, there are cases where thicknesses exceed 6 m.

(a)

(b)

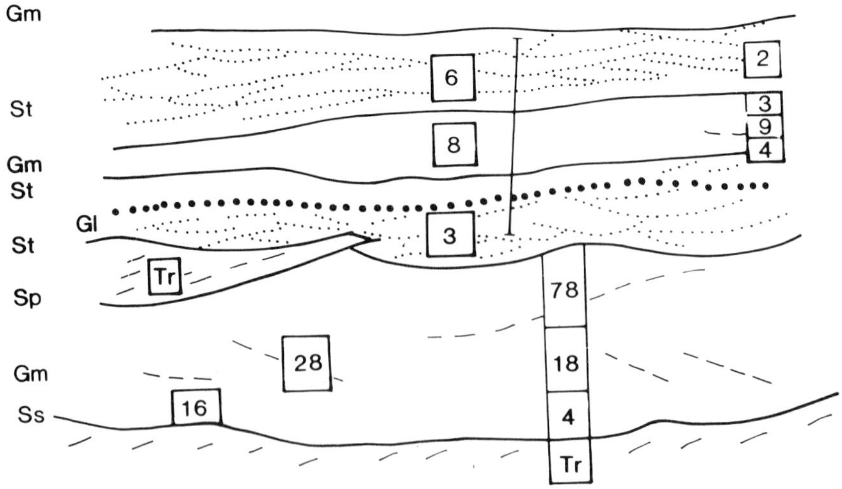

Figure 9.9 Photograph of Elsburg No. 5 placer (a) in the Klerksdorp goldfield with diagram (b) defining three main sedimentary facies found in ancient placers, namely massive gravel (Gm), planar cross-bedded sand (Sp), and trough cross-bedded sand (St) separated by scour surfaces (Ss). Relative gold contents of each facies are indicated in ppm. Scale 1 m.

The gravel facies also occur in planar and trough cross-bedded forms in a sandy matrix and the arenite is either horizontally or trough cross-bedded and rarely planar cross-bedded (Figure 9.9). Bedding planes in longitudinal exposures down the palaeodip are crudely horizontal but transverse sections indicate irregular lenticular individual beds describing bedforms preserved within channels (Smith and Minter, 1980).

9.4.3 *Deposition*

The sequence of sedimentary events recorded in the placers, from scour surface to gravel unit and to cross-bedded sand facies (Figure 9.9), indicates both rapid and extreme variations in flow velocity from erosional flood stage to accumulation during waning flow. Grain-size changes reflect lateral as well as vertical changes in the flow regime. The environment of deposition of the placer sediment appears always to have been of a fluvial nature in which unimodal transport by shallow transient streams prevailed (Smith and Minter, 1980). These processes are inferred to have operated during accumulation of all the placers discussed, whether the substrate was lithified or not.

9.4.4 *Heavy mineral fraction*

About 40 different detrital minerals have been identified in the Witwatersrand Supergroup placers (Feather and Koen, 1975). The most abundant detrital minerals are pyrite, chromite, leucoxene (after ilmenite), uraninite, and zircon. Although gold is a relatively minor component, it is the most important in an economic sense, and in this regard pyrite and uraninite are valuable by-products. In addition to these, platinum-group minerals in trace amounts are recovered on gravity tables from the Witwatersrand Supergroup placers. The Dominion Group placers contain garnet and monazite and their uranium content may exceed 2000 ppm, whereas the Witwatersrand Supergroup placers contain only 100–400 ppm uranium. In the Jacobina and Moeda placers, uranium is generally less abundant than in the Witwatersrand Supergroup placers. Traces of garnet have been reported from the Tarkwa deposits but no pyrite or uranium minerals occur, and chromite has not been reported. Monazite also occurs in Pongola placers as well as arsenopyrite, which has also been reported from the Moeda placer (Minter *et al.*, 1990). Rare diamonds have been recovered from placers at Tarkwa and in the Witwatersrand Supergroup. The Roraima Group in Venezuela, which may be equated with the Tarkwa deposits, are thought to be the source of gold- and diamond-bearing placers in recently derived alluvium (Priem *et al.*, 1973).

Hallbauer (1986) has estimated that up to 90% of Witwatersrand Supergroup gold is free and detrital and that from 5 to 40% has been redistributed during metamorphism. The average diameter of the gold particles is from 50 to 100 μm. Only 2% of the gold is enclosed within the allogenic pyrite. The gold in the Tarkwa deposits is reported to be between 10 and 15 μm in diameter and free milling. In the Moeda, most of the gold appears to be enclosed by authigenic pyrite; these particles have an average diameter of between 30 and 80 μm.

The heavy minerals are concentrated in the placers on (a) scour surfaces, particularly where trapped by pebble lags, (b) within clast-supported conglomerates

(Minter, 1978), and (c) on foreset and planar bedding surfaces in the sand facies. Pebbly bottom and top surfaces of gravel increments in the placer units are generally better mineralized, representing the lowest surface of degradation and the topmost winnowed surface.

The relative proportions of detrital placer minerals are generally related. The best documented ratios are those between gold and uranium (Figure 9.10), in which their log/log correlation coefficient is 0.81 (Smith and Minter, 1980). Their respective ratios depend on the sedimentary facies in which they occur and also on their geographical position in the depositional area. In a particular example from the Leader placer (Welkom), this relationship also applies to the gold and uranium in the carbonaceous material. Uranium contents generally increase down palaeoslopes, and in relatively distal deposits of the Steyn placer high concentrations of zircon (400 ppm), uraninite (1400 ppm) and chromite (3000 ppm) are common. The coefficients of correlation between uranium and gold (0.83), gold and zircon (0.44), and zircon and uranium (0.63) are all significant. Comparisons of the log/log coefficients of correlation between gold and uranium appear to be characteristic for each placer (Figure 9.11; Robb and Meyer, 1985). This is supporting evidence for a common placer origin and it seems likely that hydraulic sorting processes and specific original size distributions of the minerals supplied may account for the characteristics in each case.

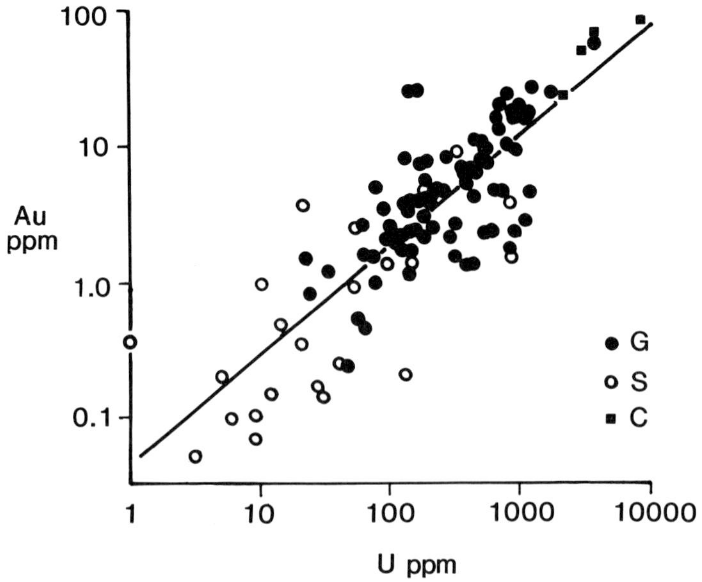

Figure 9.10 Gold versus uranium concentrations in the Leader placer, Welkom goldfield, illustrating correlation in sand (S), gravel (G), and carbon (C) facies.

On the small scale, concentrating mechanisms involve interactions between the fluid, the sediment bed, and the transported particles. The sorting mechanisms involve the free or hindered settling in usually turbulent water, the entrainment from a granular bed by flowing water, the transport of grains by flowing water, and the shearing of grains in a moving granular dispersion. The selective removal of more easily entrained grains has been emphasized by Slingerland and Smith (1986) as being particularly important. The activity of this small-scale process over intermediate-scale bars and channels and, in turn, over the scale of an entire deposit, effectively produces economically viable ore deposits (Kuhnle, 1986).

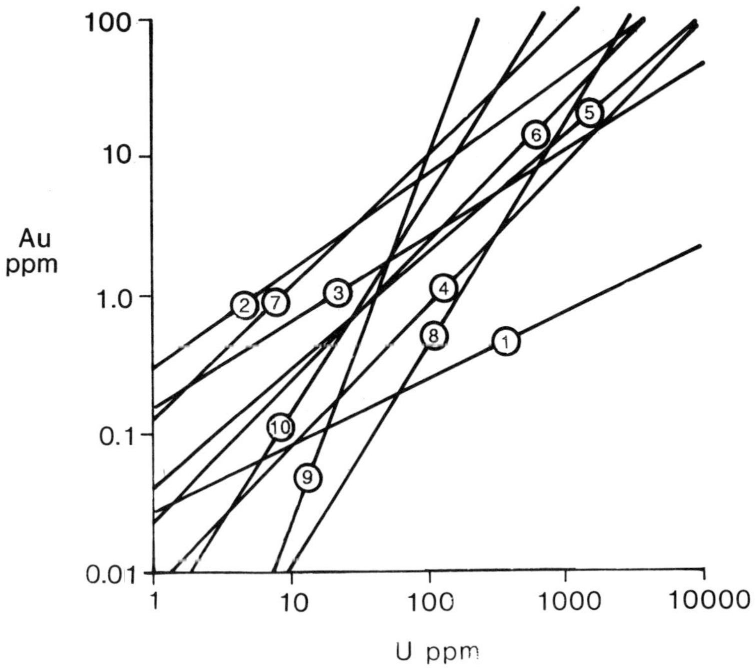

Figure 9.11 Gold versus uranium in ten different Witwatersrand placers: 1 – Promise, 2 – Basal, 3 - Dominion, 4 - White, 5 - Vaal, 6 - Leader, 7 - Venterspost, 8 - Carbon Leader, 9 - Elsburg, 10 - Kimberley. (After Robb and Meyer, 1985).

Size-sorting of the rounded pyrite grains across an entire deposit is well illustrated in the Steyn and Basal placers in the Welkom goldfield (Figure 9.12(a)). The isopleths, supported by palaeocurrent data, indicate that the pyrite is not an *in- situ* manifestation but a result of the sorting of supplied material. The concentration of gold on the same scale by sorting is illustrated in Figure 9.12(b). This structurally reconstructed simplified plan of the Steyn and Basal placer deposits clearly outlines the central mid-fan site of concentration. Detailed re-entrants in the proximal fringe and tongues in distal regions are clearly defined by the moving-average trend surface plan of gold content (Figure 9.12(c)). These areas represent broad channelways or braided belts on the palaeosurface (Minter, 1981).

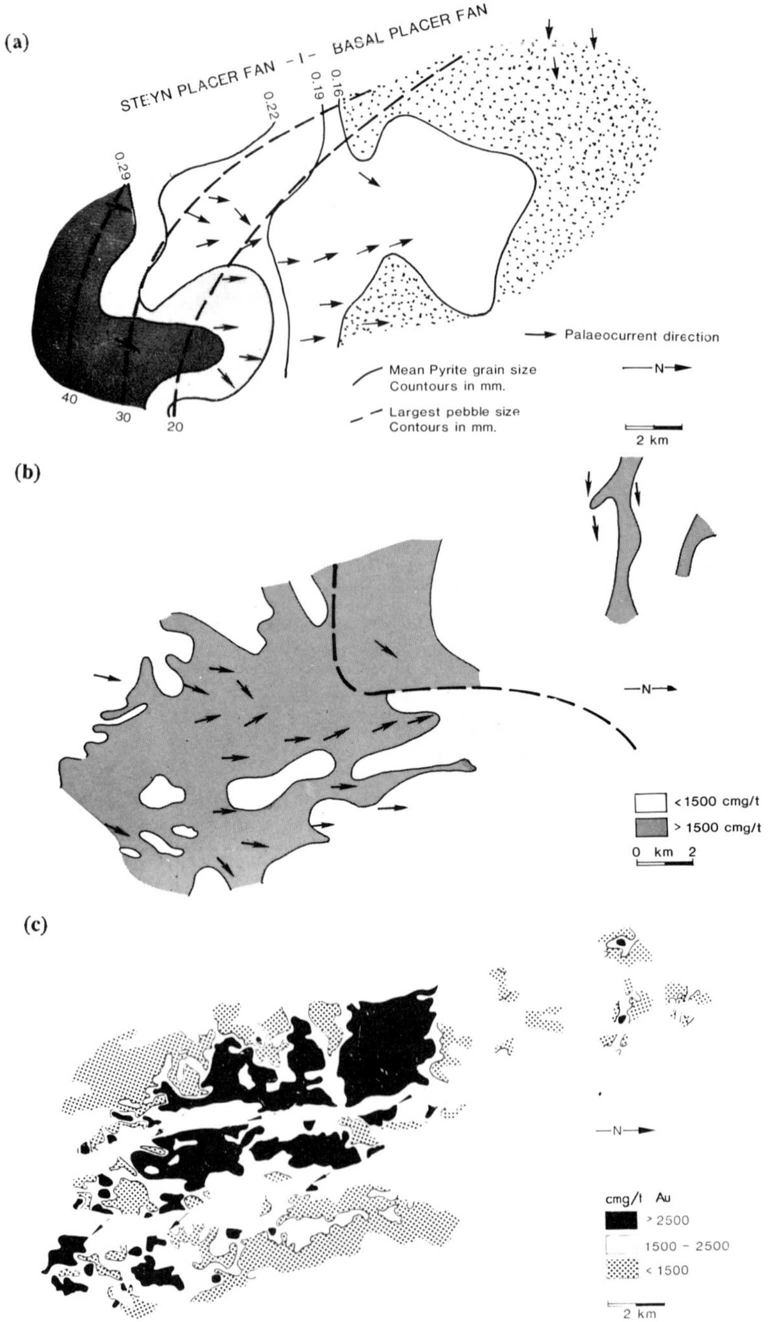

Figure 9.12 (a) Rounded-pyrite and pebble-size isopleths from Basal and Steyn placer samples in the Welkom goldfield which, in combination with palaeocurrent vectors, illustrate the dispersal of detrital components of the placers across the palaeosurface. (b) Simplified plan of gold distribution prior to faulting in the Basal and Steyn placers, Welkom goldfield, demonstrating that the main concentration is located in what is believed to have been a mid-fan area. (c) Moving-average trend surface analysis of gold contents in the Basal and Steyn placer based on over 350 000 samples grouped into 11 000 cells of 100 m. Linear pay shoots developed over considerable distances are displaced by younger major faults.

9.4.5 *Mineralogy*

Mineralogical research in recent years has focused predominantly on the composition of allogenic pyrite, chromite, uranium minerals, and gold. According to Hallbauer (1986) and Meyer *et al.* (1985), it seems that the Co/Ni ratios of detrital pyrites from the Barberton greenstone belt are an order of magnitude less than those from the Witwatersrand Supergroup placers. The Co/Ni ratios in the placer pyrites, however, are more comparable with those in pyrite from the Murchison and Pietersburg greenstone belts and also in pyrite from the basement granites.

The detrital chromites display a range in composition from low to high Cr content but the frequency distribution of these varieties in the Central Rand Group placers indicates a dominance of ultramafic source rocks (Eales and Reynolds, 1983).

The high abundance of leucoxene after ilmenite in the placers is generally overlooked, and certainly represents the 'missing' black sand component that is usually queried. The Ti-rich leucoxene played an important role in capturing remobilized uranium during the burial history of the placers but shows little sign of having been pyritized *in situ*. Although much of the pyrite is pseudomorphous after ilmenite, Ramdohr (1958) believed that it arrived in the basin in a sulphide form.

Uranium in the form of uraninite is best preserved in the kerogen. In the placer matrix, it is coated by gersdorffite or titania. These relic uraninites, authigenic uraniferous leucoxene, brannerite and coffinite all indicate that hexavalent uranium reacted with surface water or ground water (Smits, 1986). The repeated exposure of uraninite concentrates to weathering during reworking after each onlapping event in the Welkom goldfield (Minter *et al.*, 1988) demonstrates this process because the final placer of the sequence is barren of original uraninite (Figure 9.8).

The silver and mercury contents of gold particles in the placers have also aroused interest. Mercury content of between 1 and 5% in gold from the Witwatersrand Supergroup is well above the trace amounts detected in Barberton gold but resembles that from the Murchison Belt, thus supporting the conclusions concerning source-terrain made from the Co/Ni ratios in pyrite. However, Oberthur and Saager (1986) contended that the mercury content was derived from the host rocks during metamorphism and, therefore, that it does not have a bearing on provenance of the gold.

The silver content of gold particles has been used by Utter (1979) in an attempt to fingerprint source areas, but 1000 analyses of an extensive sample set across the entire Vaal Reef placer by Reid *et al.* (1988) indicated a complex heterogeneity of composition. Individual particles are homogenous with respect to both silver and mercury content, but the variance between particles, whether within or alongside pyrite, quartz or kerogen, is the same on both a local and regional scale. It will be necessary to establish what conditions prevailed within small cells before being able to understand the nature of the gold composition. Examination of gold particles from the Moeda in Brazil (Reid *et al.*, 1988) revealed similar contents of silver and mercury.

Gold has not migrated significantly away from surfaces where it was originally deposited. Examples illustrated by Hallbauer (1986) and Minter (in prep.) demonstrate how individual particles are spatially located on natural concentration surfaces like scours and cross-bedded foresets. The gold particles situated in these positions, however, are individually and collectively geochemically homogenous.

This indicates that although the gold does not seem to have been redistributed, the silver and mercury content from one particle to another has been homogenized, probably as a result of metamorphism. It is very unlikely that particles of the same composition could have been selected from probably diverse sources, undergone transport, and then been deposited together in a single cross-bed.

9.4.6 *Kerogen*

Carbonaceous material has been reported from all the ancient placers discussed, with the exception of the Tarkwa deposits. It occurs patchily as granules and as thin seams from a millimetre to a few centimetres thick. Its spatial distribution in the sediment is characteristically in the form of thin drapes or spotty accumulations on unconformities, which represent palaeochannel scour surfaces. Carbon also encapsulates pebbles and occurs on slip-faces of planar cross-beds (Minter, 1978), on coset boundaries, and over the tops of gravel bars (Smith and Minter, 1980).

The layers are composed both of vertical filamentous internal structures forming the mats and of rounded forms (Hallbauer *et al.*, 1977). This microscopic morphology and the spatial distribution of the carbon in relation to various sedimentary bedding planes has been used to argue persuasively that the carbon in related to fossil plant material (kerogen), possibly algal, and that the rounded forms may represent spores (Hallbauer *et al.*, 1977). Gold particles and detrital minerals are enclosed by the carbon fibres and gold also occurs as a fine powder on the surfaces of the carbon sheaths. Chemical analyses indicate that the carbon is composed mainly of hydrocarbons together with organic sulphur and oxygen compounds (Zumberge *et al.*, 1978).

Rich concentrates of gold, uraninite, and other detrital heavy minerals are associated with the kerogen, particularly on major scour surfaces, probably because these sites were surfaces of minimal aggradation on which heavy mineral traction carpets had had an opportunity to accumulate and upon which algal mats could develop without being killed off by rapid burial. The relative concentrations of uranium and gold in kerogen from scour surfaces correlate well with these element abundances in sand and in gravel facies (Smith and Minter, 1980). This provides supporting evidence of original sedimentary control of gold and uranium mineralization rather than a chemical origin.

Mossman and Dyer (1985) have also suggested that the kerogen was syngenetic and probably derived from prokaryotic microbial mats. They believe that bacteriological processes, such as the production of organic acids and/or sulphur-cycle intermediates, may have played a role in the solution and redistribution of gold. Kerogen not associated with major scour surfaces is not well mineralized with gold.

My preferred interpretation is that the role played by the kerogen is incidental and localized; areally, it is more prevalent in the more sandy, distal types of placers like the Vaal, Basal, and Carbon Leader placers (Minter, 1978) than in more proximal placers. Nevertheless, remnants of carbon seams in channelled depressions and as disaggregated grains do occur in coarse, gravelly, more proximal placer facies like the Ventersport and Moeda placers, and as disaggregated grains in reworked placers like the Beatrix placer (Minter *et al.*, 1986). However, the kerogen does not appear to have been as well preserved, probably because of the erosive nature of the sedimentary environments that must have prevailed.

9.5 Metamorphism

It is clear that the Witwatersrand Supergroup rocks have reached a lower greenschist facies of metamorphism. Early petrological and mineralogical studies by Young (1917) and by Fuller (1958) on shales in the sequence, and a more recent appraisal by Phillips (1988), indicated that pyrophyllite, chloritoid, chlorite, muscovite, quartz, rutile, tourmaline, and pyrite co-exist in metapelites throughout the basin. This assemblage would have required temperatures of at least 350°C during peak metamorphism. From the results of fluid inclusion studies, Hallbauer has suggested that temperatures of 250°C were reached in the orebodies. The presence of cordierite and garnet in the Dominion Group indicate that epidote–amphibolite facies conditions were reached in the deepest part of the basin. Thermally produced high grades of metamorphism of the Witwatersrand Supergroup occur locally towards the contact with the Bushveld Complex, and Schreyer (1982) has reported kyanite associated with the Vredefort structure.

The Tarkwa and Moeda deposits have also reached lower greenschist facies of metamorphism. At Jacobina, the rocks have reached a higher level of metamorphism than in any of the other placers and are at amphibolite facies. In places, the Tarkwa deposits are also reported to have reached almandine–amphibolite facies.

The effect of fluids, generated during peak metamorphism, on mineralization has not been fully assessed. Phillips (1988) has suggested that gold may have been very mobile during the conditions thought to have been prevalent during metamorphism of the Witwatersrand Supergroup but, from evidence described before, it does not seem that the primary distribution of heavy mineral grains has been changed significantly. Most alteration appears to have taken place around grains *in situ*. Certainly, in the Jacobina placers, gold has migrated on a scale of millimetres, being redeposited in tiny fractures within the orebody.

9.6 Structural control

Most of the placer deposits discussed are believed to have been associated with rifting in one way or another. The Dominion Group sediments are primarily a fluvial sandstone veneer on the Archaean granites of the craton but are followed by a great thickness of lavas that denote rifting. The Venterspost placer is similarly thin and buried by flood basalts but the terrain was evidently in an upland area and controlled by rifted half-grabens. Thicker sedimentary sequences are involved in the Pongola and Tarkwa deposits and they are also believed to lie within rifts. In the case of the Pongola, the placers occur in fluvial fan deltas in what Burke *et al.* (1985) have suggested was an intracratonic rift. In the case of the Tarkwa, Sestini (1973) believed that the placers were part of a series of alluvial fans within an intermontane setting. These are superimposed above previous grabens along a series of rifts (Hirdes and Leube, 1986). The lack of volcanics in the Moeda sequence, which fines upward into shallow marine shales, carbonates, and shelf iron formations, seems to support a cratonic continental margin setting but the structural complexity is still being researched.

Pretorius (1976) has suggested that the basin occupied by the Witwatersrand Supergroup was produced by rifting and that the Central Rand Group, in which most of the placer deposits occur, occupies a half-graben with a thicker sedimentary sequence located along the north-western flank. In recent years, plate tectonic concepts have been applied. For instance, van Biljon (1980) has suggested that the basin developed as an embayment along a suture zone between colliding Archaean continents, and Burke *et al.* (1986) have proposed that the Witwatersrand Supergroup developed in a foreland basin which developed initially on the cratonward side of an Andean-type arc. They suggest that further subsidence and deposition occurred in the basin as a result of the continental collision between the Zimbabwean and Kaapvaal Cratons. Winter (1987) also favoured a thrust model but concluded that the deformation is such that until the structural dislocations have been restored one cannot sensibly interpret the sediments.

To date, no fault-control on Central Rand Group sedimentation has been demonstrated. The faulting prevalent throughout the orebodies is all younger than middle Ventersdorp and post-dates lithification of the strata. It is clear, however, that folding took place synchronously with sedimentation (Minter *et al.*, 1986) in the Welkom goldfield; the isopachs of successive formations, each separated by an unconformity, demonstrate consecutive folding events. The tectonic transport was from the west and, during the deposition of the last Central Rand Group sediments in the Eldorado Formation, it intensified to the extent of overturning the strata. The results of ongoing reflective seismology surveys may well resolve these concepts.

9.7 Summary

There are a number of characteristics common to ancient gold placers that are related to the stage of crustal evolution, the nature of the Earth's atmosphere, the lithology of the source terrain, the degree of weathering, structural controls, and the sedimentary processes that prevailed. They are listed in Table 9.1.

Table 9.1 Common characteristics of ancient gold-bearing placers

- Ancient gold placers overlie Archaean granite-greenstone basement representing an auriferous stable cratonic source.
- The relief of the palaeotopography is low, indicating a mature landsurface. This is supported by the lithic clast assemblage of the sedimentary host rock, which is composed predominantly of quartz and chert, with a noticeable absence of granitic and volcanic components. Small proportions of felsic porphyry may occur.
- They may lie stratigraphically beneath Superior-type iron formations. This reflects an age of greater than 2.2 Ga, a cratonic margin setting, and evidence of preservation.
- They occupy quartz–arenite-filled channels on unconformities that are greater than 200 km^2 in extent. This indicates widespread erosion and sediment maturity.
- They occur in stacked tectonic/sedimentary units. This demonstrates repetition of degradation/aggradation cycles but does not preclude single clastic packages having a gold potential.
- They are associated with onlapping marginal unconformities, and palaeocurrent evidence indicates flow perpendicular to the subcrop strike of the unconformities. This indicates crustal warping to form highlands and adjacent basins.
- The palaeocurrent systems are unimodal and the deposits reflect shallow gravel-bar and channel morphology, indicating fluvial transport by braided streams on fans or braid deltas.

Table 9.1 *Continued*

- The sedimentary rock mineralogy comprises dominant quartz, chert, and feldspar, with accessory zircon, chromite, pyrite, monazite, leucoxene and uraninite concentrated in the placers. These reflect derivation from a source-rock terrain with characteristics favourable for the erosion and supply of gold. The presence of gold is certainly diagnostic but it may be confined to sedimentologically specific horizons.
- The presence of detrital sulphides and uraninite in particular is thought to reflect weathering in a low-oxygen atmosphere, and fits the age bracket for such conditions. However, there are younger examples in which hematite is a dominant detrital mineral (e.g. Tarkwa).
- The detrital heavy mineral components are concentrated in the matrix of the gravels and on scour surfaces. This demonstrates heavy mineral sorting during sedimentary processes such as transport, settling, and selective entrainment.

To appreciate the strength of the argument for a detrital origin for gold in the Witwatersrand Supergroup and for other examples of ancient gold placers, it is necessary to understand in detail the sedimentary nature of the deposits. The gross braid-delta geometry of the deposits can be recognized in modern analogues (Figure 9.13) and significant progress is being made in understanding the dynamics of sediment movement. Many of the previously empirical relationships concerning the efficiency by which very high specific gravity minerals are concentrated can now be explained (Slingerland and Smith, 1986). Without a sound sedimentological approach to ore genesis, an examination of their present mineralogy is frustrated by a confusing overprint of subsequent metamorphic events.

Estimates of gold production from Tarkwa (250 t), Jacobina (25 t), Moeda (10 t), and Pongola (0.8 t) are paltry compared to the gold deposits on the Kaapvaal Craton

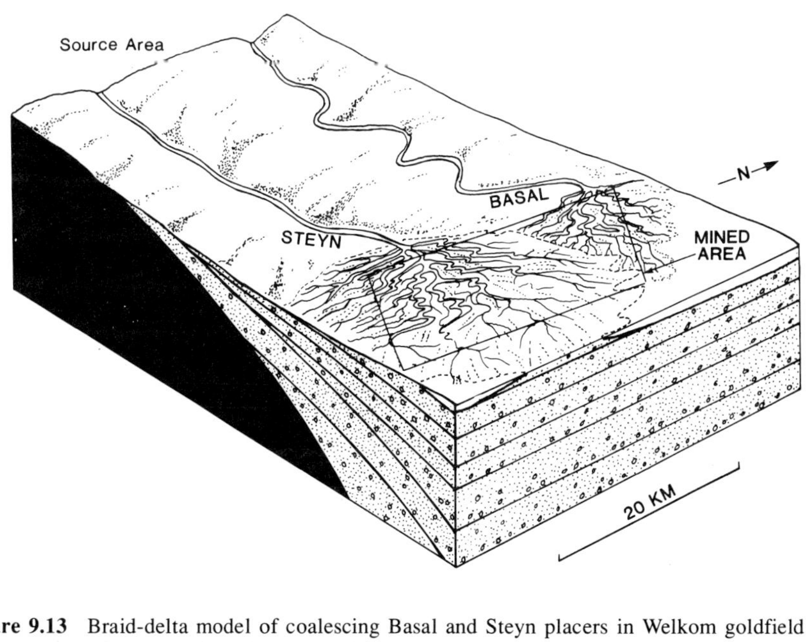

Figure 9.13 Braid-delta model of coalescing Basal and Steyn placers in Welkom goldfield is representative of many ancient placer deposits.

(42 333 t). The source of the prodigious supply of gold to the Witwatersrand Supergroup is still an enigma. The clastic composition of the placer sediments clearly indicates a provenance dominated by Archaean granitoids containing a significant component of greenstone. The variety of detrital minerals in the placers supports this concept, and the chemical composition of the associated shales also indicates that both mafic and acid igneous rocks contributed to the sediment supply.

Mineralogical research aimed at matching Witwatersrand Supergroup gold and pyrite compositions with those in the Archaean greenstones and granitoids have produced mixed results. It is not clear to what extent subsequent alteration has modified the original compositions. A notable concept that has emerged, however, is that granitoids adjacent to the basin may possibly have been altered by late hydrothermal activity during development of the Central Rand deposits and that roof zones enriched in gold and uranium were stripped by erosion and supplied to the basin (Robb and Meyer, 1985). It will be important to date such an event before being able to consider it as a possible source.

The possibility that hydrothermal fluids may have produced pyritic, cherty, gold-rich exhalative deposits on a seafloor associated with basaltic volcanism proximal to basin edge faults has been proposed by Hutchinson and Viljoen (1986) as an additional source.

The structural control of basin development in most of these placer deposits is said to be related to rifting but their real tectonic setting is simply not known. Even in the case of the most extensively explored Witwatersrand Supergroup, the rifting concept has lost favour due to lack of supporting evidence, and thrust tectonic models are now in vogue.

References

Antrobus, E.S.A., Brink, W.C.J., Brink, M.C., Caulkin, J., Hutchinson, R.I., Thomas, D.E., Van Graan, J.A. and Viljoen, J.J. (1986) The Klerksdorp goldfield. In *Mineral Deposits of Southern Africa*, eds. Anhauesser, C.R. and Maske, S. Geol. Soc. of South Africa, Vol. 1, 549–598.

Armstrong, R.A., Compston, W., Retief, E.A. and Welke, H.J. (1986) Ages and isotopic evolution of the Ventersdorp volcanics. *Abstracts Geocongress '86*, Geol. Soc. of South Africa, Johannesburg, 89-92.

Buck, S.G. and Minter, W.E.L. (1985) Placer formation by fluvial degradation of an alluvial fan sequence: the Proterozoic Carbon Leader placer, Witwatersrand Supergroup, South Africa. *J. Geol. Soc. London* **142**, 757-764.

Burke, K., Kidd, W.S.F. and Kusky, T.M. (1985) The Pongola structure of southeastern Africa: the world's oldest preserved rift? *J. Geodynamics* **2**, 35-49.

Burke, K., Kidd, W.S.F. and Kusky, T.M. (1986) Archean foreland basin tectonics in the Witwatersrand, South Africa. *Tectonics* **5**, 439-456.

Button, A. and Tyler, N. (1981) The character and economic significance of Precambrian paleoweathering and erosion surfaces in southern Africa. *Econ. Geol.*, 75th Anniversary Vol., 676-699.

de Wit, M.J., Jeffery, M., Bergh, H. and Nicolaysen, L. (1988) Geological map of sectors of Gondwana reconstructed to their disposition ~ 150 Ma. 1 : 10 000 000 Lambert equal area projection centred at 20°C, 40°E. The Amer. Assoc. Petroleum Geologists.

Eales, H.V. and Reynolds, I.M. (1983) Factors influencing the composition of chromite and magnetite in some southern African rocks. *Spec. Pub. Geol. Soc. S. Africa* **7**, 5-20.

Els, B.G. (1987) The auriferous Middelvlei Reef depositional system, West Wits Line, Witwatersrand Supergroup. Unpublished Ph.D. thesis, Rand Afrikaans Univ., Johannesburg S.A., 329.

Feather, C.E. and Glatthaar, C.W. (1987) A review of uranium bearing minerals in the Dominion and Witwatersrand placers. In *Uranium Deposits in Proterozoic Quartz-Pebble Conglomerates*, ed. Pretorius, D.A. International Atomic Energy Agency, TECDOC 427, 355-386.

Feather, C.E. and Koen, G.M. (1975) The mineralogy of the Witwatersrand reefs. *Minerals, Sci. Engng.* **7**, 189- 224.

Fuller, A.O. (1958) A contribution to the petrology of the Witwatersrand System. *Trans. Geol. Soc. South Africa* **61**, 10-50.

Hallbauer, D.K. (1986) The mineralogy and geochemistry of Witwatersrand pyrite, gold, uranium and carbonaceous matter. In *Mineral Deposits of Southern Africa*, eds. Anhauesser, C.R. and Maske, S. Geol. Soc. of South Africa, Vol. 1, 731-752.

Hallbauer, D.K., Jahns, H.M. and Beltmann, H.A. (1977) Morphological and anatomical observations on some Precambrian plants from the Witwatersrand, South Africa. *Geol. Rundschau* **66**, 477-491.

Hirdes, W. and Leube, A. (1986) The early Proterozoic (Birimian and Tarkwaian) of Ghana and some aspects of its associated gold mineralizations. *Abstracts Geocongress '86*, Johannesburg, 315-317.

Hutchinson, R.W., and Viljoen, R.P. (1986) Re-evaluation of gold source in Witwatersrand ores. *South African J. Geol.* **91**, 157-173.

Krapez, B. (1985) The Ventersdorp Contact placer : a gold–pyrite placer of stream and debris flow origins from the Archaean Witwatersrand Basin of South Africa. *Sedimentology* **32**, 223-234.

Kuhnle, R.A. (1986) Experimental studies of heavy-mineral transportation, segregation and deposition in gravel-bed streams. Unpublished Ph.D. thesis, Dept. Earth, Atmospheric and Planetary Sciences, M.I.T., 209.

McPherson, J.G., Shanmugan, G. and Moiola, R.J. (1987) Fan-deltas and braid deltas : Varieties of coarse-grained deltas. Geol. Soc. Amer., Bull. 99, 331-340.

McWha, M. (1988) The geomorphology of the Ventersdorp Contact Reef at Western Deep Levels South Gold Mine. *Abstracts, Ventersdorp Contact Reef Symposium.* Geol. Soc. of South Africa, 9–17.

Meyer, M., Oberthur, Th., Robb, L.J., Saager, R. and Stupp, H.D. (1985) Ni, Co, and Au contents of pyrites from Archaean granite–greenstone terranes and early Proterozoic sedimentary deposits in Southern Africa. *Information Circular* 176, Economic Geology Research Unit, Univ. of the Witwatersrand, Johannesburg, 11.

Minter, W.E.L. (1976) Detrital gold, uraninite and pyrite concentrations related to sedimentology in the Precambrian Vaal Reef placer, Witwatersrand, South Africa. *Econ. Geol.* **71**, 157-175.

Minter, W.E.L. (1978) A sedimentological synthesis of placer gold, uranium and pyrite concentrations in Proterozoic Witwatersrand sediments. In *Fluvial Sedimentology*, ed. Miall, A.D. Can. Soc. Petroleum Geol., Memoir 5, 801–829.

Minter, W.E.L. (1981) The Welkom goldfield exhumed: borehole predictions compared with thirty years of mining results. In *Proc. 12th CMMI Congress*, ed. Glen, H.W. South African Inst. of Mining and Metallurgy, Vol. 1, 137-145.

Minter, W.E.L. (1985) Very large ancient (2.7 Ga) fluvial bedforms in the Steyn placer, Witwatersrand, South Africa. *Abstracts 3rd Internat. Fluvial Sedimentology Conf.*, Colorado State Univ., Fort Collins, 29.

Minter, W.E.L., Feather, C.E. and Glatthaar, C.W. (1988) Sedimentological and mineralogical aspects of the newly discovered Witwatersrand deposit that reflect Proterozoic weathering, Welkom goldfield, South Africa. *Econ. Geol.* **83**, 481-491.

Minter, W.E.L., Hill, W.C.N., Kidger, R.J., Kingsley, C.S. and Snowden, P.A. (1986) The Welkom Goldfield. In *Mineral Deposits of Southern Africa*, eds. Anhauesser C.R. and Maske, S. Geol. Soc. of South Africa, Johannesburg, Vol. 1, 497-539.

Minter, W.E.L., Renger, F.E. and Siegers, A. (1990) Early Proterozoic gold placers of the Moeda Formation within the Gandarela syncline, Minas Gerais, Brazil. *Econ. Geol.*, (in press).

Mossman, D.J. and Dyer, B.D. (1985) The geochemistry of Witwatersrand-type gold deposits and the possible influence of ancient prokaryotic communities on gold dissolution and precipitation. *Precambrian Res.* **30**, 303-319.

Oberthur, T. and Saager, R. (1986) Silver and mercury in gold particles from the Proterozoic Witwatersrand placer deposits of South Africa: metallogenic and geochemical implications. *Econ. Geol.* **81**, 20-31.

Papenfus, J.A. (1957) The geology of the Government gold mining areas and new state areas, with particular reference to channel sediments and the origin of minerals of economic importance. Unpublished Ph.D. thesis, Univ. of South Africa, Pretoria.

Phillips, G.N. (1988) Widespread fluid infiltration during metamorphism of the Witwatersrand goldfields: generation of chloritoid and pyrophyllite. *J. Metamorphic Geol.* **6**, 311-332.

Pretorius, D.A. (1976) The nature of the Witwatersrand gold–uranium deposits. In *Handbook of Stratabound and Stratiform Ore Deposits*, ed. Wolf, K.H. Elsevier, Amsterdam, Vol. 7, 29-88.

Pretorius, D.A. (1986) The Witwatersrand Basin: Map of surface and subsurface geology and structure. Scale 1 : 500 000. Compiled to accompany *Mineral Deposits of Southern Africa*, published by Geol. Soc. of South Africa, Johannesburg.

Priem, H.N.A., Boelrijk, N.A.I.M., Hebeda, E.H., Verdurmen, E.H.Th. and Verschure, R.H. (1973) Age of the Precambrian Roraima Formation in northeastern South America: evidence from isotopic dating of Roraima pyroclastic volcanic rocks in Suriname. Geol. Soc. of Amer., Bull. 84, 1677-1684.

Ramdohr, P. (1958) New observations on the ores of the Witwatersrand in South Africa and their genetic significance. *Trans. Geol. Soc. South Africa* **61**, 1-10.

Reid, A., le Roex, A.P. and Minter, W.E.L. (1988) Composition of gold grains in the Vaal placer, Klerksdorp, South Africa. *Mineral. Deposita* **23**, 211-217.

Robb, L.J. and Meyer, M. (1985) The nature of the Witwatersrand hinterland: conjectures on the source-area problem. *Information Circular* 178. Economic Geology Research Unit, Univ. of the Witwatersrand, Johannesburg, 25.

Saager, R., Stupp, H.D., Utter, T. and Mattheys, H.O. (1986) Geological and mineralogical notes on placer occurrences in some conglomerates of the Pongola sequence. In *Mineral Deposits of Southern Africa*, eds. Anhauesser, C.R. and Maske, S. Geol. Soc. of South Africa, Vol. 1, 473-487.

Schreyer, W. and Bisschoff, A.A. (1982) Kyanite as a metamorphic mineral in Witwatersrand sediments at Johannesburg and Krugersdorp, South Africa. *Trans. Geol. Soc. South Africa* **85**, 211-214.

Sestini, G. (1973) Sedimentology of a paleoplacer: the gold- bearing Tarkwaian of Ghana. In *Ores in Sediments*, eds. Amstutz, G.C. and Bernard, A, J. Springer-Verlag, New York, 275- 305.

Slingerland, R. and Smith, N.D. (1986) Occurrence and formation of water-laid placers. *Ann. Rev. Earth Planet. Sci.* **14**, 113-147.

Smith, N.D. and Minter, W.E.L. (1980) Sedimentological controls of gold and uranium in two Witwatersrand paleoplacers. *Econ. Geol.* **75**, 1-14.

Smits, G. (1986) The mobility of uranium during Witwatersrand times. *Abstracts, Geocongress '86*, Geol. Soc. of South Africa, Johannesburg, 203-206.

Tankard, A.J., Jackson, M.P.A., Eriksson, K.A., Hobday, D.K., Hunter, D.R. and Minter, W.E.L. (1982) *Crustal Evolution of Southern Africa.* Springer-Verlag, New York, 115-150.

Tweedie, E.B. (1986) The Evander goldfield. In *Mineral Deposits of Southern Africa*, eds. Anhauesser, C.R. and Maske, S. Geol. Soc. of South Africa, Vol. 1, 705-730.

Utter, T. (1979) The morphology and silver content of gold from the upper Witwatersrand and Ventersdorp Systems of the Klerksdorp gold field, South Africa. *Econ. Geol.* **74**, 27-44.

Van Biljon, W.J. (1980) Plate-tectonics and the origin of the Witwatersrand Basin. In *Proc. 5th IAGOD Symposium*, ed. Ridge, J.W. Schweizerbartsche Verlagbuchhandlung, Stuttgart, Vol. 1, 796.

Walraven, F., Armstrong R.A. and Kruger, F.J. (1989) A chronostratigraphic framework for the north-central Kaapvaal craton, Bushveld Complex, and Vredefort structure. *Tectonophysics* (in press).

Watchorn, M.B. (1980) Fluvial and tidal sedimentation in the 3000 Ma Mozaan basin, South Africa. *Precambrian Res.* **13**, 27-42.

Winter, H. de la R. (1987) Cratonic foreland model for Witwatersrand basin development in a continental back-arc plate- tectonic setting. *South African J. Geol.* **90**, 409- 427.

Young, R.B. (1917) *The Banket of South African goldfields.* Gurney and Jackson, London.

Zumberge, J.E., Sigleo, A.C. and Nagy, B. (1978) Molecular and elemental analysis of the carbonaceous matter in the gold and uranium bearing Vaal Reef carbon seams, Witwatersrand sequence. *Minerals, Sci. Engng.* **10**, 223-246.

10 Geochemical exploration for gold in temperate, arid, semi-arid, and rain-forest terrains

H. ZEEGERS and C. LEDUC

10.1 Introduction

In this chapter we examine how geochemical techniques can be utilized in the exploration for different types of gold mineralization, the specific signatures of which must be identified in the various sampling media (rock, soil, etc.) used for exploration. The geochemical signature of gold mineralization depends upon several parameters, particularly the style of mineralization in question and the physico-chemical characteristics of the supergene environment. Consequently, different topographic and climatic environments will be considered, where the transposition of the characteristics of the primary mineralization to the surface sampling media may vary. This will be illustrated using a range of examples, emphasizing the procedures used as well as the results obtained. Two main components of the geochemical signature will be distinguished: Au and the pathfinder elements. By using the latter, an indication is obtained of the lithology and mineralizing processes, factors which are especially useful in terrains where outcrop is poor. However, problems specific to explorations in till-covered areas are considered by Coker and Shilts in Chapter 11.

10.2 Geochemical signatures of gold mineralization

10.2.1 *Attributes of bedrock mineralization*

The attributes of gold mineralization which define the surface or subsurface geochemical expression (signature) may be subdivided into two groups:

(i) attributes directly related to the gold concentration:
- dimensions, geometry, shape;
- gold contents (mean, peak);

(ii) attributes descriptive of the mineralization:
- Au mineralogically expressed or not;
- grain size of gold;
- fineness;
- host minerals: quartz, sulphides, oxides...
- mineral association;
- nature of country rocks;

Some of these parameters strictly depend on the type of mineralization as, for instance, the mineral association, which is a result of different processes (mostly hydrothermal) pre-dating, accompanying, or post-dating the mineralization. The main minerals characterizing hydrothermal alteration haloes are: silica (mostly quartz), tourmaline, chlorite, sericite or muscovite, K-feldspars, carbonates, barite, pyrite, As-, Ag- and Sb-bearing minerals, galena, sphalerite, and chalcopyrite, together with the element Hg. The relative abundance and the possible pathfinder roles of these components with regards to the type of mineralization are summarized in Table 10.1.

Table 10.1 Mean relative abundance and relevance (as possible pathfinders) of some specific minerals and elements for different types of mineralization

	(1)*	(2)	(3)	(4)	(5)	(6)	(7)	(8)	(9)
Silica	+–++	++–+++	+++	+++	+++	+–++	++–+++	0–++	+–++
Carbonates	+–+++	0–++	0	0–+	0	0	++–+++	0–+	++–+++
Sericite	++–+++	0–++	0–+	++–+++	0–+	++–+++	0–+	+–+++	0–++
K-Feldspar	0–++	0–+	0–+	+–+++	0	++–+++	0	0–+	0–+
Chlorite	0–++	+–++	0	+–++	0	+	0–++	+–+++	+–++
Tourmaline	0–++	0–+	0	0–+	0	0–+	0–++	0–+	0–+++
Barite	0	0	0–++	0–++	+–++	0	0	0–+++	0
Pyrite	+–+++	+–++	+–++	+	+	++–+++	++–+++	++–+++	+
As	0–+++	++–+++	+	+–++	+	0–+	++–+++	+–++	+
Pb	0–++	0–+++	+–++	++–+++	+	0–++	0–+	+–++	+
Zn	0–+	0–+	0–++	++–+++	0–+	0–+++	0–+	0–+++	0–+
Cu	0–+++	0–+	0–+++	+	0–+	+–+++	0–+	+–+++	0–+
Sb	0–+	0–+++	+–++	+–+++	0–++	0	0–+	+	+–++
Hg	0–+	0–+++	+–+++	0–+++	++–+++	0–+	0	0–+++	+–++
W	0–+	0–+++	0–+	0–++	0–+	0–++	0	0	+–++
Mn	0	0	+–++	+–+++	0–++	0–+	0–+	0	+–++
Mo	0–++	0–+	0	0	0–+	0–+++	0–+	+	0–++

**Column heading*: (1) Archaean and Proterozoic lodes; (2) Phanerozoic lodes; (3) epithermal, acid sulphate; (4) epithermal, sericite–adularia; (5) Carlin, sediment-hosted; (6) intrusion related; (7) BIF; (8) volcanogenic massive sulphides; (9) stratabound epigenetic.

0, none; +, small; ++, medium; +++, high.

Principal references: Kerrich (1983); Colvine *et al.* (1988); Groves *et al.* (1988); Bonnemaison (1986, 1987); Silberman and Berger (1985); Hayba *et al.* (1985); Heald *et al.* (1987); Sillitoe (1988); Bagby and Berger (1985); Large *et al.* (1988); Fripp (1976); Harris (1986).

Other attributes such as dimensions or geometry vary markedly from one type of deposit to the next. Large deposits may be found in the intrusion-related, epithermal, sediment-hosted or BIF-types. Lode deposits are characterized by inclined tabular orebodies commonly not more than a few metres thick, but frequently surrounded by a large alteration halo with a well-developed geochemical signature that represents a more convenient target for exploration than the mineralization itself.

A priori knowledge of the physical state of the gold (i.e. free and fine grained, free and coarse grained, small grains enclosed in sulphide minerals, or included in the lattice of minerals such as arsenopyrite), is essential to determine the respective domains of application of geochemical and heavy-concentrate (panning) techniques for gold exploration.

10.2.2 *Supergene cycle of gold*

Theoretical studies and numerous examples clearly demonstrate that gold can be hydromorphically dispersed in the supergene environment. A summary of the supergene behaviour of gold can be collated from Krauskopf (1951), Ling Ong and Swanson (1969), Goleva *et al.* (1970), Baker (1978), Cloke and Kelly (1984), Webster and Mann (1984), Mann (1984a, b), and Webster (1986) as follows:

(i) In its solid state, Au^0 is very stable; in aqueous solution, it may occur with oxidation states 1 and 3. Au^+ and Au^{+++} are unstable and thus gold tends to complex with ligands such as Cl^-, Br^-, $(CN)^{--}$, $(S_2O_2)^{--}$, or organic matter. Gold may also migrate as colloids or 'particulate' gold.

(ii) Consideration of the kinetics of the reactions possibly leading to the mobilization of gold through the formation of chlorides, thiosulphates, or other compounds (humates, thiocyanates...) demonstrates that extreme physico-chemical conditions are required. These conditions may actually occur in micro-environments, but are not likely to persist over long distances away from the source of gold. This is illustrated for instance by Freyssinet *et al.* (1987) who showed that at least part of the gold mobilized in lateritic profiles re-precipitates near to the primary gold sources.

(iii) The sulphide content of the mineralization is assumed to be of importance to the mobilization of gold, especially if the environment contains potential buffers such as carbonates; in such situations, gold can be mobilized as thiosulphate.

(iv) The presence of abundant Mn-oxides or goethite is also considered to favour the mobilization of gold in some circumstances.

Chemical processes, therefore, contribute to the formation of the secondary dispersion halo of gold, together with mechanical processes. The balance between both processes will depend upon the morphological and climatic conditions, the style (and therefore the primary attributes) of the mineralization, and the soil horizon concerned. As shown later, mechanical processes prevail in temperate regions, whereas the chemical component of the dispersion may be predominant in tropically weathered terrains. Different reactions are considered to be responsible for gold mobilization. In arid terrains such as Western Australia, chloride complexes are probably responsible for the secondary redistribution of gold in some supergene deposits (Butt, 1988; Lawrance, 1988a). In the rainforest environment of Gabon, the solubility of gold has been clearly demonstrated but the corresponding chemical processes have not been identified (Colin *et al.*, 1989). Conversely, Webster and Mann (1984) described a secondary gold occurrence in Papua New Guinea that probably resulted from mobilization of gold as a thiosulphate complex.

10.2.3 *Transposition of the primary characteristics to the supergene environment*

The geochemical signature of gold mineralization obtained in different surface and subsurface sampling media mostly depends upon the behaviour of gold and accompanying minerals in the relevant supergene environment.

In temperate climates, most of the pathfinder minerals (with the exception of some of the sulphides such as pyrite and sphalerite) are not strongly affected by supergene processes. The surface geochemical signature thus includes most of the elements present in the primary mineral association characterizing the mineralization, with eventually some mechanical reworking according to the topographic conditions. Thus, the only important problem confronting explorationists is the lack of any significant physical geochemical dispersion processes, especially in environments with poorly contrasted morphology.

In tropically weathered terrains, some constituents of the mineralization are leached during weathering. The geochemical signature of the mineralized bedrock is thus (qualitatively and quantitatively) altered in the different horizons of the weathering profile. Some elements are leached, others maintained or even enriched. As schematically illustrated in Figure 10.1, most of the pathfinder elements are immobile or only partly mobilized during weathering, so that they can be used even in deeply leached environments. However, sphalerite, chlorite and, to some extent, Cu-bearing sulphides are severely leached in such environments.

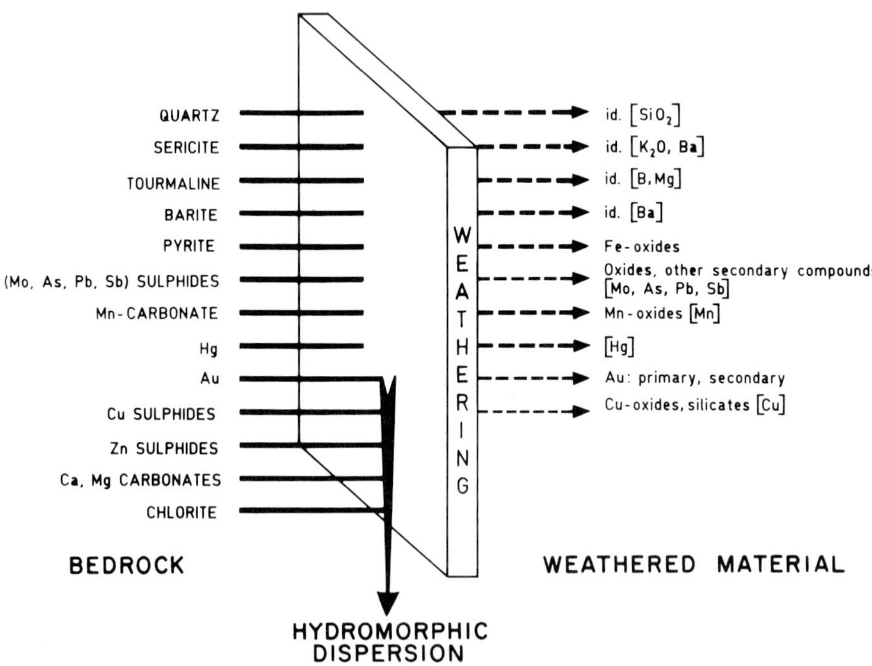

Figure 10.1 Diagram showing the geochemical behaviour during weathering of key pathfinder minerals and elements.

10.3 Examples and case histories

10.3.1 *Introduction*

Case histories of geochemical exploration for gold as well as general information on geochemical dispersion of gold and the pathfinder elements in the main geochemical sampling media (saprolite, soils and stream sediments) will be presented using the following broad climatic and morphological subdivisions:

- temperate climates, excluding till environments (see Chapter 11)
- arid to savanna terrains: cold and warm deserts, inter-tropical savanna environments, having or not having undergone lateritic type weathering
- tropical rain forest terrains.

Specific characteristics of the geochemical response of gold mineralization and their consequences for exploration will be considered for all of these groups. Two principal parameters will be emphasized: the contrast of the gold geochemical anomaly, if any, and the quality of the multi-element geochemical signature, reflecting the mineral association of the mineralization. It will be shown that the most variable parameter is the dispersion of gold itself, mostly by mechanical means in temperate climates and with a more important hydromorphic component in tropically weathered environments.

10.3.2 *Temperate terrains*

General characteristics. Temperate terrains are characterized by a mild climate, a medium rainfall and therefore weathering of low intensity. As far as known from available published data, the chemical component of dispersion of gold and most pathfinders is not very effective in these environments. This has two main consequences:

(i) The extent of the dispersion halo is minimal unless moderate to high relief promotes mechanical dispersion by processes such as solifluxion. Thus anomalies are commonly of limited dimension and therefore are detected only by a relatively close spaced sampling.
(ii) The residual component of geochemical anomalies includes most elements derived from the primary mineral association, as a large part of these minerals survive in the supergene environment. The interpretation of geochemical signatures may thus allow a reconstruction of the mineral paragenesis, and therefore an early diagnosis of the mineralization style, if multi-element analysis is used. The dimensions of the surface dispersion halo of the gold itself may not significantly exceed that of the mineralization (or of the hydrothermal alteration zone).

Nevertheless supergene processes can be rather active if considerable quantities of sulphides are present in the primary environment, resulting in a low pH during hydrolysis. Gossans commonly form under temperate climates, and their mechanical degradation is responsible for geochemical anomalies in soils or stream sediments.

Several regions currently experiencing temperate climates have previously undergone glacial climates and relics of such former climatic episodes may be found, for example as loess in the Massif Armoricain of France. Such allochthonous situations are analogous to those described in Chapter 11 for glacial environments.

Examples and case histories. Vasquez-Lopez *et al.* (1987) compared results obtained by geochemical and heavy concentrate surveys in the Massif Armoricain of France. At the regional level of exploration in terrains of low relief such as this, anomalous areas are better identified by the presence of visible gold in heavy mineral concentrates (1 sample/km^2) than by gold contents of stream sediment samples (2-3 samples/km^2), assuming the bedrock mineralization contains free gold. In geochemical samples, elements such as B, Li, and As may provide additional information and allow selection of the most promising anomalies. Detailed surveys are based on the analysis of gold and pathfinders (As, Pb, Sb, W) in soil samples, while accurate delineation of the mineralized bodies before drilling or trenching necessitates sampling in the weathered bedrock at the base of the soil profile at a very tight sampling interval (10 m). Gold contents obtained at this stage commonly exceed 1 ppm if the prospect is to approach economic viability.

At Le Bourneix (Massif Central of France), a lode-type mineralization only a few metres thick (Barbier and Wilhelm, 1978) exhibits a wide gold dispersion halo with the 200 ppb Au contour line extending over more than 100 m in the surface soil on both sides of the orebody. The development of the surface gold anomaly is interpreted as being related to solifluxion. Braux *et al.* (1989) described the geochemical response corresponding to a lode-type mineralization having already produced 11 t of Au at Le Châtelet, Massif Central, France (Bache, 1982). The distribution of gold contents in

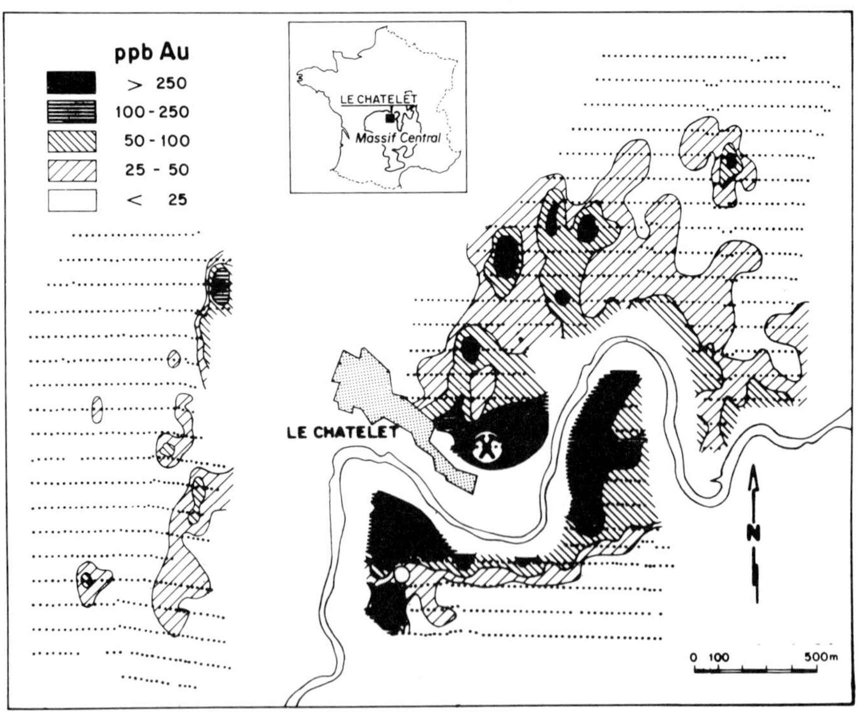

Figure 10.2 Le Châtelet gold prospect (Massif Central, France). Distribution of Au in soil samples around an old mine. (From Braux *et al.*, 1989.)

soils collected on a grid of 100 × 25 m (Figure 10.2) provided a fair image of the extent of the mineralized system. Indeed, several contrasted Au anomalies (up to 500 ppb) indicated not only the known mineralization but also new drilling targets. Multi-element analysis of soil samples provided information about the hydrothermal paragenesis, and principal component factor analysis revealed an association of Au with B (tourmaline ?), Li (muscovite ?), and As (arsenopyrite), with respective anomalous thresholds of 80 ppm (B), 100 ppm (Li), and 250 ppm As.

Some aspects of exploration for lode gold deposits in the Hercynian and the Archaean basements of France and Canada respectively are discussed by Bonnemaison (1986) and Bonnemaison and Marcoux (1987), who emphasized that gold introduced during such early stages of the evolution of the shear zones may be very fine grained and therefore not detectable by panning. At the detailed scale of exploration, pathfinders specific to the rich ore zones have to be identified: Pb, As, and Sb are commonly mentioned, but Au probably remains its own best pathfinder.

Janatka and Moravek (1987) gave examples of the geochemical response of quartz stockwork-type gold mineralization in Bohemia, with rather extensive As (up to several thousand ppm) and Hg anomalies surrounding smaller Au anomalies (0.1–1.0 ppm) in the fine (< 200 mesh) size fraction of soil samples collected in the B horizon. At the regional scale of exploration, good results were obtained by combining heavy concentrate sampling at a density of 1 sample/km^2 (with identification of visible gold) and soil multi-element geochemistry.

In South Carolina, USA, Kinkel and Lesure (1968) reported that some residual concentration of gold may occur in the upper horizons of the soil profile due to the removal of soil particles during run-off of surface water. A better geochemical contrast may thus be expected at the surface compared with deeper horizons. Also, some mobility of gold is demonstrated by the presence of gold (up to 2.9 ppm) in freshly precipitated limonite in a weathered mineralized environment, with the gold initially being associated with pyrite in quartz veins.

Under the drier climate of central California (500–700 mm annual rainfall), Chaffee and Hill (1989) have described a good surface geochemical response for Mother Lode-type deposits. Soil samples collected in the B-horizon at a depth of 20–30 cm and at a 10 m interval displayed multi-element anomalies with the largest haloes obtained for Au, Ag, As, Hg, Sb, and W. Depletion of Ca and Mg was also observed, reflecting the alteration halo around the mineralized faults.

In summary, in temperate climates, there is no evidence of an important role for hydromorphic processes in the dispersion of gold. Therefore, the geochemical anomalies for gold or other elements can be expected to be mostly mechanical and thus dependent on the topography. Predominantly mechanical dispersion also suggests the possible use of heavy-concentrate surveys with identification of free gold being important for locating targets containing some coarse free gold. Such an approach is especially efficient in terrains of low relief, as panning enhances low gold contents in alluvial sediments. In terms of the multi-element signature, most of the elements characterizing the minerals of the primary paragenesis are stable in the supergene zone of temperate terrains. Therefore, the transposition of the geochemical information to the surface is interference-free and the geochemical signature is mostly reliable.

10.3.3 *Arid and semi-arid terrains*

General characteristics. Within arid to semi-arid environments (i.e. annual rainfall less than 600 mm) we here consider warm and cool deserts and tropical savannas. In these terrains, very different conditions of geochemical dispersion exist. Indeed, the key factors are the morphological and climatic history of the region and, more especially in the inter-tropical zone, the presence of weathered material such as laterites representing relics of former climatic episodes (Butt and Zeegers, 1989). Two main subdivisions will thus be considered, depending on whether or not a pre-existing weathering profile is present.

Arid terrains where the pre-existing lateritic weathering profile is truncated or lacking. In such environments, the pedogenic activity, mostly physical, is not sufficient to develop a mature soil profile. The material resulting from degradation of rocks or mineralization is mechanically dispersed, either by water or by the wind.

In desert terrains with low relief, outcrop conditions are commonly very poor, so that indirect exploration techniques such as geochemistry have to be employed. The examples described by Salpéteur and Sabir (1989) from Saudi Arabia, lead to the general conclusion that, under the conditions of the central Arabian pediplain, the fine size fraction (< 80 μm) of soil or stream sediment samples gives the best contrast and the longest dispersion train for gold. This is in contradiction with recommendations by other authors such as Barbier (1987) to use the coarser fractions for geochemical purposes in order to get rid of aeolian contamination. Salpéteur and Sabir (1989) interpreted the good response obtained in the fine fraction of surface material as a result of aeolian deflation of clay or sand particles, with a relative enrichment of the heavy fraction, including gold grains. The coarse fraction of soils may also show anomalies (up to several ppm Au) but these anomalies are mostly related to old workings and not directly to the main bedrock mineralization. For regional exploration, the authors recommend a rather high sampling density (2 samples/km^2), especially in low-relief terrains where the dispersion of gold and pathfinders is limited. Samples should best be collected in medium size tributaries, under the upper, sandy, aeolian layer. For follow-up surveys, soil sampling (< 80 μm fraction of the upper argillaceous horizon) is recommended where relief is poorly developed while, in hilly terrains, stream sediment sampling at high density (up to 10 samples/km^2) may represent a first stage for defining an anomaly.

In the arid environment of Nevada (annual rainfall of about 300 mm), Wargo and Powers (1978) described the results of geochemical sampling on the Saddle gold prospect. The mineralization consisted of fine-grained Au (2–8 μm) disseminated in cherty and carbonaceous rocks of the Slaven Chert (Devonian). The best geochemical response of gold mineralization in the skeletal and colluvial soil was obtained in the coarse (>1.168 mm) fraction, either for Au or As and Hg. It can be inferred that such a dispersion is mostly achieved by mechanical processes, enhanced by the well-developed topography.

> Thus, conditions of geochemical dispersion in non-lateritic arid terrains may, to some extent, be compared to those prevailing under temperate climates, with the role of physical processes exceeding that of chemical processes. Depending on the local conditions, it was shown that identification of dispersion haloes for

exploration may be achieved by selecting for analysis either the fine or the coarse size fraction. Orientation surveys, therefore, are strongly recommended in such arid environments to ensure selection of the optimum procedures.

Pre-existing lateritic weathering profile partly preserved. Here, conditions for the dispersion of gold and the pathfinder elements are different. Chemical (or palaeochemical) processes of dispersion are active and so the size of the dispersion haloes of gold (and of some pathfinders) is significantly increased. Such conditions are met in West Africa (savanna climate), Australia, India, part of Brazil, and in some Gulf countries where remnants of former lateritic surfaces survive. In general, the lateritic weathering profile, summarizing from Nahon (1976), Leprun (1979) and Tardy and Nahon (1985), comprises (from bottom to top):

(i) *the saprolite* (30–100 m thick) where most rock-forming minerals have been transformed into secondary kaolinite and Fe-oxides (goethite, hematite); the original rock fabric is preserved;
(ii) *the mottled zone* (1–3 m) where the rock structures are progressively destroyed and ferruginous spots and nodules form, producing the mottled appearance; total Fe content is higher than in the saprolite;
(iii) *the ferruginous cuirasse* (1–7 m) consisting of a massive, hardened Fe-accumulation (hematite or goethite) horizon. The cuirasse, also commonly referred to as 'iron oxide cap' or 'ferricrete', can be pisolithic, lamellar, or tubular. At the surface, degradation may lead to the formation of a nodular soil.

Such profiles are observed in part of the Red Sea Hills province of Sudan (Fletcher, 1985) where a thick lateritic profile is described above gold mineralization. Fletcher (*op. cit*) remarked that dispersion of gold was mostly detrital where the pre-existing lateritic profile had been eroded, while in deeply weathered situations both mechanical and chemical processes were active. The effects of deep weathering were shown to be responsible for gold mobilization in the oxide zone and concentration as a zone of secondary enrichment at the water table. Remobilization of gold can lead to the formation of supergene gold deposits, as is the case in most of the deposits of Western Australia (Butt, 1988; Lawrance, 1988a)

In situations where both preserved and truncated lateritic profiles occur at the prospect scale, geochemical results may be misleading. Thus, in the same Red Sea Hills region of Sudan (Fletcher, *op. cit.*), gold soil anomalies discordant to the mineralization were observed, indicating that lateritic gold has undergone lateral transport away from mineralization due to former intensive weathering conditions. In both cases, gold dispersion seems to be independent of the present-day steep-sided topography and relates to an older, gently sloped, weathered surface. Fletcher (*op. cit.*) considered that surface reworking may be intense, so that false gold geochemical anomalies in the present-day active stream sediments may be derived from gold-rich palaeosols not directly related to bedrock mineralization. In general, however, soil sampling is very effective and gold anomalies obtained near mineralization in the < 177 μm fraction range from less than 10 ppb up to 3000 ppb, often extending several hundred metres from the mineralized source. At the regional scale, it is considered that panning stream sediment samples followed by identification of gold and other heavy minerals is the most effective exploration technique.

In other dry or seasonally humid terrains partly covered by a lateritic regolith (Western Australia, West Africa, Brazil, India), the mobility of gold in the supergene environment has been clearly demonstrated (Butt, 1988; Lawrance, 1988a, b; Freyssinet *et al.*, 1987, 1989; Vasconcelos and Kyle, 1988; Nair *et al.*, 1987; Davies *et al.*, 1989). The dispersion processes are interpreted in terms of evolution of the landscape and a succession of different climates. The highest mobility of gold seems to be achieved during dry climatic periods, when saline groundwaters may mobilize gold as chloride (Mann, 1984a; Webster and Mann, 1984; Lawrance, 1988a). Gold may thus be mobilized and leached from some horizons of the weathering profile and further redistributed in the saprolite, forming wide dispersion haloes (Figure 10.3). It can also be mechanically dispersed, either free or associated for instance with Fe-rich pisoliths (Mann, 1984a).

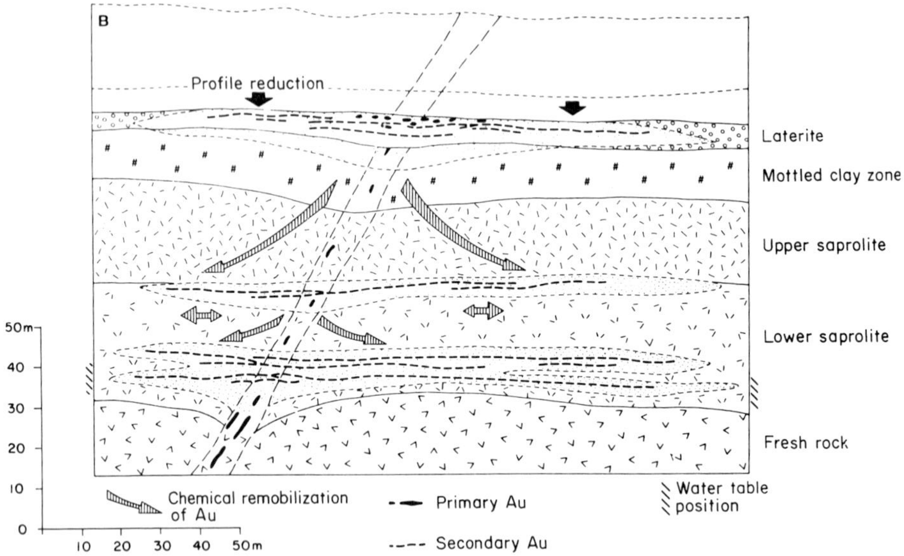

Figure 10.3 Schematic representation of chemical and physical remobilization of gold with increasing aridity (from Lawrance, 1988a).

Pathfinder elements will behave differently depending on their mobility during the lateritic weathering processes. They will be retained, or not, depending on whether (a) they are incorporated in resistate minerals (e.g. Sn in cassiterite or Zn in willemite) or whether (b) they are incorporated in one or several of the main secondary mineral phases (e.g. Fe- or Al-oxides, kaolinite, smectites, carbonates from calcretes), or whether (c) they form supergene, stable minerals such as Pb-phosphates.

Geochemical procedures must be adapted to each specific environment; the selected examples briefly discussed hereafter concern different environments, exploration targets, and adaptations of the geochemical techniques.

Freyssinet *et al.* (1987, 1989) studied the dispersion of gold in lateritic weathering profiles in Mali. They showed that gold is fairly well retained (residual or secondary) in any horizon of the profile, even in the uppermost Fe-rich cuirasse, which can therefore be used for geochemical sampling.

Smith *et al.* (1989) used pisolithic laterite as a sampling media for low-density regional exploration programmes. Most of the pathfinder elements used in the exploration for gold and massive sulphide deposits are to some extent retained in such Fe-rich material, which may thus be considered as a 'memory' of the bedrock and contained mineralization. Sampling entails collecting about 1 kg of lateritic material over a radius of 5–10 m at each site, with a sampling density as low as 1 sample/km^2. Samples are crushed and about 30 elements (including Au) are determined using different analytical techniques. Anomalies of chalcophile or partly chalcophile elements (As, Bi, Sb, Mo, Ag, Sn, Ge, W, Se, and Au) are considered to reflect bedrock features defining favourable environments for mineralization.

Carver *et al.* (1987) recommended 'lag' sampling (coarse particles collected from unconsolidated surface material) for the geochemical exploration for gold in arid regions. For instance, in Western Australia both lags and soils display contrasted anomalies, but lags show more extensive lateral dispersion, making possible the use of larger sampling intervals.

The Boddington Au deposit, in Western Australia, was discovered following up an As, Cu, Pb, Mo, and Zn anomaly, with gold contents up to 1.6 ppm in lateritic soil samples (Davy and El-Ansary, 1986). Most of the deposit is covered by bauxitic laterites lying on top of a deep weathering profile. The patterns of trace-element distribution between surface horizons – uppermost laterite zone – and fresh bedrock were compared on two traverse lines. In surface horizons, Au, Cu, Mo, and W anomalies either overlie or are displaced no more than 100–150 m away from the mineralization. Lateral transport of gold could be more important since the occurrence of secondary gold up to 500 m from any known primary mineralization is described by Monti (1988). However, surface geochemistry in such deeply weathered terrains may be used to delineate the potential areas of interest.

> The presence of remnants of pre-existing lateritic-type weathering profiles thus implies a significant increase in the lateral extent of gold dispersion haloes due to active chemical processes. An important consequence, from a practical point of view, is that gold mineralization can be located even by using very large sample intervals. For instance, the Loulo orebody (Mali), which is not more than 6–7 m in width, was discovered after sampling lateritic soil on a 1600 × 500 m grid (Dommanget *et al.*, 1985, 1987).

10.3.4 *Rain forest environments*

General characteristics. Rain forest environments, corresponding to high annual rainfall and elevated average temperatures, are typified by extensive and deep development of the weathering profile, whether or not the region has undergone weathering related to a former climate. As defined for dryer climates, complete geochemical dispersion models can only be established by considering the possible existence of relics of pre-existing weathering surfaces. Another important parameter is the degree of leaching or transformation undergone by the pre-existing profile after its formation (Butt and Zeegers, 1989).

Examples and case histories. A first group of examples concern stone-line profiles, as described by Lecomte (1988). These profiles, commonly found for instance in central

Africa or in the rain forests of South America, are considered as residual and consist of three main horizons (Figure 10.4):

(i) a friable, sandy-clay, yellow upper horizon (0.5–7 m thick);
(ii) an accumulation horizon made up of coarse fragments (quartz, lithorelics, Fe-rich nodules or pisoliths), the so-called 'stone-line' (0.5–2 m thick); depending on the abundance of coarse Fe-rich material, considered as inherited from former weathering episodes, two submodels are distinguished – *lateritic stone-line* and *quartz stone-line* profiles;
(iii) the saprolite (up to 70 m thick).

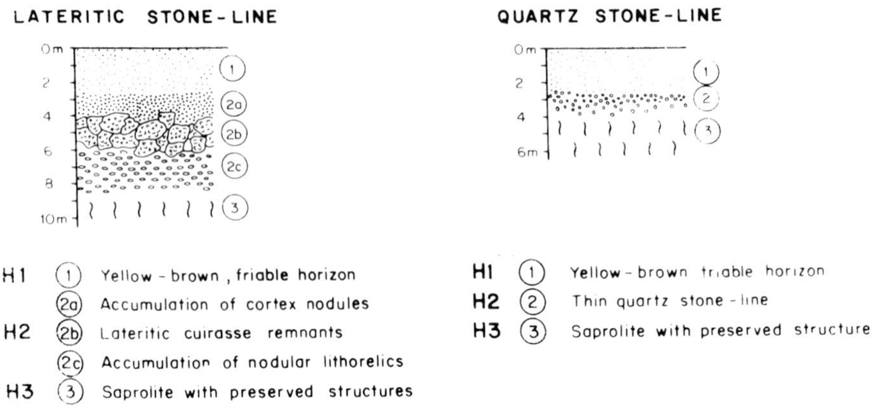

Figure 10.4 Idealized lateritic and quartz stone-line weathering profiles. (From Lecomte, 1988.)

The Mebaga gold prospect was discovered during a regional exploration programme carried out in Gabon (Barthélémy *et al.*, 1987) involving geochemical sampling (stream sediments), a heavy concentrate survey, and geological mapping. The regional anomaly consisted of one stream sediment sample having 100 ppb Au and three heavy concentrates with visible gold, and was followed up by soil sampling on a grid of 100 x 200 m (Colin and Lecomte, 1988). The local geology comprises amphibolites, amphibolitic gneisses, and ultrabasic rocks, all cross-cut by quartzo-feldspathic veins. The weathering profile is typically of the stone-line type, with an upper, friable homogenous horizon up to 7 m thick.

The soil gold anomaly (Figure 10.5), weakly contrasted (30–70 ppb) but well defined, was auger-drilled, with systematic sampling in the saprolite beneath the stone-line. Along the traverses, the auger drillholes were 5 m apart. The corresponding samples were analysed for gold and for 35 elements by ICP. The results obtained in the saprolite are compared in Figure 10.5 (Au) and Figure 10.6 (selected major and trace elements). Gold contents of several grams per tonne were recorded locally, and subsequently were shown by deep drilling to correspond to minor mineralization. The other elements may be strongly depleted in the upper horizon, but the principal content-variations along the traverse, related to major changes of the lithology, are still evident in soil samples. For instance, the presence of ultramafic rocks is well indicated at the surface by low SiO_2 and Al_2O_3 contents, but Ni and Cr, having very high contents in the saprolite, show strongly smoothed anomalies in the soil samples.

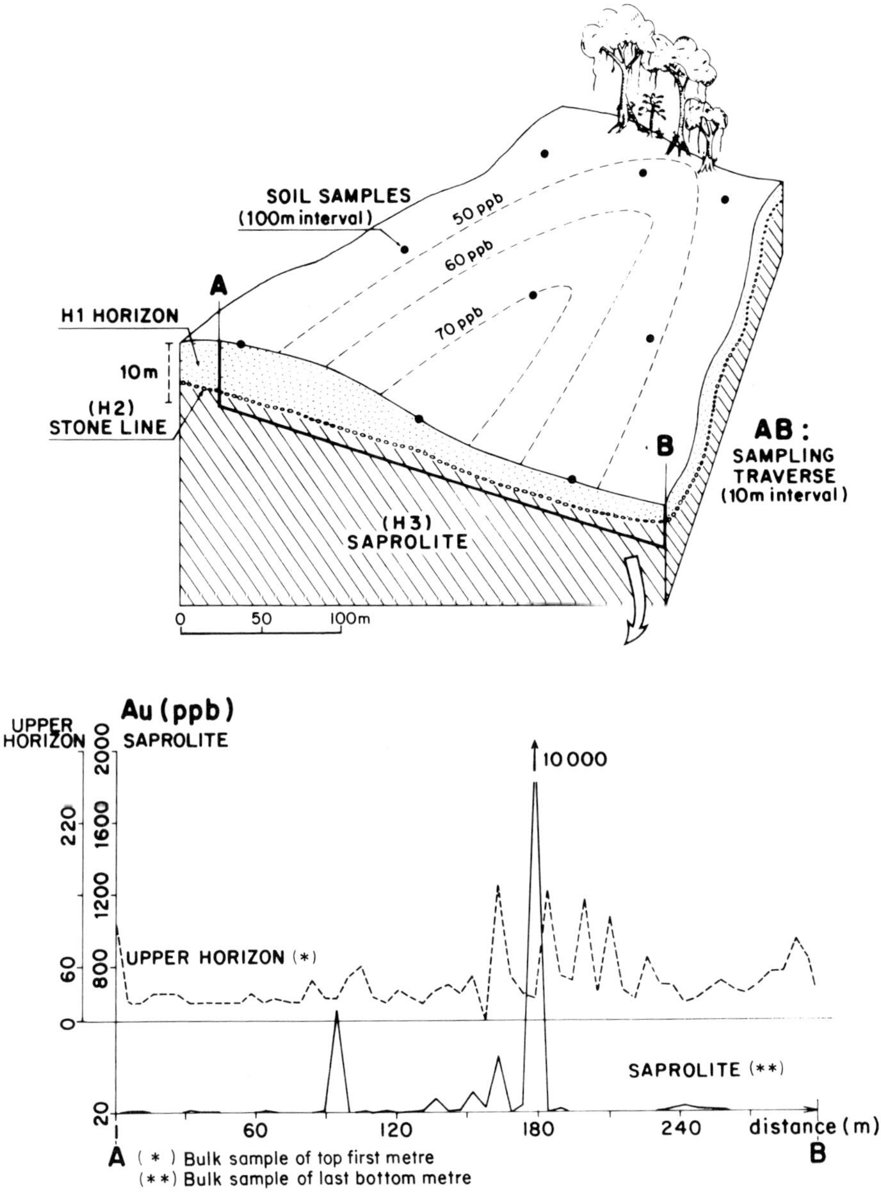

Figure 10.5 Mebaga prospect (Gabon). Diagram showing the soil gold anomaly and the location and results of the auger sampling at the surface and in the saprolite. (From Colin and Lecomte, 1988.)

This Mebaga example thus demonstrates that minor gold mineralization corresponds to weak gold anomalies in soil samples, whereas strong leaching is observed for several trace or major elements.

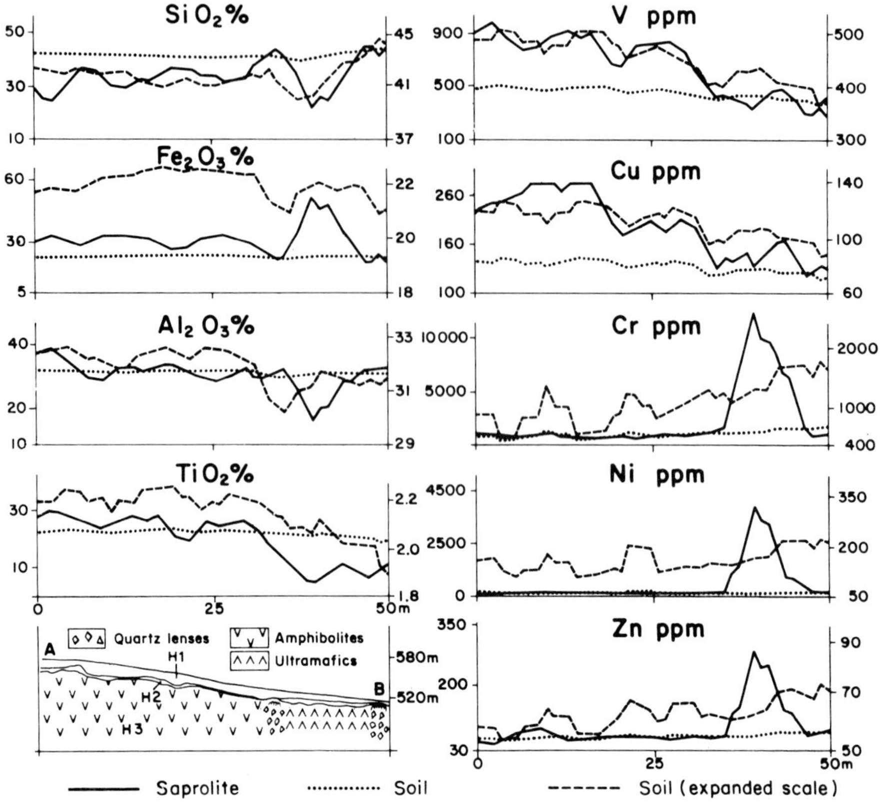

Figure 10.6 Mebaga prospect (Gabon). Distribution of selected major and trace element contents in the H1 upper horizon and in the H3 (saprolite) horizon, along traverse AB (see Figure 10.5). (From Colin and Lecomte, 1988.)

At Dondo Mobi (Gabon), the dispersion of gold in the weathering profile above bedrock mineralization was investigated by careful sampling, chemical analysis, and investigation of the morphology of gold grains (Lecomte and Colin, 1989; Colin *et al.*, 1989). The morphoscopic study of gold grains demonstrated that part of the primary gold was dissolved, but no evidence was found to indicate re-precipitation of secondary gold in the supergene environment. The gold dispersion mushroom (Figure 10.7), therefore, has an important residual component, but the presence of very fine-grained secondary gold cannot be excluded. The surface dispersion halo of the gold is about 100 m wide at the 0.5 ppm level and more than 200 m at the 0.15 ppm level. Part of the gold present in that halo is probably a residue of the profile-reduction by chemical weathering.

Stone-line environments thus correspond to conditions leading to an important surficial dispersion of gold, but the relative contributions of mechanical and hydromorphic processes to that dispersion are not fully understood. The anomalous concentrations obtained at the surface may be as high as several grams per tonne just

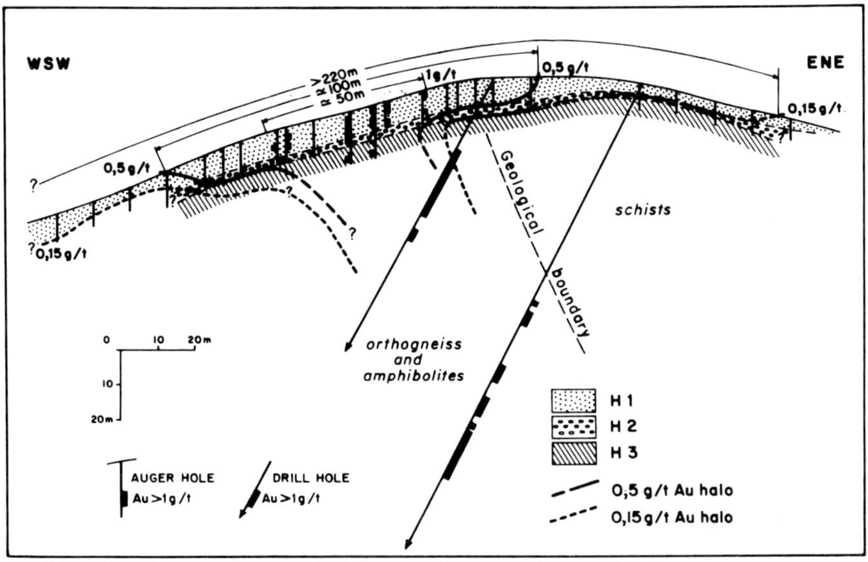

Figure 10.7 Dondo Mobi gold prospect (Gabon). Mushroom-like dispersion pattern of gold across a vertical section. H1: upper, friable horizon; H2: stone-line; H3: saprolite. (From Lecomte, 1988.)

above the mineralization (Dondo Mobi), indicating that gold is not significantly leached in such environments.

At Dorlin, French Guiana, geochemical sampling followed by multi-element analysis also proved suitable for delineating gold mineralization in deeply weathered environments (Taylor *et al.*, 1989; Zeegers, 1987). Gold contents at the surface reflected the bedrock mineralization and the hydrothermal alteration was also clearly indicated in saprolite samples collected at a few metres depth. Abundance of sericite corresponded to high K_2O contents, B and MgO to tourmaline, and As, Sb, Cu to sulphides. On the contrary, chlorite, abundant in drillhole samples, was completely leached from the saprolite (Table 10.2). Thus, 'blind' sampling of the saprolite indicates the presence of gold mineralization at depth and can define the hydrothermal alteration assemblages.

Table 10.2 Dorlin (French Guiana) gold mineralization. Variation in contents of some selected elements, corresponding to the hydrothermal paragenesis, from bedrock to near-surface saprolite (from Zeegers, 1987)

		Bedrock		Saprolite
Sericite	K_2O	1.9%	======>	1.3%
Chlorite	MgO	6%	======>	0.7%
Pyrite	Fe_2O_3	20%	======>	27%
Arsenopyrite	As	900 ppm	======>	700 ppm
Tourmaline	B	$n \times 1000$ ppm	======>	$n \times 1000$ ppm
	MgO	$n \times 1\%$	======>	$n \times 1\%$

$1 \leq n \leq 10$.

In high-relief terrains, the pre-existing profile, if any, has generally been removed by erosion, and allochthonous material is common on the slopes. Locally, stripping has resulted in removal of the entire soil profile, and the exposure of fresh rock. The mechanical component of geochemical dispersion is important, and it can be quite difficult to precisely locate the origin of geochemical anomalies obtained in residual soils or colluvial material. Chemical processes of dispersion of gold can also be active, as demonstrated by Webster and Mann (1984) from the Morobe Goldfield, Papua New Guinea. There, primary gold occurs as epithermal mineralization, commonly as gold-rich quartz–rhodocrosite veins. In the weathering zone, the Mn-carbonates are replaced by oxides, and gold and silver may be enriched as secondary Au or electrum. Complexing of gold and silver is assumed to be achieved through thiosulphate ligands formed by the weathering of sulphides, the pH being controlled by the carbonate minerals.

10.4 Operating procedures

10.4.1 *Sampling and sample preparation*

Collecting geochemical samples for gold exploration is not fundamentally different to exploration for other types of ore deposits. However, anomalous samples may have rather low gold contents (commonly less than 0.2 ppm), with important consequences for the mass of sample to be analysed in order to incorporate a minimum number of gold particles (Harris, 1982; Day and Fletcher, 1986). Large grain size and low contents of gold may lead to erratic and unreproducible analytical data for geochemical samples. It should also be remembered that gold is malleable and that pulverizing samples is not very effective on gold particles (Harris, 1982). Sampling media and intervals, and sample preparation procedures used by several authors and corresponding to different topographic and climatic environments are presented in Table 10.3.

At the regional stage, sampling for gold exploration does not require unique or unusual procedures, especially if stream sediments are collected. Such samples are generally prepared by drying and sieving. As for other elements, a fine fraction is generally retained for analysis. In many instances, a common practice in gold exploration involves the collection of 500 g of sample and the recovery by sieving of the –80 mesh (< 177 μm) fraction (Nichol, 1986). In Gabon, gold was determined in the < 125 μm fraction of stream sediment samples (Barthélémy *et al.*, 1987) collected during regional exploration. In the desert environment of Saudi Arabia, Salpéteur and Sabir (1989) recommended the use of the –80 μm fraction for determining gold in stream sediment and soil samples. Heavy concentrates are also commonly used as a sampling medium, followed by the determination of visible gold (see page 327 of this chapter).

In situations where soil sampling is required, for example at the regional stage in low relief terrains or for follow-up surveys, selecting the most appropriate sampling media may be more difficult, especially in deeply weathered, lateritic, environments. There, gold or the pathfinder elements may be incorporated in specific mineral phases, corresponding to different grain size material. If, for instance, the Fe-oxides are considered to retain gold or other elements, sampling should focus on either the

Table 10.3 Geochemical exploration for gold: summary of different sampling media and intervals, size fractions retained for analysis, Au geochemical responses obtained, and significant pathfinder elements, in relation to various climatic and morphological environments

Ref.	Climate	Relief	Stage of exploration	Sampling media	Sampling Interval	Size fraction	Au response	Path-finders
(1)	Temperate	Low	REGIO	SS	2–3/km^2	<125 µm	None	(+)As, Li, B
(1)	Temperate	Low	REGIO	HC	1/km^2	–	Good	
(1)	Temperate	Low	DETAIL	SO	50×100 m, 50×200 m,...	<125 µm	Good	(+) As, Pb, Sb, W
(1)	Temperate	Low	PRE–D	WR, RO	5, 10, 20 m	Total	Good	
(2)	Temperate	Moderate	DETAIL	SO	100×25 m	<125 µm	Good	(+)As, Li, B
(3)	Temperate	Moderate	REGIO	HC	1/km^2		Good	
(3)	Temperate	Moderate	DETAIL	SO		<63 µm	Good	(+)As, Hg
(4)	Temperate	Moderate	DETAIL	SO	10 m		Good	(+)As, Ag, Hg, Sb, W (–)Ca, Mg
(5)	Arid	Low	REGIO	SS	2/km^2	<80 µm	Good	
(5)	Arid	Low	DETAIL	SO	50, 100 m	<80 µm	Good	
(6)	Arid	Low	REGIO	PL	1/km^2	Total	Good	(+)As, Bi, Sb, Mo, Ag, Sn, Ge, W Se
(7)	Savanna	Low	REGIO	SO, SS	1600×500 m	<125 µm	Good	(+)B
(8)	Savanna	Low	REGIO	SO	300×500 m	Total	Good	(+)Cu, Zn, Mo
(9)	Rainforest	Moderate	REGIO	SS	1.5/km^2	<125 µm	Good	
(9)	Rainforest	Moderate	REGIO	HC	0.5/km^2		Good	
(10)	Rainforest	Moderate	DETAIL	SO	100×200 m	>125 µm	Poor	
(10)	Rainforest	Moderate	PRE–D	SA	10 m	Total	Good	
(11)	Rainforest	Moderate	PRE–D	SA	5, 10 m	Total	Good	(+)Ag, Mo, Pb, Si (–)Fe, Ti, Sr
(12)	Rainforest	Moderate	PRE–D	SA	10, 20 m	Total	Good	(+)As, B, Sb, Cu, K, Mg

Exploration stage: REGIO, regional; DETAIL, detailed; PRE–D, pre-drilling.

Sampling media: SS, stream-sediments; HC, heavy concentrates (with determination of visible gold); SO, soils; PL, pisolitic laterite; SA, saprolite; WB, weathered bedrock; RO, bedrock.

Pathfinder elements: (+) enriched; (–) depleted.

Reference: (1) Vasquez-Lopez *et al.* (1987); (2) Braux *et al.* (1989); (3) Janatka and Moravek (1987); (4) Chaffee and Hill (1989); (5) Salpéteur and Sabir (1989); (6) Smith *et al.* (1989); (7) Dommanget *et al.* (1987); (8) Ouedraogo (1988);(9) Barthélémy *et al.* (1987); (10) Colin and Lecomte (1988); (11) Zeegers (1987); (12) Taylor *et al.* (1989).

ferruginous cuirasse itself or derived products such as pisoliths (Smith *et al.*, 1989) or lag (Carver *et al.*, 1987). Concentration of the selected phase from the raw material can be achieved by sieving and recovering the coarse fraction.

Saprolite is commonly sampled at the detailed exploration stage, and can be considered as an alternative to lithogeochemistry in deeply weathered terrains. Taylor *et al.* (1989) have used saprolite sampling on an intensive scale at Dorlin, French Guiana. They were able to delineate alteration patterns and to distinguish subtle characteristics such as the Fe-content of tourmalines from multi-element analysis of the saprolite.

10.4.2 *Sample analysis*

The considerable interest in gold exploration over the past decade has resulted in the improvement of analytical methods for determining the gold contents of surface materials. Hoffman and Brooker (1982) and Hall and Bonham-Carter (1988) reviewed and discussed the different sample-decomposition procedures for solid materials and the main analytical techniques used for the determination of gold (Table 10.4).

Table 10.4 Methods in common use to determine gold in geological materials

Decomposition	Analysis	Detection limit (ppb)	Mass (g)
Pb–FA	F–AAS	5	10–30
	DCP– or ICP–ES	1	20
	NAA	1	20
	ICP–MS	1	10
	ICP–AFS	2	20
NiS–FA	NAA	1	25
	ICP–MS	1	25
AR or HBr–Br_2	GF–AAS	1	10
	INAA	5	10–30

Explanation: Pb–FA, lead fire assay; NiS–FA, nickel sulphide fire assay; AR, aqua regia; F–ASS, flame atomic absorption spectrometry; FG–AAS, graphite furnace atomic absorption spectrometry; ICP, inductively coupled plasma; DCP, direct current plasma; ES, emission spectrometry; MS, mass spectrometry; AFS, atomic fluorescence spectrometry; NAA, neutron activation analysis; INAA, instrumental neutron activation analysis. (From Hall and Bonham-Carter, 1988.)

These instrumental techniques are designed to detect low concentrations of gold (ppb range) in most geological materials, and the classical fire assay with gravimetric estimation of the gold content is now reserved for ore-grade determinations due to sensitivity limitations (ppm range). New methods of gold exploration (hydro- and biogeochemistry) require analytical techniques with very low sensitivity (0.1 ppb in plants by INAA; 1 ppt in water samples by NAA and GF–AAS) as well as specific sample preparation techniques (Brooks *et al.*, 1981; Hoffman and Brooker, 1986; Hall and Bonham-Carter, 1988; McHugh, 1988).

Mention must also be made of a special method of gold determination currently used in the desert terrains of Australia – BLEG (Bulk Leaching of Extractable Gold), which consists of cyanide leaching of a large (0.5–5 kg) sample of bulk material. Very subtle (ppb level) gold anomalies have been shown to indicate mineralization where conventional techniques failed (Sharpe, 1988).

10.5 Alternative sampling techniques

10.5.1 *Lithogeochemistry*

Lithogeochemical techniques applied to gold exploration are not really specific, compared to applications in the search for other types of ore deposits. Most of the published data are issued from orientation surveys undertaken in known mineralized environments. Two different scales of lithogeochemical surveys can be distinguished and are considered below:

At the regional scale, some specific lithofacies, reputed to host gold mineralization, may be sought by the identification of some of their chemical attributes. Such an approach, consisting more of predictive metallogeny than exploration, is especially suited to syngenetic types of deposits, e.g. massive sulphides. The presence of enhanced gold contents in some lithologies, thus suggesting a potential for mineralization, is mentioned in a general way by Boyle (1979); more specifically Roslyakov and Roslyakova (1975, in Boyle, 1979) described extensive primary haloes of varied size (12–600 km^2) around hydrothermal gold deposits in the USSR, where the gold contents exceeded at least twice the regional background. Moravek and Pouba (1984, 1987) suggested that the gold of different types of gold-bearing mineralization in the Bohemian Massif was derived from Proterozoic and Phanerozoic volcano-sedimentary rocks showing enrichment of gold (2.8–7.5 ppb on average) compared to the surrounding Palaeozoic sedimentary and crystalline rocks, which averaged approximatively 1 ppb Au.

At the detailed scale, numerous examples of primary haloes around gold orebodies have been described. Lode- or vein-type deposits are generally characterized by gold haloes (Boyle, 1979, 1982), the dimensions and shapes of which are similar to those of haloes developed around base-metal deposits (Govett, 1983). In the Archaean basement of Canada, Perrault and Trudel (1984) and Beaudouin *et al.* (1987) showed that primary gold haloes, developed up to 400 m distant from some lode deposits, can be used to assess the potential of mining properties. Other types of deposits, such as Carlin type, may develop leakage haloes related to pervasive hydrothermal alteration (Boyle, 1979), or other primary haloes (Crone *et al.*, 1984; Ronkos, 1986).

10.5.2 *Heavy-mineral concentrate geochemistry*

Exploration programmes commonly make use of alternative sampling techniques involving the separation of fractions of surface material (soil, alluvium) according to specific gravity or magnetic susceptibility. In gold exploration, different approaches are possible using heavy-mineral concentrates obtained by panning or by heavy liquid separation:

(i) optical determination and estimation of the content of visible gold and identification of possible pathfinder minerals;

(ii) chemical analysis with determination of gold content of the non-magnetic, heavy-mineral fraction (AAS).

Numerous examples of such applications are available, corresponding to different topographic and climatic enviroments:

- mountainous terrain in Yukon (Boyle, 1979) or British Columbia (Barakso and Tegart, 1982);
- subtropical: north Queensland, Australia (Watters, 1983);
- arid, lateritic: Sudan (Fletcher, 1985), Western Australia and Botswana (Farrel, 1984);
- temperate: Brittany, France (Vasquez-Lopez *et al.*, 1987); Czechoslovakia (Janatka and Moravek, 1987).

Most examples demonstrate the efficiency of heavy mineral surveys for delineating high-potential areas for gold at the regional scale of exploration. Where compared with conventional sampling (< 177 μm fraction of stream sediments), the results obtained from heavy concentrates show an improved anomaly contrast and a greater extent. The sampling interval can then be optimized, resulting in cost-saving. Moreover, as the anomalous contents are much greater (commonly ppm instead of ppb), the confidence level attached to the results is also high (Nichol, 1986).

Sampling requires some precautions, as the sample representativity can be hampered by the scarcity of gold particles in alluvium. This can be minimized by collecting and processing a large volume of material. Also, the hydraulic energy level corresponding to each sampling site should be taken into account, as this clearly influences the total content of heavy minerals in the sample (Day and Fletcher, 1986, 1989).

To summarize, heavy concentrate surveys for gold exploration seem to be most useful for regional exploration. Of course, optimum responses can be expected where physical processes of dispersion are most active. As mentioned previously (see page 315), it should be kept in mind that some styles of gold mineralization (in which gold occurs in the crystal lattice of sulphides, or free but very fine grained) can be missed by heavy-concentrate techniques.

10.5.3 *Hydrogeochemistry*

Apart from its use in the USSR (Chernyaev *et al.*, 1969; Volkov and Shakhbazova, 1975) direct hydrogeochemical prospecting for gold has been largely ignored. Only a limited amount of work has been done in the western world to assess the effectiveness of the technique and the results are somewhat diverse depending on the scale of the surveys.

Natural waters (spring, stream, and mine drainage) draining gold-barren and gold-enriched terrains in a large area of Colorado were filtered and assayed for their particulate (> 0.1 μm), solute, and total gold contents by Gosling *et al.* (1971). Total gold contents (NAA determination) ranged from less than 0.001 ppb up to 0.150 ppb, but the highest contents did not appear to be indicative of specific gold deposits.

Turner and Ikramuddin (1982) collected and determined Au, Ag, As, and Si in filtered stream-water samples from mineralized areas in Washington where Au–Ag-bearing quartz veins are known to occur. Concentrations of Au ranged between 0.009 and 0.140 ppb, and in this case the samples in which element contents exceeded the anomalous thresholds (determined by using Au/Si, Ag/Si, As/Si ratios) mostly corresponded to mineralized areas.

Hamilton *et al.* (1983) compared particulate and dissolved gold contents of surface-water samples collected upstream, within, and downstream of well-defined gold-bearing areas in New South Wales, Australia. The use of particulate gold ($> 1.2\,\mu m$) was found unpromising for locating the gold anomalies, whereas anomalous dissolved gold contents (up to 0.130 ppb) where shown to be related directly to gold mineralization.

McHugh (1988) investigated the amount of gold present in natural water by summarizing and critically evaluating data published since 1969, and presented gold analyses for various filtered waters collected throughout the world. Particularly interesting was the fact that the average gold contents of waters collected in unmineralized and mineralized areas were comparable for most of the examples considered – 0.002 ppb and 0.101 ppb respectively.

Thus, under favourable conditions, bedrock gold mineralization may develop hydrogeochemical gold-dispersion haloes in surface waters. Nevertheless, no examples of systematic and successful applications of hydrogeochemical techniques to gold exploration are available. Thus the technique is still to be considered to be at the development stage.

10.5.4 *Atmogeochemistry*

Atmogeochemical surveys for mineral deposits are based on the analysis of gases and volatile compounds associated with the mineralization types. The analysis of such compounds trapped in surficial material (mostly soils) is, therefore, designed for the search for hidden, often deeply buried, deposits. However, there is no specific atmogeochemical method designed for prospecting for gold deposits.

The main gases or volatile compounds associated with mineral deposits (Beus and Grigorian, 1977; Degranges *et al.*, 1983; McCarthy, Jr., 1986) are:

(i) gases directly associated with ore-formation processes (halogens, CO_2);
(ii) gases generated by supergene weathering of sulphide minerals or by secondary reactions with carbonate gangue minerals (SO_2, H_2S, COS, CO_2);
(iii) volatile compounds of Hg, I, As, Sb;
(iv) primary gases (He, H_2), detected along deep-seated faults, that can be useful in exploration if the mineral deposits are controlled by such faults.

The migration processes of gases in the supergene environment and the main sampling and analytical techniques have been reviewed and discussed by McCarthy, Jr. (1986) and Degranges *et al.* (1983). In terms of gold exploration, the following applications should be considered;

- CO_2, COS, Hg for gold-bearing volcanogenic massive sulphide deposits;
- Hg, As and Sb for epithermal deposits; volatile hydrides such as AsH_3 and SbH_3 may be formed if H_2 is available (McCarthy, Jr., *op. cit.*);
- He, As, Hg for lode-type deposits.

The use of atmogeochemistry for gold exploration should be considered in conjunction with other techniques but may be particularly appropriate to the search for buried deposits. Optimum conditions of use are met in dry climates ; in humid terrains, the presence of ground or surficial water may hamper the migration of gases to the surface.

10.5.5 *Biogeochemistry*

Biogeochemistry was first used for gold exploration at an operational level in the USSR (Brooks, 1982). Subsequently, the technique was extensively used in North America where forests grow on glacial till or lacustrine deposits. Humus or mull-litter sampling should also be considered as an alternative biogeochemical technique.

Just as for other elements, biogeochemical prospecting of gold faces specific problems, as the gold contents vary according to the species, the part, and the age of the part of the floral species used, and also the season of sampling. Some examples of such variations were presented by Erdman and Olson (1985).

Few examples are available concerning the use of biogeochemistry for gold exploration in non-till environments. In the Mojave Desert of California, the analysis of selected bushes revealed anomalous gold contents (2–5 ppb compared to a background of < 2 ppb, on dry weight basis) above epithermal gold mineralization covered by some 30 m of allochthonous overburden (Busche, 1989).

In Western Australia, Smith and Keel (1984) compared results obtained by shallow drilling in a deep weathering profile above a Au-bearing shear zone at Norseman with biogeochemical data. Geochemical sampling only revealed significant gold contents (>10 ppb) in the upper part of the weathering profile. These weakly contrasted anomalies, which were interpreted as possibly resulting from decayed plant litter, did not allow delineation of the trace of the underlying mineralization. Conversely, a marked gold anomaly (up to 230 ppb in ash compared to a background of 60 ppb) was obtained by biogeochemical sampling at the surface projection of the mineralization.

In Tasmania, Baker (1986) reported highly contrasted gold anomalies (up to 76 ppm in ash) above alluvial gold concentrations overlain by a thick transported overburden.

Thus, from a general viewpoint, biogeochemical techniques may be useful for exploration for gold where conventional techniques, best adapted for residual environments, are hampered by the presence of overburden. However, using soil, stream sediments, saprolite, and rocks is far easier than collecting, processing and analysing organic material such as leaves, twigs or bark.

10.6 Conclusions

Temperate climates. The chemical mobility of gold is low and mechanical processes dominate. Panning for visible gold is the most pragmatic aproach to regional exploration, although some styles of mineralization, particularly sulphide-impregnated shear zones, may not yield haloes of detrital gold. In such situations, multi-element analysis is strongly recommended.

Arid terrains. Mechanical dispersion is predominant and so panning is recommended for regional exploration. In some circumstances, geochemical sampling and analysis of the fine fraction (e.g. -80 mesh) may be useful in both regional and detailed surveys.

Semi-arid terrains with relics of lateritic weathering. Chemical dispersion is more important and large dispersion haloes can develop. Economically viable supergene deposits can develop but the secondary gold is commonly very fine grained and not detectable by panning.

Rain-forest environments. Gold is strongly dispersed by chemical and mechanical processes resulting in haloes extending several hundred metres away from the primary mineralization. Net loss of gold from horizons overlying deposits is often low and so chemical analysis for gold or appropriate pathfinder elements is reliable.

In most lateritic environments, selection of the correct horizon and/or size fraction is of critical importance; a common and fairly reliable procedure consists of sampling a ferruginous coarse size-fraction. Chemical analysis of samples for gold, sometimes accompanied by panning, is commonly sufficient to detect most types of bedrock gold mineralization. However, multi-element analysis offers the additional potential advantages of identifying the style and therefore the potential of the target mineralization at an early stage in the exploration programme.

Alternative sampling media such as rocks ('predictive metallogeny' – p.327), water, volatiles, or plants may be considered in specific conditions where conventional techniques are hampered, particularly in areas of transported overburden.

References

Bache J.J. (1982) Les gisements d'or dans le monde. Essai de typologie quantative. Mém. BRGM 118.

Bagby, W.C. and Berger, B.R. (1985) Geologic characteristics of sediment-hosted, disseminated precious-metals deposits in the Western United States. In *Geology and Geochemistry of Epithermal Systems*, eds. Berger, B.R. and Bethke, P.M. Reviews in Economic Geology, 2, Soc. of Econ. Geol., 169–202.

Baker, W.E. (1978) The role of humic acid in the transport of gold. *Geochem. Cosmochim. Acta* **42** 645-649.

Baker W.E. (1986) Gold in vegetation as prospecting in Tasmania, Australia. In *Mineral Exploration : Biolological Systems and Organic Matter*, Rubey Volume V, eds, Carlisle, D., Berry, W.L., Kaplan, I.R. and Watterson, J.R. Prentice-Hall, Englewood Cliffs, N.J., 150–158.

Barasko, J.J. and Tegart, P. (1982) Revised stream sediment geochemistry at selected precious metal deposits in British Columbia. *Western Miner*, April, 53-65.

Barbier, J. (1987) Arid environment geochemical exploitation in Saudi Arabia. The problem of aeolian contamination. *J. Geochem. Explor.* **27**, 29-51.

Barbier, J. and Wilhelm, E. (1978) Superficial geochemical dispersion around sulphide deposits: some examples in France. *J. Geochem. Explor.* **10**, 1-38.

Barthélémy, F., Kassa Mombo, M. and Labbé, J.F. (1987) Inventaire minier transgabonais: campagnes géochimiques régionales et détaillées 1979-1986. Acquisition des données sur le terrain. In *Abstracts of the 12th IGES*, Orléans, France, April 23-26.

Beaudoin, A., Perrault, G. and Bouchard, M. (1978) Distribution of gold, arsenic and tungsten around the Dest-Or orebody, Noranda district, Abitibi, Quebec. *J. Geochem. Explor.* **28**, 41-70.

Beus A.A. and Grigorian S.V. (1977) *Geochemical Exploration Methods for Mineral Deposits.* Applied Publishing, Illinois.

Bonnemaison, M. (1986) Les 'filons de quartz aurifère: un cas particulier de shear zone aurifière. *Chron. Rech. Min.* **482**, 55-66.

Bonnemaison, M. and Marcoux, E. (1987) Les zones de cisaillements aurifères du socle hercynien francais. *Chron. Rech. Min.* **488**, 29-42.

Boyle, R.W. (1979) The geochemistry of gold and its deposits (together with a chapter on geochemical prospecting for the element). Geol. Surv. of Can., Bull. 280.

Boyle, R.W. (1982) Geochemical methods for the discovery of blind mineral deposits. Part 1. *CIM Bull.* **75**(844), 123- 142.

Braux, C., Bonnemaison, M., Prévot, J.C., Maurin, G. and Zeegers, H. (1989) Le Châtelet gold deposit, France, Creuse: gitology and geochemical exploration. *Gold '89 in Europe*, Toulouse, France, May 23-25 Terra Abstracts **1**, 19.

Brooks, R.R. (1982) Biological methods of prospecting for gold. *J. Geochem. Explor.* **17**, 109-122.

Brooks, R.R., Chatterjee, A.K. and Ryan, D.E. (1981) Determination of gold in natural waters at the parts-per-trillion (pg cm^{-3}) level. *Chem. Geol.* **33**, 163-169.

Busche, F.D. (1989) Using plants as an exploration tool for gold. *J. Geochem. Explor.* **32**, 199-209.

Butt, C.R.M. (1988) Genesis of lateritic and supergene gold deposits in the Yilgarn Block, Western Australia. In *Proc. of Gold '88 Symposium*, Melbourne, 359-364.

Butt, C.R.M. and Zeegers, H. (1989) Classification of geochemical exploration models for tropically weathered terrains. *J. Geochem. Explor.* **32**, 65-74.

Carver, R.N., Chenoweth, L.M., Mazzucchelli, R.H., Oates, C.J. and Robbins, T.W. (1987) 'Lag' – A geochemical sampling medium for arid regions. *J. Geochem. Explor.* **28**, 183-199.

Chaffee, M.A. and Hills, R.H. (1989) Soil geochemistry of Mother Lode-type deposits in the Hodson mining district, central California, USA (extended abstact). *J. Geochem. Explor.* **32**, 53-55.

Chernyayev, A.M., Chernyayeva, L.E., Yeremeyeva, M.N. and Andreyev, M.I. (1969) Hydrogeochemistry of gold. *Geochem. Int.* **6**, 348-358.

Cloke, P.L. and Kelly, W.C. (1964) Solubility of gold under inorganic supergene conditions. *Econ. Geol.* **59**, 259-270.

Colin, F. and Lecomte, P. (1988) Etude minéralogique et chimique du profil d'altération de Mébaga Mvomo (Gabon). *Chron. Rech. Min.* **491**, 55-65.

Colin, F., Lecomte, P. and Boulange, B. (1989) Dissolution features of gold particules in a lateritic profile at Donbo Mobi, Gabon. *Geoderma* (in press).

Colvine, A.C., Fyon, J.A., Heather, K.B., Marmont, S., Smith, P.H. and Troop, D.G. (1988) Archean lode gold deposits in Ontario. Ontario Geol. Surv., Misc. Paper 139.

Crone, W., Larson, L.T., Carpenter, R.H., Chao, T.T. and Sanzolone, R.F. (1984) A comparison of iron oxide-rich joint coatings and rock chips as geochemical sampling media in exploration for disseminated gold deposits. *J. Geochem. Explor.* **20**, 161-178.

Davies, T.C., Friedrich, G. and Wiechowski, A. (1989) Geochemistry and mineralogy of laterites in the Sula Mountains greenstone belt, Lake Sonfon gold district, Sierra Leone. *J. Geochem. Explor.* **32**, 75-98.

Davy, R. and El-Ansary, M. (1986) Geochemical patterns in laterite profile at the Boddington gold deposit, Western Australia. *J. Geochem. Explor.* **26**, 199-144.

Day, S. and Fletcher, K. (1986) Particule size and abundance of gold in selected streams sediments, southern British Columbia, Canada, *J. Geochem. Explor.* **26**, 203-214.

Day, S. and Fletcher, K. (1989) Effect of valley and local channel morphology on the distribution of gold in stream sediments from Harris Creek, British Columbia, Canada. *J. Geochem. Explor.* **32**, 1-16.

Degranges, P., Baubron, J.-C. and Leleu, M. (1983) Où en est l'atmogéochimie appliquée à la prospection des gîtes? *Chron. Rech. Min.* **473**, 67-73.

Dommanget, A., Diallo, M. and Guilloux, L. (1985). Un nouveau type de gisement d'or: Loulo (Mali). *Chron. Rech. Min.* **481**, 5-18.

Dommanget, A., Traore, H. and Zeegers, H. (1987) Discovery of the Loulo gold deposit (Mali). In *Abstracts of the 12th IGES*, Orléans, France, April 23-26.

Erdman, J.A. and Olson, J.C. (1985) The use of plants in prospecting for gold : a brief overview with a selected bibliography and topic index. *J. Geochem. Explor.* **24**, 281-304.

Farrell, B.L. (1984) The use of 'loam' concentrates in geochemical exploration in deeply weathered arid terrains. *J. Geochem. Explor.* **22**, 101-118.

Fletcher, R.J. (1985) Geochemical exploration for gold in the Red Sea Hills, Sudan. I.M.M. Spec. Issue, *Prospecting in Areas of Desert Terrain*, 79-94.

Freyssinet, P., Zeegers, H. and Tardy, Y. (1987) Néoformation de l'or dans les cuirasses latéritiques : dissolution, migration, précipitation. *C.R. Acad. Sci. Paris* **305**, Série II, 867-874.

Freyssinet, P., Zeegers, H. and Tardy, Y. (1989) Morphology and geochemistry of gold grains in lateritic profiles of southern Mali. *J. Geochem. Explor.* **32**, 17-31.

Fripp, R.E.P. (1976) Stratabound gold deposits in archean banded iron-formation, Rhodesia. *Econ. Geol.* **71**, 58-75.

Goleva, G.A., Krivenkov, V.A. and Guttz, Z.G. (1970) Geochemical trends in the occurrence and migration forms of gold in natural water. *Geoch. Int.* **7**, 518-529.

Gosling, A.W., Jenne, E.A. and Chao, T.T. (1971) Gold content of natural waters in Colorado. *Econ. Geol.* **66**, 309- 313.

Govett, G.J.S. (1983) *Rock Geochemistry in Mineral Exploration.* Elsevier, Amsterdam.

Groves, D.I., Ho, S.E., McNaughton, N.J., Mueller, A.G., Perring, C.S., Rock, N.M.S. and Skwarnecki, M. (1988) Genetic models for Archaean lode-gold deposits in Western Australia. In *Advances in Understanding Precambrian Gold Deposits*, Vol. II, eds. Ho, S.E. and Groves, D.I. Geol. Dept. Univ. Extension, Univ. of W. Austr, Pub. 12, 1-22.

Hall, G.E.M. and Bonham-Carter, G.F. (1988) Review of methods to determine gold, platinum and palladium in production-oriented geochemical laboratories, with application of a statistical procedure to test for bias. *J. Geochem. Explor.* **30**, 255-286.

Hamilton, T.W., Ellis, J., Florence, T.M. and Fardy, J.J. (1983) Analysis of gold in surface waters from Australian goldfields : an investigation into direct hydrogeochemical prospecting for gold *Econ. Geol.* **78**, 1335-1341.

Harris, D.C. (1986) Mineralogy and geochemistry of the main Hemlo gold deposits, Hemlo, Ontario, Canada. In *Proc. of Gold '86 Symposium*, Toronto, 297-310.

Harris, J.F. (1982) Sampling and analytical requirements for effective use of geochemistry in exploration for gold. In *Precious Metals in the Northern Cordillera*, Assoc. of Expln. Geochemists, Special Vol. 10, 53-67.

Hayba, D.O., Bethke, P.M., Heald, P. and Foley, N.K. (1985) Geologic, mineralogic andd geochemical characteristics of volcanic-hosted epithermal, precious-metal deposits. In *Geology and Geochemistry of Epithermal Systems*, eds. Berger, B.R. and Bethke, P.M. Reviews in Economic Geology, 2, Soc. of Econ. Geol., 129-167.

Heald, P., Foley, N. and Hayba, D.O. (1987) Comparative anatomy of volcanic-hosted epithermal deposits: acid-sulfate and adularia- sericite types. *Econ. Geol.* **82**, 1-26.

Hoffman, E.L. and Brooker, E.J. (1982) The determination of gold by neutron activation analysis. In *Precious Metals in the Northern Cordillera*. Assoc. of Expln. Geochemists, Special Vol. 10, 69-77.

Hoffman, E.L. and Brooker, E.J. (1986) Biogeochemical prospecting for gold using instrumental neutron activation analysis with reference to some canadian gold deposits. In *Mineral Exploration: Biolological Systems and Organic Matter*, Rubey Volume V, eds. Carlisle D., Berry W.L., Kaplan I.R. and Watterson J.R. Prentice-Hall, Englewood Cliffs, N.J., 159-169.

Janatka, J. and Moravek, P. (1987) Geochemical exploration in the Célina Mokrsko gold district in the Jilove Belt, Bohemian massif, Czechoslovakia. In *Abstracts of 12th IGES*, Orléans, France, April 23-26.

Kerrich, R. (1983) Geochemistry of gold deposits in the Abitibi greenstone belt. CIM. Spec. Vol. 27, 1-75.

Kinkel, A.R. Jr. and Lesure, F.G. (1968) Residual enrichment and supergene migration of gold, southeastern United States. US Geol. Surv. Prof. Paper 600-D, D174-D178.

Krauskopf, K.B. (1951) The solubility of gold. *Econ. Geol.* **46**, 858-870.

Large, R., Huston, D., McGoldrick, P., McArthur, G and Ruxton, P. (1988) Gold distribution and genesis in Paleozoic volcanogenic massive sulphide systems, Eastern Australia. In *Proc. of Gold '88 Symposium*, Melbourne, 121-126.

Lawrance, L.M. (1988a) Behaviour of gold within the weathering profile in the Yilgarn block, Western Australia. In *Advances in Understanding Precambrian Gold Deposits*, Vol. II, eds. Ho, S.E. and Groves, D.I. Geol. Dept. Univ. Extension, Univ. of W. Austr, Pub. 12, 335-351.

Lawrance, L.M. (1988b) Geochemical dispersion anomalies in transported overburden around the quartz vein system at the Mount Pleasant gold mine, Western Australia. In *The 2nd Int. Conf. on Prospecting in Arid Terrain*, Perth, W. Australia. Aust. IMM, Symposia Series 57, 87-93.

Lecomte, P. (1988) Stone line profiles: importance in geochemical exploration. *J. Geochem. Explor.* **30**, 35-61.

Lecomte, P. and Colin, (1989) Gold dispersion and size fraction distribution in a tropical rain forest weathering profile at Dondo-Mobi, Gabon. *J. Geochem. Explor.* (in press).

Leprun, J.C. (1979) Les cuirasses ferrugineuses des pays cristallins de l'Afrique Occiddentale sèche. Genèse-Transformations-Dégradation. Mém. Sci. Géol., Strasbourg, 58.

Ling Ong, H. and Swanson, V.E. (1969) Natural organic acids in the transportation, deposition and concentration of gold. *Colorado School Mines Quarterly* **64**, 395-425.

McCarthy, J.H. Jr. (1986) The uses of gases in geochemical exploration. *Anal. Chem.* **43**, 171-180.

McHugh, J.B. (1988) Concentration of gold in natural waters. *J. Geochem. Explor.* **30**, 85-94.

Mann, A.W. (1984a) Mobility of gold and silver in lateritic weathering profiles. Some observations from Western Australia. *Econ. Geol.* **39**, 38-49.

Mann, A.W. (1984b) Redistribution of gold in the oxidized zone of some Western Australian deposits. The Aust. IMM, Perth and Kalgoorlie branches, Regional Conf. on Gold Mining, Metall. and Geol., 1-12.

Monti, R. (1988) The Boddington lateritic gold deposit, Western Australia: a product of supergene enrichment processes. In *Recent Advances in Understanding Precambrian Gold Deposits*, Vol II eds. Ho, S.E. and Groves, D.I. Geol. Dept. & Univ. Extension, Univ. W. Aust., Publ. 11, 355-368.

Moravek, P. and Pouba, Z. (1984) Gold mineralization and granitoids in the Bohemian Massif, Czechoslovakia. In *Gold '82 : The Geology Geochemistry and Genesis of Gold Deposits*, ed. Foster R.P. Balkema, Rotterdam, 713-729.

Moravek, P. and Pouba, Z. (1987) Precambrian and Phanerozoic history of gold mineralization in the Bohemian massif. *Econ. Geol.* **82**, 2098-2114.

Nahon, D. (1976) Cuirasses ferrugineuses et encroûtements calcaires au Sénégal occidental et en Mauritanie. Systèmes évolutifs : géochimie, structures, relais et coexistence. Mém. Soc. Géol., Strasbourg, 44.

Nair, N.G.K., Santosh, M. and Mahadevan, R. (1987) Lateritisation as a possible contributor to gold placers in Nilambur Valley, Southwest India. *Chem. Geol.* **60**, 309-315.

Nichol, I. (1986) Geochemical exploration for gold: a special problem. In *Applied Geochemistry in the 1980s : Proc. Meeting to Honour the Contribution of Professor John S. Webb to Applied Geochemistry*, eds. Thornton, I. and Howarth, R.J. Graham and Trotman, London, 60-85.

Ouedraogo, M.F. (1988) The Kwademen gold deposit: a polymorphic mineralization within a Precambrian (Birrimian) greenstone belt in Burkina Faso, West Africa : a case history (Abstract). In *The 2nd Int. Conf. on Prospecting in Arid Terrain*, Perth Western Australia. Aust. IMM, Symposia Series 57, p. 105.

Perrault, G. and Trudel, P. (1984) Auriferous halos associated with the gold deposits at Lamaque mine, Quebec. *Econ. Geol.* **79**, 227-238.

Prian, J.P., Siméon, Y., Johan, V., Ledru, P., Piantone, P., Coste, B. and Eko N'Dong, J. (1988) Valorisation géologique de l'inventaire minier de l'Archéen et du Protérozoique inférieur des feuilles de Mitzig, Booué et Mouila, à 1/200 000 (Gabon central). *Chron. Rech. Min.* **491**, 67-104.

Ronkos, C.J. (1986) Geology and interpretation of geochemistry at the Standard mine, Humbolt county, Nevada. *J. Geochem. Explor.* **25**, 129-137.

Salpéteur, I. and Sabir, H. (1989) Orientation studies for gold in the central pediplain of the Saudi Arabian shield. *J. Geochem. Explor.* (in press).

Sharpe, W.V. (1988) BLEG: a geochemical tool in the Great Sandy Desert. In *The 2nd Int. Conf. on Prospecting in Arid Terrain*, Perth, Western Australia. Symposia Series 57, 117- 119.

Silberman, M.L. and Berger, B.R. (1985) Relationship of trace- element patterns to alteration and morphology in epithermal precious-metal deposits. In *Geology and Geochemistry of Epithermal Systems*, eds. Berger, B.R. and Bethke, P.M. Reviews in Economic Geology, 2, Soc. of Econ. Geol., 203-232.

Sillitoe, R.H. (1988) Environments, styles and origins of gold deposits in western Pacific island arcs. In *Proc. of Gold '88 Symposium*, Melbourne, 127-138.

Smith, B.H. and Keele, R.A. (1984) Some observations on the geochemistry of gold mineralization in the weathered zone at Norseman, Western Australia. *J. Geochem. Explor.* **22**, 1-20.

Smith, R.E., Birrell, R.D. and Bigden, J.F. (1989) The implications to exploration of chalcophile corridors in the Archaean Yilgarn Block, Western Australia, as revealed by laterite geochemistry. *J. Geochem. Explor.* **32**, 169-184.

Tardy, Y. and Nahon, D. (1985) Geochemistry of laterites, stability of Al–goethite, Al–hematite and Fe-kaolinite in bauxites and ferricretes: an approach to the mechanism of concretion formation. *Amer. J. Sci.* **285**, 865-903.

Taylor, G.H., Coste, B., Lambert, A. and Zeegers, H. (1989) Geochemical signature (bedrock and saprolite) of gold mineralization and associated hydrothermal alteration at Dorlin, French Guyana (extended abstract). *J. Geochem. Explor.* **32**, 59-60.

Turner, L.D. and Ikramuddin, M. (1982) Electrothermal atomic absorption determination of gold, silver, and arsenic in stream waters and their relationship to gold–silver occurrences in the Republic Graben, N.E. Washington. In *Precious Metals in the Northern Cordillera*, Assoc. of Expln. Geochemists, Special Vol. 10, 79-88.

Vasconcelos, P. and Kyle, J.R. (1988) Gold geochemistry during semiarid weathering : a case study of the Fazenda Brasileiro deposit, Bahia, Brazil. In *Proc. of Gold '88 Symposium*, Melbourne, 645-647.

Vasquez-Lopez, R., Lefur, Y., Chevance, H., Bellivier, F., Brosset, R., Callier, L. and Kerjean, M. (1987) La recherche de l'or primaire dans le massif Armoricain (France): approches géochimiques et minéralogiques combinées. In *Abstracts of the 12th IGES*, Orléans, France, April 23- 26.

Volkov, G.A. and Shakhbazova, L.V. (1975) Some data on geochemistry of gold in natural waters of the Beskessk ore field. *Internat. Geology Rev.* **17**, 142-148.

Wargo, J.G. and Powers, H.A. (1978) Disseminated gold in Saddle prospect, Lander county, Nevada. *J. Geochem. Explor.* **19**, 236-241.

Watters, R.A. (1983) Geochemical exploration for uranium and other metals in tropical and subtropical environments using heavy mineral concentrates. *J. Geochem. Explor.* **19**, 103- 124.

Webster, J.G. (1986) The solubility of gold and silver in the system Au–Ag–S–O_2–H_2O at 25°C and 1 atm. *Geochim. Cosmochim. Acta* **50**, 1837-1845.

Webster, J.G. and Mann, A.W. (1984) The influence of climate, geomorphology and primary geology on the supergene migration of gold and silver. *J. Geochem. Explor.* **22**, 21-24.

Zeegers H. (1987) Remaniements de surface et prospection géochimique de l'or. *Chron. Rech. Min.* **488**, 55- 61.

11 Geochemical exploration for gold in glaciated terrain

W.B. COKER and W.W. SHILTS

11.1 Introduction

In regions that were glaciated in the Quaternary, mineral exploration can be hampered by the complexity of the surficial sediments, which are largely allochthonous in relation to the bedrock they overlie. In the context of almost totally glaciated landscapes, which includes most of North America north of 40°N, Greenland and Iceland, most of Europe north of 50°N, extensive parts of Asia north of 60°N, parts of South America, and Antarctica, the sediments have particular characteristics that influence the selection of sample media, sampling design and interpretation of data. Foremost among these sediments is till or its recycled derivatives. Till represents the texturally and compositionally heterogeneous debris eroded by and carried directly within or on a glacier. Once deposited, till has distinctive provenance features: (1) it is a first-derivative sediment; (2) it is widespread; and (3) it serves as the parent material for most other glacial and non-glacial surficial sediments.

11.2 Glacial dispersal

The nature of glacial dispersal and the resultant dispersal trains, particularly as related to mineral exploration, have been described in general by a number of authors in the last decade (e.g. Shilts, 1975, 1976, 1982a, 1984; Hirvas, 1977; Minell, 1978; Miller, 1984; Geological Surveys of Finland, Norway and Sweden, 1986a–f; Salonen, 1986a, b, 1987; Strobel and Faure, 1987; Clark, 1987; and Coker and DiLabio, 1989–to cite a few examples).

During the Quaternary, till was produced by the glacial erosion, transport, and deposition of fresh and weathered unconsolidated sediments and bedrock. Till is, therefore, a geologically young sediment which at any given site is not an *in-situ* weathering product, but a lithological summation of source units up-ice from the site. Debris from any size of source unit is dispersed down-ice to produce a ribbon-shaped or fan-shaped dispersal train, comprising a body of till that is enriched in debris from the source relative to the till surrounding the train. Shilts (1976) has shown that, ideally, a plot of the abundance of glacially dispersed debris vs. distance down-ice approximates a negative exponential curve (Figure 11.1), in which the concentration of a component reaches a peak near its source (i.e. within a zone referred to as the 'head' of dispersal) and then declines exponentially to background levels down-ice

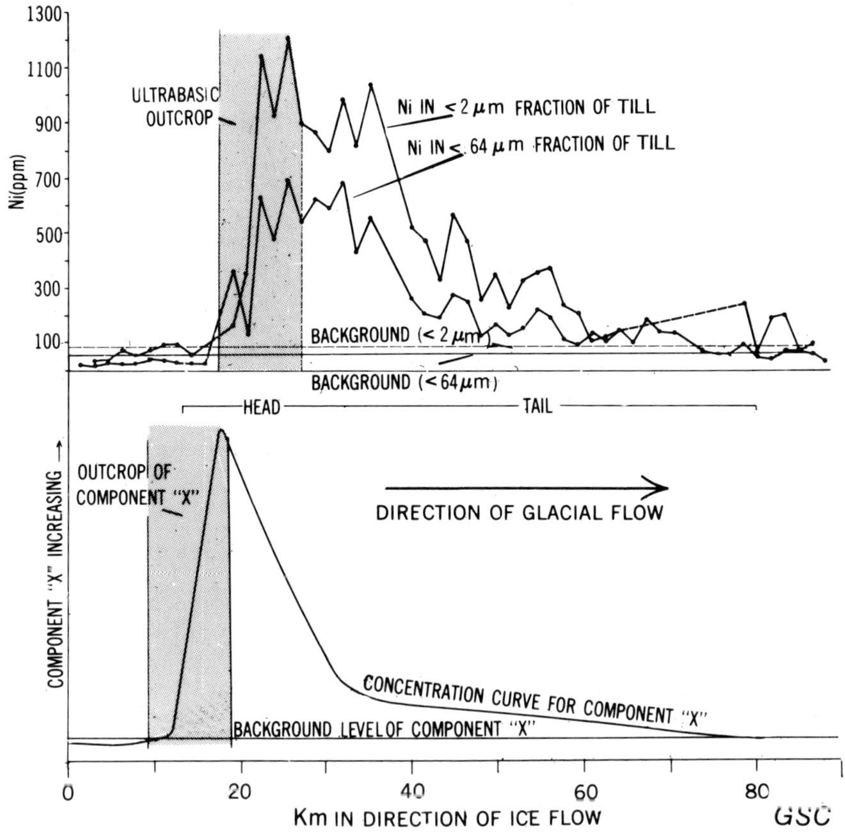

Figure 11.1 Dispersal curves for nickel in till, Thetford Mines area, Quebec. Actual (top) and idealized (bottom) curves show the relationship of the head and tail of a negative exponential curve. (After Shilts, 1976.)

(i.e. in a much larger zone termed the 'tail' of a dispersal train). The dispersal tail is generally many times larger than the head, and as a result it is generally the part of a dispersal train first detected by till sampling programmes. An advantage of prospecting in glaciated terrain is that, if it can be detected, a dispersal train of distinctive boulders, minerals, trace and/or major elements, or radioactive components may enhance the size of mineral exploration targets by several orders of magnitude. A major objective of till geochemistry, then, is simply to detect the tail of a dispersal train, trace it back to its head, and locate its source.

The composition of a till sample at a given site is truly like a fingerprint – no two samples are exactly alike, despite the well-known propensity for glaciers to homogenize bedrock signals through thorough mixing of drift components during transport and deposition. The sediment composition is potentially the net result of mixing all bedrock lithologies from the sample site along a flow line to the dispersal centre for the glacier that deposited it. This complicated melange is further influenced by the width of outcrop of each lithology traversed, the topography of the bed, and the changing dynamic conditions of the ice itself during entrainment, transport, and release of the particles of which the sediment is composed. These, among many other factors, make the likelihood remote that one sample will be strictly comparable in all compositional aspects to another, even if they are collected in close proximity. Thus, the 'case history' approach to developing exploration strategies in glaciated terrain has often foundered; it is a much more useful strategy in areas of residual soil where only easily observable local factors influence soil composition.

In glaciated terrain, the interpretation of the composition of samples and their intercomparisons can be done most effectively by understanding glacial sedimentation principles, history of the depositing and previous ice sheets, and nature of post-glacial weathering, periglacial disturbance, fluvial erosion, etc. In other words, comparison of drift prospecting data gathered on similar mineralization in contrasting glaciated settings is likely as not to lead to false interpretations. Evaluation on the basis of known glacial and post-glacial history and on principles of glacial sedimentation, while considerably more difficult than direct comparison and requiring input from a geologist trained in glacial geology and/or drift prospecting, is far more likely to produce viable exploration targets in glaciated terrane.

In glaciated terrain, the composition of a till sample may be the composite of many overlapping dispersal trains. Most of the individual dispersal trains are not identifiable, however, because they are too small or are composed of rocks or minerals that are not distinctive. The size and shape of the dispersal train are controlled by the orientation of the source relative to ice flow, by the size and susceptibility to erosion of the source, and by the topography of the source and dispersal areas, which can trap trains in valleys, break them into disjointed segments in rough terrane, or even truncate them against hills or mountains.

Dispersal can occur at a variety of scales ranging from continental (100s of kilometres), to regional (100 to 10s of kilometres), to local (< 10 km), to small-scale (final stages of mineral exploration in the 100s to 10s of metres) (Shilts, 1984a). Continental-scale dispersal trains are found throughout the area originally covered by the Laurentide Ice Sheet in Canada and the United States of America. Among the most typical and prominent of these is a train of red till and matrix that extends eastwards from Dubawnt Group rocks into and across northern Hudson Bay (Shilts *et al.*, 1979; Shilts, 1982b; Figure 11.2). Continental-scale trains can be detected only when a characteristic lithological component of the train is present in adequate amounts or is particularly distinctive against background rock types in the dispersal area. For drift prospecting purposes, these large trains are significant in that the exotic lithology of the till can mask the lithology and geochemistry of mineralized debris eroded from local sources, for example the Palaeozoic carbonate trains that extend across greenstone belts south-west of Hudson Bay (Geddes and Kristjansson, 1986; Gleeson and Sheehan, 1987). Large trains such as these can be detected by sampling at densities as low as one till sample per 100 km^2, the sort of sampling carried out during surficial geology or bedrock geology mapping in Canada.

Regional-scale dispersal trains are more likely to be detected in the preliminary stages of mineral exploration programmes and may reflect mineralization or bedrock environments suitable for mineralization. At this stage of exploration, one till sample per km^2 will define which parts of a favourable bedrock unit are most metalliferous and may even detect the tails of dispersal trains derived from small sources. This sampling density may provide geochemical targets that should be sampled at a

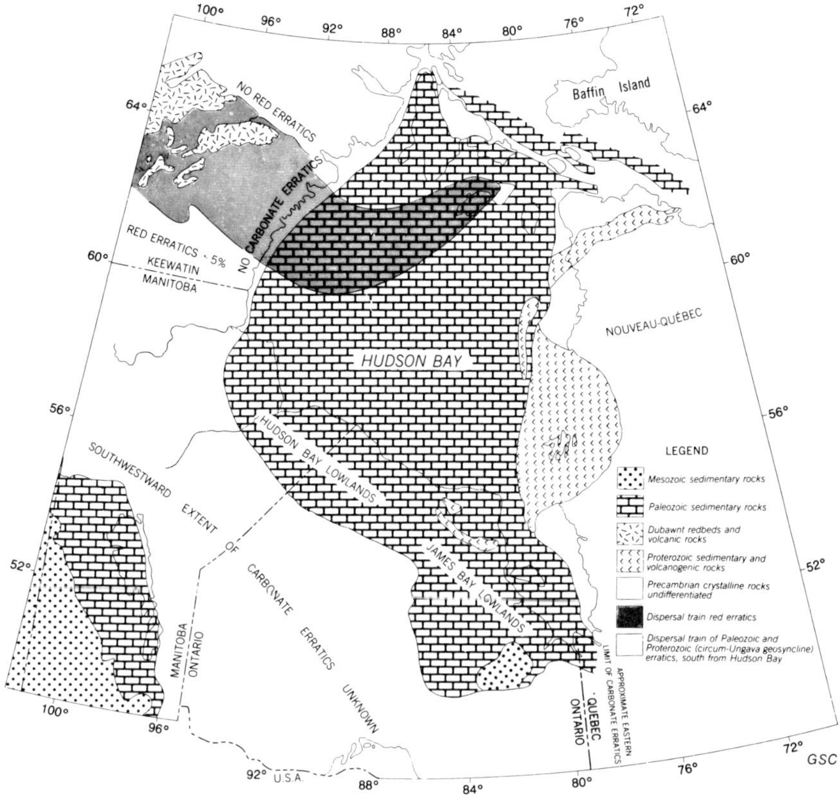

Figure 11.2 Major dispersal trains around Hudson Bay. (Redrawn and adapted from Shilts, 1982b, and Kaszycki and DiLabio, 1986.)

detailed scale to differentiate areas of high background metal levels from those resulting from several overlapping small trains derived from areas of mineralized bedrock.

'Detailed' sampling, in which sample spacing is on the order of 10s to 100s of metres, is designed to locate heads of dispersal trains. This sample spacing is designed to trace dispersal trains up-ice or to test geophysical anomalies and/or favourable geological structures and contacts. An idealized glacial dispersal model (Miller, 1984) serves to illustrate some of the characteristic features of 'normal' dispersal trains (Figure 11.3): they are generally ribbon-, fan- or flame-shaped in outline; they have abrupt lateral edges with the surrounding barren till; and the concentration of the distinctive component within a train decays rapidly down-ice. 'Normal' dispersal trains are formed near or at the base of glaciers by repeated cycles of erosion, entrainment, abrasion, deposition and re-entrainment. At the detailed scale of sampling, post-glacial mobilization of trace elements in groundwater and soil water may spread the dispersal train downslope, partially obscuring its original shape.

Recently, another type of dispersal train, formed by late glacial ice streams, has been recognized. Trains formed by ice-stream mechanisms are often long and have abrupt lateral and distal terminations. Component concentrations may decay

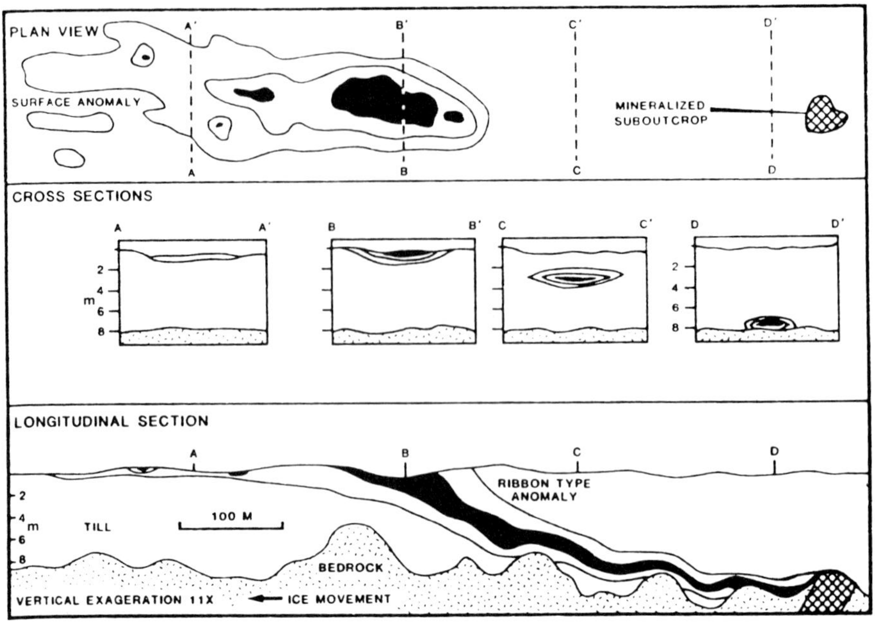

Figure 11.3 Idealized glacial dispersal model. (Redrawn and adapted from Miller, 1984.)

insignificantly down-ice, and they may not obey the exponential curve criterion (Hicock, 1988; Thorliefson and Kristjansson, 1988).

Only in recent times has use been made of samples collected stratigraphically in areas of deeper overburden, using various drilling techniques to produce three-dimensional data sets (i.e. Thompson, 1979; Averill and Zimmerman, 1986; Bird and Coker, 1987; Sauerbrei *et al.*, 1987; Harron *et al.*, 1987; Smith and Shilts, 1987; Brereton *et al.*, 1988; Coker *et al.*, 1988). Variations in these data reflect vertical changes in stratigraphy, sedimentary facies, and diagenesis – particularly weathering during non-glacial intervals such as the present.

11.3 Glacial stratigraphy and ice-movement directions

Till can be thought of as the first derivative of bedrock (Shilts, 1976). Sediments resulting from the reworking of till or other unconsolidated sediments (i.e. stratified drift) are second-derivative sediments; they have been subjected to sorting and have undergone an episode of transport in air or water, often along a different path from the direction of the original ice movement. In this way, glaciofluvial gravel and sand represent the coarse fractions, and glaciolacustrine silt and clay represent the fine fractions derived directly from the texturally heterogeneous debris at the base of the glacier or entrained in ice near its base. Indirectly they may be formed by meltwater erosion of already-deposited till. It is difficult to interpret the provenance of these sediments because they have travelled along transport paths consisting of at least two vectors, transported first by ice, then by water. Till is clearly the optimum glacial sediment type to use in mineral exploration, because it has the least complicated source-transport-deposition history.

Data on ice-movement directions may be obtained from a variety of glacial features including: striations; glacially moulded, streamlined landforms; dispersal trains; and fabric of glacial diamictons (till) or palaeocurrent measurements of associated ice-contact fluvial sediments. Ice-flow directions, estimated by measurement of striae, are not always the most significant flow directions in terms of drift transport (Shilts, 1984a). It has been noted at several sites that the bulk of the till was deposited by movement of ice in a direction different from the ice-flow direction indicated by the youngest set of striae (Veillette, 1986; Kaszycki and DiLabio, 1986a).

The past decade has seen a substantial increase in our knowledge of the relationship of glacial sedimentation, dispersal patterns, stratigraphy and ice-flow directions to practical problems in mineral exploration. Several examples of the integration of the principles of glacial geology and mineral exploration can be cited:

(i) The Nordkalott Project, which included a regional surficial geochemistry and mapping component, was carried out by the Geological Surveys of Finland, Norway and Sweden north of latitude 66°N (250 000 km^2) on the Fennoscandian peninsula (Geological Surveys of Finland, Norway and Sweden, 1986a–f, 1987). Ice-movement directions were systematically recorded, revealing areas of simple unidirectional ice flow in coastal areas and multiple directions of ice flow inland. Trenching and drilling, using techniques developed in major regional sampling projects carried out by the Geological Survey of Finland in the late 1960s and 1970s, provided data on the Quaternary stratigraphy. Data on glacial geology were used to interpret the regional geochemical patterns that emerged.

(ii) In Labrador, Klassen and Thompson (1987) identified ice-flow patterns that are simple near the coast, becoming complex inland, reflecting the complicated ice-flow history of the shifting Labrador–Nouveau Quebec Ice Divide. These ice-flow patterns cause dispersal trains to be ribbon-shaped near the coast and fan-shaped or amoeboid inland (Figure 11.4; Klassen and Thompson, 1989).

(iii) In Nova Scotia, Stea *et al.* (1988) found, as in Labrador, that complex dispersal is recorded in areas of shifting centres of glacial outflow. These authors were able to classify different areas of Nova Scotia as to their expected sequence of ice-flow events based on dispersal of various indicators from lithologically distinctive source outcrops.

(iv) Veillette (1986, 1989) identified three ice-flow events in the Abitibi region of western Quebec (Figure 11.5), and showed that the intermediate one was responsible for the bulk of the drift transport. In an area of active exploration by drift prospecting, this interpretation was immediately useful in exploration.

11.4 Sampling and analytical methods

The most important aspect of data collection, and the resultant sample treatment and geochemical analyses of glacial overburden, starts in the field or at the drill site, where it is essential to make the best possible identification of the type of glacial sediment

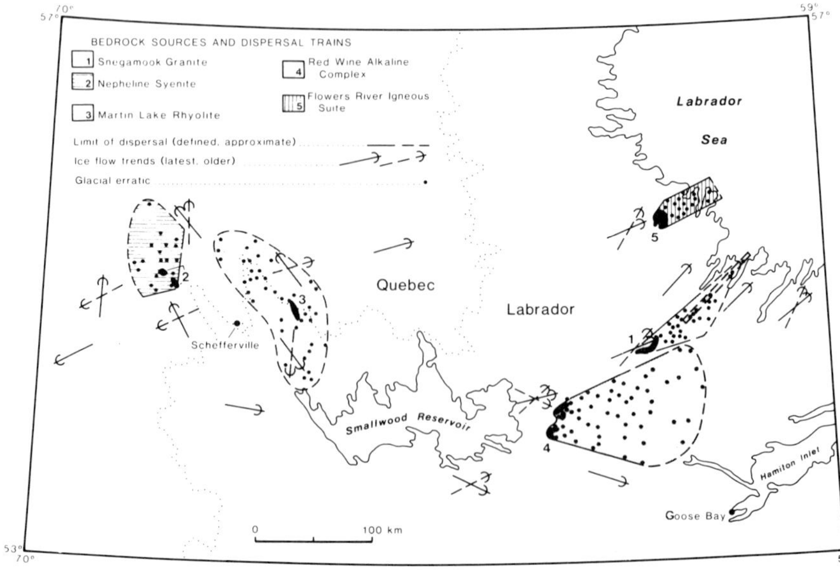

Figure 11.4 Dispersal trains in Labrador (from Klassen and Thompson, 1989). Broad fan shapes reflect transport in two or more phases of ice flow.

being sampled. Appropriately educated and trained scientists or technicians, glacial sedimentologists or geologists, must be employed to ensure that glacial sediment samples are adequately identified and logged. In Fennoscandia, the use of glacial geologists/applied geochemists on overburden geochemical programmes is accepted and routine. In Canada, this is not the case, since most overburden drilling and sampling is being done by personnel trained in neither glacial geology nor applied geochemistry.

Correct identification of the genetic class of the glacial sediment is the key to tracing geochemically anomalous overburden back to a bedrock source. Where overburden is thick and consists of deposits from more than one glaciation, the stratigraphic position of each till sample must be determined properly. In particular, it is important to determine to which sedimentary package and which ice-movement direction a till belongs. It is only when a diamicton is correctly identified as till, and when its stratigraphic position and associated ice-movement direction are understood, that one can use the pattern of geochemically anomalous till samples to find a bedrock source. Equally important is the requirement to assess the extent to which post-depositional weathering may have altered the labile minerals that are a common component of unoxidized till. Without these controls, even the most sophisticated sample preparation, analyses, and interpretation of data derived from overburden may be inadequate or, at best, inefficient for locating mineralization.

Samples used in regional geochemical/Quaternary mapping programmes are usually collected at or near the surface from hand-dug pits (< 2 m), from holes drilled by a hand-held auger, from holes made by percussion drills with flow-through samplers, or from holes excavated by back hoes (< 5 m) (see Table 11.1). Areas of deep overburden

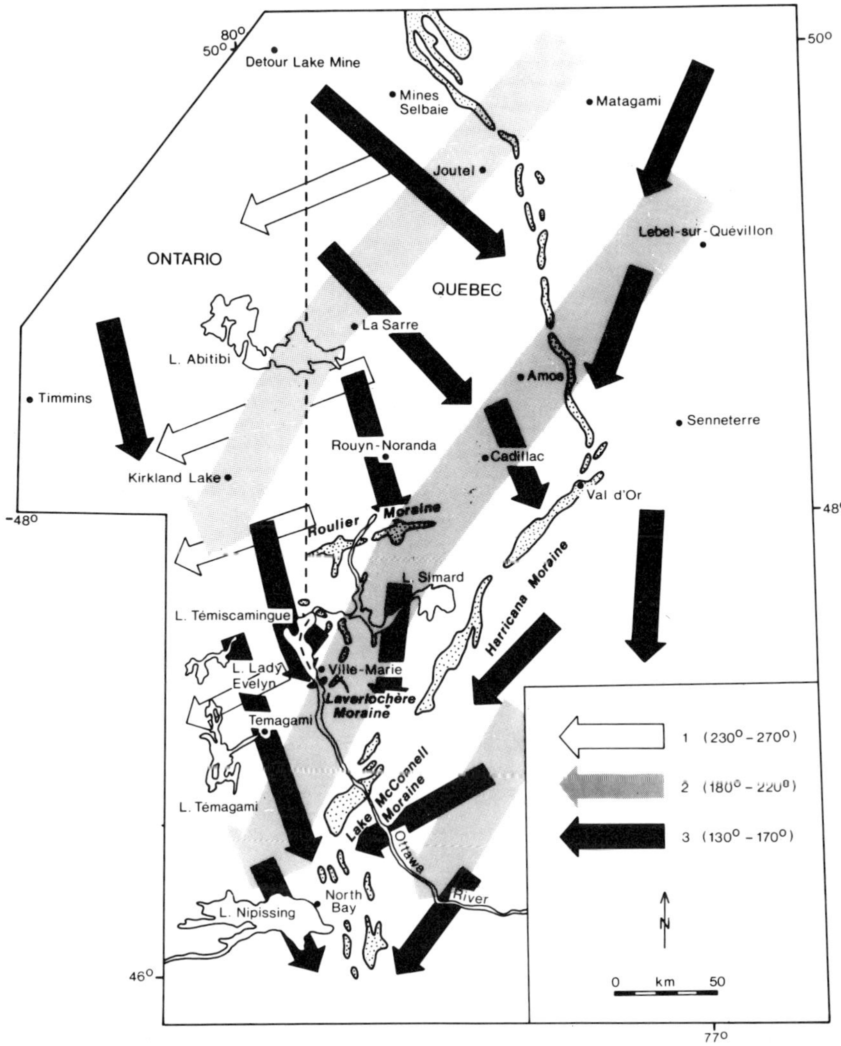

Figure 11.5 The three ice-flow directions recorded in the Abitibi–Timiskaming area of Quebec and Ontario. (After Veillette, 1986.)

are commonly sampled using power augers, percussion drills, reverse-circulation rotary drills (RCD), and rotasonic drills (Table 11.1). Percussion drills have been widely used in Canada and in Fennoscandia, but during the past ten years the greatest experience and success in Canada has been with reverse-circulation drills (Coker and DiLabio, 1989). In recent years, rotasonic drills have started to play a more significant role, particularly in stratigraphic drilling programmes designed to calibrate the crude stratigraphy derived from disaggregated samples produced by reverse-circulation

Table 11.1 Features of various overburden drilling systems (1985 data)[1]

	Reverse circulation drills (Longyear or Acker) (Nodwell mounted)	Rotasonic drills (Nodwell or truck mounted)	Small percussion and vibrasonic drills (various)	Auger drills (various)
1. Production cost estimate per:				
day (10 h)	$ 1800–$ 2000	$ 3000–$ 4000	$ 500–$ 1000	$ 800–$ 1500
metre	$ 25–$ 40	$ 50–$ 80	$ 20–$ 40	$ 25–$ 50
2. Penetration depth	Unlimited (125 m?)	Unlimited (125 m?)	10–20 m (greater ?)	15–30 m (boulder-free)
3. Environmental damage	5 m - wide trails (may have to be cut in areas of larger trees)	5 m - wide cut trails	nil	2–3 m–wide cut trails (Nodwell, muskeg; all terrain vehicle-mounted quite manoeuverable)
4. Size of sample	5 kg (wet)	Continuous core	300 g (dry), or continuous core	3–6 kg (dry or wet)
5. Sample of bedrock	Yes (chips)	Yes (core)	Yes (chips) if reached	Unlikely, if hollow auger, split spoon sampler can be used for chips
6. Sample recovery				
(a) till	Good	Excellent	Good	Good
(b) stratified drift	Moderate	Excellent	Good	Poor to moderate
7. Holes per day (10 h)	4 @ 15–20 m 1 @ 60–80 m	4 @ 15–20 m 1 @ 60–80 m	5 @ 6–10 m	1–3 @ 15–20 m
8. Metres per day (10 h)	60–80 m	60–80 m	30–50 m	20–60 m
9. Time to pull rods	10 min @ 15 m	10 min @ 15 m	30–60 min @ 15 m	20–40 min @ 15 m
10. Time to move	10–20 min	15–30 min	30 min	15–60 min
11. Negotiability	Good	Moderate	Good (poor if manually carried on wet terrain)	Good to reasonable
12. Trails required	Yes, may have to be cut in areas of larger forest	Yes, must be cut	No	Yes and no
13. Ease in collecting sample	Good	Excellent, continuous core	Sometimes difficult to extract from sampler	Good (contamination ?)
14. Type of bit	Milltooth or tungsten carbide tricone	Tungsten carbide ring bits	Flow through sampler, continuous coring	Auger with tungsten carbide teeth

15. Type of power	Hydraulic-rotary	Hydraulic-rotasonic	Hydraulic percussion (gas engine percussion, vibrasonic)	Hydraulic-rotary
16. Method of pulling rods	Hydraulic	Hydraulic	Hydraulic jack, hand jack or winch	Winch or hydraulics
17. Ability to penetrate boulders	Excellent	Excellent, cores bedrock	Poor	Poor to moderate
18. Texture of sample	Slurry (disturbed sample)	Original texture (core can be shortened, lengthened and/or contorted)	Original texture	Original texture (dry) to slurry (wet)
19. Contamination of sample	Nil, fines lost (tungsten)	Nil (tungsten)	Nil (tungsten)	Nil to high (tungsten)

[1] Costs in Canadian dollars; in 1985 C$ 1.00≈US$ 1.00.

drilling. To date, however, no cost-effective drilling system for recovering large till samples at intermediate depths, from 10 to 20 m, has been devised.

In reverse-circulation drilling, water, sometimes used in conjunction with compressed air, is pumped down the outer tube of a system of dual tube rods. The water mixes with the cuttings at a tricone bit (tungsten-carbide buttons) and the slurry is forced to the surface through the inner tube. The sample slurry discharges into a cyclone, to reduce the velocity of the discharge material, and empties through a 2 mm (10 mesh) sieve into a series of sample buckets. Logging is carried out by the Quaternary geologist on the drill. The geologist sees a washed and disturbed sample, which makes accurate logging difficult for qualified personnel and impossible for unqualified personnel. There is only one chance to log the sample material as it goes by. In addition, the fine fraction of the material, such as the ore minerals and including some forms of gold, is generally lost (Shelp and Nichol, 1987) and becomes cross-contaminated by the recirculated water.

The rotasonic drill uses high-frequency resonant vibration produced by a rotating eccentric cam (averaging 5000 vpm) and rotation of the drill bit to obtain continuous solid cores. Sediments are cored with tungsten-carbide-tipped bits. The cores are extruded, in 5 ft lengths, into plastic sleeves and placed in core boxes. Logging and sampling of the cores can be carried out on-site or at a later stage. Rotasonic drills may produce cores that are longer, shorter, or equal in length to the interval sampled. These variations appear to be due mainly to the manner in which different sediment types react to the drilling stresses. Deformation due to the drilling also has been observed in rotasonic drill core (Smith and Rainbird, 1987), and can interfere with interpretations of sediment origin, particularly where deformations may be confused with structures caused by glacial overriding or proglacial, subaerial, or sublacustrine mud flows.

For gold analysis, sand-sized heavy mineral concentrates (HMCs) are the only reliable media that can be prepared routinely from samples obtained by reverse-circulation drilling. Although any grain size can be analysed selectively from rotasonic core, the material must first be disaggregated, which, in the case of overconsolidated tills, can be difficult and time consuming.

The heavier materials are separated from the lighter materials by agitation in water on a shaker table (see Sivamohan and Forssberg, (1985) and Stewart (1986) on the principles of tabling). Once again, fine materials, probably fine gold, are lost (Shelp and Nichol, 1987). The tabled heavies are dried and the magnetic fraction removed (using a magnet) and stored. For gold mineralization associated with iron formation, the usefulness of the removal of the magnetic fraction should be carefully considered. The non-magnetic fraction is further concentrated using a heavy liquid (e.g. methylene iodide, SG = 3.3) separation technique. The samples are cleaned and dried, and in some instances crushed and ground to homogenize them, before being analysed. In many projects, the preconcentration on a shaker table is omitted, the sand-silt-sized material being fractionated directly by submersion in a heavy liquid. Using this method, little is lost but concentrating samples large enough to avoid the 'nugget effect' in gold analysis is time consuming.

With rotasonic-core or surface-till samples, the fine fractions (e.g. clay-sized material (< 2 μm) and/or silt and clay sizes combined (–250 mesh; < 63 μm)) as well as the sand-sized heavy minerals, can be separated and analysed. If a till is sufficiently cohesive to come up the reverse-circulation drill in balls or lumps, it is also possible to collect, in the sieve above the collection bucket, a relatively undisaggregated sample from which to obtain the fine fraction.

HMCs and the other size fractions of tills are commonly analysed for a wide but varied suite of elements, determined by the type of mineralization and nature of deposit being sought. Whether heavy minerals are used is determined by the oxidation state of the till and by the presumed resistance to weathering of the minerals being sought (Shilts, 1975, 1984; Shilts and Kettles, in press). All analytical work should be quality- controlled using reference control and duplicate sample analyses. Analytical techniques generally include various combinations of fire assay (FA) – dissolution – atomic absorption spectrometry (AAS)/direct current plasma or inductively coupled plasma emission spectrometry (DCP–ES or ICP–ES) or mass spectrometry (MS) methods as discussed by Hall and Bonham-Carter (1989). This series of analytical techniques usually involves sample-splitting, which, in the case of analysis for gold, particularly in HMCs, may lead to spurious or non-comparable results, due to the 'nugget effect' (Clifton *et al.*, 1969). These techniques also commonly involve destruction of the sample. Neutron activation allows non-destructive analysis of the whole sample, but one must be aware of the type of irradiation used, because some samples, depending on their matrices and/or chemistry, will be rendered permanently too radioactive to handle. Non-destructive analysis and subsequent recovery of the whole HMC sample facilitates later mineralogical work on anomalous samples to gain some insight on the nature of the mineralization, its geological environment, and the distance of transport of ore fragments.

Contamination by elements such as W and Co, due to fragments from tungsten-carbide bits, also can be detected during examination of HMCs. Although some laboratories offer to analyse grain morphology, caution must be exercised in utilizing the shape of gold grains, or any other mineral grain for that matter, as an indication of distance of transport. Variability in the original morphology of the grains and the manner in which they were glacially transported are far more important in determining grain shape than distance of transport. In the case of style of transport, the grain shape has radically different implications, depending on whether the particle was transported in the active basal zone or over long distances in the passive englacial zone of the ice.

All too often larger clasts (i.e. those > 1mm) are routinely discarded from samples. These clasts are rock fragments, and when counted can give a useful picture of the petrology of the lithologies across which the depositing glacier flowed (see Bird and Coker, 1987). In addition, sand-sized or finer labile minerals, which can be destroyed by weathering, may be enclosed within siliceous, coarser rock fragments from which they can release a geochemical signature on crushing and analysis.

11.5 Occurrence of gold in till and soil and the effects of weathering

In recent years, more attention has been placed on trying to understand the comminution behaviour of ore minerals during glacial erosion and deposition, the residence sites of metals in tills, and the effects of weathering on trace metal contents (Shilts, 1975, 1984). Studies have shown that the mineralogy, petrography and major element chemistry of tills are clearly dependent on till-forming processes as well as on bedrock variations (Haldorsen, 1977, 1983; Taipale *et al.*, 1986).

Fractionation experiments on gold-bearing till (DiLabio, 1982a, 1985, 1988; Guindon and Nichol, 1983; Nichol, 1986; Shelp and Nichol, 1987; Coker *et al.*, 1988)

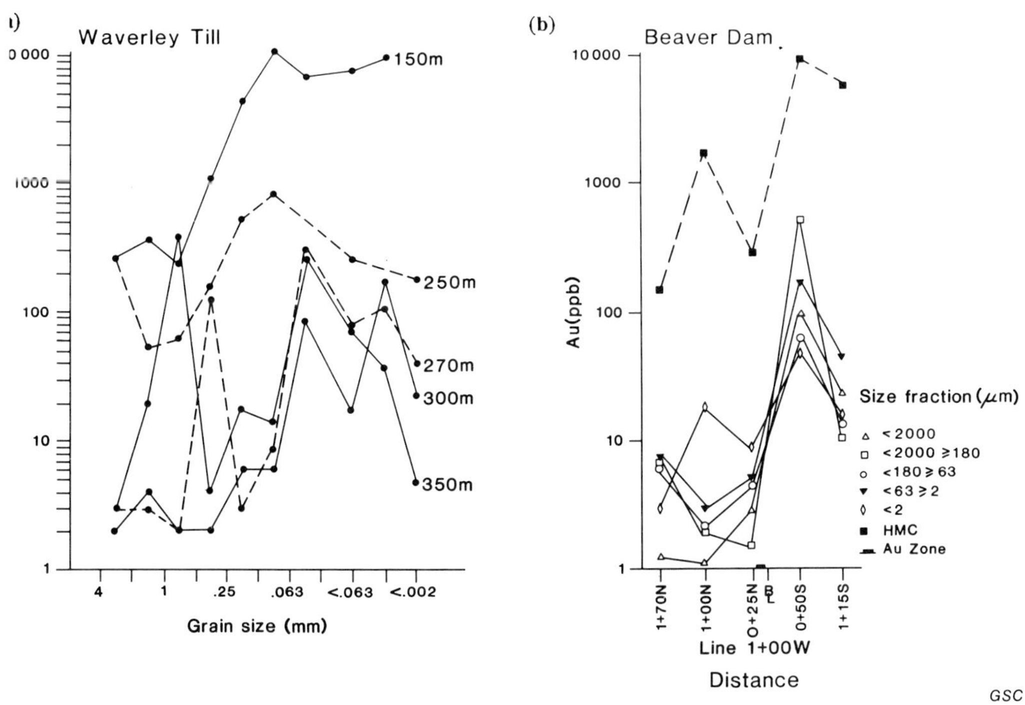

Figure 11.6 Abundance of gold vs. grain size of analysed fraction of till at (a) different distances down-ice from a gold deposit at Waverley, Nova Scotia (after DiLabio, 1982a, 1985, 1988), and (b) along a sample line traversing the Beaver Dam gold deposit, Nova Scotia. (Coker *et al.*, 1988.)

indicate that gold distributions in various particle sizes of till are complex, because of the combined effects of (a) glacially comminuted detrital particulate gold; (b) the grain size of re-precipitated gold released during the weathering of sulphides; (c) the grain size of precipitated or adsorbed gold; and (d) the original grain size of native gold at its source. In general, till is richest in gold in its finer size ranges (Figure 11.6), although coarse fractions may also be auriferous (Figure 11.6).

At the base of the thick (> 30 m) Quaternary sequence at the Owl Creek gold deposit near Timmins, a green till and its oxidized equivalent are preserved. The unoxidized

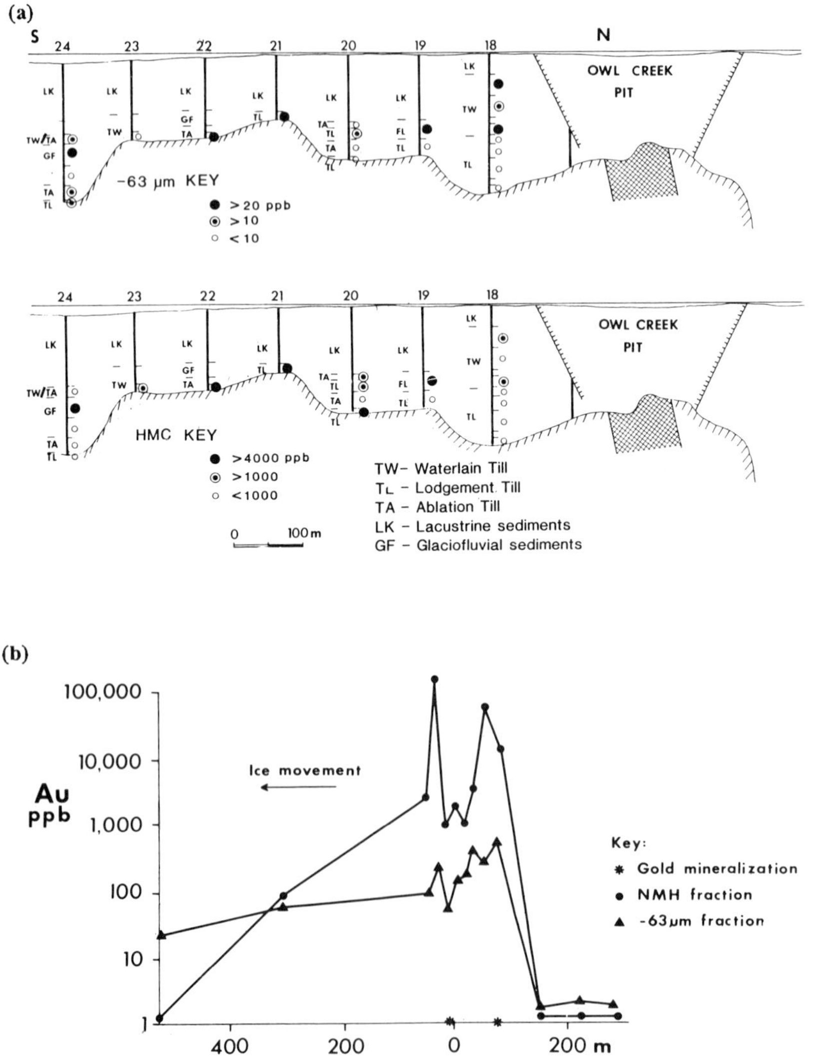

Figure 11.7 Comparison of the gold distribution in the heavy mineral concentrate (HMC) (NMH = non-magnetic heavy mineral concentrate) and the < 63 μm fractions of till at (a) the Owl Creek gold mine, Timmins, Ontario, and (b) associated with gold mineralization at Hemlo, Ontario. (From Shelp and Nichol, 1987.)

green till contains most gold and abundant fresh pyrite in the fine sand sizes (DiLabio, 1985, 1988). In the overlying oxidized layer, pyrite has been altered to earthy limonite–goethite that ranges in size and habit from sand-sized pseudomorphs after pyrite to amorphous silt- and clay-sized grains. As the oxidized till is much more auriferous in all size fractions than the unoxidized till, it was concluded that gold had been added to the limonite–goethite grains, perhaps from groundwater.

Work by Shelp and Nichol (1987) at the Owl Creek gold mine shows that both the gold content of the HMCs and of the < 63 μm (clay and silt) material clearly depict anomalous glacial dispersal trains (Figure 11.7(a)). The lengths of the anomalous dispersal trains are similar in both cases, although the contrast and absolute levels of gold in the < 63 μm fraction are lower. For Hemlo-type mineralization (Figure 11.7(b)), they found that HMCs anomalously enriched in gold are restricted to drift lying directly over the deposit and show little evidence of dispersal. The < 63 μm fraction shows significant dispersal. Thus, analyses of fine fractions and HMCs can provide different, often complementary, information.

It should be apparent that the suitability of the use of the conventional -80 mesh material for till geochemical surveys should be carefully evaluated on a project by project basis. If possible, orientation surveys should be conducted for each project area to determine the optimum size fraction to use. If it is not possible to carry out a meaningful orientation survey, fine-grained materials (< 63 μm) should be used, because they are easiest to obtain, and in many cases they are the best size fraction for reflecting gold mineralization. The nature of the weathering history of the material also determines the appropriateness of the grain size to be used. In tills that show evidence of weathering, the use of data on the geochemistry of the fine and HMC fractions to complement each other may well increase the effectiveness and reliability of gold exploration.

11.6 Drift prospecting for gold

Drift prospecting is the use of data on the geochemistry and lithology of glacial sediments (mainly till) to identify economically significant components in the sediments and to trace them up-ice to their bedrock source (DiLabio, 1989). The concept of predictable patterns within dispersal trains (e.g. exponential decay, as shown in Figure 11.1, and the shape of trains, as shown in Figure 11.3), when considered during the design of a geochemical exploration programme, will influence the choice of sample types, the sampling plan, the analytical scheme, and the interpretation of the data. Compared to geochemical anomalies in unglaciated areas, once a part of a glacial dispersal train has been detected, it can be traced more easily up-ice to its source because simple clastic dispersal is the main mechanism involved in the formation of a train. This assumes, in the case of gold, that the gold observed visually is from a bedrock or discrete source and is not formed *in situ* by the weathering of sulphides carrying trace amounts of gold.

In the last decade, many studies of gold dispersal trains have been documented within Fennoscandia and Canada. Details of most of these studies/and case histories are summarized and referenced in Coker and Dilabio (1989). It is impractical to go into much depth of discussion for all of these examples. The focus has been on studies which are most familiar and which best make the points wc wish to emphasize, such as understanding: (a) ice-movement directions; (b) glacial stratigraphy and bedrock topography as they control the nature of glacial dispersal; (c) distinguishing gold that was dispersed clastically from a 'mother lode' from gold that was concentrated on surface or in buried placers, or formed by diagenetic (weathering) processes; and (d) gold dispersal trains.

Geochemical orientation surveys were carried out across the Beaver Dam gold deposit in Nova Scotia to determine the most effective sample media and those elements most suitable for indicating gold mineralization (Coker *et al.*, 1988). Fractionation studies indicated that gold was preferentially partitioned into the coarser fractions of the till (see Figure 11.6). In this study, the till matrix (< 2 mm ground to < 75 μm), analysed for gold (Figure 11.8) and arsenic, may well be the most effective and least costly fraction to use in exploring for gold in this geological environment.

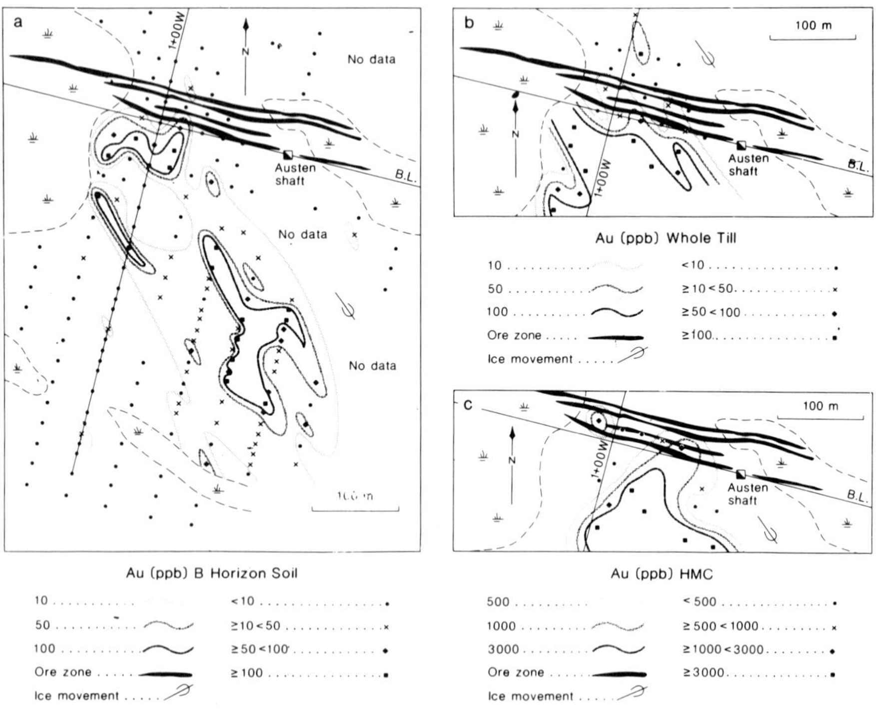

Figure 11.8 Beaver Dam gold deposit, Nova Scotia. The distribution of gold in (a) B-horizon soil; (b) whole till; and (c) heavy mineral concentrates. (From Coker *et al.*, 1988.)

A reverse circulation overburden drilling programme carried out at the Golden Pond gold deposit, Casa Berardi, Quebec, by Sauerbrei *et al.* (1987) identified a thin lower till containing anomalous geochemical concentrations of gold and abundant gold grains. Of note was the discovery that glacial dispersal of gold was very limited (200–400 m). This was because the ice that deposited the lower till moved subparallel to the strike of both the mineralized structure, itself recessive, and a bedrock trough, which further confined the dispersal train.

In Timmins, Ontario, a reverse-circulation and rotasonic drilling programme was carried out by Bird and Coker (1987) at the Owl Creek gold mine. This revealed a deep and complex overburden stratigraphy with up to four glacial sediment packages, each with different ice-movement directions (Figure 11.9). In the lowest (Older) till, on bedrock, dispersal is very local being truncated against a bedrock ridge. The highest gold concentrations in till are adjacent to the subcropping gold mineralization. Although the overlying Matheson till has not been in contact with mineralization or bedrock, it contains gold reworked from the lower (Older) till. Matheson dispersal is longer, approximately 600 m, and the area of maximum gold concentration in till is displaced some 300 m down-ice from the subcropping gold mineralization.

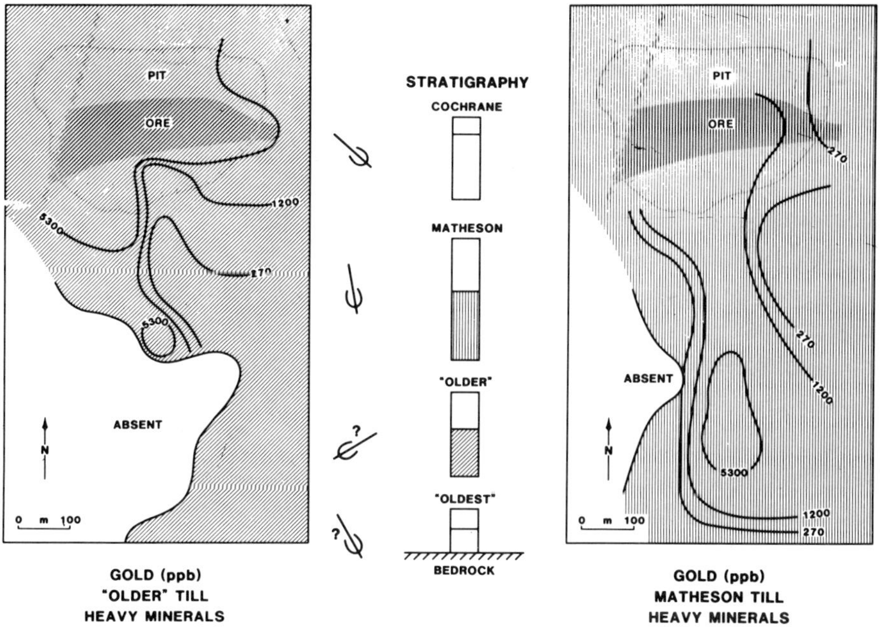

Figure 11.9 Gold contents in heavy mineral concentrates from the 'Older' and Matheson Tills at the Owl Creek gold mine, Timmins, Ontario. (Redrawn and adapted from Bird and Coker, 1987.)

In the Hemlo area, at the Page-Williams gold deposit 'A' zone, Gleeson and Sheehan (1987) sampled till using percussion drills with 'flow through' bits. They found that the upper 'exotic' calcareous till gave little indication of the gold mineralization (Figure 11.10). This till is probably analogous to the ice-stream tills (Hicock, 1988; Thorliefson and Kristjansson, 1988) of the Beardmore–Geraldton area. In areas underlain by this type of till, the geochemical and mineralogical signature of the underlying bedrock is almost totally masked. The underlying, locally-derived limonite-rich till gave good response to the gold mineralization (Figure 11.10) in all size fractions and HMCs, using gold, arsenic, antimony, molybdenum, mercury, tungsten, and barium. Once again, dispersal was short (i.e. 200 m), partly because the deposit lies in the lee of a bedrock-high and is protected, and partly because dispersal is truncated against a bedrock ridge down-ice.

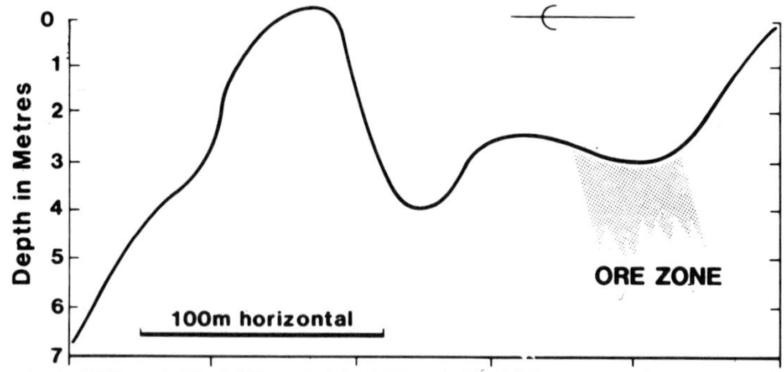

Metal Values in Basal Till (−250 mesh)		L15+50E	L16+00E	L16+50E	L17+00E	L17+50E	L18+00E
ppb	Au	12	13	45	300	+15,000	11
ppb	Hg	70	155	790	1,390	297,000	30
ppm	As	2	3	11	25	700	4
ppm	Sb	2	<1	7	46	804	<1
ppm	Mo	3	3	6	7	98	1
ppm	W	2	4	6	4	34	<2
ppm	Ba	560	8,413	1,420	788	1,404	572

Figure 11.10 Page–Williams gold deposit, 'A' zone, Hemlo, Ontario: bedrock topographic profile and geochemistry across the mineralization. (From Gleeson and Sheehan, 1987.)

A rotasonic overburden drilling programme carried out by Averill and Zimmerman (1986) in northern Saskatchewan located a dispersal train in which the HMCs from the till contained abundant native gold, gold–silver, and copper, as well as galena, chalcocite, and pyromorphite, which led to the discovery of the EP gold zone at Waddy Lake. This classic dispersal train is ribbon shaped with sharp edges (Figure 11.11). It is noteworthy that the trend of the glacial dispersal train is 15° off the direction for ice movement indicated by glacial striae in the area.

11.7 Source of placer gold in glaciated terrain

Perhaps one of the most frustrating and commonly unresolved problems of gold exploration in glaciated terrain is determining whether high gold concentrations in drift or preserved pre-glacial regolith or alluvium are the result of clastic dispersal from a discrete bedrock source ('mother lode') or whether the gold was formed and concentrated by secondary, low-temperature, geochemical processes acting on a diffuse, non-economic source. In the latter case, further complications arise in determining whether low-temperature processes might have been associated with prolonged weathering over a broad area or whether hydrothermal action caused gold to be precipitated in discrete zones. Concentration of geochemically immobile gold formed by these processes could have been by placer-forming processes or by intense weathering over a time period sufficient to remove much of the silicate part of the soil.

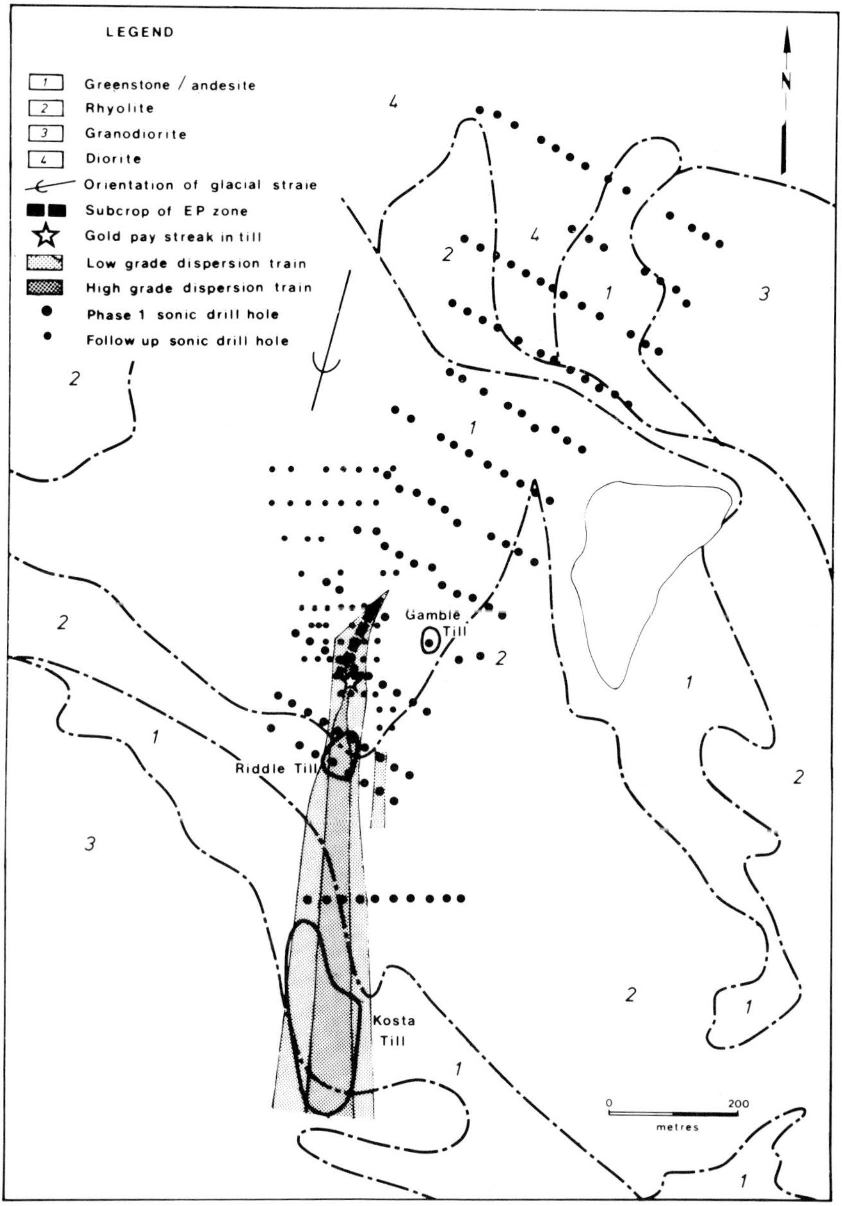

Figure 11.11 Glacial dispersal train from the EP gold zone, Waddy Lake, Saskatchewan. (After Averill and Zimmerman, 1986.)

For over a century, in both the Cordillera and south-eastern Quebec, these questions have intrigued and frustrated prospectors working in placer deposits. The voluminous literature on the problem of placer gold in the Yukon and British Columbia is beyond an adequate discussion in this chapter, but the interested reader is referred to recent work by Morrison and Hein (1987) and Eyles and Kocsis (1988).

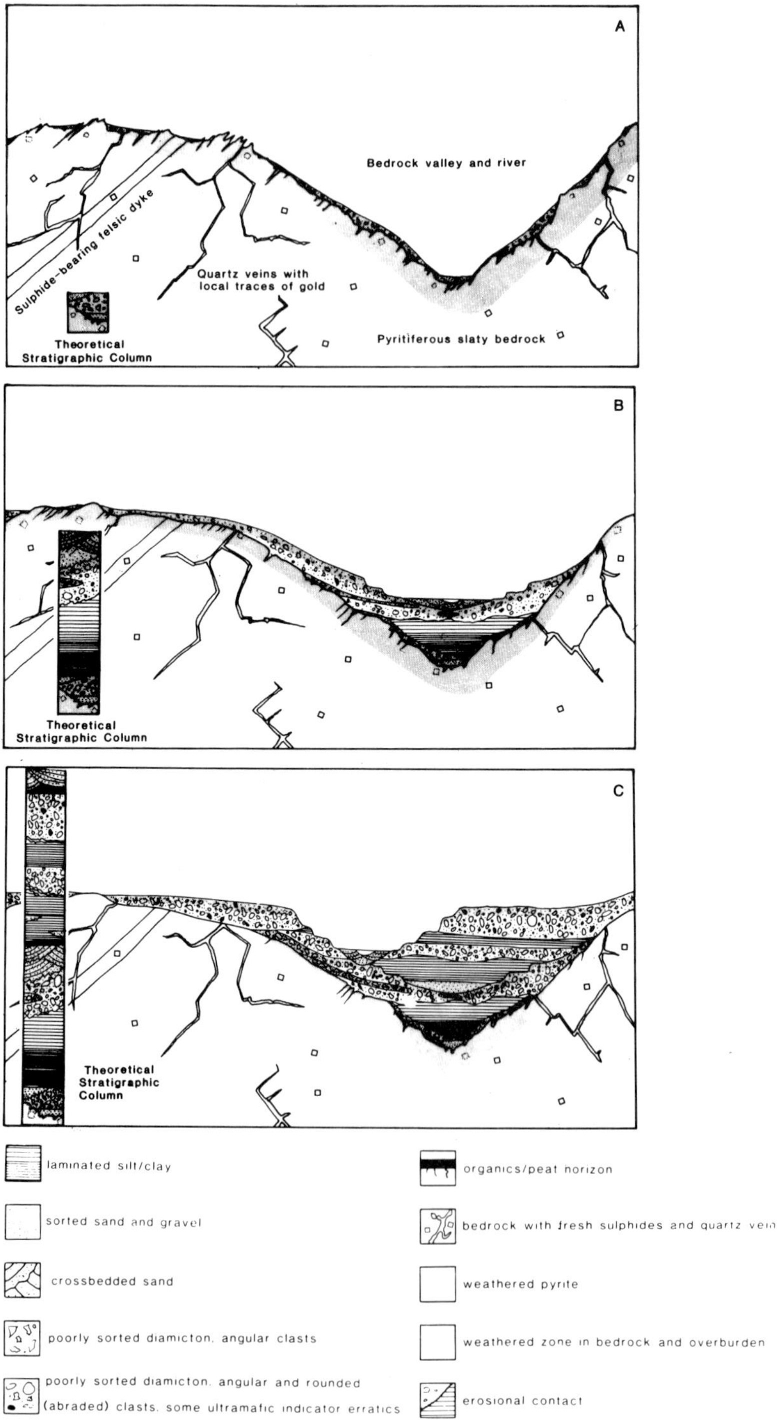
A
Bedrock valley and river
Sulphide-bearing felsic dyke
Quartz veins with local traces of gold
Theoretical Stratigraphic Column
Pyritiferous slaty bedrock
B
Theoretical Stratigraphic Column
C
Theoretical Stratigraphic Column
laminated silt/clay
sorted sand and gravel
crossbedded sand
poorly sorted diamicton, angular clasts
poorly sorted diamicton, angular and rounded (abraded) clasts, some ultramafic indicator erratics
organics/peat horizon
bedrock with fresh sulphides and quartz veins
weathered pyrite
weathered zone in bedrock and overburden
erosional contact

In south-eastern Quebec, a research programme (Shilts and Smith, 1986a,b; Smith and Shilts, 1987; Shilts and Smith, 1988) was carried out from 1984 to 1987 using rotasonic stratigraphic drilling through complex glacial sequences overlying pockets of pre-glacial, gold-bearing regolith and gravel, that have been mined sporadically for over a century (Figure 11.12). The objectives of the programme were: (a) to find the source of the gold in the regolith; (b) to find ways to predict where other occurrences of the regolith might be found; and (c) to model the pathway of gold enrichment from either bedrock or the pre-glacial deposits to modern streams and soils across a span of geological time encompassing at least three glaciations (Shilts, 1981). At the heart of the century-old problem of the origin of gold in this earliest (1820) of the Canadian gold rush terrains is the question most commonly asked in areas of alluvial placers: 'Is the gold transported and concentrated by normal glacial and fluvial sedimentation processes from discrete bedrock sources, or is the gold produced more or less ubiquitously by weathering of rock with background gold concentrations that are enhanced by concomitant gold precipitation and concentration by water and/or down-slope movement by mass wasting processes ?'

In the Quebec study, it was concluded that the pyritiferous, flyschoid sediments of the Appalachian Mountains supplied gold in solution as their sulphides were weathered over a long period prior to glaciation. The weathered kaolinitic regolith served as a host for gold precipitated from solution, and the gold was transported down the steep local slopes to valley bottoms, primarily by mass wasting processes. Streams further concentrated the gold in placer deposits. Some of the individual nuggets from this region have the appearance of large concretions, and two weighed

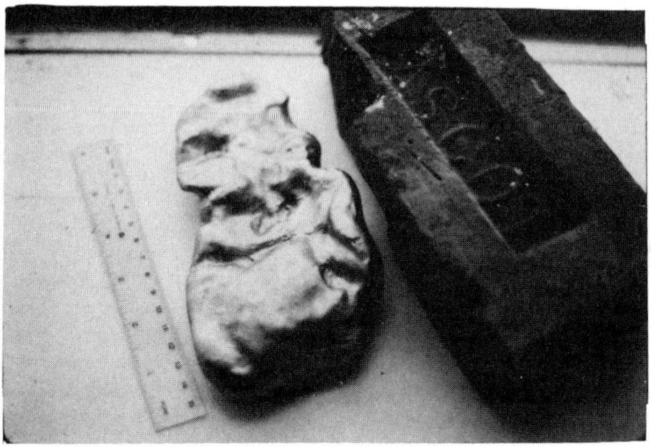

Figure 11.13 A model of the MacDonald nugget containing 1500 g of gold.

Figure 11.12 Interpretation of borehole records across Rivière Gilbert Valley, Québec, in which gold-bearing regolith has been protected from glacial erosion and buried under a complex glacial sedimentary sequence. Periods of pre-glacial and inter-glacial weathering, indicated by shading, were times of gold concentration by chemical destruction of pyrite and concomitant precipitation of gold. Pre-glacial, inter-glacial and post-glacial streams concentrated chemically precipitated and glacially reworked gold into placers. (A) Development of pre-glacial, gold-bearing regolith and alluvial placers by prolonged weathering; (B) inter-glacial weathering and accumulation of gold after first glacial cycle (note that glaciers block drainage, forming lakes when they enter or leave this region); (C) post-glacial weathering, fluvial erosion, and accumulation of gold after several glacial cycles. (Modified from sketches by Sharon Smith in Shilts and Smith, 1988.)

more than 1100 g (Figure 11.13). Repeated glaciations (at least three) removed most of this weathered regolith and diluted it with fresh, glacially ground bedrock. In the steeper reaches of some valleys oriented perpendicular to the direction of glacial flow, pockets of palaeoplacers and the regolith were preserved and mined in the nineteenth and twentieth centuries.

Gold, which is now widespread in modern alluvium and soil from the same area, comes from at least two sources: (a) gold reworked through at least three glaciations by glacial mixing of the pre-glacially-widespread regolith with glacially eroded detritus derived from local bedrock; and (b) gold which appears to be forming at present in the solum on till, and presumably also during inter-glacial or inter-stadial intervals as a result of low-temperature geochemical processes associated with weathering (Figure 11.12). This is a particularly rapid process compared to pre-glacial rates of gold formation, because glaciers have crushed, on average, one or more metres of bedrock to sand and finer sizes, liberating sulphides as discrete grains, and increasing the chemical reaction surface by several orders of magnitude over that available on the exposed and fracture surfaces of consolidated bedrock. From experience, this can translate to bulk, near-surface till samples with 6–10 grains of visible gold per kilogram in a till containing pyrite averaging less than 50 ppb gold (Table 11.2).

Table 11.2 Gold concentrations in pyrite from the Eastern Townships, Quebec (from Shilts and Smith, 1988)

Locality	Gold (ppb)	Details
Samson River	105	Streambed placer, 20 g, 1 mm–1 cm pyrite cubes (Locat, pers. comm.)
Samson River	94	Streambed placer, 20 g, 1 mm–1 cm pyrite cleaned of surface tarnish
Mining Brook	33	Panned, oxidized pyrite, <10 g
Mining Brook	123	Panned, unoxidized pyrite, <10 g
Various	5–65	Unoxidized pyrite from till samples (range from 11 samples)

11.8 Conclusions and future trends

The studies described above illustrate the special knowledge of glacial geology required to address fundamental problems in gold exploration in glaciated terrain. Although glacial geological principles have not resolved centuries-old controversies and hypotheses of gold genesis in glaciated terrain, they have relatively recently begun to bring some order to the chaos that the peculiar fever or enthusiasm of gold discovery provokes.

A number of points have been emphasized in this chapter. Many are areas in much need of further work and research. These include the need for:

(i) Regional studies of till provenance, including the effects of exotic drift on local geochemistry.
(ii) Correlation of tills in areas of complex stratigraphy and assignment of ice-flow directions to tills.
(iii) Investigation of the comminution and weathering behaviour of ore and ore-generating (sulphide) minerals, particularly gold and platinum group elements, in order to design better sampling and analytical schemes.

(iv) Development of more cost-effective drilling systems, particularly in areas of intermediate overburden thickness (i.e. 10–20 m).

(v) Education and training of geologists in geochemistry and Quaternary geology and a commitment by senior explorationists and exploration managers to the use of qualified people to carry out surficial geochemical and overburden drilling programmes in till- covered terrains.

References

Averill, S.A. and Zimmerman, J.R. (1986) The riddle resolved: the discovery of the Partridge Gold Zone using sonic drilling in glacial overburden at Waddy Lake, Saskatchewan. *Can Geol. J. Can. Inst. Mining Metall.* **1**, 14-20.

Bird, D.J. and Coker, W.B. (1987) Quaternary stratigraphy and geochemistry at the Owl Creek gold mine, Timmins, Ontario, Canada. In *Geochemical Exploration 1985*, part 1, ed. Garrett, R.G. *J. Geochem. Explor.* **28** (1/3), 267-284.

Brereton, W.E., Briggs, D.N. and Rollinson, J.P. (1988) Till prospecting in the area of the Farley Lake gold deposits, Northwestern Manitoba, Canada. In *Prospecting in Areas of Glaciated Terrain-1988*, ed. MacDonald, D.R. Can. Inst. of Mining and Metallurgy, 225–240.

Clark, P.U. (1987) Subglacial sediment dispersal and till composition. *J. Geol.* **95**, 527-541.

Clifton, H.E., Hunter, R.E., Swanson, F.J. and Phillips, R.L. (1969) Sample size and meaningful gold analysis. US Geol. Surv., Prof. Paper 625-C. 17pp.

Coker, W.B., Sexton, A., Lawyer, I. and Duncan, D. (1988) Bedrock, till and soil geochemical signatures at the Beaver Dam gold deposit, Nova Scotia, Canada. In *Prospecting in Areas of Glaciated Terrain – 1988*, ed. MacDonald, D.R. Can. Inst. of Mining and Metallurgy, 241-254.

Coker, W.B. and DiLabio, R.N.W. (1989) Paper 30, Geochemical exploration in glaciated terrain: geochemical responses. In *Proc. Exploration '87: Third Decennial Internat. Conf. on Geophysical and Geochemical Exploration for Minerals and Groundwater*, ed. Garland, G.D. Ontario Geol. Surv., Special Vol. 3, 336-383. 960pp.

DiLabio, R.N.W. (1982) Gold and tungsten abundance vs. grain size in till at Waverley, Nova Scotia. In *Current Research*, Part B. Geol. Surv. of Can., Paper 82-1B, 57-62.

DiLabio, R.N.W. (1985) Gold abundances vs. grain size in weathered and unweathered till. In *Current Research*, Part A. Geol. Surv. of Can. Paper 85-1A, 117-122.

DiLabio, R.N.W. (1988) Residence sites of gold, PGE and rare lithophile elements in till. In *Prospecting in Areas of Glaciated Terrain - 1988*, eds. MacDonald, D.R. Can. Inst. of Mining and Metallurgy 121-140.

DiLabio, R.N.W. (1989) Terrain geochemistry. Chapter 10 in *Quaternary Geology of Canada and Greenland*, eds. Fulton, R.J., Heginbottom, J.A. and Funder, S. Geol. Surv. of Can. *Geology of Canada*, Vol. 1.

Eyles, N. and Kocsis, S.P. (1988) Gold placers in Pleistocene glacial deposits; Barkerville, British Columbia. *Can. Inst. Mining and Metall. Bull.* **81** (916), 71-79.

Geddes, R.S. and Kristjansson, F.J. (1986) Quaternary geology of the Hemlo area: constraints on mineral exploration. *Can. Geol. J. Can. Inst. Mining Metall.* **1** (1), 5-8.

Geological Surveys of Finland, Norway and Sweden (1986a) Map of Quaternary Geology, Sheet 2: Glacial Geomorphology and Paleohydrography, Northern Fennoscandia. ISBN-91-7158-378-5.

Geological Surveys of Finland, Norway and Sweden (1986b) Map of Quaternary Geology, Sheet 3: Ice Flow Indicators, Northern Fennoscandia. ISBN-91-7158-379-3.

Geological Surveys of Finland, Norway and Sweden (1986c) Map of Quaternary Geology, Sheet 4: Quaternary Stratigraphy, Northern Fennoscandia. ISBN-91-7158-380-7.

Geological Surveys of Finland, Norway and Sweden (1986d) Map of Quaternary Geology, Sheet 5: Ice Flow Directions, Northern Fennoscandia. ISBN-91-7158-381-5.

Geological Surveys of Finland, Norway and Sweden (1986e) Geochemical Interpretation Map, Northern Fennoscandia. ISBN-91- 7158-382-3.

Geological Surveys of Finland, Norway and Sweden (1986f) Geochemical Atlas of Northern Fennoscandia. ISBN-91-7158-383-1.

Geological Surveys of Finland, Norway and Sweden (1987) Map of Quaternary Geology, Sheet 1: Quaternary Deposits, Northern Fennoscandia. ISBN-9-7158-377-7.

Gleeson, C.F. and Sheehan, D.G. (1987) Humus and till geochemistry over the Doyon, Bousquet and Williams gold deposits. *Can. Inst. Mining Metall. Bull.* **80** (898), 58-66.

Guindon, D.L. and Nichol, I. (1983) Speciation of free gold in glacial overburden. Ontario Geol. Surv., Open-File Report 5444. 15pp.

Haldorsen, S. (1977) The petrology of tills - a study from Ringsaker, South-Eastern Norway. *Norges Geologiske Undersokelse* **336**, 1-36.

Haldorsen, S. (1983) Mineralogy and geochemistry of basal till and their relationship to till forming processes. *Norsk Geologisk Tidsskrift* **63**, 15-25.

Hall, G.E.M. and Bonham-Carter, G.F. (1988) Review of methods to determine gold, platinum and palladium in production-oriented laboratories, with application of a statistical procedure to test for bias. *J. Geochem. Explor.* **30**, 255-286.

Harron, G.A., Middleton, R.S., Durham, R.B. and Philipps, A. (1987) Geochemical and geophysical gold exploration in the Timmins area, Ontario; a case history. *Can. Inst. Mining Metall. Bull.* **80** (898), 52-57.

Hicock, S.R. (1988) Calcareous till facies north of Lake Superior, Ontario: Implications for Laurentide ice streaming. *Geog. Phys. Quater.* **42** (2), 120-135.

Hirvas, H. (1977) Glacial transport in Finnish Lapland. In *Prospecting in Areas of Glaciated Terrain - 1977*, ed. Davis, G.R. Inst. Mining and Metall., London, 128-137.

Kaszycki, C.A. and DiLabio, R.N.W. (1986) Surficial geology and till geochemistry, Lynn Lake–Leaf Rapids Region, Manitoba. In *Current Research*, Part B. Geol. Surv. of Can. Paper 86-1B, 245-256.

Klassen, R.A. and Thompson, F.J. (1987). Ice flow history and glacial dispersal in the Labrador Trough. In *Current Research*, Part A. Geol. Surv. of Can., Paper 87-1A, 61-71.

Klassen, R.A. and Thompson, F.J. (1989) Glacial history and drift prospecting in Labrador. In *Drift Prospecting – 1987*, eds. DiLabio, R.N.W. and Coker, W.B. Geol. Surv. of Can., Paper 89-20 21-40.

Miller, J.K. (1984) Model for clastic indicator trains in till. In *Prospecting in Areas of Glaciated Terrain - 1984*, Inst. of Mining and Metallurgy, London, 69-77.

Minell, H. (1978) Glaciological interpretations of boulder trains for the purpose of prospecting in till. *Sveriges Geologiska Undersokning*, Series C, Number 743. 51pp.

Morrison, S.R. and Hein, F.J. (1987) Sedimentology of the White Channel gravels, Klondike area, Yukon Territory: Fluvial deposits of a confined valley. Soc. of Econ. Paleontol. and Mineral. Special Pub. 39, 205-216.

Nichol, I. (1986) Geochemical exploration for gold deposits in areas of glaciated overburden: Problems and new developments. In *Prospecting in Areas of Glaciated Terrain - 1986*, Inst. of Mining and Metallurgy, London, 7-16.

Salonen, V.-P. (1986a) Glacial transport distance distributions of surface boulders in Finland. Geol. Surv. of Finland, Bull. 338. 57pp.

Salonen, V-P. (1986b) Length of boulder transport in Finland. In *Prospecting in Areas of Glaciated Terrain – 1986*, Inst. of Mining and Metallurgy London, 211-215.

Salonen, V-P. (1987) Observations on boulder transport in Finland. In *INQUA Till Symposium, Finland 1985*, eds. Kujansuu, R. and Saarnisto, M. Geol. Surv. of Finland, Special Paper 3, 103-110.

Sauerbrei, J.A., Pattison, E.F. and Averill, S.A. (1987) Till sampling in the Casa-Berardi area, Quebec: A case history in orientation and discovery. In *Geochemical Exploration 1985*, Part 1, ed. Garrett, R.G. *J. Geochem. Explor.* **28**, 297-314.

Shelp, G.S. and Nichol, I. (1987) Distribution and dispersion of gold in glacial till associated with gold mineralization in the Canadian Shield. In *Geochemical Exploration 1985*, ed. Garrett, R.G. *J. Geochem. Explor.* **28**, 315-336.

Shilts, W.W. (1975) Principles of geochemical exploration for sulphide deposits using shallow samples of glacial drift. *Can. Inst. Mining Metall. Bull.* **68** (757), 73-80.

Shilts, W.W. (1976) Glacial till and mineral exploration. In *Glacial Till: an Interdisciplinary Study*, ed. Legget, R.F. Roy. Soc. of Can. Special Pub. 12, 205-224.

Shilts, W.W. (1981) Surficial geology of the Lac Megantic area, Quebec. Geol. Surv. of Can., Memoir 397. 102pp.

Shilts, W.W. (1982a) Glacial dispersal – principles and practical applications. *Geoscience Canada* **9** (1), 42- 47.

Shilts, W.W. (1982b) Quaternary evolution of the Hudson/James Bay region. *Le Naturaliste Canadien* **109**, 95-117.

Shilts, W.W. (1984) Till geochemistry in Finland and Canada. *J. Geochem. Explor.* **21**, 95-117.

Shilts, W.W. and Smith, S.L. (1986a) Stratigraphy of placer gold deposits; overburden drilling in Chaudiere Valley, Quebec. In *Current Research*, Part A. Geol. Surv. of Can. Paper 86-1A, 703-712.

Shilts, W.W. and Smith, S.L. (1986b) Stratigraphic setting of buried gold-bearing sediments, Beauceville area, Quebec. In *Current Research*, Part B. Geol. Surv. of Can., Paper 86- 1B, 271-278.

Shilts, W.W. and Smith, S.L. (1988) Glacial geology and overburden drilling in prospecting for buried gold placer deposits, Southern Quebec. In *Prospecting in Areas of Glaciated Terrain - 1988*, ed. MacDonald, D.R. Can. Inst. of Mining and Metallurgy, 141-169.

Shilts, W.W. and Kettles, I.M. (in press) Geochemical/mineralogical profiles through fresh and weathered till. In *Handbook of Glacial Indicator Tracing*, eds. Kujansuu, R. and Saarnisto, M. Balkema, Rotterdam.

Shilts, W.W., Cunningham, C.M. and Kaszycki, C.A. (1979) Keewatin Ice Sheet - Re-evaluation of the traditional concept of the Laurentide Ice Sheet. *Geology* **7**, 537-541.

Sivamohan, R. and Forssberg, E. (1985) Principles of tabling. *J. Mineral Proc.* **15**, 281-295.

Smith, S.L. and Rainbird, R.H. (1987) Soft-sediment deformation structures in overburden drill core, Quebec. In *Current Research*, Part A. Geol. Surv. of Can., Paper 87-1A, 53-60.

Smith, S.L. and Shilts, W.W. (1987) Quaternary stratigraphy of Noire River cores, Beauceville area, Quebec. In *Current Research*, Part A. Geol. Surv. of Can., Paper 87-1A, 159-164.

Stea, R.R., Turner, R.G., Finck, P.W. and Graves, R.M. (1988) Glacial dispersal in Nova Scotia: a zonal concept. In *Prospecting in Areas of Glaciated Terrain - 1988*, ed. MacDonald, D.R. Can. Inst. of Mining and Metallurgy, 57-79.

Stewart, R.A. (1986) Routine heavy mineral analysis using a concentrating table. *J. of Sediment. Petrol.* **56**, 555-556.

Strobel, M.L. and Faure, G. (1987) Transport of indicator clasts by ice sheets and transport half-distance. A contribution to Prospecting for Ore Deposits, *J. Geology* **95**, 687-697.

Taipale, K., Nevalainen, R. and Saarnisto, M. (1986) Silicate analyses and normative compositions of the fine fraction of till: examples from eastern Finland. *J. Sediment. Petrol.* **56**, 370-378.

Thompson, I.S. (1979) Till prospecting for sulphide ores in the Abitibi Clay Belt of Ontario. *Can. Inst. Mining Metall. Bull.* **72** (807), 65-72.

Thorliefson, L.H. and Kristjansson, F.J. (1988) Stratigraphy and visible gold content of till in the Beardmore–Geraldton area, Northern Ontario. In *Current Research*, Part C. Geol. Surv. of Can., Paper 88-1C, 217-221.

Veillette, J.J. (1986) Former southwesterly ice flows in the Abitibi–Timiskaming region: implications for the configuration of the Late Wisconsinan ice sheet. *Can. J. Earth Sci.* **23**, 1724-1741.

Veillette, J.J. (1989) Ice movements, till sheets and glacial transport in Abitibi–Timiskaming, Quebec and Ontario. In *Drift Prospecting - 1987*, eds. DiLabio, R.N.W. and Coker, W.B. Geol. Surv. of Can., Paper 89-20 139-154.

12 Geophysical exploration for gold

N.R. PATERSON and P.G. HALLOF

12.1 Introduction

The world-wide interest in gold that has characterized mineral exploration in the 1980s has had a profound effect on the focus and style of geophysical prospecting. In the 1950s and 1960s, direct detection methods (chiefly airborne and ground electromagnetics (EM)) were applied widely in the search for massive sulphide-hosted base-metal deposits (Paterson, 1967). At the same time, induced-polarization (IP) methods were developed for the direct detection of porphyry coppers and other disseminated sulphide ores (Hallof, 1967). In the 1970s, refinements were made to extend the search by electrical methods under deeply weathered rocks (Sumner, 1977; Ward 1977; Hallof, 1980). In all of this direct search for sulphides, there also lay the possibility of finding gold. Hannington *et al.* (this volume) estimate that about 5% of massive sulphide deposits world wide contain mineable gold, and some of the richest gold deposits have been found while exploring for base metals (e.g. Crone, 1984; Dowsett and Krause, 1984)

In the 1980s, however, these highly focused campaigns have given place to the ingenious application of a wide variety of methods, some new, but mostly tried and proven, generally in combination, not to search directly for gold but to locate environments in which gold is most likely to occur.

The indirect use of geophysics in exploration is not new. Airborne methods have been applied extensively for geological mapping since the 1940s (Paterson, 1962; Boyd, 1984). Magnetic and electrical methods have long been used to delineate faults and shears, iron formations, and igneous intrusions that may have an indirect association with precious metals, base metals, uranium, chrome, nickel, asbestos, tin, diamonds, and a host of other minerals. In the course of this work, gold deposits have indeed been found (e.g. Kelly, 1957b; Coggan, 1984).

The recent campaigns of the 1980s differ partly in focus, but mainly in scale and scope. Whereas airborne surveys were conducted widely to map geology (e.g. Reeves, 1989), now they are carried out specifically to locate structures and environments favourable for gold. Where previous exploration programmes for gold were generally confined to a particular geological structure or camp (e.g. Koulomzine and Brossard, 1957), recent campaigns have covered huge tracts of land such as the entire Casa Berardi belt and adjacent areas in Quebec, where a combined helicopter electromagnetic, magnetic and VLF EM survey (Podolsky, 1986) covered 7700 km^2 in which there were only three known gold occurrences. Programmes are currently underway in Nevada and in Australia, covering areas almost as large.

In this chapter, the authors examine the characteristics that constitute the typical signatures of environments in which gold deposits commonly occur. Geophysical methods and exploration strategies are described, and illustrated by examples from a variety of gold occurrences. A short glossary of geophysical terms is provided at the end of this chapter for those readers less familiar with geophysical techniques and terminology.

12.2 Geological and geophysical models

The hardest task for the geophysicist is to ascribe meaningful geophysical signatures to the wide variety of geological models proposed for gold deposits in the literature and described comprehensively in this volume. Without such a geophysical 'model', however, the job of looking for a favourable gold environment would be quite impossible.

The problem stems from the diversity of geological characteristics that may be associated with a single class of gold deposit. For example, Groves and Foster (this volume) list in their Table 3.2 the main characteristics of three types of Archaean lode-gold deposit. Host rocks of these deposits vary from sediments and pyroclastics, through felsic volcanics to tholeiitic basalts, dolerites, and ultramafics. Enclosing structures appear to be mainly shear zones but iron formations are also represented. Thirteen different alteration minerals are mentioned, of which there are no less than 71 alternative combinations, each varying to some degree in physical properties. Seventeen ore minerals are listed, occurring in a correspondingly larger variety of combinations.

The regional settings of gold deposits are equally diverse, although attempts have been made (e.g. Mitchell and Garson, 1981) to simplify or categorize them. Guides, even as broad as those given by Berger (this volume) for Carlin-type deposits, are essential for the geophysicist who is interpreting aeromagnetic and/or gravity data on a country-wide scale.

In Table 12.1, the authors have grouped gold deposits into seven classes based mainly on mode of occurrence rather than genesis, age or regional setting. This is done to simplify the formulation of geophysical models and exploration methods, since these vary more according to mode (e.g. veins vs. disseminated deposits) than from one genesis to another (e.g. young epithermal to Archaean lode). Differences between the geological characteristics listed in this table and those given in other chapters of this volume can be explained mainly by an attempt to focus on those that most affect the geophysical response.

It is clear from the table that the geological environments of most gold deposits can be recognized geophysically. In some cases, the ore zone itself is detectable, usually through accessory minerals such as pyrite or pyrrhotite. In very few instances, however, can the ore zone be distinguished from zones of similar mineral assemblages, structure and alteration, that do not carry economic gold.

The challenge facing geophysicists is to reduce the odds in gold prospecting by helping to direct the search to favourable environments and, once these have been located, to pin-point features with signatures that match those of the gold deposit model. As in every other kind of exploration programme, the selection of targets for ultimate drilling must involve the synthesis of all available geological, geophysical,

Table 12.1 Gold deposit models and geophysical signatures

	Host	Structure	Ore minerals	Alteration	Other characteristics	Geophysical model, signature
Veins, stockworks, lodes	Greenstones, slates, tuffs, iron formation; intruded by granitoids, gabbros	Faulting, fracturing, shearing; rifting; drag folds, horsetails favourable	Pyrite, quartz, calcite, arsenopyrite, typical; sometimes graphite, scheelite, stibnite, tellurides	Silica, chlorite, ankerite, sericite, argillite (epithermal)	Granitization frequent; metamorphic grade important locally	Weak conductor over fault or shear, sometimes extending into altered zone; silicification produces local resistivity high; pyrite may cause minor IP anomaly; host, structure and, metamorphic grade generally interpretable by magnetics/gravity/ radiometrics ; piezo-electricity may locate qtz. veins; scheelite produces UV luminescence; local K radioactivity possible
Skarn deposits	Carbonates and metacarbonates, usually adjacent to granitoids	Granite contacts; brecciation common; mineralization sometimes fault controlled	Pyrite, pyrrhotite, arsenopyrite usually abundant	Ca–Mg–Fe silicates, magnetite, hematite	High-grade metamorphism; injection of small granite bodies; large, irregular ore zones	Strong magnetic response common; possibly high U, K radioactivity; sulphides produce strong IP anomalies and local conductivity highs

Auriferous volcanogenic sulphides	Usually with felsic facies of volcanic piles; local granitoid or gabbroic intrusion common	Shearing common	Chalcopyrite, pyrite, galena, sphalerite; often magnetite, graphite	Chlorite, ankerite, sericite, etc.	Sulphide host usually massive, lens-like; iron-formation bands nearby; metamorphism commonly greenschist facies	Strong local conductivity from massive sulphide body, sometimes augmented by graphite; strong IP anomaly from disseminated sulphides; direct magnetic anomaly in some cases from magnetite or pyrrhotite; adjacent weak magnetic anomalies common from iron formation; conductive zone typically wide
Auriferous granitoids	Felsic, coarse-grained, discordant plutons	Shattering, brecciation	Chalcopyrite, molybdenite (porphyries); quartz, silver and sulphides (non-porphyries)	Biotite, anhydrite, magnetite, sericite, kaolinite, anhydrite, hematite, fluorite, calcite	Ore occurs as disseminations, veins, stockworks in the intrusive, often related to adjacent skarns and replacement deposits	Intrusive is usually detectable magnetically, by K, Th radioactivity, and by negative density contrast in most environments; shattering, brecciation commonly detectable as VLF conductor and resistivity low; ore zone may give IP response

Table 12.1 *Continued*

	Host	Structure	Ore minerals	Alteration	Other characteristics	Geophysical model, signature
Dessiminated deposits in igneous, volcanic, and sedimentary rocks	Typically stratabound in felsic intrusives, volcanics, tuffs, chemically favourable seds (carbonates preferred)	Faulting, fracturing drag-folding	Quartz, pyrite; often calcite, barite, stibnite, fluorite, arsenopyrite; sometimes graphite	Depending on genesis and host rock, siliceous or argillic alteration; often hematite, kaolinite, alunite, sericite, adularia, etc.	Ore usually closely tied to silicification, with or without pyrite; sometimes pyrite; sometimes flanked by clay alteration	Magnetics, radiometrics useful in locating igneous and structural controls; host beds may give direct resistivity, magnetic or gravity signatures; clay alteration, fracturing provide VLF and resistivity signatures; sometimes magnetic lows; ore zones commonly detectable by resistivity highs with or without coincident IP anomalies; scheelite produces UV luminescence

Palaeoplacers: pebble conglomerates, quartzites	Conglomerates, quartzites, sedimentary breccias	Eroded palaeo-surfaces; old stream, beach, fan, flood plain deposits, at base of sedimentary pile	Magnetite, pyrite, ilmenite, uranium, monazite (other heavy minerals)	Low-grade metamorphism; post-depositional oxidation of pyrite to SO_4; hematite, kaolinite, hydrocarbons frequent	Most deposits are Precambrian age and are confined to margins of ancient shields; slow subsidence; sedimentary hiatus important	Magnetics and gravity useful in delineating boundaries and configurations of basins; magnetic marker beds in sediments can be mapped to guide drilling, seismics; unconformity at base of sediments is sometimes detectable magnetically and usually by seismics; radioactive anomalies sometimes found over outcropping conglomerates beds with uranium, thorium; direct magnetic anomalies from magnetite in ore zone seldom detectable

Table 12.1 *Continued*

	Host	Structure	Ore minerals	Alteration	Other characteristics	Geophysical model, signature
Placers	Coarse sediments	All associated with alluvial, eluvial or beach deposits on eroded basement surface	Magnetite, ilmenite and other heavy minerals (similar to palaeoplacers)	Depending upon location and age, varying degrees of surface weathering -oxidation	Commonly Tertiary–Quaternary; shape and size determined by stream dynamics	Buried stream channels usually detectable by seismics, often by radar, resistivity, gravity, deposits in marine environments can sometimes be delineated by sub-bottom seismic profiling; if sufficient magnetite present the deposit can be followed by ground magnetics

and geochemical data. The non-uniqueness inherent in geophysical exploration for gold can thereby be substantially reduced.

12.3 Exploration strategy and methods

12.3.1 *Reconnaissance*

When exploration for gold is starting in a new physiographic, metallogenic, or tectonic terrane, it may be premature to commence a systematic survey, however regional in scope, without obtaining some basic knowledge of the geophysical responses in the area. This phase is often referred to as 'orientation'. Ideally it will include the determination of geophysical signatures of the significant lithostratographic units and structures in the area and the physical properties (e.g. magnetic susceptibility, density, electrical conductivity) of representative lithologies and, where appropriate, ores. Orientation airborne and/or ground surveys should be followed by close field examination of outcrops in the vicinities of selected, representative anomalies – a procedure often called 'ground truth'. Based on this knowledge, a sensible geophysical programme can be designed.

12.3.2 *Regional*

Regional surveys are normally those designed to locate environments of regional scale: e.g. greenstone belts, intracratonic basin margins, rift zones, plate boundaries, and regional fault patterns. Such information is still unavailable in many parts of the world and must be determined before a gold exploration campaign can be launched. Aeromagnetic and gravity surveys are most helpful at this stage (Boyd, 1984; Paterson and Reeves, 1985; Reeves, 1989). Such data are available over large areas of relatively unmapped terrane, or can be obtained at an acceptable cost if necessary (Reford, 1980; Paterson, 1983). Typical of such a programme was the aeromagnetic survey of the Kalahari area of Botswana (Paterson *et al.*, 1979; Reeves and Hutchins, 1982; Reeves, 1985) which followed a regional gravity survey, and was followed, itself, by a deep drilling programme (Meixner and Peart, 1984) to test and validate interpretations of buried lithologies and structures.

Often, however, regional surveys are regional only in scale, but fairly detailed in scope. For example, the systematic coverage of Canada by aeromagnetics (Hood *et al.*, 1985) has produced maps at 1 : 50 000 scale that are capable of pin-pointing features of 1–2 km dimension, such as individual faults, iron-formation units, and small intrusive bodies. The granite stocks that control the distribution of gold mines in the Red Lake camp of north-west Ontario (Figures 12.1(a)–(c)) show clearly on both the aeromagnetic and gravity maps of the area. In Africa, combined aeromagnetic and radiometric surveys of entire countries have been conducted at a traverse interval of 500 m (Grant *et al.*, 1980; Batterham and Bullock, 1983), and these have been used to prepare interpretations of quite detailed character. Target areas for surface exploration for gold have been identified and followed up in Ivory Coast and elsewhere as a direct result of these surveys.

Regional surveys are often conducted with combined magnetic and EM survey equipment, with the dual objective of directly locating massive sulphide bodies and

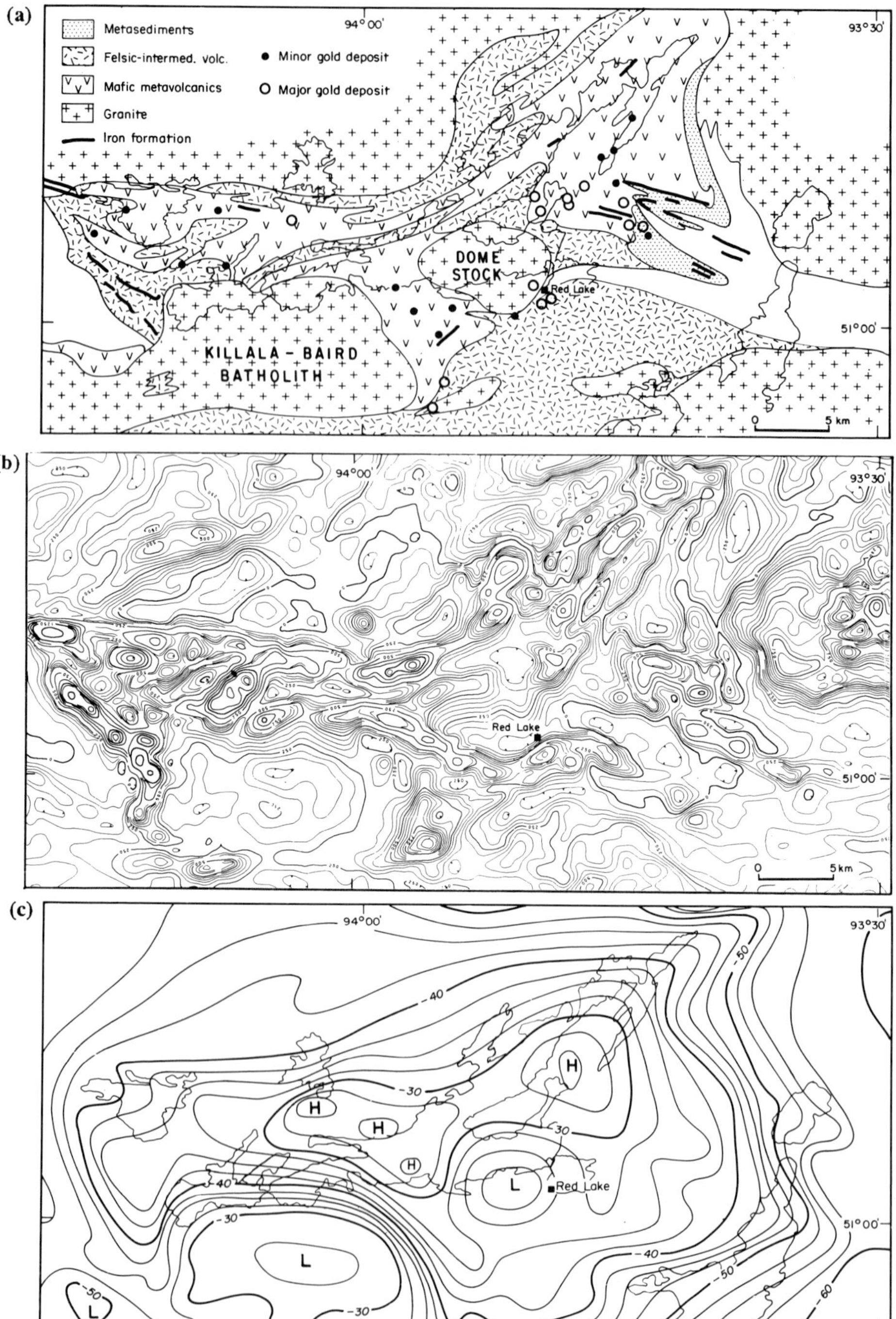

Figure 12.1(a) Simplified geology and gold deposits, Red Lake Mining District, Ontario. (b) Aeromagnetic contours, same area as (a). (c) Bouguer gravity contours, same areas as (a).

mapping geological structure and lithology. Palacky (1988, 1989) has demonstrated the role of airborne EM in identifying lithologies by virtue of the conductivity signatures produced by surface weathering. Reed (1989) has shown how conductive (usually graphitic) marker formations can help trace favourable stratigraphic zones for large distances under a thick layer of overburden. Wide-band EM systems, both fixed-wing and helicopter-mounted, are now used routinely to locate conductive features such as major shears and zones of argillic alteration. The discovery of the rich Hishikari epithermal gold deposit illustrates (Figure 12.2) the successful application of this technique.

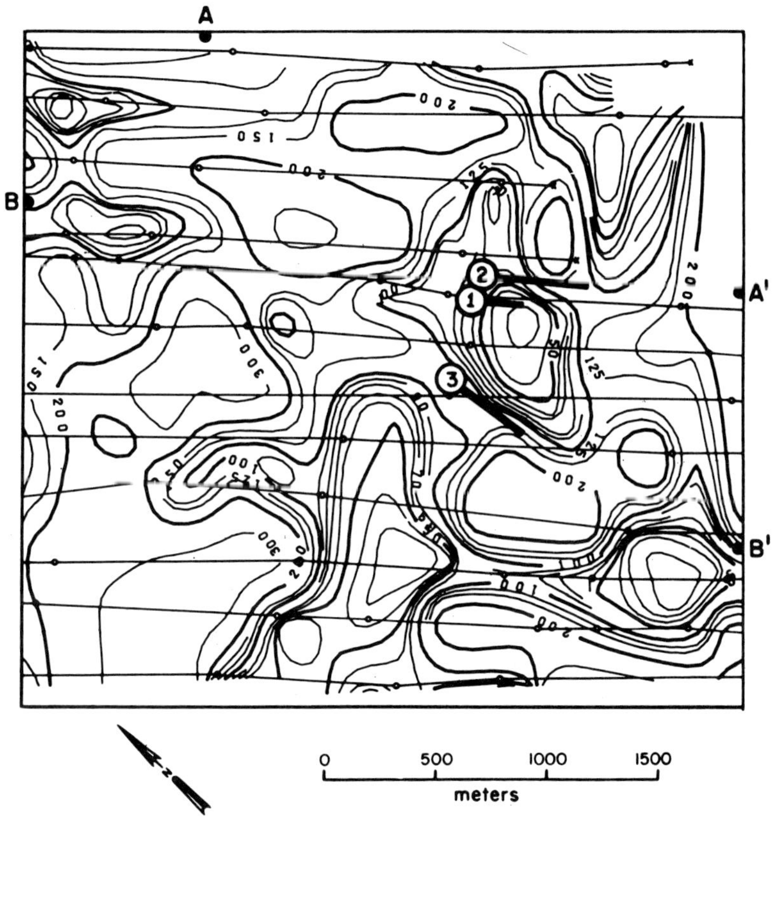

Figure 12.2 Apparent resistivity contours at 735 Hz from helicopter EM survey, Hishikari gold deposit; with discovery drillholes 1, 2, and 3. (After Johnson and Fujita, 1985.)

An effective first step in locating gold targets is the combined aeromagnetic and VLF EM survey, which can be conducted on a regional or detailed scale, usually at the small traverse interval of 100 m. The relatively low cost of this approach has led to its widespread application in areas where the VLF response is not masked by high surface conductivity (Whiteway, 1985).

12.3.3 *Detailed*

Detailed exploration differs from regional mainly in the scale of the search, typical programmes being of property size (a few km^2 to a few tens of km^2). Surveys may be airborne or ground depending upon the techniques used, the terrane, and the target. Typical airborne programmes are combined helicopter aeromagnetic, EM and VLF EM surveys of 300–1000 line km. On the ground, IP, gravity and controlled source audio magneto-telluric (CSAMT) surveys of the order of 100–200 line km are not uncommon. The final phase of all of these programmes is almost invariably a detailed ground geophysical survey, followed by trenching or drilling.

Because of the frequent association of gold with sulphides (Table 12.1), IP and resistivity methods have been favoured in many terranes (Doyle, 1986; Reed, 1989). Some gold occurrences, however, lack a sulphide association and can be located only by their relationship to structure and/or hydrothermal alteration. Examples of detailed geophysical signatures over the seven classes of gold deposits described in Table 12.1 are given in the following case histories.

12.4 Examples

12.4.1 *Veins, stockworks and lodes*

This is not strictly a single class of gold deposits but a collection of deposit types falling outside the other six classes that have been defined in Table 12.1. The examples chosen to illustrate geophysical exploration for this diverse group are divided into subgroups primarily according to age, but also regional tectonic environment.

1. Deposits in Archaean greenstone belts. Intensive gold exploration programmes have been undertaken in the shield areas of Canada and Australia, usually on a fairly detailed scale. The signatures of vein and stockwork deposits in greenstone environments are generally subtle, and the methodology of locating them is not easily transferrable from one gold camp to another. The following examples illustrate three approaches, each useful in a different geological environment.

Century Mine, Elbow Lake, Manitoba, Canada. Gold veins hosted by mafic volcanics are often characterized by linear magnetic lows, sometimes related to hematitization of magnetite and sometimes to the intrusion of narrow feldspar porphyry dykes or massive quartz veins. Shears are the dominant structural control. The combination of ground magnetometer and VLF EM surveys has been credited with the discovery of many small gold occurrences.

The ground geophysical and geochemical surveys illustrated in Figure 12.3 were carried out by Ram Petroleums Limited in 1980. They centred on the Webb and Garbutt veins, comprising the main ore zone of the Century Mine, on which development work was carried out in the 1930s. The veins are in sheared intermediate-mafic metavolcanics, adjacent to a feldspar porphyry dyke. The zone is generally altered and brecciated, and carries both quartz and sulphides locally.

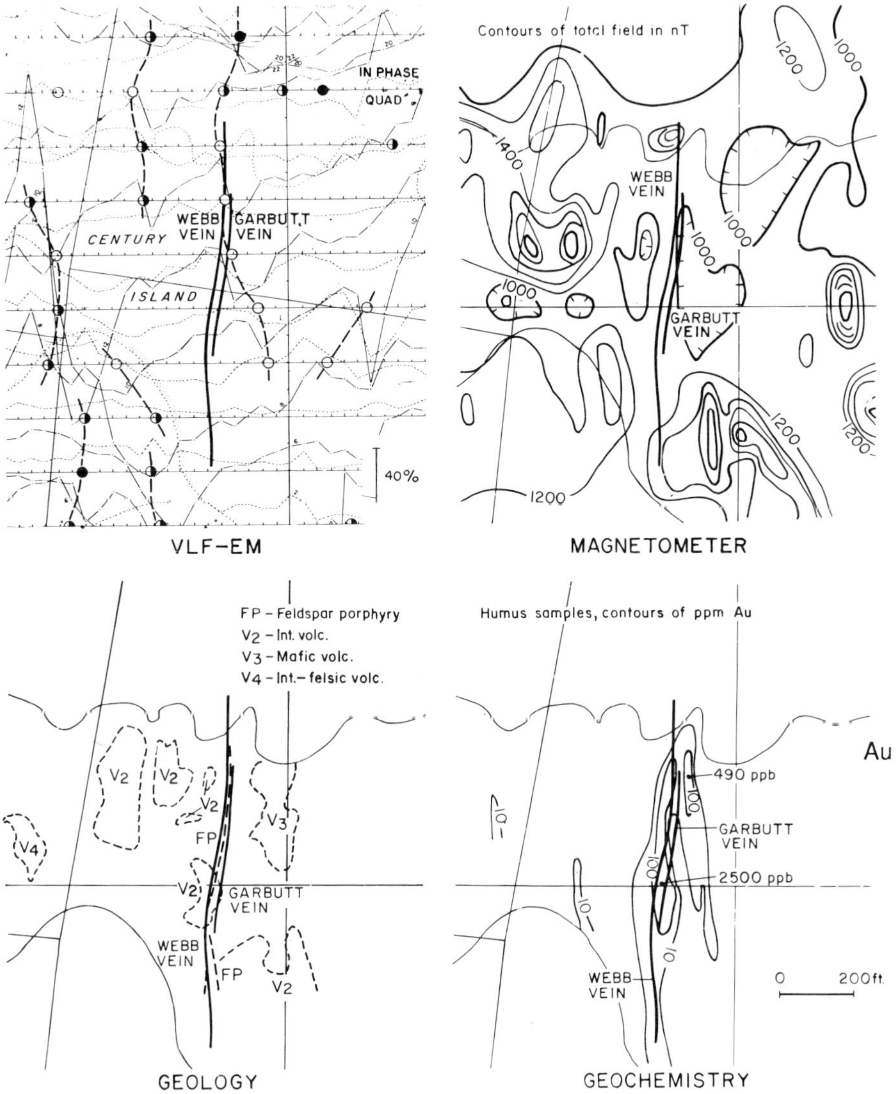

Figure 12.3 Geophysical and geochemical signatures of the Century Mine, Elbow Lake, Manitoba. (Courtesy Ram Petroleum Ltd.)

The magnetometer survey shows a picture characteristic of metavolcanics, with the predominant magnetic grain following the NW–SE direction of foliation. The abrupt, transgressive magnetic low near the Webb and Garbutt veins was thought to reflect a north–south trending structure, and possibly the feldspar porphyry. The VLF EM survey showed quite discrete anomalies, one of which followed the veins locally but continued to the north and branched eastwards to the south.

Humus samples taken at 25-ft centres showed gold concentrations up to 2500 ppb over the veins and weaker but significant values over some of the other VLF EM anomalies in the area. Coincident magnetic lows and VLF conductors were found in most cases to represent zones of shearing and alteration, several of which gave good geochemical responses.

Barraute gold zone, Val D'or, Quebec, Canada. Where the gold is closely associated with disseminated sulphides, either in the ore zone or in the altered wallrock, IP surveys have been applied quite effectively.

The ore-grade, gold- and sulphide-bearing quartz vein at the Barraute property near Val D'or, Quebec, is exposed in an open pit. The vein itself is about 5 feet (1.5 m) in width, and contains considerable sulphide mineralization. In addition, there is 1% to 3% pyritic mineralization that extends a few feet into the Archaean-age andesitic wallrocks.

The open pit lies to the east of Line 72 + 50E and, at the west end of the pit, the gold–quartz vein is offset by a fault. It was felt that, somewhere to the west of the fault, it would be possible to find the western extension of the vein.

Line 72+50E was the easternmost of several lines that were surveyed to the west of the known position of the fault. Phase IP and resistivity measurements were made using a frequency of 1.0 Hz. There was not much glacial overburden present, and, due to the narrow width of the source being sought, a short electrode interval was used. The dipole–dipole electrode configuration was employed, with dipole separation (X)=15ft (4.6m), and distance between dipoles (n) = 1, 2, 3, 4, 5 times the dipole separation.

The pseudosection plots for the measurements made on Line 72 + 50E (Figure 12.4) show an apparent resistivity, apparent phase shift, and apparent Metal Factor anomaly centred at approximately 5 + 50S to 5 + 35S. Since the measurements made for $n = 1$ were not anomalous, there is some depth indicated to the top of the source of the anomalies.

There was little or no surface expression of the vein and so an angle drillhole was planned. The drillhole intersected the quartz vein with metallic mineralization at a depth of about 10 ft (3.1 m).

Central Patricia property, Pickle Lake, Ontario, Canada. Vein gold deposits in banded iron formations may have both electrical and magnetic signatures. Airborne and ground magnetometer surveys are used extensively to focus IP/resistivity surveys on the most favourable structures – usually drag folds or faults cutting siliceous iron-formation bands in Archaean metasedimentary and metavolcanic belts. The target is usually the pyrite which is commonly associated with gold mineralization in these environments.

Ore deposits in the Pickle Lake area are found both as quartz veins cross-cutting the iron formation (e.g. Pickle Crow) or as sulphide 'stringers' occupying fractures or drag folds in the iron formation. At the Central Patricia deposit, there is a close association of gold with pyrite and arsenopyrite. Pyrrhotite is pervasive in both the chert–magnetite and sulphide facies of iron formation but is seldom associated directly with the gold mineralization.

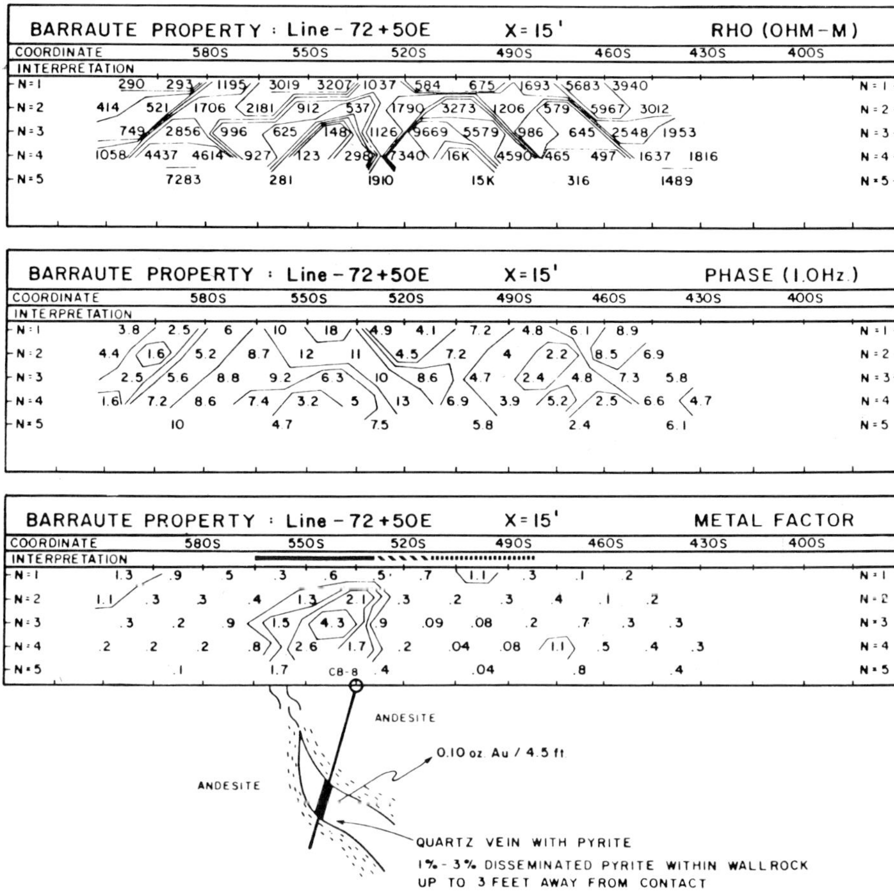

Figure 12.4 Dipole–dipole IP/resistivity pseudosection on Line 72+50E, Barraute Property, Val d'Or, Quebec. (Courtesy Barexor Minerals Inc.)

Exploration in the Pickle Lake area is guided by careful mapping of the iron formation using both airborne (Figure 12.5) and ground magnetics. The SW Powderhouse zone of the Central Patricia deposit was found by Noramco Explorations Inc. by drilling a strong IP anomaly (Figure 12.6) on the nose of a folded iron-formation band. The ore zone is a narrow lens of pyrite–arsenopyrite in a drag fold at the hanging wall of the structure. Gold values average about 6 g/t over typical widths of 1–2 m.

Strong resistivity lows (Figure 12.6) are indicative of the massive pyrrhotite in the sulphide facies of iron formation. At this deposit, however, gold appears to be structurally confined to the outer chert–magnetite zone.

2. *Deposits in Palaeozoic fold belts.* Regional exploration for these deposits usually involves studies of aeromagnetic and gravity data to define major structures and outline intrusive and metamorphic units. Robinson *et al.* (1985) and Tenison-Woods and Webster (1985) described studies of this type in the Carolina Slate Belt of eastern

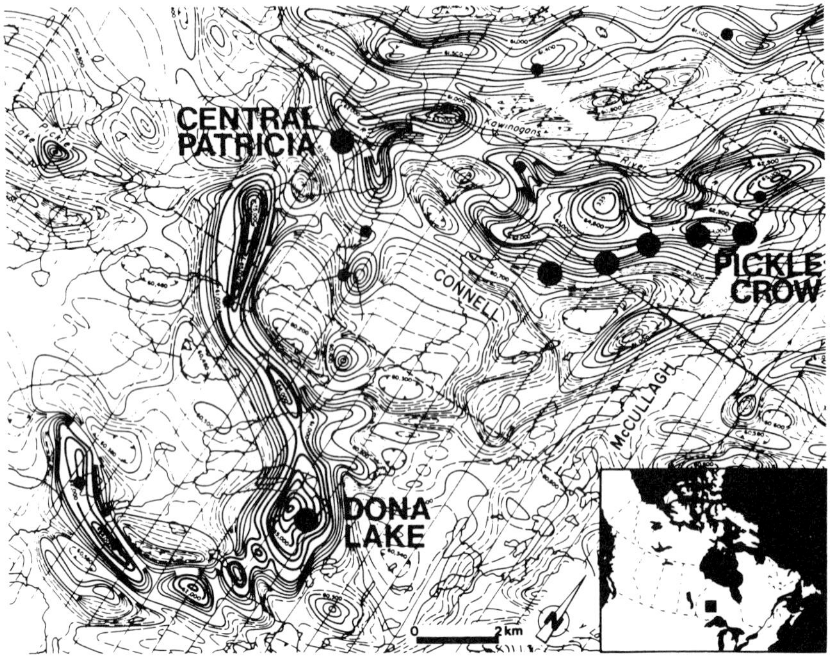

Figure 12.5 Aeromagnetic contours of the Pickle Lake Mining District, Ontario, with gold deposits. (After Reed, 1989.)

USA and the Lachlan Fold Belt of eastern Australia respectively. The deposits themselves appear to be characterized by a more widespread halo of alteration than is common in Archaean rocks, and can be explored effectively by a combination of IP, VLF EM, resistivity and magnetic methods.

In the Drake mineral field in north-eastern New South Wales (Staltari, 1985) IP/resistivity surveys were successful in outlining the hydrothermal alteration associated with gold–silver mineralization in the Drake volcanics.

3. Palaeozoic-Tertiary volcanic/subvolcanic granitoid-related deposits. Geophysical exploration for these deposits centres on the identification (regional) and delineation (detailed) of the granitoids, usually by magnetics or a combination of magnetics, gravity, and gamma-ray spectrometry. The orebodies are usually found as stockworks in breccia zones or fissures systems, often related to the intrusion of the granite. Magnetite may be depleted (through conversion to pyrite, ankerite or hematite) or it may have been introduced with the mineralizing solutions. As in the case of skarn deposits, detailed ground magnetic surveys may or may not be of direct use in targeting gold mineralization. Sulphides, usually pyrite, are almost always present, however, and the IP method has been used successfully in several mining camps.

Cripple Creek Mining District, Colorado, USA. The Cripple Creek mining district, though not strictly granitoid-related, presents a good example of the use of magnetic and gravity surveys, not only to outline regional structure but also to pin-point zones

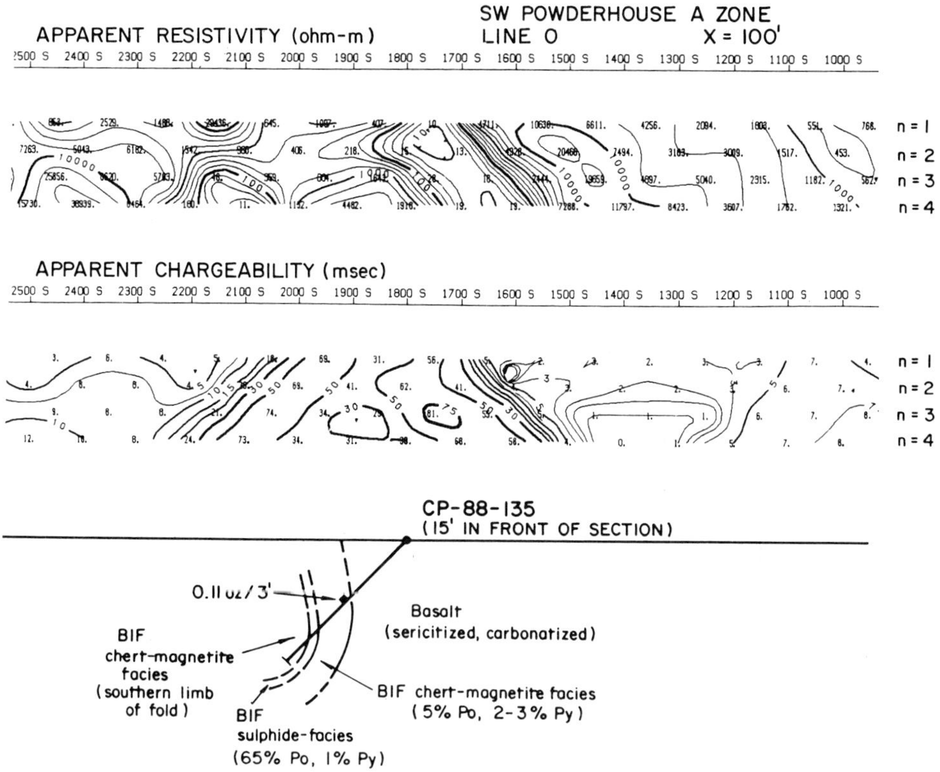

Figure 12.6 Dipole dipole IP/resistivity pseudosection and idealized geology, Line 0, SW Powderhouse ore zone, Central Particia gold deposit, Ontario. (Courtesy Noramco Explorations Inc.)

of intense brecciation and alteration where gold has been found. The Cripple Creek gold deposits occur in a Tertiary volcanic-subsidence basin within Precambrian granite, gneiss and schist. Magnetic surveys successfully outline the basin (Kleinkopf et al., 1970) and display magnetic lows over the more intensely altered volcanics. The basin is also well delineated by gravity surveys, and sharp gravity lows reflect fracture systems and associated higher porosity. Strong potassic alteration is mentioned, suggesting that shallow deposits of this sort might be recognizable radiometrically.

Mt. Milligan Mine, B.C., Canada. The intrusive bodies in the Quesnel Belt of British Columbia are alkaline syenite, monzonite, syenodiorite, monzodiorite, diorite, and syenogabbro batholiths, stocks, plugs, and dykes. Many lie along linear trends, believed to reflect fault zones, which form discernible patterns in the aeromagnetics. The plutons are usually recognizable by a zoned magnetic signature.

The ore at the Mt. Milligan deposit occurs as replacement veins and stockworks both in the mainly basaltic volcanic host rocks and in the monzonite porphyry itself. Ore minerals include pyrite and chalcopyrite. Magnetite is present in a breccia zone adjacent to the Mt. Milligan pluton. Alteration is characterized by both silicate and propylitic assemblages but no distinct resistivity patterns are observed.

Section 9500 (Figure 12.7), crossing the MBX zone of the Mt. Milligan deposit, shows a typical IP response from the pyrite mineralization and a strong magnetic anomaly associated with magnetite in the contact zone. IP has been used effectively to outline other pyrite concentrations in the vicinity of the pluton but no simple relationship has been found between gold and pyrite concentrations. VLF EM is reported to have assisted in delineating fault zones that have been important controls for the gold mineralization in the deposit (R. Dickinson, pers. comm.)

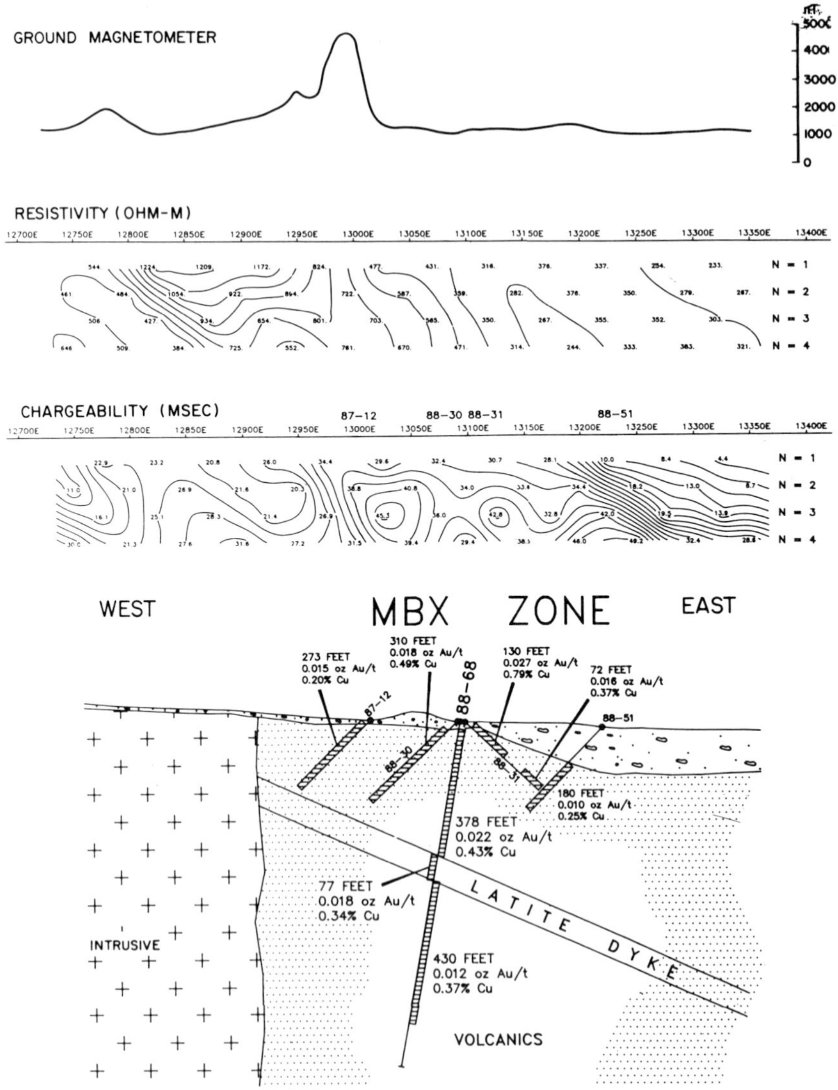

Figure 12.7 Dipole–dipole/resistivity pseudosection, and idealized geological section on Line 9500, MBX Zone, Mt. Milligan, British Columbia. (Courtesy Continental Gold Inc. and BP Resources Canada Limited.)

4. Mesozoic-Quaternary epithermal deposits. Although currently the target of widespread and intensive exploration in Nevada and elsewhere, this class of deposits is one of the most difficult to locate or delineate by geophysical methods. Favourable environments are extensional fault systems but frequently there are insufficient magnetic susceptibility or density contrasts in the country rocks to allow effective use of magnetic or gravity methods. Sulphides, common in other lode deposits, are either frequently absent or completely oxidized, rendering IP ineffective.

The most useful characteristic, geophysically, is the intense silicification that usually accompanies the gold mineralization, particularly in the Carlin deposits of Nevada. The strong, local resisitivity highs over these zones are often detectable by CSAMT or other electrical methods.

At the Hishikari gold deposit, Japan, volcanic rocks have been subjected to clay alteration and are recognizable (Figure 12.2) as local resistivity lows (Johnson and Fujita, 1985). The same is said to be true of epithermally altered shaley limestones in the Getchell Trend of Nevada (D. Hoover, pers. comm.).

Electrical and EM methods have also been used effectively to map structure and thereby localize drilling programmes. Some examples follow.

Marigold Mine – 8-South ore zone. The 8-South orebody at the Marigold Mine, in Humboldt County, Nevada, is one of several that have been outlined on the mine property. Since the area is almost entirely covered by a few tens of metres of Tertiary Carlin sediments, the zones of mineralization were found by a programme of pattern drilling to test the Palaeozoic clastic and calcareous sediments that form the basement.

The mineralized gold-bearing zones in the 'dirty limestone' sediments in the basement are entirely oxidized and highly altered. The geologic mapping, based on the results of the detailed drilling necessary to evaluate the ore zone, indicated that several faults extended in an approximately north–south direction through the orebody. This has been confirmed by detailed mapping, now that the overburden has been removed preparatory to mining.

Before the overburden removal began, a CSAMT survey was completed on a large grid that included the 8-South ore zone. The purpose of the survey was to search for bedrock conductors that might be due to the zones of increased porosity associated with the faulting. The measurements were made using an E-field dipole of 200 ft (61 m), and twelve frequencies in the range 1.0–2048 Hz.

The apparent resistivity data in Figure 12.8 show the increased thickness of conductive overburden that accompanies the near-vertical conductor (fault) at E0200E that has displaced the higher resisitivity basement rocks downwards to the east. This conducting feature correlates exactly with one of the faults that have been mapped as passing through the orebody. A second fault is interpreted at E1600E. The other known faults can be detected only as weak conductors, with shorter E-field dipoles (not shown).

A helicopter EM survey of the same area (Figure 12.9) showed a large zone of low resistivity in the south-east corner of Section 8, almost certainly representing the down-faulted Carlin sediments. The NW-trending gradient flanking this feature correlates well with the two faults found by the CSAMT survey.

Dee gold mine – Capstone Extension, Elko County, Nevada. The Capstone orebody is currently being prepared for production by Newmont Gold Inc. The silicified

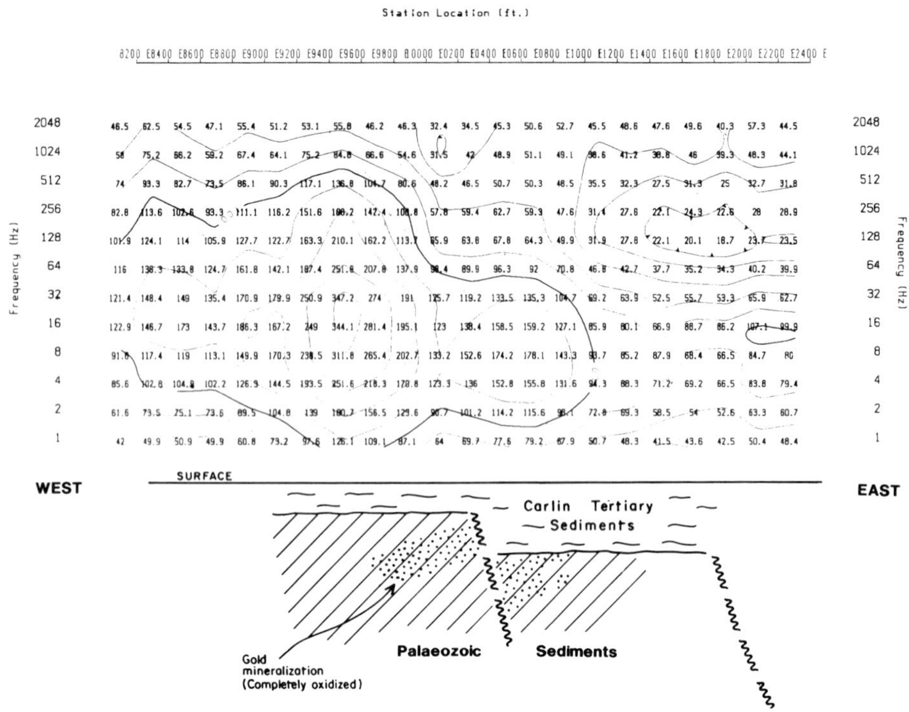

Figure 12.8 CSAMT apparent resistivities and idealized geologic section, Line 20,400E, 8-South ore zone area, Marigold Mine, Nevada. (Courtesy Marigold Mines Inc.)

gold–sulphide ore zone lies a few hundred feet south of the south-east boundary of the Dee gold mine property. At its north end, the Capstone orebody plunges to the north and disappears beneath the Tertiary Carlin sediments and volcanics that lie at the surface. A CSAMT test survey (not shown) conducted over the Carlin sediments, at the edge of the outcrop, indicated that the ore zone itself was a distinct resistivity high (greater than 1000 ohm-m) due to the intense silicification and reduced porosity.

A later CSAMT survey was conducted along Line 64+00N on the Dee gold mine claim boundary, approximately 800 ft (244 m) to the north. At this location, an unknown thickness of the Carlin sediments and volcanics was present. The apparent resistivity pseudosection on Line 64+00N (Figure 12.10) shows a narrow apparent resistivity high at depth (i.e. detected only at the lower frequencies used).

The results were interpreted using an approximate, plane field and two-dimensional computer inversion programme. The result suggested that a ridge more than 100m high might exist on the basement. Further, the 'best-fit' solution required that the ridge have a higher true resistivity (less porosity) than the surrounding bedrock. Later drilling has shown that the bedrock configuration is very much like that predicted by the computer inversion of the pseudosection. A non-porous, silicified hill is present in the bedrock, and significant gold values are present. The geology suggests that it is the northern continuation of the Capstone orebody.

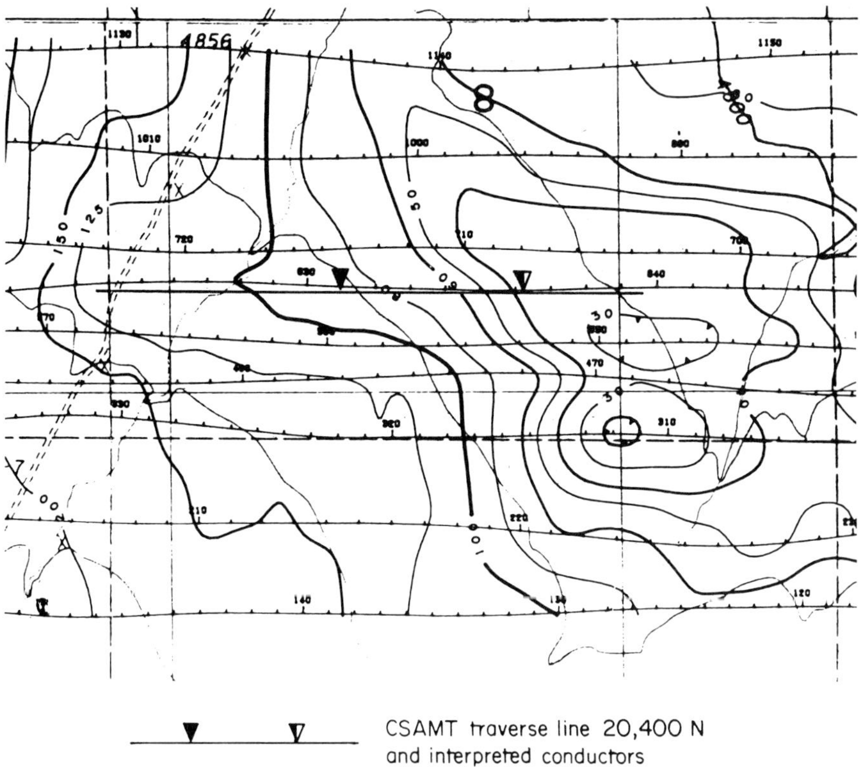

Figure 12.9 Apparent resistivity contours at 900 Hz from helicopter EM (DIGHEM IV) survey, 8-South ore zone area, Marigold Mine, Nevada. (Courtesy Dighem Surveys & Processing Inc. and Teck Resources Inc.)

12.4.2 *Skarns*

On a regional scale, magnetic and gravity surveys can be very helpful in locating and delineating intrusive bodies. Radiometric data can assist in identifying the intrusives and the nature of the country rock. Still at a regional scale, extensive zones of magnetic and/or radiometric activity over calcareous sediments, adjacent to or possibly overlying roofed intrusions, can be recognized as possible skarns. Examples of such occurrences are widespread in Malaysia and Thailand, often associated with Triassic granodiorite. Not all skarns, however, produce detectable magnetic or radiometric responses. At the Nickel Plate orebody near Hedley, B.C., the gold mineralization is closely associated with disseminated pyrrhotite but the ore zones are not recognizable by either ground or airborne magnetics. IP can detect the mineralization but is unable to discriminate between the gold-bearing and barren sulphide zones. VLF EM has been used to interpret structure, which is thought to control some of the ore.

Fortitude deposit, Battle Mountain, Nevada, USA. The Fortitude gold–silver deposit (Wotruba *et al.*, 1988) is related to a 'wallrock' copper porphyry system developed

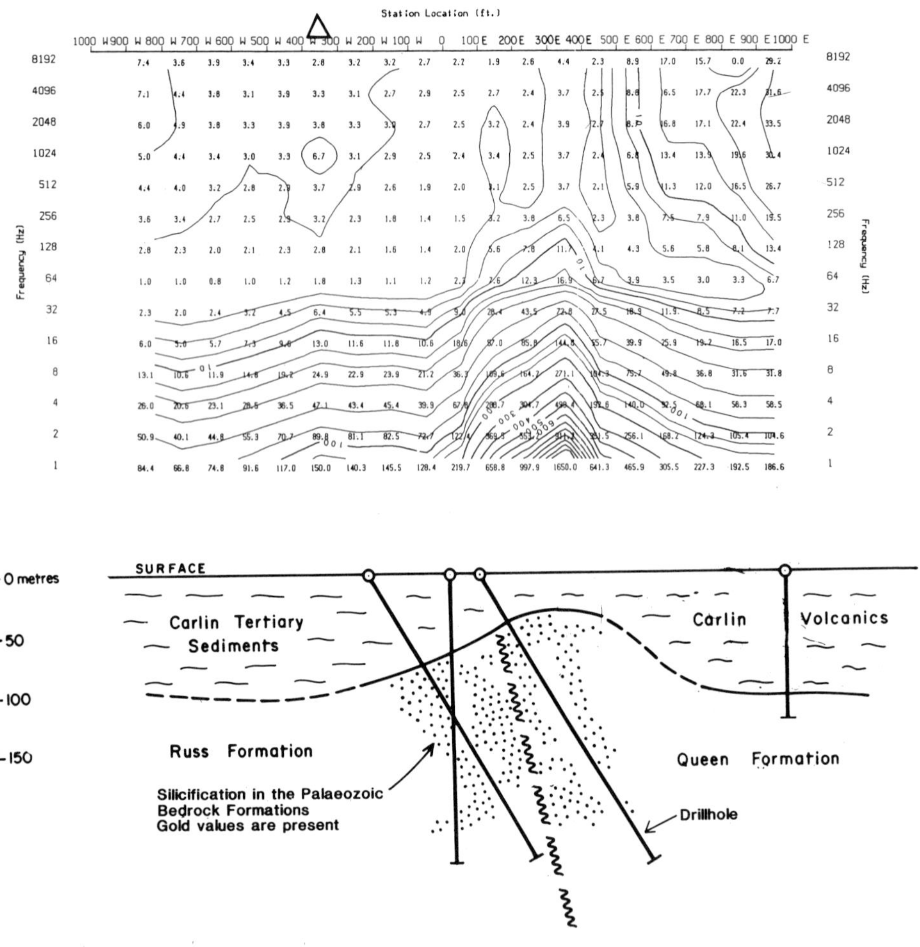

Figure 12.10 CSAMT apparent resistivities and idealized geologic section, Line 64 + 00N, Dee gold mine Nevada. (Courtesy Dee Gold Mining Co.)

within Middle Pennsylvanian to Permian carbonate and coarse clastic rocks adjacent to an altered Tertiary granodiorite stock. The orebodies occur with disseminated and massive sulphide-replacement mineralization of skarn-like or calc-silicate-rich units. A major north-trending normal fault is believed to have controlled the hydrothermal fluids responsible for the mineralization.

Figure 12.11 shows the generalized geology and the corresponding magnetic anomaly map of the Copper Canyon area. The granodiorite stock and the central Cu+Au+Ag metal zonation halo fall within a broad magnetic low, probably resulting from pervasive alteration and a relative lack of secondary pyrrhotite. To the north, however, a magnetically high area surrounds the Lower Fortitude, Upper Fortitude, West and Northeast orebodies. Widespread pyrrhotite replacement of early-formed calc–silicate minerals is responsible. Gold is closely associated with this phase of the cycle. The Lower Fortitude orebody, which contains the highest concentration of pyrrhotite (up to 50%, locally, of the rock volume), gives the most pronounced magnetic anomaly.

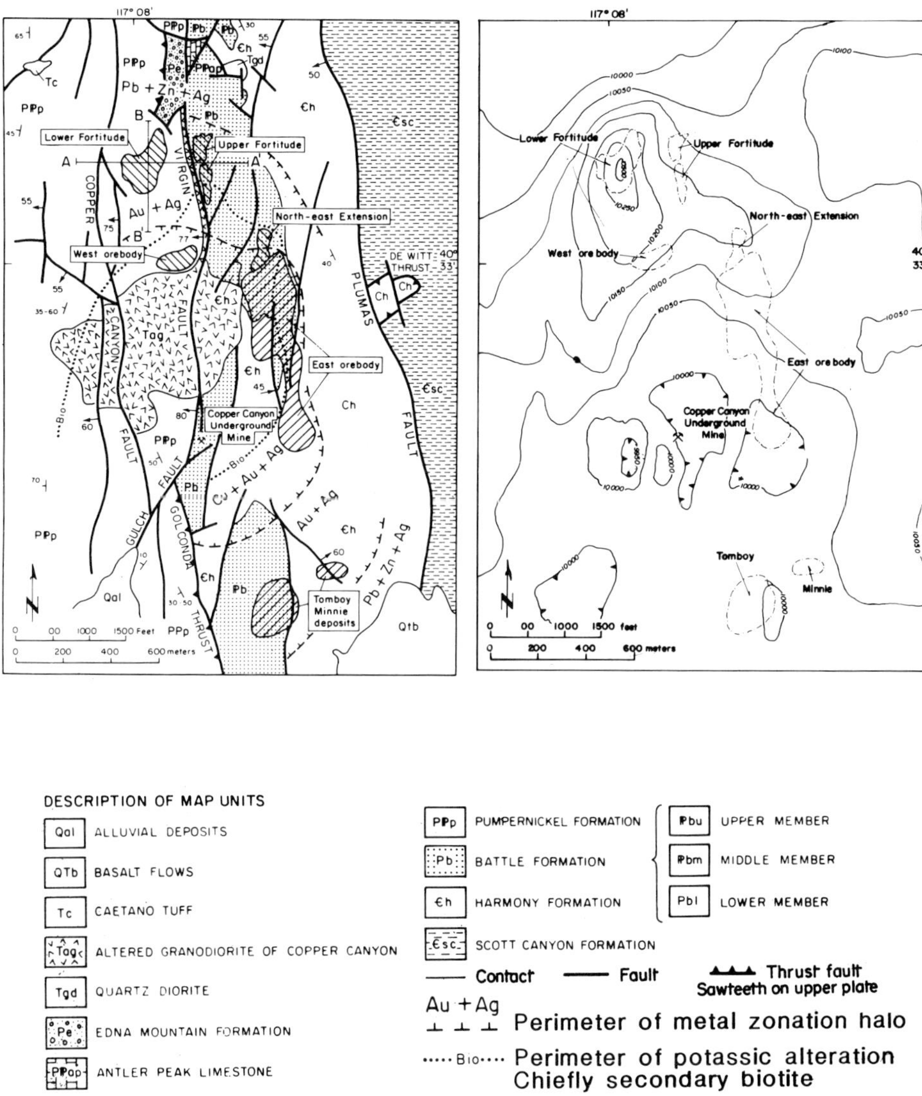

Figure 12.11 Geology (left) and corresponding residual magnetic anomaly map (right): Fortitude gold–silver deposit, Nevada. (After Wotruba *et al.*, 1988.)

Airborne and ground magnetics, which were instrumental in the discovery of the Lower Fortitude orebody, have served a useful role in ongoing exploration in the Copper Canyon area.

12.4.3 *Auriferous volcanogenic sulphides*

The statistics of finding mineable gold by exploring for massive sulphide bodies are not favourable (Paterson, 1967; Hannington *et al.*, this volume). Nevertheless, some very important gold occurrences have been discovered in this way (Reed, 1989). Given a close association of gold with sulphides in a given deposit or camp (e.g. Noranda, Quebec; Mt. Magnet, W.A.), geophysical techniques can be used effectively by delineating the massive sulphide bodies in three dimensions (e.g. Lindeman, 1984).

The discovery of gold-bearing massive sulphide bodies has been mainly the result of systematic airborne EM surveys carried out primarily for base-metal exploration. The Rouez orebody (France) was discovered as the result of a regional airborne magnetic/EM (INPUT)* survey in 1975 (Despointes, 1984). The roughly 100-million-tonne, large, pyritic deposit contains an average of only 1.5 g/t gold but is mineable in the oxidized zone which carries up to 10 g/t. The traditional geophysical techniques for delineating massive sulphides, namely gravity, IP, EM and resistivity, have been used successfully.

The Kerry Road orebody at Gairloch, Scotland (Bowker and Hill, 1987) is a stratabound, massive copper–zinc–gold sulphide body in Lewisian age hornblende and quartz–mica schists. Magnetic, EM, IP, self-potential and radiometric methods were successfully applied in combination. The radiometric and magnetic data were useful indirectly in mapping lithology and structure.

Detour Lake orebody, north-eastern Ontario, Canada. Crone (1984) described in detail the discovery and subsequent exploration by ground geophysics of the Detour Lake zinc–copper–gold–silver deposit. The main orebodies occur in a cherty, felsic tuff flanking an ultramafic intrusion within a steeply dipping sequence of Archaean basalts, tuffs and agglomerates. The ore minerals consist of pyrrhotite, chalcopyrite and minor pyrite, averaging 10–15%, but locally reaching 30% total sulphides by volume. Gold, which is not visible in the drill core, averaged 5.5 g/t in the zone initially tested.

The discovery was made as a result of drilling a moderate to weak ground EM and VLF EM conductor, with a coincident magnetic anomaly (Figures 12.12 and 12.13). These were obtained during the follow-up of a fixed-wing EM (INPUT)/magnetometer survey conducted by Amoco Canada Petroleum Company (Mineral Division) in 1974 primarily to search for base metals. The airborne anomaly (Figure 12.14) was rated as second priority, owing to its relatively low apparent conductance (1–2 S). Crone (1984) credited the discovery to the policy of thorough drill-testing of all interesting conductors, rather than 'high-grading' them, which is the practice of some exploration companies. He also points out the importance of assaying for gold, even when the primary target is base metals.

Further gold exploration in the area has centred naturally on testing moderate to weak EM conductors in association with mafic to ultramafic intrusions. Unfortunately the footwall intrusive at the Detour Mine is virtually non-magnetic (Figure 12.13), presumably because of the pervasive talc–carbonate alteration. However, large-spacing, vertical-loop, ground EM has been successful in mapping the contact between the ultramafic and cherty tuff units.

* INPUT: Registered trademark of Barringer Research Ltd.

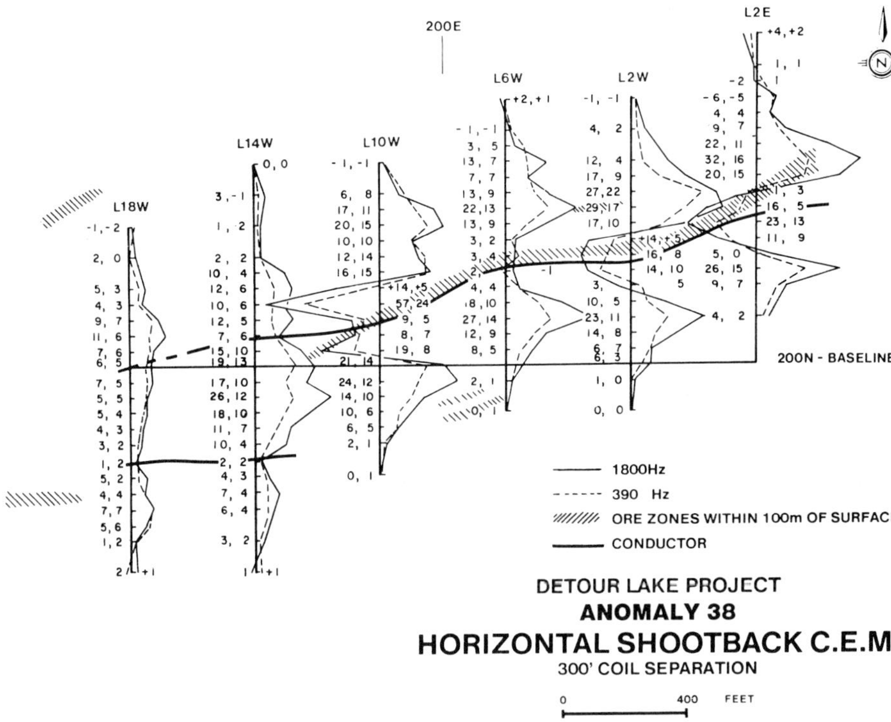

Figure 12.12 Ground EM (discovery) profiles and main ore zones, Detour Lake gold deposit, Ontario. (After Crone, 1984.)

12.4.4 *Auriferous granitoids*

The contacts of granitic intrusions with older volcanic and sedimentary rocks are favourable environments for mineralization of many types. These contacts can usually be mapped geophysically by an appropriate combination of magnetics, gravity and gamma-ray spectrometry (see, for example, Figure 12.1). The granitoids themselves can be either magnetic or non-magnetic, depending on composition. In the former case, mineralized zones may be reflected as magnetic lows. Within the granitoids, gold mineralization is usually found in shears and fracture zones, commonly accompanied by intense local alteration. Such structures can be detected simply, in many cases, by VLF EM (see Ferderber gold deposit below) or by a combination of VLF EM and magnetics.

Porphyry gold deposits are usually associated with disseminated pyrite, and can be explored by IP methods. Local potassium concentrations have been found by gamma-ray spectrometry over some Cu–Au porphyry deposits in Triassic granodiorites in South-East Asia.

Ferderber gold deposit, Val d'Or, Quebec, Canada. The Ferderber deposit is one of two known shear-hosted gold orebodies on the property of Belmoral Mines Ltd. near

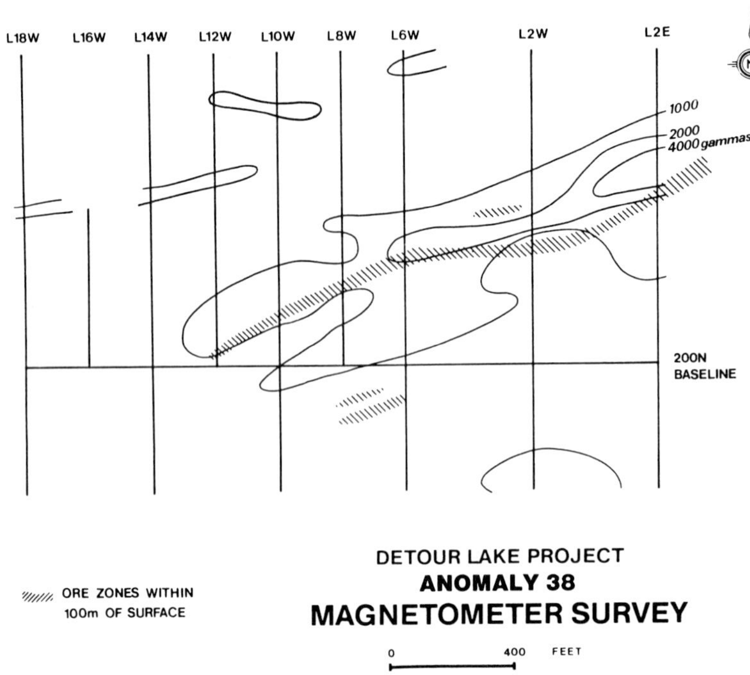

Figure 12.13 Simplified total field magnetic contours and main ore zones, same area as Figure 12.12. (After Crone, 1984.)

Val d'Or, Quebec (Darling *et al.*, 1986). Unlike most of the mines in the immediate area, which follow the contact of the Bourlamaque granodiorite batholith with adjacent Archaean volcano-sedimentary rocks, the Ferderber deposit lies near the centre of this 8.5 km × 4 km intrusive body.

The ore zones consist of quartz veins with carbonates and up to 25% combined pyrite and chalcopyrite. The pyrite–gold association is so close that mine geologists use the ratio 35 g/t Au per 10% pyrite to estimate gold grade. The veins are spread over a shear zone varying in width from 1 to 15 m that has been traced along strike for 3 km and is still open in both directions. The Ferderber Mine has published reserves (including ore already mined) of 0.79 Mt averaging 7.5 g/t Au. Smaller orebodies, such as the Ferderber West Zone, have been discovered on the same and parallel shear zones.

The deposits were discovered in 1974 by ground magnetometer and VLF EM surveys carried out over a large part of the Bourlamaque batholith (Bergmann, 1980). Weak magnetic anomalies suggested an extensive zone of shearing striking in an east–west direction across the batholith. VLF EM conductors (Figure 12.15) were found to coincide with this zone and suggested a pattern of branching shears more than 100 m wide and several kilometres long. Initial drilling, carried out on the B-

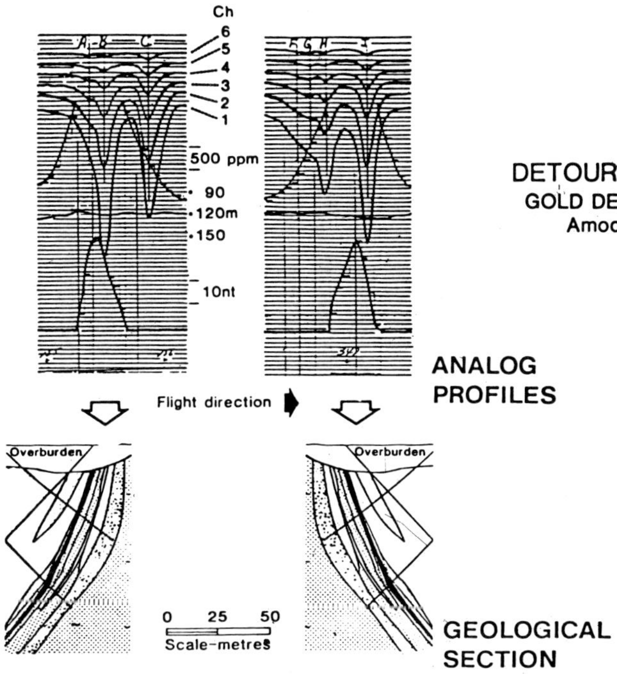

Figure 12.14 Airborne EM (INPUT) discovery profiles and simplified geologic section, Detour Lake gold deposit, Ontario. (Courtesy Questor Surveys Limited.)

Zone of strong VLF EM anomalies, discovered the main ore zone shown in Figure 12.15.

Conductivity responsible for the extensive VLF EM anomalies is probably related to porosity caused by shearing, together with the presence of minor clay-type alteration products. The magnetic anomalies may be related to the reported sporadic pyrrhotite mineralization.

12.4.5 *Disseminated deposits in igneous, volcanic and sedimentary units*

This class of deposits lends itself ideally to the dual approach: (a) indirect exploration for the appropriate stratigraphic or structural environments; and (b) direct exploration for the associated ore minerals (e.g. pyrite) or alteration zones (e.g. siliceous or argillic). The ore zones themselves are generally larger in volume than other lode deposits and are therefore more easily detected.

Wynn and Luce (1984) described the use of VLF EM, EM, magnetic and IP methods to map gold-bearing metasediments at the Haile gold mine in South Carolina. The mineralization is characterized by a zone of siliceous alteration more than 300 m wide that is easily identifiable by the VLF resistivity method.

A comprehensive suite of geophysical surveys was conducted over the Hope Brook gold deposit near Chetwynd, Newfoundland (Reed, 1989). Ground magnetics and VLF EM assisted both in mapping structure and tracing a stratabound pyrite zone

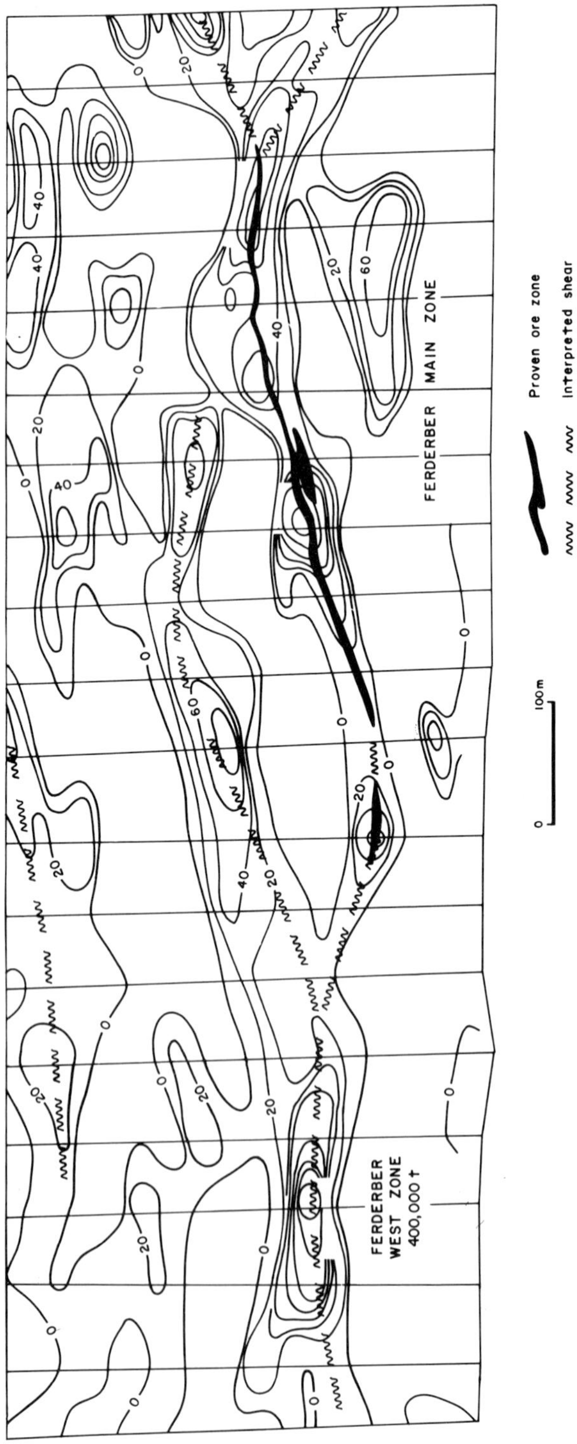

Figure 12.15 Filtered VLF EM contours (discovery data) and proven ore zone, Ferderber gold deposit, Val d'Or, Quebec. (Courtesy Belmoral Mines Ltd.)

which is subparallel to the orebody. IP surveys detected the pyrite body but also produced a weak response from the sulphides associated with the ore zone.

Holt-McDermott Mine, Harker & Holloway Twps., Ontario. The Holt-McDermott deposit is contained within a band of Archaean sedimentary and/or rhyolitic rocks that are believed to follow a branch of the famous Porcupine-Destor Fault Zone that controls several orebodies in the Abitibi Belt in northern Ontario and Quebec. The ore zone currently being mined (at the Holt-McDermott Mine) was discovered by an extensive drilling programme that extended at least a mile to the east from an outcropping zone of gold-bearing sulphide mineralization.

The development of the ore by drilling progressed slowly since, to the east, the position of the favourable structure and the mineralized rock units were covered by a layer of conductive, 'clay-belt', glacial lake sediments. During the development drilling it was found that a carefully controlled Phase IP survey could be used to trace the position of the gold-bearing sulphide mineralization, even beneath the conducting overburden layer.

The dipole–dipole measurements with $X = 40$ m (Figure 12.16) show no apparent resistivity anomaly from the mineralization in the basement, perhaps due to the presence of the conducting glacial sediment layer. The Phase IP measurements gave rise to a low-magnitude, but quite distinct, anomalous pattern on the pseudosection. The significant depth to the top of the source is indicated by the fact that the maximum anomalies were measured for $n = 3$ and $n = 4$. With larger values of X, the anomaly was detected by the $n = 1$ measurement.

Golden Hope – East Ore Zone, Estrades Township, Quebec, Canada. The Golden Hope orebody was one of the first discovered in the Casa Berardi area of north-western Quebec. The gold values were discovered within a sulphide- and graphite-rich band of Archaean volcanic and sedimentary rocks. The conductivity of the zone was sufficient that the ore zone was one of several horizontal-loop EM anomalies detected, following airborne magnetic and EM surveys. The initial discovery, the West Ore Zone, was centred at about Line 24W (not shown). At this position, the zone had a width of about 10 m. There were enough conducting minerals present that the zone also gave rise to a low-magnitude, but distinct, anomaly on the apparent resistivity pseudosection from a CSAMT survey.

The East Ore Zone was discovered by drilling, on Line 17W (Figure 12.17), at a location where no EM anomaly had been detected. There is less metallic mineralization at this point, and the source has less width. The CSAMT apparent resistivity pseudosection on this line (Figure 12.18) shows only the layered pattern due to the 35 m of conducting glacial overburden that is present. This layer of conducting overburden is also evident from the apparent resistivity results measured with the dipole–dipole electrode configuration and $X = 25$ m. There is a slight indication of the increased conductivity due to the metallic mineralization that is present, but this apparent resistivity anomaly is not as definite as that measured on Line 24W with the same electrode interval.

The variable frequency IP anomaly measured with $X = 25$ m was as easily recognizable on Line 17W as on Line 24W. The ore has subsequently been found to follow the IP anomaly.

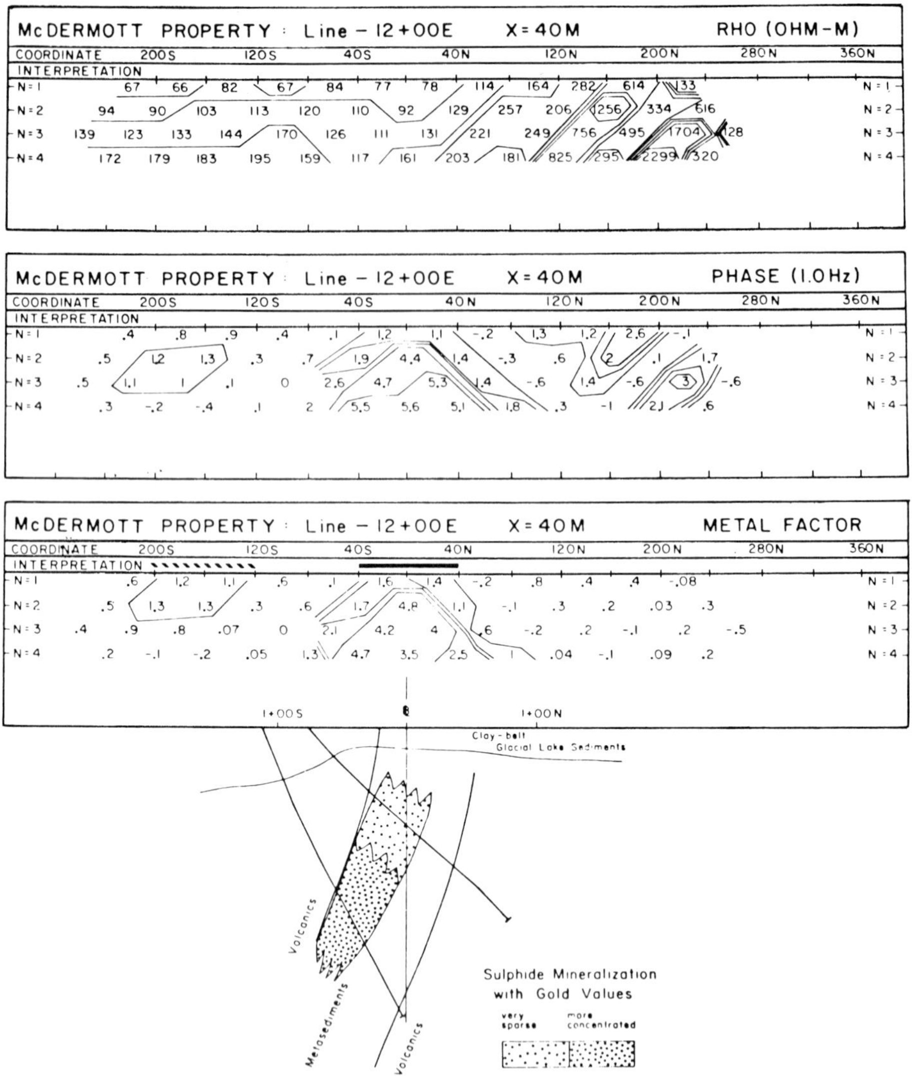

Figure 12.16 Dipole–dipole IP/resistivity pseudosection and idealized geology, Holt-McDermott gold deposit, Ontario. (Courtesy American Barrick Resources Corporation.)

Madsen Mine, Red Lake, Ontario, Canada. Gold ore at the Madsen Mine occurs mainly as disseminations of pyrite and pyrrhotite in Archaean tuffaceous, andesitic beds in a predominantly mafic volcanic sequence. Primary magnetite is probably present also. Figure 12.19, taken from an exploratory magnetometer/self-potential (SP) survey on the property, shows distinct magnetic and SP responses over the gold-bearing beds. In plan form, the magnetic and SP contours closely delineate the surface traces of the beds.

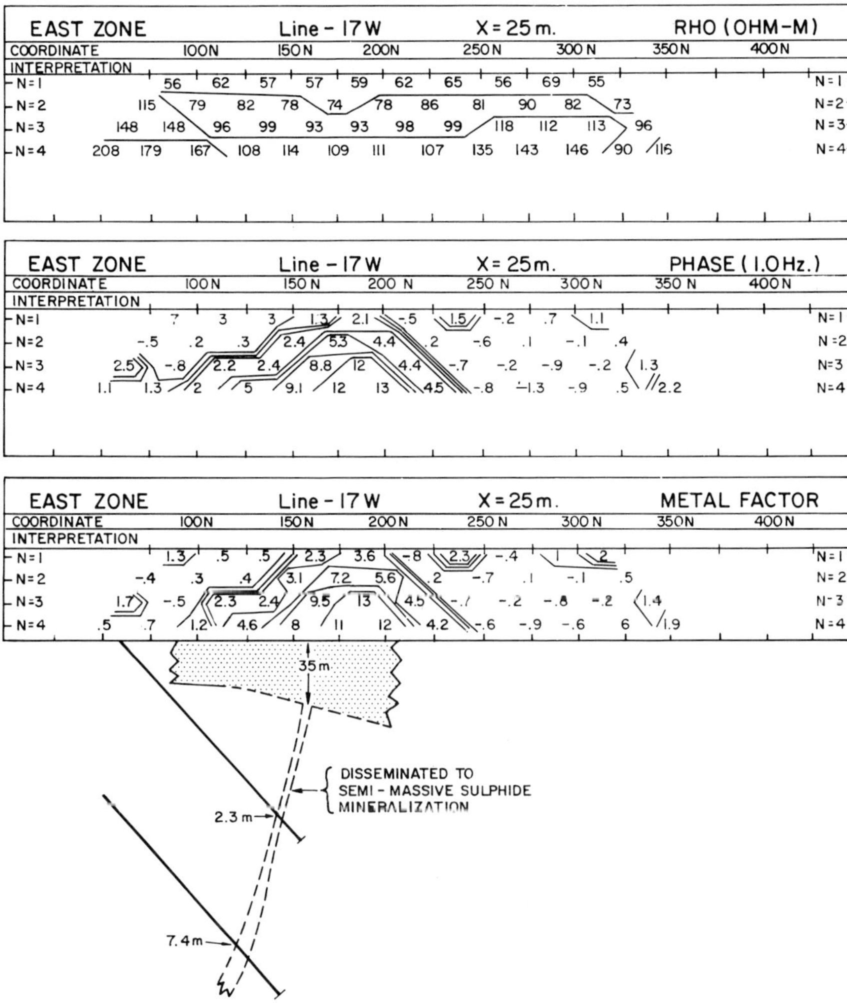

Figure 12.17 Dipole–dipole IP/resistivity pseudosection and idealized geology, East Zone, Golden Hope gold deposit, Quebec. (Courtesy Teck Explorations Ltd.)

Elsewhere on the Madsen property it was found that the magnetic anomalies could not be separated from the background responses of the country rocks. Here, VLF resistivity methods were used effectively to delineate the mineralized beds.

12.4.6 *Palaeoplacers*

Gold and other heavy mineral deposits can be found in a large number of continental basins, usually of Archaean and Proterozoic age. The deposits are mostly located in basal conglomerates or quartzites and the distribution is governed, to a considerable extent, by the pattern of fluvial and lacustrine sedimentation during the early

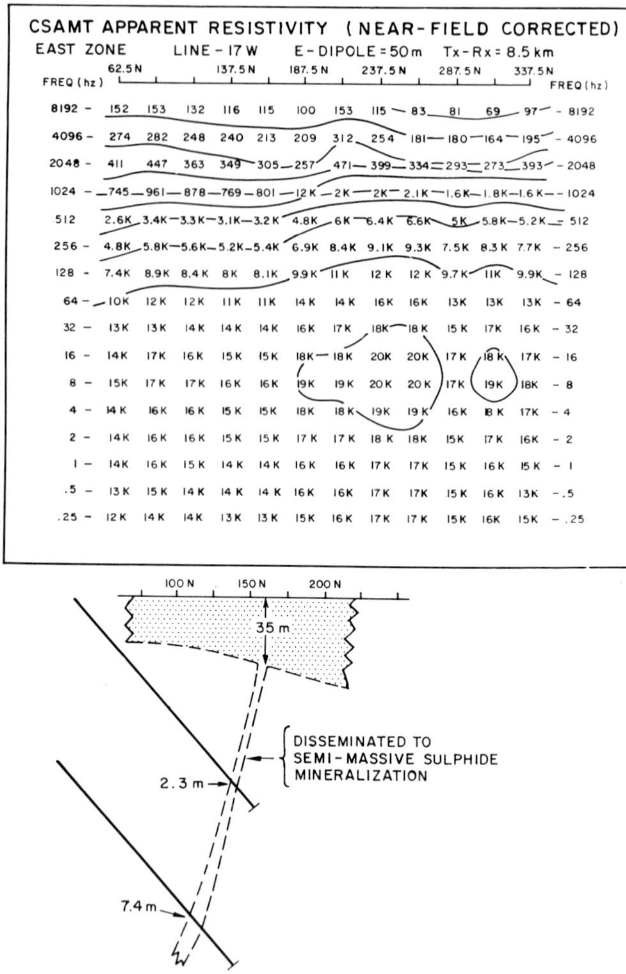

Figure 12.18 CSAMT apparent resistivities and idealized geology, same area as Figure 12.17. (Courtesy Teck Explorations Ltd.)

subsidence of the basin. Exploration for these deposits again becomes a two-stage process: (a) to determine the configuration and depth of the basin; and (b) to locate directly, if possible, the host beds of the gold deposits.

Undoubtedly the Witwatersrand Basin, Republic of South Africa, is the most important and the best studied of such gold-bearing structures. Doyle (1986) reviewed the history of geophysical exploration in the basin, describing both reconnaissance (magnetic and gravity) methods of mapping major structure, and the much higher resolution seismic methods that have recently taken over the detailed exploration role.

Mapping the configuration of Precambrian intracratonic basins is not generally as simple as mapping sedimentary basins in petroleum exploration. The structures are

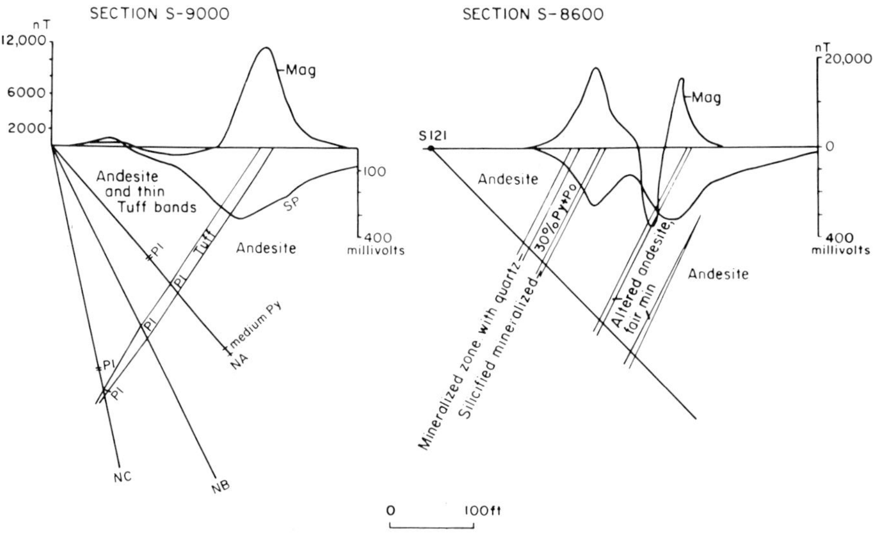

Figure 12.19 Magnetometer and SP profiles over gold- bearing zones, Madsen Mine, Red Lake, Ontario.

usually more complicated and, more importantly, the basin rocks are frequently metamorphosed and may contain not only volcanic units but also syn- and post-tectonic intrusions. All of these tend to distort and even obliterate the signature of the basement unconformity. They do, however, produce information (e.g. post-basin faulting) that may be used directly in the exploration programme.

Two recent papers (Corner and Wilsher, 1989; and Pretorius *et al.*, 1989) have described in some detail the gravity-magnetic and seismic approaches, respectively, to exploration in the Witwatersrand Basin. Corner and Wilsher illustrated effectively the application of qualitative and quantitative interpretation, including the mapping of marker beds within the basement, to the determination of regional structure. Pretorius *et al.* (1989), without denying the role of magnetics and gravity, demonstrated dramatically the better precision obtainable from high-resolution, continuous seismic profiling, particularly in the deeper parts of the basin. Figures 12.20 and 12.21 are reproduced here with their permission.

The relatively high cost of seismics prohibits its use as a reconnaissance tool in relatively unexplored basins. Magnetic, gravity and AMT methods continue to produce useful results in such areas as the Proterozoic Basins of the Northwest Territories, Canada, and the Karroo Basins of Southern Africa (Reeves, 1985).

12.4.7 *Placers*

Shallow seismic refraction and gravity methods have been used for many years to locate and delineate palaeochannels beneath semi-consolidated and unconsolidated sediments. Seismic velocity and density contrasts at this interface are normally enhanced by the coarse nature of the detrital beds that are the object of exploration.

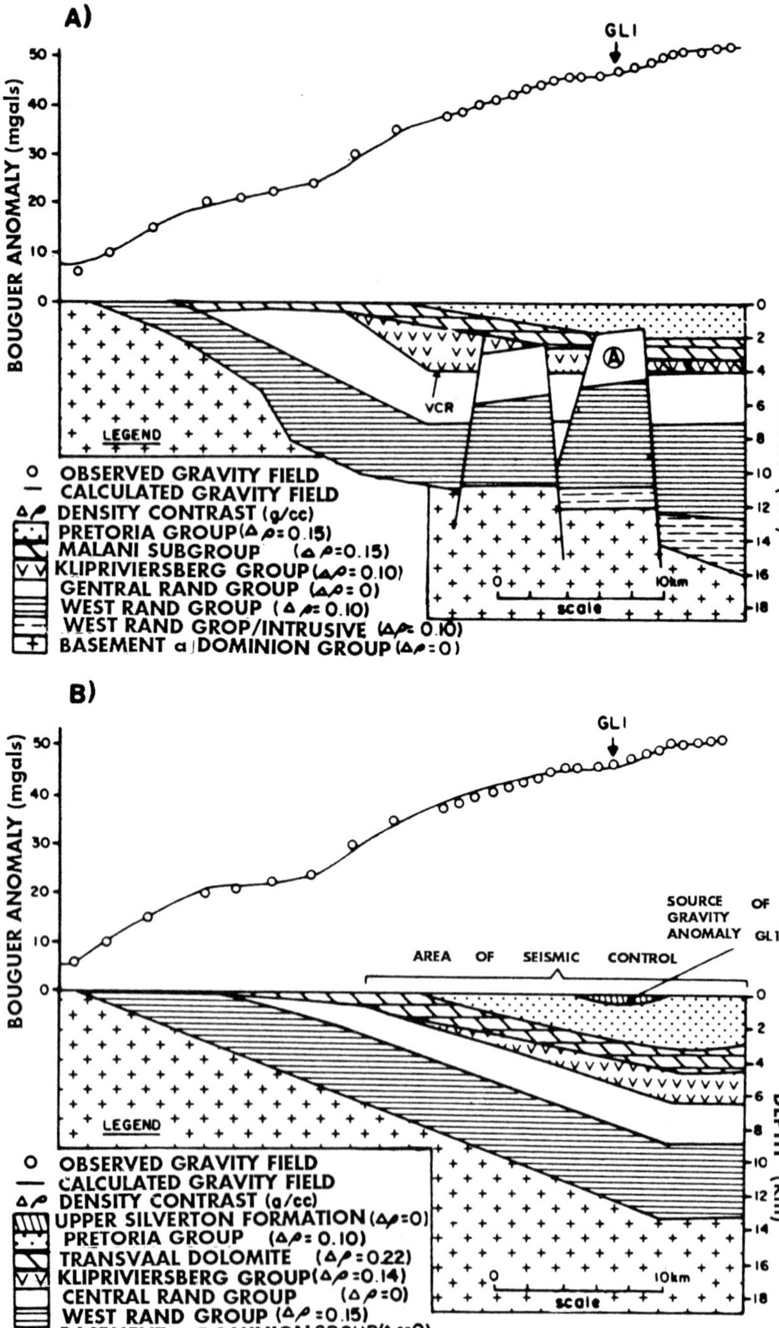

Figure 12.20 (A) Early gravity modelling exercise, without seismic control; (B) gravity modelling exercise with seismic control; Witwatersrand Basin, South Africa. (After Pretorius *et al.*, 1989.)

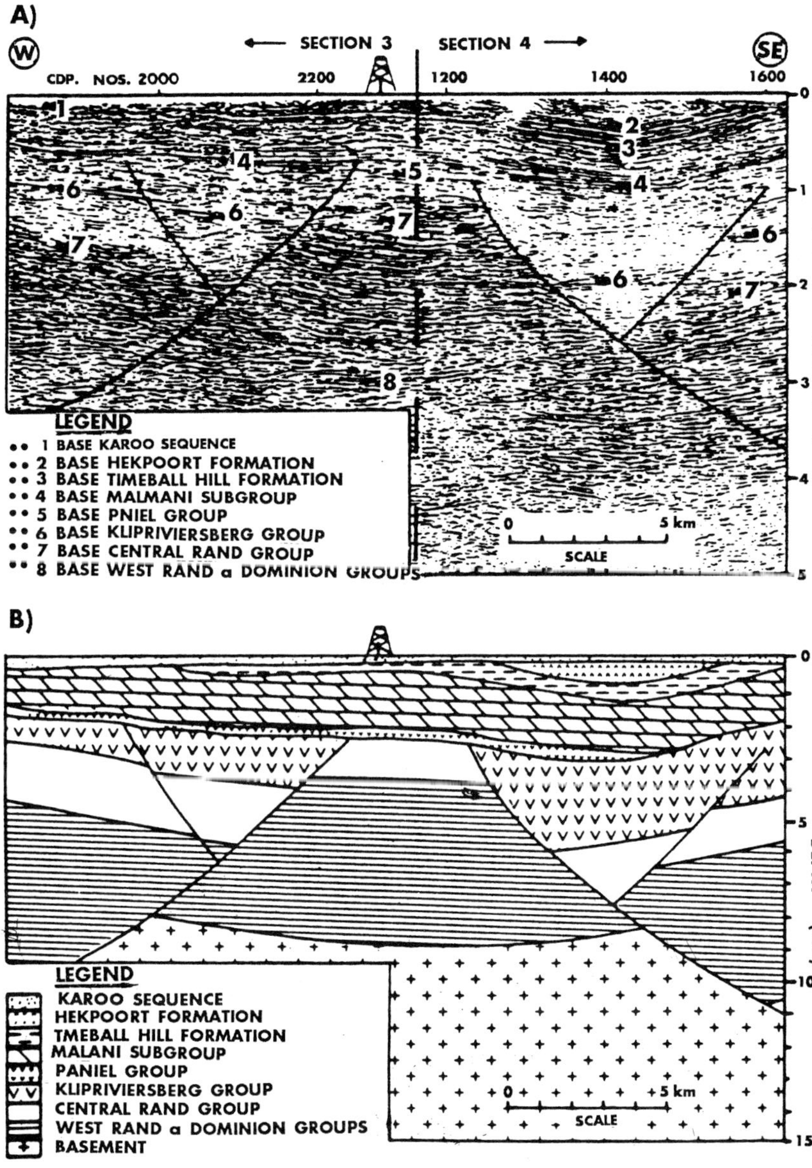

Figure 12.21 (A) Interpretation of seismic sections 3 and 4; (B) depth-converted interpretation of seismic sections 3 and 4; Witwatersrand Basin, South Africa. (After Pretorius *et al.*, 1989.)

Chapman *et al.* (1980) found a density contrast of 0.5 between basement rocks and auriferous gravels in Tertiary channels in California. Hunter *et al.* (1984) have applied seismic reflection methods to the mapping of buried river channels since the late

1960s but have encountered difficulties in obtaining good reflections in coarse gravels over relatively shallow bedrock. Davis *et al.* (1987) described placer gold exploration using a combination of seismic and radar methods, and report penetration of up to 30 m in sands and gravel, with a spatial resolution of about 1 m. The technique is a useful complement to reflection seismics which has lower resolution but works well with deeper reflectors and finer sediments. Magnetite is a common accessory mineral in gold placers. Schwartz and Wright (1987) described total magnetic field and vertical magnetic gradient surveys over buried placers along the Chaudière River, Quebec. The results showed that sediments with slightly higher magnetite content than the surrounding alluvium could be detected by closely spaced traverses. The greatest problem stems from the difficulty of separating anomalies from the placers themselves and from those originating in the underlying bedrock.

Waterborne continuous seismic profiling is widely used in exploration for placer tin deposits, and is currently being applied to gold exploration in Alaska and Nova Scotia.

12.5 Conclusions

Geophysical methods have been helpful in discovering and/or delineating gold deposits of all types and in many geological settings.

Table 12.2 Methods used for exploration of gold deposits

Deposit type	Exploration method	Example
Veins stockwork lodes		
1. Archaean greenstone belts	IP, VLF EM, magnetic, resistivity	Century Mine Barraute, Central Patricia
2. Palaeozoic fold belts	IP, VLF EM, magnetic, resistivity	Drake Field, Colorado Slate Belt, Lachlan Fold Belt
3. Palaeozoic–Tertiary volcanic/subvolcanic granitoid-related	Magnetic, gravity, IP, VLF EM, resistivity	Cripple Creek, Mount Milligan
4. Mesozoic–Quaternary epithermal	Helicopter EM, CSAMT, IP/resistivity, magnetic	Hishikari, Marigold, Dee gold mine
Skarns	Magnetic, radiometric, IP/resistivity, VLF EM	Fortitude, Nickel Plate
Auriferous volcanogenic sulphides	Airborne and ground EM, IP, magnetic, gravity	Rouez, Kerry Road, Detour Lake
Auriferous granitoids	VLF EM, magnetic, IP, radiometric	Ferderber deposit
Disseminated in igneous volcanic and sedimentary rocks	IP, VLF EM, EM, magnetic, CSAMT, SP	Haile Mine, Holt-McDermott Mine, Hope Brook Mine, Golden Hope, Madsen Mine
Palaeoplacers	Seismic reflection magnetic, gravity, AMT	Witwatersrand Basin
Placers	Seismic refraction, radar, magnetic,	Chaudière River

For regional exploration, to locate appropriate geological environments, aeromagnetic, gravity, radiometric, airborne EM and ground CSAMT methods have all been found to provide useful information. In palaeoplacer exploration, the seismic reflection method has been useful as both a regional and detailed exploration tool.

Table 12.2 lists the methods used most commonly for detailed exploration, citing examples mentioned in this paper. The table gives only a few of the many hundreds of cases where geophysical surveys are routinely playing an important role in worldwide gold exploration. Improved knowledge of the physical properties associated with the metamorphic, hydrothermal and structural processes accompanying gold mineralization will add further to the value of geophysics.

12.6 Glossary of geophysical terms

Apparent conductivity Calculation of ground conductivity by geophysical measurement, usually based on assumption of uniform or two-layer earth (see bibliography, Palacky, 1989).

Audio magneto-telluric (AMT) Same as CSAMT, using natural fields as source (Goldstein and Strangway, 1975).

Controlled source audio magneto-telluric (CSAMT) Simultaneous ground measurements of magnetic and electrical fields created by an applied (induced or galvanic) source (Goldstein and Strangway, 1975)

Dipole–dipole array Electrode system often used in IP/resistivity surveys (Sumner, 1977).

Dipole distance Distance *nX* between centres of current and potential dipoles (Sumner, 1977).

Dipole separation Distance *X* between current or potential electrodes (the same in the dipole–dipole array) (Sumner, 1977).

Electrical conductivity Property of rock/soil to conduct electricity; units: S/m (Palacky, 1988, 1989).

Electrical resistivity and apparent resistivity Inverse of conductivity.

Electromagnetic (EM) Measurements of ground conductivity by inductive methods, ground or airborne (Ward, 1977; Palacky, 1989).

Gravity Measurements of the natural gravitational field; units milligals (mgal) (Paterson and Reeves, 1985).

Induced polarization (IP) Measurements of polarization effect in rocks by applied (ground) galvanic currents; units Percent Frequency Effect (PFE %) or milliseconds (ms) (Sumner, 1977; Hallof, 1980).

INPUT™ 'Induced pulse transient' method of airborne EM surveying; registered trademark of Barringer Research Ltd., Toronto (Becker, 1977).

Magnetic/magnetometer Measurements of the natural magnetic field, ground or airborne; units: nano-Tesla (nT) (Hood *et al.*, 1977; Paterson, 1983; Paterson and Reeves, 1985).

Magnetic susceptibility Property of rock to concentrate magnetic flux; units: emu; 10^{-2} emu is roughly equivalent to 4% Fe_3O_4 by volume (Paterson, 1983).

Metal factor Measurement of polarizability relative to ground resistivity, used to estimate concentration of metallic mineralization (Sumner, 1977; Hallof, 1980).

Phase IP Measurements of IP effect through phase changes at a single frequency (Hallof, 1980).

Radiometric Gamma-ray spectrometer measurements, airborne or ground; usually expressed as K (%), U (ppm) and Th (ppm) equivalents, and total count as uranium equivalent (ur) (Paterson, 1983; Darnley and Ford, 1989).

Seismic reflection Measurements of reflection time and amplitude of sound waves generated at surface and reflected by geologic (usually stratigraphic) boundaries (Pretorius *et al.*, 1989; Hunter *et al.*, 1984).

Seismic refraction Measurements of travel time and amplitude of sound waves generated at surface and refracted along geologic (usually stratigraphic) boundaries (Davis *et al.*, 1987; Hunter *et al.*, 1984).

Self-potential (SP) Measurements of natural electric potential of ground; affected by local oxidizing agency (e.g. sulphides), or concentration of natural currents by conductivity variations (Kelly, 1957).

VLF EM 'Very low frequency' (radio) passive electromagnetic measurements, airborne or ground (Paterson and Ronka, 1971; Palacky, 1989).

Note For a more complete explanation of geophysical terms, the reader is referred to: Sheriff, R.E. (1984). *Encyclopedic Dictionary of Exploration Geophysics*. Soc. of Explor. Geophys., Tulsa. 323pp.

References

Batterham, P.M. and Bullock, S.M. (1983) Tanzania: integrated interpretation of aeromagnetic and radiometric maps for mineral exploration. *Trans. Inst. Min. Metall.* (Sec. B: Appl. Earth Sci.) **92**, 83-92.

Becker, A. (1977) Airborne electromagnetic methods. In *Geophysics and Geochemistry in the Search for Metallic Ores*, ed. Hood, P.J. *Proc. of Exploration '77*. Geol. Surv. of Can., Econ. Geol. Rep. 31, 33-44.

Bergmann, H.J. (1980). The Belmoral Story. Presented at the Annual Meeting of the Prospectors and Developers Association, Toronto, Ont., 11 Mar., 1980.

Bowker, A.M. and Hill, I.A. (1987) Geophysical study of Kerry Road orebody, Gairloch. *Trans. Instn. Min. Metall.* (Sect. B: Appl. Earth Sci.), **96**, B213–B220.

Boyd, D. (1984) Aeromagnetic surveys and gold. In *Geophysical Exploration for Precambrian Gold Deposits*, ed. Doyle, H.A. Dept. Geol. & Extension Service, Univ. of W. Australia, Pub. 10, 81-96.

Chapman, R.H., Clark, W.B. and Chase, B.W. (1980) A geophysical approach to locating Tertiary gold channels. *Calif. Geol.* **33**, 173-180.

Coggan, J.H. (1984) Geophysics as an aid to gold exploration in the Yilgarn Block. In *Geophysical Exploration for Precambrian Gold Deposits*, ed. Doyle, H.A. Dept. Geol. & Extension Service, Univ. of W. Australia, Pub. 10, 113-138.

Corner, B. and Wilsher, W.A. (1989) Structure of the Witwatersrand Basin derived from interpretation of aeromagnetic and gravity data. In *Proc. of Exploration '87*, ed. Garland, G.D. Ontario Geol. Surv., Spec. Vol. 3, 532-546.

Crone, J.D. (1984) Amoco's Detour Gold Discovery, Detour Lake area, N.E. Ontario. (Abstract) Can. Inst. Min. Metall., *Geophysics for Gold Symposium*, Abstracts, Vol., p.47.

Darling, R., Vu, L. Dussalt, C., Waitzenegger, B. and Popov, V. (1986) Geology of the Ferderber gold deposit, Val d'Or. *Can. Min. J.*, August, 25-29.

Darnley, A.G. and Ford, K.L. (1989) Regional Airborne Gamma-Ray Surveys: A Review. In *Proc. of Exploration '87*, ed. Garland, G.D. Ontario Geol. Surv., Special Vol. 3, 229-240.

Davis, J.L., Annan, A.P. and Vaughan, C.J. (1987) Placer exploration using radar and seismic methods. *Can. Inst. Min. Bull.* **80**, 67-72.

Despointes, T.H. (1984) La decouverte du gite de Rouez. Can. Inst. Min. Metall., *Geophysics for Gold Symposium*, Abstracts, 79.
Dowsett, J.S. and Krause, B.R. (1984) Geophysics of the Casa Berardi area. Can. Inst. Min. Metall., *Geophysics for Gold Symposium*, Abstracts, 50-63.
Doyle, H.A. (1986). Geophysical exploration for gold – A review. *Explor. Geophys.* **17**, 169-180.
Goldstein, M.A. and Strangway, D.W. (1975) Audio-frequency magnetotellurics with a grounded electric dipole source. *Geophys.* **40**, 669-683.
Grant, F.S., Reeves, C.V., Misener, D.J. and Angoran Y. (1980) An airborne magnetometer and gamma-ray spectrometer survey of Ivory Coast, West Africa. *EOS* 61, 416. Paper presented to the Amer. Geophys. Union, Spring Meeting, Toronto, 1980.
Hallof, P.G. (1967) The use of induced polarization measurements to locate massive sulphide mineralization in environments in which EM methods fail. In *Mining and Groundwater Geophysics/1967*. Geol. Surv. of Can., Econ. Geol. Rep. No. 26, 302-309.
Hallof, P.G. (1980) Grounded electrical methods in geophysical exploration. In *Practical Geophysics for the Exploration Geologist*, ed. van Blaricom, R. Northwest Mining Assoc., Spokane, Washington.
Hood, P.J., Holroyd, M.T. and McGrath, P.H. (1977) Magnetic methods applied to base metal exploration. In *Geophysics and Geochemistry in the Search for Metallic Ores*, ed. Hood, P.J. *Proc. of Exploration '77.* Geol. Surv. of Can., Econ. Geol. Rep. 31, 77-104.
Hood, P.J., McGrath, P.H. and Teskey, D.J. (1985) Evolution of Geological Survey of Canada magnetic-anomaly maps. In *The Utility of Regional Gravity and Magnetic Anomaly Maps*, ed. Hinze W.J. Soc. Expl. Geoph., Tulsa, 62-68.
Hunter, J.A., Pullan, S.E., Burns, R.A., Gagné, R.M. and Good, R.L. (1984) Shallow seismic reflection mapping of the overburden-bedrock interface with the engineering seismograph – some simple techniques. *Geophys.* **49**, 1381- 1385.
Johnson, I.M. and Fujita, M. (1985) The Hishikari gold deposit: an airborne EM discovery. *Can. Inst. Min. Metall. Bull.* **78**, 61-66.
Kelly, S.F. (1957a) Spontaneous polarization, or self-potential method. In *Methods and Case Histories in Mining Geophysics*, ed. de Wet, J.P. Sixth Commonwealth. Min. Metall. Cong., Mercury Press, Montreal, 53-58.
Kelly, S.F. (1957b). Resistivity and magnetic surveys in 1936 in the Broulan-Porcupine gold prospect, South Porcupine, Ontario. In *Methods and Case Histories in Mining Geophysics*, ed. de Wet, J.P. Sixth Commonwealth. Min. Metall. Cong., Mercury Press, Montreal, 283-289.
Kleinkopf, M.D., Peterson, D.L. and Gott, G. (1970) Geophysical studies of the Cripple Creek Mining District, Colorado. *Geophys.* **35**, 490–500.
Koulomzine, T. and Brossard L. (1957) Magnetometer surveys in the area of the Bourlamaque batholith and its satellites. In *Methods and Case Histories in Mining Geophysics*, ed. de Wet, J.P. Sixth Commonwealth. Min. Metall. Cong., Mercury Press, Montreal, 176-184.
Lindeman, F.W. (1984) Geophysical case history of Water Tank Hill–Mt. Magnet, W.A. In *Geophysical Exploration for Precambrian Gold Deposits*, ed. Doyle, H.A. Dept. Geol. & Extension Service, Univ. of W. Australia, Pub. 10, 97-112.
Meixner, H.M. and Peart, R.J. (1984) The Kalahari drilling project. Geol. Surv. Botswana, Bull. 27. 224pp.
Mitchell, A.H.G. and Garson, M.S. (1981) *Mineral Deposits and Global Tectonic Settings.* Geology Series, Academic Press, London. 405pp.
Palacky, G.J. (1988) Resistivity characteristics of geologic targets: Electromagnetic methods. In *Applied Geophysics*, **1** ed. Nabighian, M.N. Soc. of Explor. Geophys., Tulsa, 29-91.
Palacky, G.J. (1988) Resistivity characteristics of geological targets: Electromagnetic methods. In *Applied Geophysics*, ed. Nabighian, M.N. Soc. Expl. Geoph., Tulsa, 29-91.
Palacky, G.J. (1989). Advances in geological mapping with airborne electromagnetic systems. In *Proc. Exploration '87*, ed. Garland, G.D. Ontario Geol. Surv., Spec. Vol. 3, 137- 152.
Paterson, N.R. (1962). Geological mapping by magnetometer surveys. *Proc. Benedum Earth Magnetism Symp.* Univ. of Pittsburgh Press, 139-157.
Paterson, N.R. (1967). Exploration for massive sulphides in the Canadian Shield. In *Mining and Groundwater Geophysics/1967.* Geol. Surv. of Can., Econ. Geol. Rep. No. 26, 275-289.
Paterson, N.R. (1983) Exploration geophysics: airborne. In *AGID Guide to Mineral Resources Development*, eds. Woakes M. and Carman, J. Report Series 10, Assoc. of Geoscientists for Int. Dev., Bangkok, Thailand, 121-152.

Paterson, N.R. and Reeves, C.V. (1985) Applications of gravity and magnetic surveys: The state of the art in 1985. *Geophys.* **50**, 2558-2594.

Paterson, N.R., Reeves, C.V. and Reford, M.S. (1979) The reconnaissance aeromagnetic survey of Botswana – I: Some novel techniques of survey execution. Botswana Geol. Surv., Bull. 22, 31-65.

Paterson, N.R. and Ronka, V. (1971) Five years of surveying with the VLF EM method. *Geoexploration* **9**, 7-26.

Podolsky, G. (1986) Eye in the sky. *Northern Miner Mag.* **1**(2). Northern Miner Press, Toronto.

Pretorius, C.C., Jamison, A.A. and Irons, C. (1989) Seismic exploration in the Witwatersrand Basin, Republic of South Africa. In *Proc. of Exploration '87*, ed. Garland, G.D. Ontario Geol. Surv., Special Vol. 3, 241-253.

Reed, L.E. (1989) Geophysics in gold exploration. In *Proc. of Exploration '87*, ed. Garland, G.D. Ontario Geol. Surv., Spec. Vol. 3, 473-485.

Reeves, C.V. (1985) The Kalahari Desert, central southern Africa: a case history of regional gravity and magnetic exploration. In *The Utility of Regional Gravity and Magnetic Anomaly Maps*, ed. Hinze, W.J. Soc. Expl. Geoph., Tulsa, 144-153.

Reeves, C.V. (1989) Geophysical mapping of Precambrian granite–greenstone terranes as an aid to exploration. In *Proc. of Exploration '87*, ed. Garland, G.D. Ontario, Geol. Surv., Special Vol. 3, 254-266.

Reeves, C.V. and Hutchins, D.G. (1982) A progress report on the geophysical exploration of the Kalahari in Botswana. *Geoexploration* **20**, 209-224.

Reford, M.S. (1980) Magnetic method. *Geophys.* **45**, 1640-1658.

Robinson, E.S., Poland, P.V., Glover, L. and Speer, J.A. (1985). Some effects of regional metamorphism and geological structure on magnetic anomalies over the Carolina Slate belt near Roxboro, North Carolina. In *The Utility of Regional Gravity and Magnetic Anomaly Maps*, ed. Hinze, W.J. Soc. Explor. Geophys., Tulsa, 320-324.

Schwartz, E.J., and Wright, N. (1987). Buried placers in Chaudière River sediments indicated by ground magnetometer survey, Eastern Townships, Quebec. In *Current Research*, Part A. Geol. Surv. of Can., Paper 87-1A, 423-428.

Staltari, G. (1985). The use of IP in mapping alteration and related Au–Ag mineralization in the Drake Mineral Field, NSW. *Explor. Geophys.* **16**, 279-283.

Sumner, J.S. (1977) The induced-polarization exploration method. In *Geophysics and Geochemistry in the Search for Metallic Ores*, ed. Hood, P.J. *Proc. of Exploration '77*. Geol. Surv. of Can., Econ. Geol. Rep. 31, 123-134.

Tenison-Woods, K. and Webster, S.S. (1985) Geophysical signature of gold and porphyry copper mineral deposits in the Lachlan Fold Belt, NSW. *Explor. Geophys.* **16**, 279-283.

Ward, S.H. (1977) Ground electromagnetic methods and base metals. In *Geophysics and Geochemistry in the Search for Metallic Ores*, ed. Hood, P.J. *Proc. of Exploration '77*. Geol. Surv. of Can., Econ. Geol. Rep. 31, 45-62.

Whiteway, P. (1985) Entrepreneurs exploit market niche. *The Northern Miner*, 7 March 1985, B18.

Wotruba, P.R., Benson, R.G. and Schmidt, K.W. (1988) Geology of the fortitude gold-silver skarn deposit, Copper Canyon, Lander County, Nevada. In *Bulk Mineable Precious Metal Deposits of the Western United States*. Geol. Surv. of Nevada, Reno, 159- 171.

Wynn, J.C. and Luce, R.W. (1984) Geophysical methods as mapping tools in a strata-bound gold deposit: Haile Mine, South Carolina Slate Belt. *Econ. Geol.* **79**, 382-388.

13 Economics of gold deposits

B.W. MACKENZIE

13.1 Introduction

The primary role of gold mining in the economy is wealth creation. Wealth is generated by finding, delineating, and developing economic gold deposits and, then, mining, recovering and selling gold from them. Thus, economic deposits are the focal point of the gold-supply process. The economic characteristics of gold deposits are shaped by their geological features. Gold deposits are initially unknown and, once discovered, are fixed in size and highly variable in quality.

The gold-supply process starts at the exploration stage, where there is not only a long period of investment but also a high risk of total loss through failure to discover an economic deposit. This chapter examines one of the main components of the cost–risk–return relationship which underlies gold exploration – the return or prize realized as the result of success. Economic deposits are the motivation for gold exploration. At the beginning of the supply process, gold deposits constitute exploration targets. The economic characteristics of deposits which have already been found can tell us something useful about this undiscovered potential.

The demand for gold establishes a market price. The price of gold provides the incentive for exploration and development. The resulting production from gold mines supplies the market. An apparently simple cyclical process is made exceedingly uncertain and challenging by the geological environments in which economic gold deposits occur.

We start off by having a brief look at the gold market and, in particular, gold prices. Then, world-wide trends in gold-mine production are examined, portraying the aggregate historical record of the discovery and depletion of gold deposits. This leads us to establish an analytical procedure for the evaluation of gold deposits. Finally, the characteristics of economic gold deposits are illustrated by applying this procedure to lode-gold deposits discovered in Canada during a 40-year period.

13.2 Market setting

For over 100 years, the US dollar value of gold was fixed under various forms of gold standard. Then, in 1934, the United States government devalued the dollar by increasing the gold price from the US$20.67 per ounce, which had prevailed since 1879, to US$35 per ounce. This revaluation, as we shall see, had a highly stimulating global effect on gold-mine production. Following the initial impetus, however, the fixed nominal price combined with cost inflation through the 1950s and 1960s to force a steady decline in gold-mining activity.

By the 1960s, several factors had reduced the world's confidence in the ability of the United States to redeem its currency and debt in gold. A 'two-tier' gold price system was consequently established in 1968 where the official US gold price was restricted to monetary transactions involving government-owned gold, and an open-market gold price, free to fluctuate with supply and demand, was recognized for other transactions. Subsequently, the flow of gold from the US treasury to other governments was stemmed in 1971 by a presidential order ending the convertibility of US dollars into gold. Since that time the free market has flourished.

The spectacular rise in free market gold prices since the early 1970s, from the fixed US$35 per ounce price to a peak average annual price of US$613 per ounce in 1980, has provided a world-wide stimulus for gold supply. Levels of investment in gold mining during the 1980s are unprecedented.

A new wave of exploration for gold is lifting production to heights that few, if any, forecasters had imagined possible a few years ago. In the 1980s, gold has been the principal target of most mining company exploration programmes. There are even concerns that new gold discoveries coming on stream over the next decade may oversupply the market and depress the gold price.

According to Jastram (1977), long-term historical evidence indicates that the real price of gold has displayed a flat trend over four centuries. Despite prolonged deviations in the underlying gold supply and demand conditions across decades and centuries, the price tends to return to this long-term average, sooner or later. This 'golden constant' average price in 1989 terms lies between US$250 and $330 per ounce.

The 1970s witnessed an extraordinary increase in investor demand for gold. Initially, the bullion price rose dramatically to its 1980 record but, thereafter, it has eased considerably. There are two reasons for this decline:

- some of the underlying economic conditions which had fuelled the increase in demand – such as high inflation and negative real interest rates – reversed themselves.
- the prospect of substantial increases in new mine production began to make an impression on investors.

Because investor demand has probably been permanently augmented, and because the high levels of mine production achieved during the 1980s will most likely be difficult to sustain during the 1990s (Mackenzie and Doggett, 1989), it is hard to imagine the gold price dropping below Jastram's long-term mean.

According to Anders (1987), only one long-term gold forecast has remained valid over time, and that is the one made by Einzig (1987). Einzig concluded that the price would settle into a range of around US$140 per ounce in 1970–71 terms. This is the equivalent of between $420 and $440 per ounce in 1989 terms.

When examining the economics of gold deposits, the appropriate time-frame for projecting the price of gold extends far out into the future. The relevant gold prices are the ones which will prevail when the economic gold deposits currently awaiting discovery and development are in a position to produce and sell gold, most likely from the mid-1990s to the turn of the century and beyond. Based on the foregoing market perceptions, a long-term average gold price in the range of US$300–$400 per ounce would seem to be reasonable for this purpose.

13.3 Gold-mine production

The ultimate measure of the economic viability of gold deposits is the physical recovery of bullion from the ores mined. This section examines global trends in gold-mine production. The production records of literally thousands of economic gold deposits are aggregated in these historical statistics. Particular attention is given to the output from the major gold miners in the non-communist world, in order of their historical ascendancy: Brazil, the United States, Australia, South Africa, and Canada.

The histograms presented in Figures 13.1–13.6 plot the averages of annual gold-mine production for five year intervals up to 1981–85. The cumulative totals in Table 13.1 and the outline of trends during the 1980s include production up to 1988.

Mine production statistics since the 1960s have been taken from the annual reviews published by Consolidated Gold Fields. See, for example, Milling-Stanley (1989). Historical statistics before that time and the cumulative totals have been compiled from a wide range of sources, some of which are cited in the reference list.

13.3.1 *World-wide trends*

Approximately 114 000 tonnes of bullion have been produced from gold ores to 1988. Trends in the annual rate of world production from 1800 are illustrated in Figure 13.1. About 90% of the total has been mined since then, with 50% extracted from 1950. More recently, world production has increased from 1212 tonnes in 1980 to 1863 tonnes in 1988.

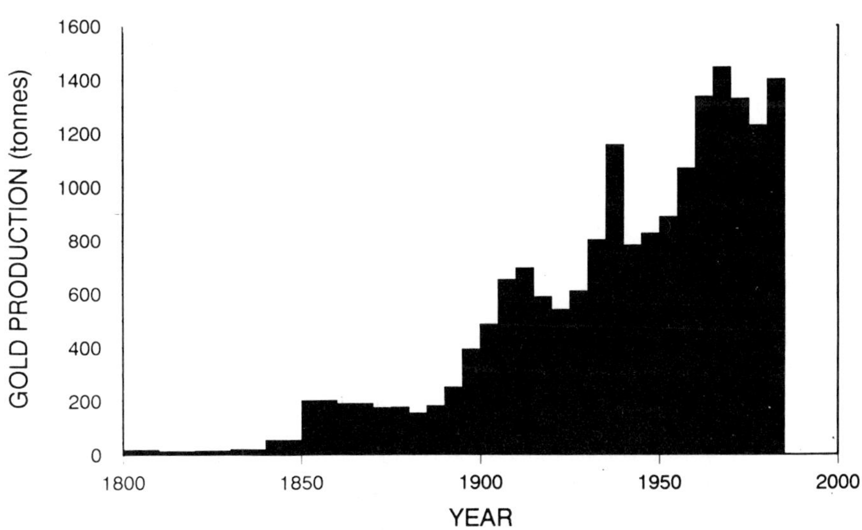

Figure 13.1 World gold-mine production.

As summarized in Table 13.1, South Africa has produced 37% of the world total, followed by the United States (9%), Australia and Canada (6% each), and Brazil (2%). A diverse group of other nations produced the remaining 40%, most notably the USSR which accounts for 15% of total output.

Table 13.1 Gold-mine production from antiquity to 1988

Region	Total production to 1988 (tonnes)	Share of World total (%)
World	113 743	100
Australia	6 506	6
Brazil	2 132	2
Canada	7 255	6
South Africa	42 042	37
United States	10 673	9
Other countries	45 135	40

While controlled by overall market forces, the world-wide distribution of gold production shown in Figure 13.1 is nevertheless a composite of distinct patterns of discovery and depletion which have characterized the major gold-mining nations. We now take a more specific look at production trends within these countries based on the references cited at the end of the chapter.

13.3.2 *Brazil*

Over the past 300 years, gold mining in Brazil has led the way. During the period 1700–1850, Brazil was the number one producer of gold in the world.

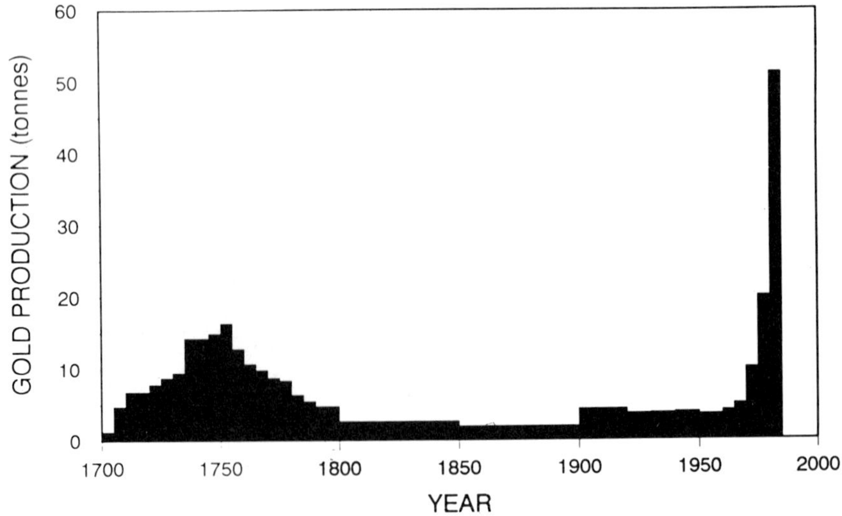

Figure 13.2 Gold mine production in Brazil.

As shown in Figure 13.2, Brazilian gold production (totalling 2100 tonnes) exhibited sustained growth in the first half of the eighteenth century, reaching its highest point at an average annual production of 16 tonnes during the period 1750–54. This increase resulted from the discovery of important alluvial and superficial deposits of gold, particularly in a region now referred to as the Quadrilatero Ferrifero (Iron Quadrangle) of Minais Gerais. Then followed a period of declining production to an average of 4.5 tonnes per year by the end of the century. This prominent eighteenth century gold cycle is referred to in Brazilian literature as the 'Seculo do Ouro' or Century of Gold. The importance of Brazil in terms of world gold production continued through the first half of the nineteenth century, after which it was eclipsed by the great gold discoveries in California and elsewhere.

Brazil's most prominent and prolific lode gold mine, Morro Velho, was the country's dominant producer from the 1830s to the 1970s. The discovery of the Serra Pelada deposit in the late 1970s sparked a remarkable increase in Brazil's gold output from 22 tonnes in 1978 to 112 tonnes in 1988 (unofficial estimates of total production provided by Departmento Nacional da Producao Mineral, Brasilia). Most of this growth is attributable to the work of several hundred thousand garimpeiros (as prospectors are called), mainly active in the Amazon Basin. In addition, international and Brazilian mining companies are gaining prominence with the development of major deposits such as Crixas, Fazenda Brasileiro, Jacobina, São Bento, Morro do Ouro, and Cuiaba.

13.3.3 *United States*

The time-distribution of gold-mine production in the United States, totalling approximately 10 700 tonnes, is presented in Figure 13.3.

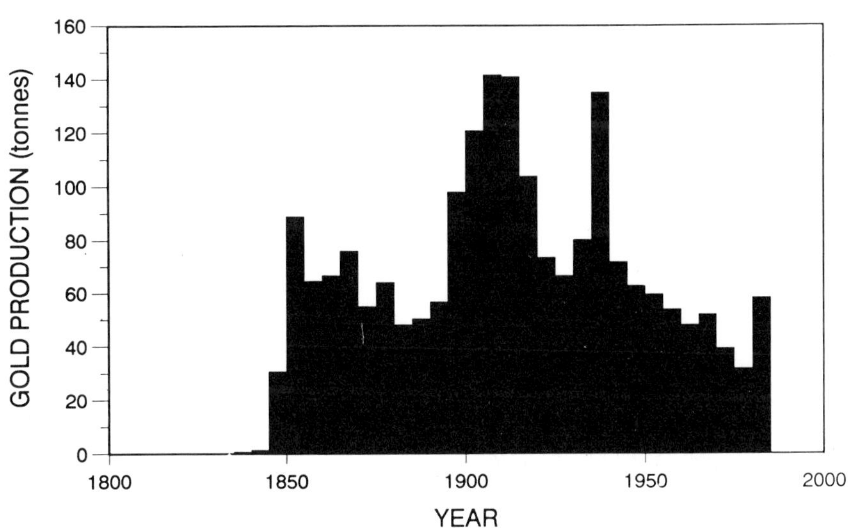

Figure 13.3 Gold-mine production in the United States.

Gold was produced from small-scale placers and weathered bedrock in the Appalachian region in the early eighteenth century. With the news of the discovery of gold at Sutter's Mill in northern California in 1848, 100 000 men headed west. The California deposits proved to be extremely rich, and production climbed to an average of 89 tonnes per year during the 1851–55 period, representing about 45% of world output.

By the late 1850s, the smaller claims had been worked out and many of the '49-ers' began to move into the interior of the western United States in search of new goldfields. This migration resulted in the opening up of the Comstock Lode and Tonopah districts in Nevada, the Clear Creek deposit in Colorado, and numerous other discoveries, mostly in Idaho and Montana, during the 1860s.

Up to the 1870s, most gold production came from easily mined placer deposits but from then on underground mining of lode deposits began to play an important role, especially in Colorado. The Homestake Mine in South Dakota started production in 1877.

From 1890, a notable climb in annual gold production occurred (Figure 13.3), to a record level of 142 tonnes per year in 1906–10. This rise was the result of placer production from the Alaskan gold rush, coupled with extensive development of the deposits discovered in the west during the 1860s and 1870s.

A lengthy decline in gold-mine production followed, sharply but temporarily arrested in the late 1930s by the effects of gold's initial rise to $35 per ounce. In the 1960s, the only significant gold mine remaining was Homestake, although by- product gold was recovered from the porphyry copper mines.

The development of the Carlin and Cortez deposits in Nevada led the way into the modern era of gold mining in the south-western United States. Thirty new gold mines have opened in Nevada alone during the last few years. These producers are mainly very low-grade deposits of disseminated gold that can be mined as open pits, often amenable to heap-leaching technology (see Berger and Bagby, this volume). New gold-mine developments, mainly in Nevada, California, and Montana, have raised gold production in the United States from 31 tonnes in 1980 to 205 tonnes in 1988.

13.3.4 *Australia*

About 6500 tonnes of bullion have been recovered from Australian gold mines to date. The pattern of production is shown in Figure 13.4.

Gold was initially discovered in eastern Australia in the 1820s. However, it was not until 1851 that the significance of these rich alluvial deposits was appreciated through the efforts of Edward Hargraves in the Bathurst district of New South Wales, and James Esmond at Clunes in Victoria, precipitating Australia's first gold rushes. It is interesting that both these men gained their experience on the California gold diggings.

The discoveries of alluvial gold were followed in 1852 by the discovery of 'deep lead' alluvials, first at Ballarat in Victoria, and then on other goldfields in Victoria and New South Wales. Small underground gold mines were established at these locations but the main deposits of lode gold were at Bendigo in Victoria where more than a thousand companies developed mines to work a field only 11 km long by 3 km wide. These mining activities lifted Australian gold production to an average of 84 tonnes per year during 1856–60.

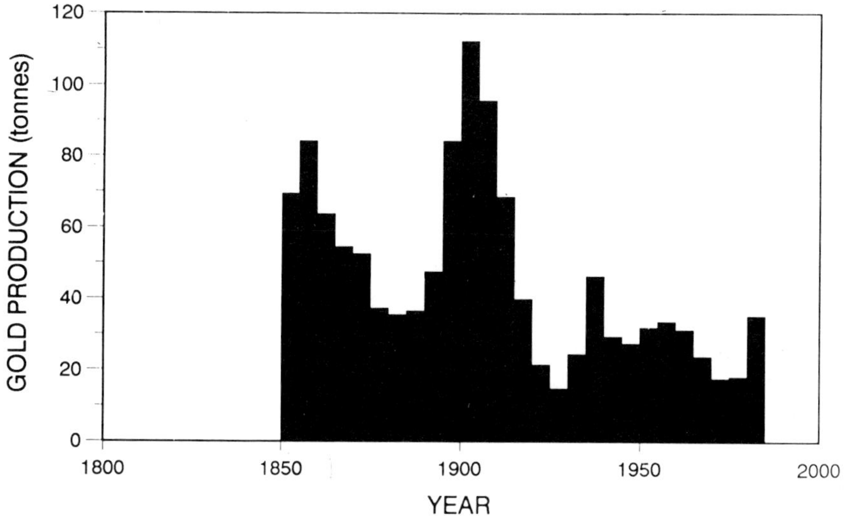

Figure 13.4 Gold-mine production in Australia.

Depletion and a steady decline in production followed this peak period. At the same time, the search for gold deposits spread anti-clockwise around Australia to Queensland in the 1860s and 1870s, and then on to the Northern Territory and Western Australia. Spectacular discoveries of gold in the Kalgoorlie region in the 1890s greatly accelerated the flow of mining capital to Australia. Between 1894 and 1896, 690 Western Australian gold-mining companies were floated in London. Record Australian production averaging 112 tonnes per year was achieved during the 1901–05 period.

Gold mining in Australia plummeted during and after the 1914–18 war (Figure 13.4). While an appreciable growth in production followed the gold price increase in the early 1930s, gold mining declined once again after the Second World War as the fixed gold price was squeezed by rising costs. By the mid-1970s, only four significant Australian gold mines had survived.

In Australia as elsewhere, the gold-mining boom of the late 1980s was undoubtedly triggered by the high gold price of 1980. The Eastern Goldfields region of Western Australia, centred around Kalgoorlie and over 500 km long, is presently the site of hundreds of gold mines, mills, and development projects. There has also been notable expansion in Queensland and the Northern Territory. Consequently, gold-mine production in Australia has grown from 17 tonnes in 1980 to 152 tonnes in 1988.

13.3.5 *South Africa*

Approximately 42 000 tonnes of gold have been mined in South Africa, almost 40% of all the bullion produced in the world. The time-distribution of this unique contribution is illustrated in Figure 13.5.

The first discovery and small-scale production of gold in South Africa occurred at Pilgrims Rest and Barberton in the eastern Transvaal during the 1870s and early-1880s. These activities were soon overshadowed by the singular event in the history of world gold mining – the discovery of the Main Reef of the Witwatersrand Basin by George Harrison (who had previously worked on the Australian goldfields) in 1886.

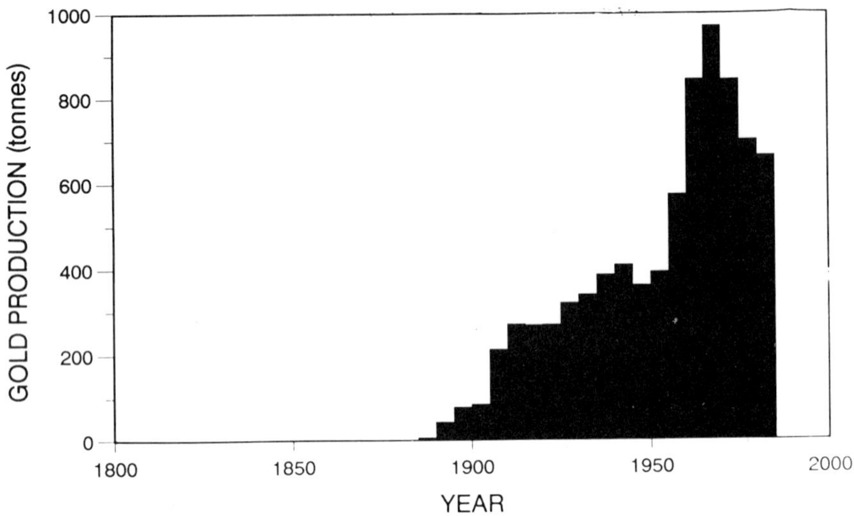

Figure 13.5 Gold-mine production in South Africa.

Full-scale mining on the Central Rand began in 1887 and accelerated during the 1890s with the introduction of the cyanide process to treat the sulphide ores. Production increased to an average of 78 tonnes per year during 1896–1900. Gold mining on the West Rand and East Rand followed the extension of the outcrop of the Main Reef from the Central Rand. From 1886 to 1930, development and production in the Witwatersrand Basin was concentrated on mines in these three areas, lying in a more or less continuous arc from Randfontein, 30 km north-west of Johannesburg, to Nigel 50 km to the south-east of Johannesburg. Gold production rose steadily to an average of 321 tonnes per year during the 1926–30 period.

The increased gold price, coupled with advances in exploration geophysics during the 1930s, resulted in the discovery of gold-bearing reefs on the West Wits Line (also known as the Far West Rand) south-west of Randfontein and, in due course, the establishment of several major new mines. Gold production from the Klerksdorp field was also brought on stream at this time.

It was not until the late 1940s and early 1950s that the Orange Free State field was discovered and development began. The last of the major areas to be brought into production on the Witwatersrand was the Evander (or Far East Rand) goldfields, about 110 km south-east of Johannesburg, which was discovered and developed in the late 1950s and early 1960s.

As the result of these discoveries around the rim of the Basin, and the subsequent development of many new gold mines, gold production in South Africa following the Second World War climbed dramatically to its highest level of 970 tonnes per year in 1966–70 (Figure 13.5). Since that time, gold exploration and development on the Witwatersrand has concentrated on downdip extensions of the reefs being mined, the potential for mining other reefs on lease areas covered by existing operations, and on-strike extensions of reefs to close the 'gaps' between the known goldfields. Despite high levels of investment in mine development, South African production has declined from its record level to 675 tonnes in 1980 and 621 tonnes by 1988.

13.3.6 *Canada*

Canada has yielded approximately 7300 tonnes of gold. Of the five major producers, Canada was the last to command international attention. The pattern of production is presented in Figure 13.6.

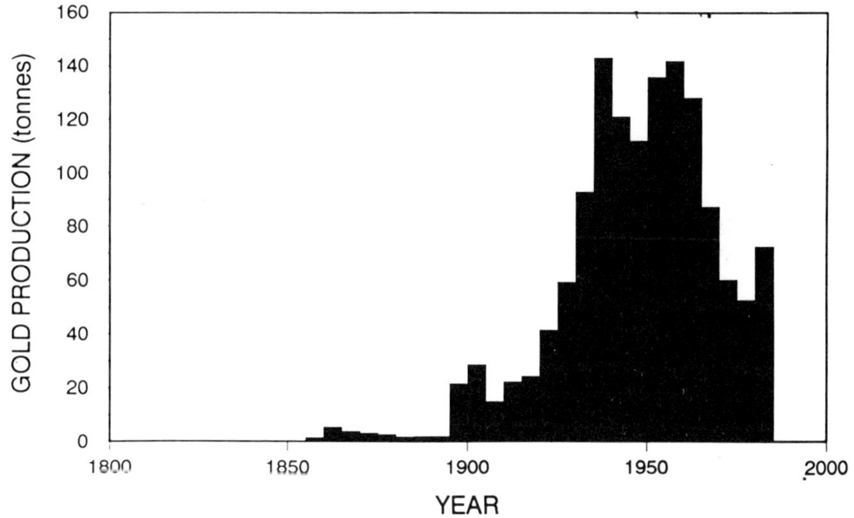

Figure 13.6 Gold-mine production in Canada.

The first recorded discovery of placer gold in Canada was in the Eastern Townships of Quebec during the 1830s. Saddle reef gold deposits were found in Nova Scotia in 1862. Significant production resulted from placer deposits found in British Columbia, on the Fraser River in 1858 and in the Cariboo district during the 1860s. Prospectors and miners, who had drifted northward following the California gold rush, played a role in the discoveries. Despite these developments, production of gold in Canada averaged only 2–3 tonnes per year until the mid-1890s.

In 1896, the Klondike gold rush began with the discovery of gold on the banks of a tributary of the Yukon River. Coming in the midst of a world depression, thousands took to the trail. The zenith was reached in 1901–05 when an average of 29 tonnes of gold per year was recovered from creeks in the area. Once the easily accessible gravels had been exhausted, production rapidly declined.

A great expansion of mining in eastern Canada was associated with the discovery of silver deposits at Cobalt, Ontario, in 1903. The prospectors involved and the wealth created at Cobalt formed the basis for development of the Ontario and Quebec gold-mining industries.

The Porcupine goldfield was found in 1908 and development proceeded rapidly with the establishment of the Dome and Hollinger mines, both of which became world-class producers. Following success at Porcupine, exploration interest turned to Kirkland Lake which developed rapidly after 1915.

In Quebec, production of gold on a significant scale commenced in 1927 as a by-product from the Horne copper mine at Noranda. The increase in the gold price in the early 1930s stimulated the discovery and development of major new mines at Malartic, Cadillac, and Val d'Or. Many other gold deposits were discovered at this time, most notably in the Red Lake area of north-western Ontario, and at Yellowknife in the Northwest Territories. Development of these camps extended into the 1940s.

Record gold production of 143 tonnes per year was attained in 1936–40. More than 100 gold mines were operating in Canada at this time. Production rates were maintained at over 100 tonnes per year until the mid-1960s but, after that, output declined rapidly to an average of 53 tonnes per year during the 1976–80 period. Only 17 gold mines were operating by 1977–78.

In contrast to the United States and Australia, where expansion during the 1980s has come mainly from a host of small open-pit mines, the gold boom in Canada is associated with hard rock underground mines. Canadian gold production has increased from 52 tonnes in 1980 to 129 tonnes in 1988. The dominant features of this growth have been the three major new mines which are developing the Hemlo deposit, new developments and expansions in the known gold mining camps, chiefly in Ontario and Quebec, and the Lupin Mine in the Arctic.

13.3.7 *The 1980s*

We are witnessing an important world-wide expansion in gold-mine production during the 1980s as well as fundamental changes in its geopolitical make up.

As summarized in Table 13.2, the annual global growth rate has averaged 5%. However, the acceleration of production has been much higher than this in Australia and the United States and, to a lesser degree, in Brazil and Canada. On the other hand, gold-mine production in South Africa is declining. Consequently, the shares of world output accounted for by Australia, Brazil, Canada, and the United States have risen from below their historical averages in 1980 (Table 13.1), to well above their historical averages in 1988. Over the same period, South Africa's share has fallen from 56% to 33%.

Table 13.2 Gold-mine production, 1980–1988

Region	Total production 1980–1988 (tonnes)	Average annual growth rate (%)	Percentage share of world total	
			1980	1988
World	13 525	5	100	100
Australia	529	32	1	8
Brazil	620	14	2	6
Canada	773	12	4	7
South Africa	5 900	–1	56	33
United States	807	27	3	11
Other countries	4 896	6	33	35

In looking ahead towards the 1990s, there are two key questions:

- how long can these exponential growth rates be sustained?
- at what point will the respective production shares of South Africa and the other four majors reach a new equilibrium?

The answers to these questions are to be found in the varied dynamic relationships which exist between the discovery and depletion of economic gold deposits within each country.

13.4 Economic evaluation of gold deposits

Gold-mine production, as illustrated above, is the consequence of finding and delineating gold deposits, followed by decisions to invest in their development. This section considers the factors that have to be taken into account in appraising the economic potential of a gold deposit, the evaluation procedure that is followed, and the types of economic criteria that can be assessed. Here, it is assumed that a mining company anticipates being or already is successful in bringing a deposit to the point of making a mine development decision.

13.4.1 *The decision process*

The starting point for the economic evaluation of a gold deposit is the generation and assembly of relevant experience and information. This provides the basis for estimating the future conditions anticipated if a decision is made to proceed with the investment. Relevant conditions include geological estimates of ore reserve size and grade, market price projections, the selection of capacity and the methods to be used for mining and processing, capital expenditures and operating costs, and government policies relating to such factors as taxation, environmental controls, and infrastructure.

Evaluation techniques are then applied to translate these estimates of possible future conditions into measures of the economic attractiveness of the contemplated mine development. These measures portray the quantitative dimensions of the investment being considered in terms of expected value, sensitivity, and risk criteria. At the same time, attention must be given to examining the consequences of non-quantifiable or 'intangible' factors.

The economic measures evaluated are used to support the decision being taken on whether or not to develop the gold deposit to production.

13.4.2 *Economic evaluation techniques*

The geological, market, engineering, and government policy estimates of future conditions are the basis for assessing the time distribution of cash flows for the project over its planned life. Cash flows are initially estimated on a before-tax basis to appraise the total potential of the gold deposit. Relevant taxation policies are then overlain on these before-tax values to determine after-tax cash flows or, in other words, to see how the project looks from the more specific viewpoint of the enterprise involved.

Several types of economic criteria can be directly determined from the estimated before-tax and after-tax cash-flow profiles. These measures include the total profit anticipated, the size of the project, and project costs expressed per ounce of gold. These 'cash-flow criteria' are facets of the overall economic picture required for informed decision-making.

The next step in the economic evaluation is to time-adjust the cash flows to allow for the cost of capital. The concepts of cash flow and time value are combined in various ways to evaluate 'discounted cash flow (DCF) criteria' such as net present value, present value ratio, and rate of return.

Single-point estimates of expected future conditions for the relevant 'input variables' are combined to assess expected values for the cash flow and DCF indicators. This first step represents the 'base case' of a deposit evaluation.

Then, sensitivity analysis is used to examine the sensitivity of the economic measures to possible variations in input variables above and below their expected values. Sensitivity analysis can also be applied to determine the break-even conditions required to justify investment. This is a helpful approach when a high degree of uncertainty is associated with one or more key variables (such as the price of gold).

Risk analysis translates perceived uncertainties concerning the input variables into probability distributions of possible values for the various economic indicators. In this way, it is possible to evaluate the risks associated with realizing the expected value outcomes. What is the insurable lower-limit value? What is the probability of economic loss if development of the deposit is undertaken?

At the end of the day, the expected value, sensitivity, and risk results, together with an appreciation of non-quantifiable intangible factors, are used to support decisions concerning the economics of gold deposits.

13.4.3 *Estimation of cash flow*

The general economic conditions which provide the overall framework for estimating the time distribution of cash flows for development of a gold deposit include:

- specification of currency to be used, usually the currency of the country in which the project is located;
- specification of money values, usually constant money units at the time of evaluation;
- future price projections for gold and any associated by-products, usually expressed in US dollars;
- exchange rate projection to convert US dollar prices to the domestic currency;
- general rate of inflation to correctly assess tax allowances and payments.

A basic understanding of the geological, marketing, engineering, and governmental factors associated with a deposit provides the basis for selecting project specifications for the evaluation. These decision variables include: the development, mining, and processing methods to be utilized; mine and mill capacities; and cut-off grade. The purpose of such a preliminary feasibility study is to optimize the decision variables by analysing a full range of technically feasible development alternatives available and, on the basis of comparative results, recommending a best set of specifications.

Development establishes productive mining and mineral processing capacity. Since processing is required to upgrade the ore mined to a gold concentrate or impure bullion that can be marketed, the construction of processing facilities is carried out in conjunction with mine development. A mill may be installed at the mine site, or a common processing facility may be used to treat ores from a number of mines in a region.

For cash-flow appraisal, estimates are necessary for the pre-production capital expenditures required to develop the specified capacity. These expenditures include: mine development; mine plant and machinery; processing facilities; infrastructure requirements for power, water, housing, town site, and road facilities; and working capital. An estimate of the time required to bring the deposit into production, and a projected schedule of pre-production capital expenditures over this development period are also needed.

The mining stage of production may include stripping waste for open pits, preparing stopes (underground excavations for mining), developing ore reserves, drilling, blasting, transporting materials to the processing facilities, filling mined-out stopes, and associated technical and planning facilities. For gold ores, processing typically comprises crushing, grinding, cyanidation, carbon-in-pulp adsorption, filtering, tailings disposal, and smelting. Alternatively, if conditions are suitable, the ore may be heap leached.

The geological size and grade of ore reserves is assessed based on delineation results and the cut-off grade specified. Part of this geological reserve inevitably has to be left behind in pillars and remnants. Also, the geological grades are diluted by the mining of barren or low-grade waste rock adjacent to the deposit itself. Thus, the recoverable ore reserves delivered to the mill will usually be significantly lower in grade and metal content than the geological deposit estimates. Mine recovery and dilution factors are applied to the initial estimates of geological ore reserves to assess the recoverable mill-feed ore reserve size and grade.

Mineral processing results in the incomplete recovery of gold (and any other metals) contained in the mill feed in the final product. Processing losses are taken into account in cash flow estimation by applying an appropriate mill recovery factor. For example, a 6.5 g/t gold ore may be treated to obtain a 98% gold bullion, with the loss of 6% of the gold content in the waste product. In this case, a mill recovery factor of 94% would be applied.

The productive mine life is determined by dividing the estimated recoverable size of ore reserves by the annual capacity. Annual revenue is estimated by multiplying capacity, the recoverable grade of ore reserves, the mill recovery factor, and the price of gold. Adjustments may be required to standardize physical units, to convert the gold price to domestic currency, and to credit any recoverable by-products which may be present. Also, a charge must be applied to cover the insurance, transportation, refining and marketing costs associated with converting the impure gold product at the mine site to refined metal in the market.

Annual production costs are next assessed, including: the capital cost of major modifications, for example, the conversion of an open pit to an underground mine; sustaining capital requirements for equipment replacement; and operating costs. Operating costs are normally subdivided into mining, milling, and overhead components.

Year-by-year differences between revenue on the one hand and the pre-production capital expenditures and production costs on the other portray the time distribution of before-tax cash flows for the deposit being evaluated. This distribution will be negative during the development period when the investment is being placed and, hopefully, positive over the productive mine life.

Taxation payments then have to be calculated and deducted. Here, the effect of inflation should be considered. The deduction of tax payments from the before-tax cash flows determines the after-tax cash flow stream.

The estimated time distributions of cash flows represent the anticipated before-tax and after-tax returns on investment.

13.4.4 *Cash-flow criteria*

A number of worthwhile economic criteria can be assessed from the estimated cash-flow characteristics, including the three types of measure outlined below.

Project size. The size of a mining project is often a relevant consideration in decision-making. A venture may be too small to be of interest to a major mining group even if it fulfils other corporate investment requirements. Size may be assessed in various ways. The total revenue that a mine is anticipated to generate over its life provides an economic measure of project size which is distinct from profitability considerations. Total revenue is simply the summation of annual revenues estimated over the mine life.

Total profit. Profit, for purposes of economic analysis, is measured by cash flow. The summation of projected annual cash flows (either before- or after-tax) represents the 'total profit' anticipated. While this criterion makes no allowance for the cost of capital, it is nevertheless a useful indicator of economic potential.

Relative cost. Because the future price of gold is so uncertain, it is of interest to determine the break-even price requirement and the relative position of a proposed development on a cumulative cost curve of competitor deposits. The project's location on this curve establishes its relative susceptibility to adverse price changes. For this purpose, costs are evaluated per ounce of gold recovered. The costs to be included in this type of competitive position evaluation depend on the time-frame considered. In the short term, the focus is on operating mines and production costs. For medium-term evaluation of development potential, pre-production capital expenditures should also be incorporated.

13.4.5 *The cost of capital*

The cost of capital is the cost associated with the investment funds required for mine development. The cost of capital represents the minimum acceptable return required on the investment. The concept of time value is used in evaluation practice to allow for the cost of capital, the one cost which is not incorporated in the estimation of cash flows.

The cost of capital is a weighted average cost, composed of the costs of individual sources of funds. Sources of funds may be simply grouped into two types – debt and equity.

The cost of debt capital is an explicit interest cost. Adjusting for taxation and inflation effects, debt capital is a low-cost source of funds, with constant money after-tax costs typically in the range of 1–3%.

The cost of equity funds is an implicit opportunity cost. An opportunity cost arises whenever the funds available for investment are limited relative to the number of economic opportunities which may be pursued. Thus, for example, investment of equity funds in a new gold-mine development should be charged with the return foregone on some alternative opportunity which could otherwise be realized. The cost of equity funds normally varies from 8 to 15%.

The financial structure of a mining company – basically how much debt and how much equity it utilizes – is applied to cost estimates for the individual sources of funds to assess the weighted average cost. Weighted average costs of capital for mining company investment normally range from 6 to 12%.

The main purpose of an economic evaluation is to compare the return on investment, as reflected by the time distribution of cash flows, on the one hand, with the cost of capital on the other. Investment is economically justified when, in the light of the analysis carried out and the experience at hand, management judges that there is a 'reasonable chance' that the return on investment will be greater than the cost of capital.

13.4.6 *DCF criteria*

Discounted cash-flow (DCF) methods overlay time-value considerations on the estimated cash-flow profile for a project in a variety of ways in order to compare the cost of capital and the return on investment. The three main DCF measures are described below.

Net present value. The net present value criterion converts the anticipated time distribution of cash flows for a mine development project into an equivalent value at the start of development. To do this, each cash flow is discounted back to this point using the cost of capital rate. The present value components are then summed to yield the net present value. Thus, net present value is the difference between the discounted positive cash flows from production and the discounted pre-production capital expenditures. Net present value measures the monetary value or economic worth of a project. It represents the anticipated return on investment over and above the minimum required return. Using this criterion, the minimum acceptable condition for an economic deposit is a net present value equal to zero. Net present value is a consequence of both the size and the profitability characteristics of a project.

Present value ratio. The present value ratio method measures the net present value per unit of investment. It is determined by dividing the net present value for a project by the absolute value of the discounted negative cash flows (or, in other words, by the time-adjusted investment). The present value ratio appraises the profitability of a project independently of size considerations. Since the cost of capital has been charged to the project in time adjusting the cash flows, the minimum economic condition of acceptability is a ratio of zero.

Rate of return. In economic terms, rate of return measures the average annual percentage return on investment that a project is anticipated to yield over its total life. It is defined as the discount rate which equates the present value of the positive cash flows from production with the present value of the pre-production capital expenditures. In other words, rate of return is the discount rate which produces a zero net present value. Unlike the other DCF methods, rate of return is determined solely on the basis of the estimated time distribution of cash flows for the project. Using this criterion, the minimum acceptable condition for an economic deposit is a rate of return equal to the cost of capital.

13.5 Characteristics of economic gold deposits

Based on the evaluation procedure and economic criteria set out, we now illustrate the characteristics of economic gold deposits. The case study assesses the economic potential of gold deposits discovered in Canada.

13.5.1 *Canadian case study*

The illustrative characteristics are derived from the evaluation of 134 significant gold deposits discovered in Canada during the 1946–85 period (Mackenzie *et al.*, 1989). A discovery is considered to have been made when the deposit, which ultimately forms the basis for mine development, was first encountered, usually by drilling. Deposits are treated individually if their discovery required essentially independent primary exploration programmes. The intention is to include all known deposits of at least 200 000 tonnes that offer medium-term potential for development.

Money values are expressed in constant 1987 Canadian dollars (the average 1987 exchange rate of US$ 0.75 = C$1.00 can be used to convert results to US dollars). The evaluation is carried out on a before-tax basis using single-point expected value estimates.

One aspect of the evaluation is to determine which of the 134 gold deposits discovered in the past would be economic under present conditions. An economic deposit is defined here as a deposit that realizes a total revenue of at least $20 million and a rate of return of at least 8%. The $20-million minimum size ensures that any deposit of significance is included in the evaluation. The 8% rate-of-return investment margin has been selected as a best estimate of the weighted average cost of capital for gold-mine investment funds.

Revenues are estimated using an expected gold price of $500 per ounce (equivalent to US $375 per ounce). For free milling gold ores, a fixed charge of $15 per ounce of gold contained in impure bullion is levied for transportation, refining, and marketing. In the case of refractory ores, $40 per ounce of gold recovered in concentrate is charged. By-product silver is credited using an expected price of $8 per ounce.

Deposit-specific estimates of recoverable ore reserves, mine and mill capacities, the length of the pre-production period, pre-production capital expenditures including the working capital requirement, a stripping ratio for open pits, the cost of major modifications, sustaining capital, operating costs, mill recovery factor, and product specifications, are made for each of the 134 gold deposits, based on present-day economic and technological conditions. In other words, we are trying to determine which of the historical discoveries would be economic to develop if found today. To apply this type of appraisal for current policy and planning purposes, we are making the assumption that deposits yet to be found will resemble, in economic terms, those that have been discovered to date. These deposit estimates mainly utilize actual historical data and feasibility study estimates. Otherwise, general Canadian costing relationships are applied, reflecting an order-of-magnitude level of accuracy.

The development and production phase characteristics for each of the gold deposits are evaluated by combining the general market estimates with the deposit-specific estimates. A number of measures of economic worth are derived from the resulting cash-flow distributions, including total revenue and rate of return. Discoveries that satisfy the minimum specified conditions of size and profitability are considered to be economic deposits.

On evaluation, 81 of the 134 gold deposits were found to satisfy the economic deposit conditions. We now examine the characteristics of these 81 economic deposits.

13.5.2 *Presentation format*

The illustrations which follow include selected input variables, cash-flow criteria, and DCF criteria for the case study set of 81 deposits. Results are presented in the form of histograms, key statistics, and scatter diagrams.

The histograms group results for each deposit parameter into an appropriate number of standard class intervals, showing the relative frequency of each interval. The relative frequency is the number of economic deposits with values within an interval divided by the total number of economic deposits evaluated (81). The distribution of values is a reflection of the underlying geological variability that exists among economic gold deposits.

Key statistics are derived from the cumulative frequency distribution for each parameter to characterize the central tendency, variability, and skewness of the overall distribution. For example, a cumulative frequency of 0.10 specifies the lower-decile statistic; i.e. 10% of the economic deposits have values less than or equal to the lower decile and 90% have higher values. The median (or middle) deposit value is designated by a 0.50 cumulative frequency, i.e. half the deposits have lower values than the median, while the other half have higher values. The mean deposit value is the arithmetic average of all deposit values. On the upside of the distribution, a cumulative frequency of 0.90 specifies the upper-decile statistic; i.e. 90% of the economic deposits have values less than or equal to the upper decile and 10% have higher values.

Scatter diagrams plot the relationship between any two specified parameters based on the 81 economic deposits evaluated. Each deposit is represented by a point on the scatter diagram.

13.5.3 *Deposit input variables*

Recoverable ore reserve size. A highly skewed distribution, as shown by the histogram (Figure 13.7) and the following key statistics:

Lower decile	*Median*	*Mean*	*Upper decile*
0.4 Mt	1.4 Mt	3.9 Mt	9.0 Mt

Although about 70% of the economic deposits have recoverable reserves of less than 2 million tonnes, 10% are larger than 9 million tonnes, including the Hemlo deposit with reserves in excess of 75 million tonnes. Thus, the mean value of the distribution is almost three times greater than the median.

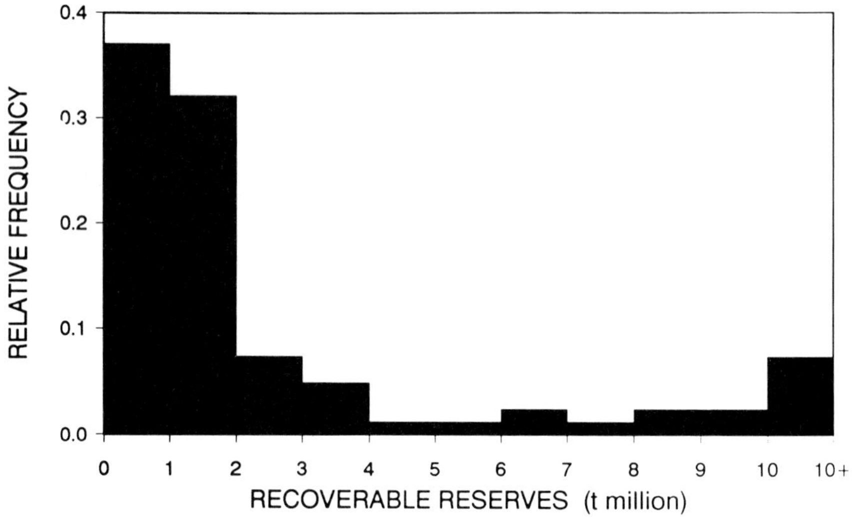

Figure 13.7 Recoverable ore reserve size.

Recoverable ore reserve grade. The high grade nature of economic gold deposits in Canada (Figure 13.8) reflects, in part, the predominance of underground mines with relatively high capital and operating costs:

Lower decile	*Median*	*Mean*	*Upper decile*
4.6 g/t	7.4 g/t	9.7 g/t	19.1 g/t

The low-grade deposits are open-pit mines. The distribution has a high-grade tail. The average recoverable grade of one-tenth of the economic deposits exceeds 19 g/t.

Size–grade relationship. The correlation between recoverable ore reserve size and grade for the 81 economic deposits is illustrated in Figure 13.9 (a few deposit points fall outside the boundaries of the scatter diagrams). The deposits with average grades above 12 g/t are quite small. Otherwise, there does not appear to be a systematic relationship between the size and grade of economic gold deposits.

Pre-production capital expenditures. The investment needed to develop a new gold mine to production is typically in the range of $10–30 million (Figure 13.10):

Lower decile	*Median*	*Mean*	*Upper decile*
$11 million	$24 million	$37 million	$57 million

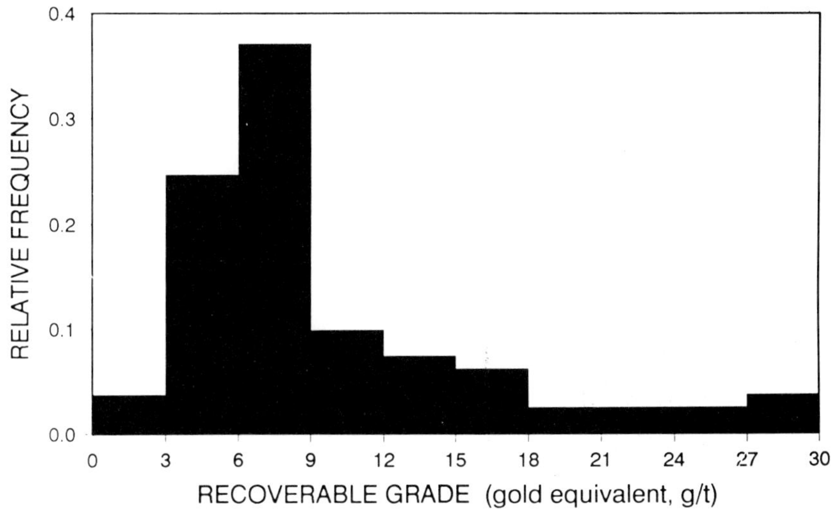

Figure 13.8 Recoverable ore reserve grade.

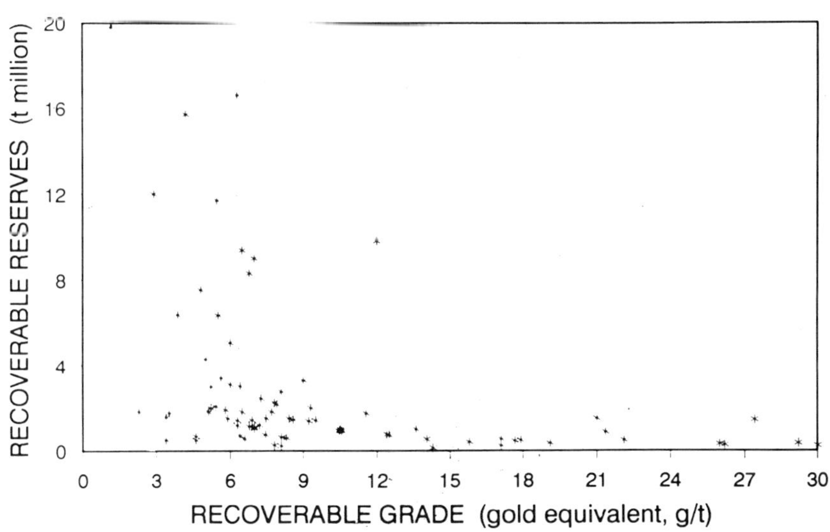

Figure 13.9 Relationship between recoverable ore reserve size and grade.

However, the mine capacities justified for the larger deposits demand considerably higher capital requirements, exceeding $57 million for 10% of the economic deposits, pulling the average investment up to $37 million.

Production costs. Unit production costs per tonne of ore mined and milled are generally high and are shown to vary over a wide range (Figure 13.11). Most of the economic deposits have production costs between $30 and $75 per tonne:

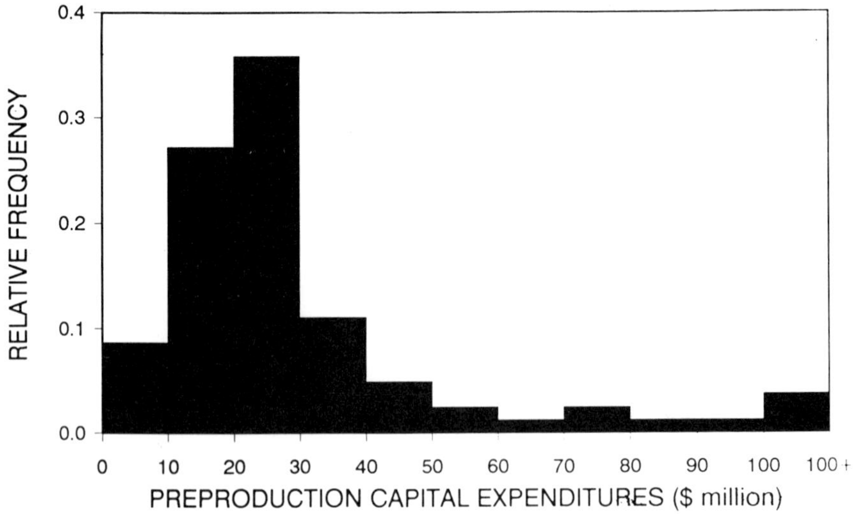

Figure 13.10 Pre-production capital expenditures.

Lower decile	*Median*	*Mean*	*Upper decile*
$33/t	$59/t	$61/t	$104/t

The production cost distribution is reasonably symmetrical. Deposits with costs at or below the lower decile are open-pit mines. On the other hand, deposits with costs above the upper decile value of $104 per tonne are characteristically small, narrow, deep, and/or high grade.

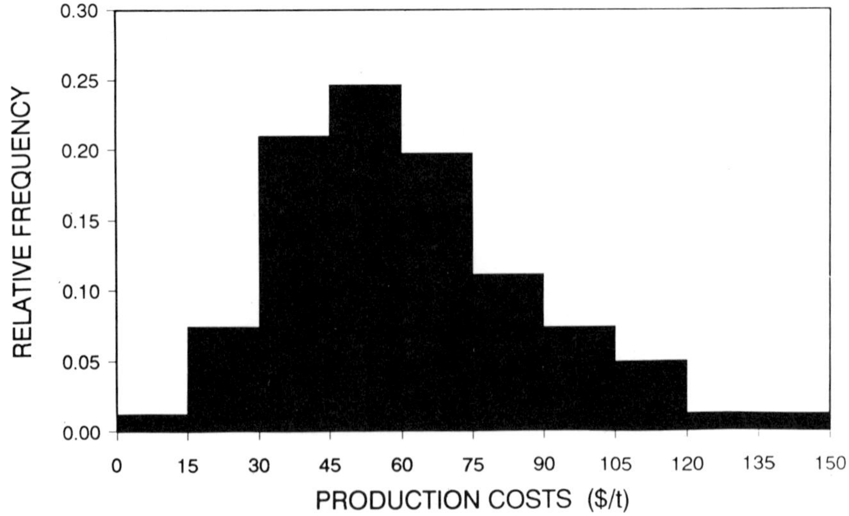

Figure 13.11 Production costs.

13.5.4 *Cash-flow criteria*

Total revenue. The size of the economic gold deposits as measured by total sales revenue (Figure 13.12) is a reflection of the ore reserve size–grade characteristics presented in Figure 13.9:

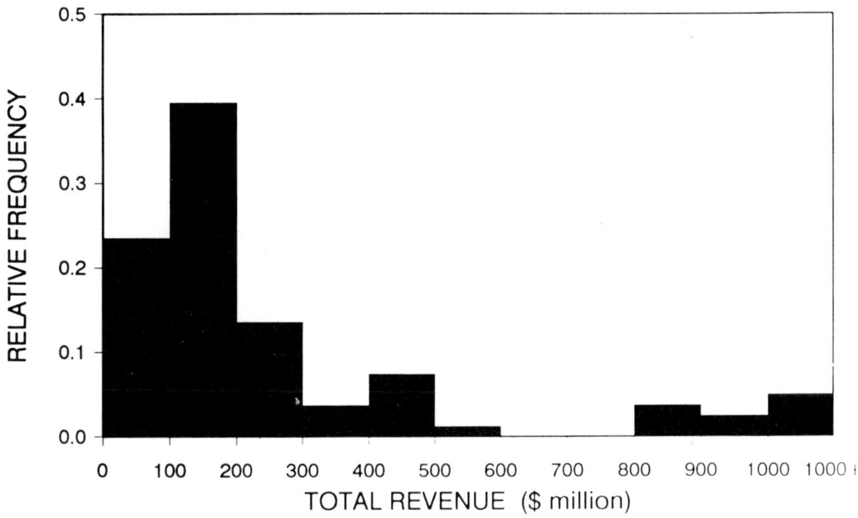

Figure 13.12 Total revenue.

Lower decile	*Median*	*Mean*	*Upper decile*
$57 million	$151 million	$377 millon	$836 million

The variability in ore reserve size is illustrated by the 15-fold difference between the lower decile and upper decile statistics. The total revenue distribution is also very skewed with a discontinuity occurring between $600 and $800 million.

Total cash flow. The total profit generated (Figure 13.13) is impressive considering that 70% of the economic deposits have recoverable reserves of less than 2 million tonnes:

Lower decile	*Median*	*Mean*	*Upper decile*
$10 million	$54 million	$169 million	$ 260 million

Once again, the total cash-flow distribution is characterized by extreme variability and skewness. Ten percent of the economic deposits generate cash flows in excess of $260 million.

Production costs per ounce. Production costs expressed per ounce of gold recovered – derived by combining the cost per tonne (Figure 13.11) and grade (Figure 13.8) parameters with mill recovery factors and any necessary time-value adjustments – represent the break-even gold price requirement for operating mines. As illustrated in

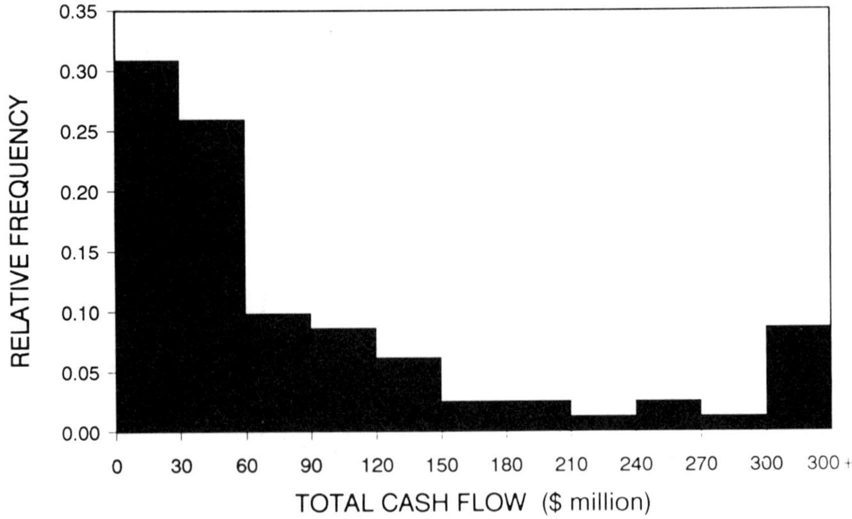

Figure 13.13 Total cash flow.

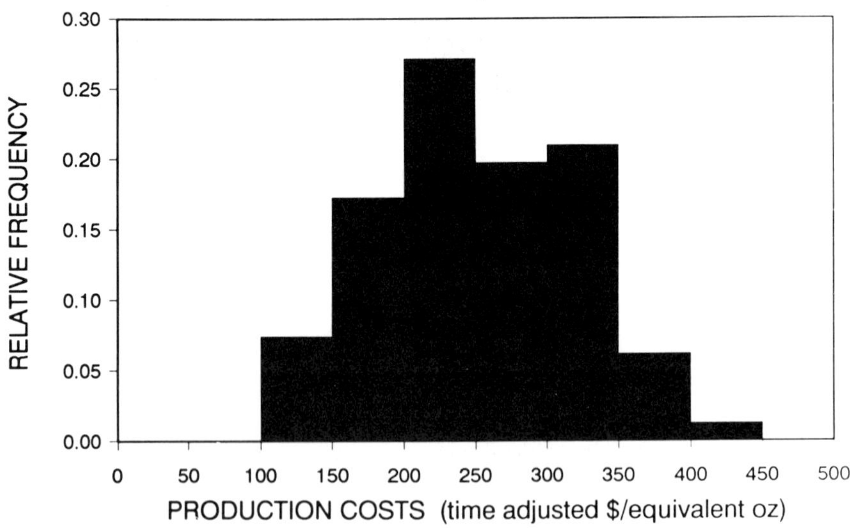

Figure 13.14 Production costs per ounce.

Figure 13.14, production costs for the 81 economic gold deposits are mainly distributed through the range of $150 to $350 per ounce:

Lower decile	*Median*	*Mean*	*Upper decile*
$162/oz	$243/oz	$250/oz	$335/oz

It is interesting that the mean production cost of $250 per ounce is exactly half of the $500 per ounce gold price used in the evaluation.

Total costs per ounce. Determination of the total cost per ounce includes both production costs (Figure 13.11) and pre-production capital expenditures (Figure 13.10). The 8% cost of capital is used to account for time-value differences between the years in which expenditures are incurred and the years when the associated gold content is recovered and sold. The total cost per ounce represents the break-even gold price required to justify investment in new mine development. The total cost per ounce shown in Figure 13.15 varies over a wide range for the 81 economic deposits from less than $50 per ounce to $500 per ounce:

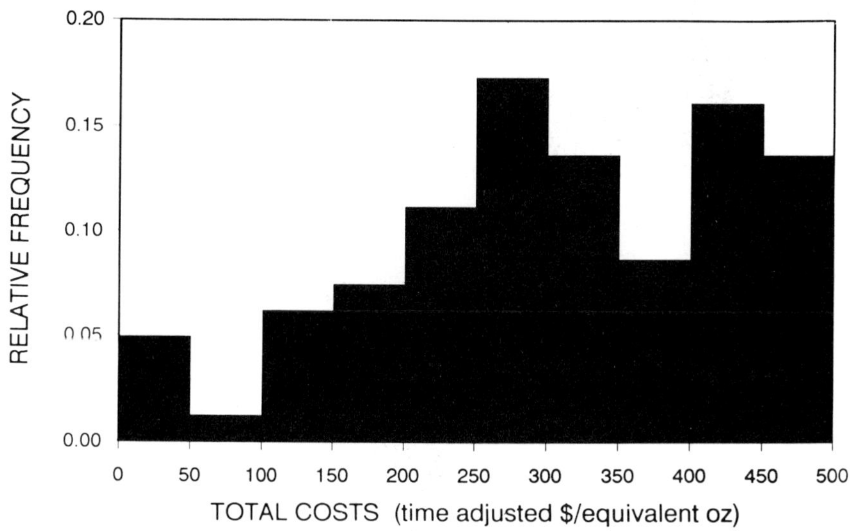

Figure 13.15 Total costs per ounce.

Lower decile	*Median*	*Mean*	*Upper decile*
$248/oz	$365/oz	$363/oz	$470/oz

The $137 per ounce difference between the assumed gold price of $500 per ounce and the mean total cost of $363 per ounce reflects the average profitability of the 81 economic gold deposits above the 8% cost-of-capital threshold. (It should be noted that only those deposits which are found to be economic when evaluated at the assumed $500 price are included in the above compilation. Therefore, after allowing for the 8% cost of capital, the combined development and production cost of these economic deposits cannot by definition exceed $500 per ounce.)

13.5.5 *DCF criteria*

Net present value. The histogram for the net present value results, presented in Figure 13.16, illustrates the wealth-creating capability of economic gold deposits:

Lower decile	*Median*	*Mean*	*Upper decile*
$3 million	$24 million	$61 million	$85 million

The mean value for the 81 economic deposits is \$61 million. The variability of the return given an economic discovery is represented by the 28-fold difference in value between the lower-decile and upper-decile economic deposits. The net present value

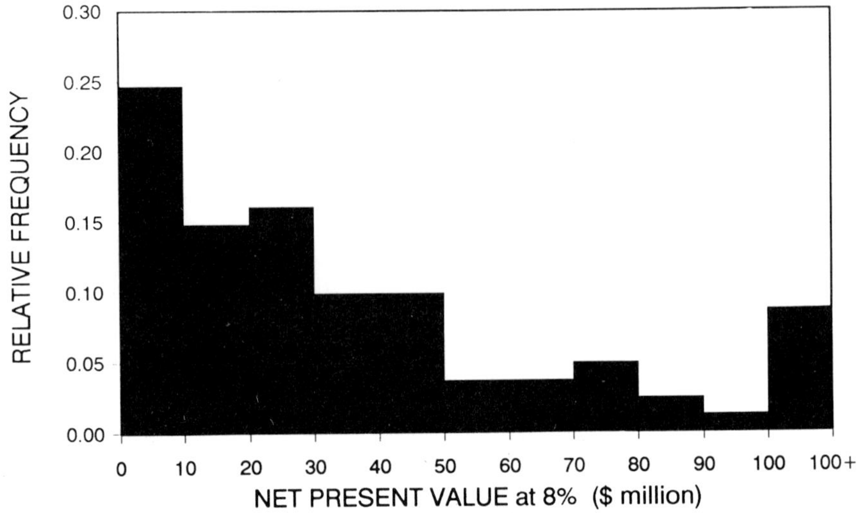

Figure 13.16 Net present value.

distribution is also highly skewed. There are many low-value economic deposits and relatively few of extremely high value (including Hemlo). Thus, the mean value is almost three times greater than the median or middle value of the distribution.

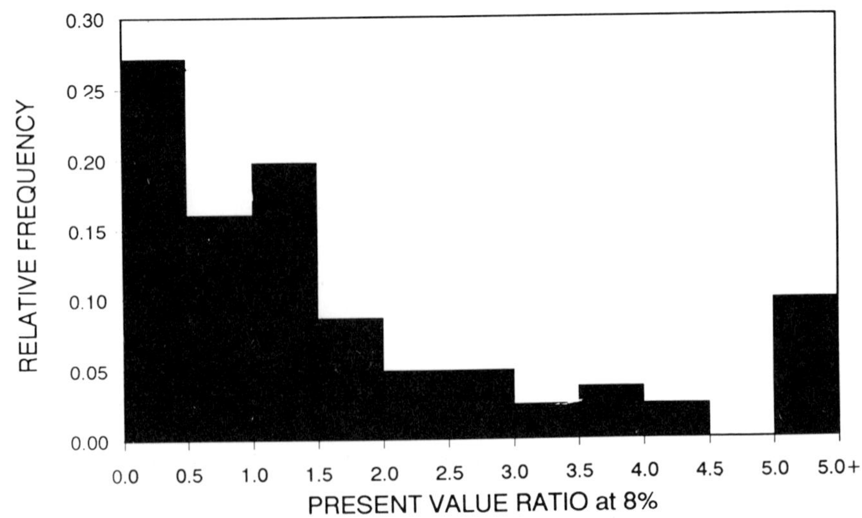

Figure 13.17 Present value ratio.

Present value ratio. The present value ratio measures the profitability of investment in the economic gold deposits (Figure 13.17). For about two-thirds of the deposits, the net present value is less than $1.50 per dollar invested, after the recovery of all costs including the 8% cost of capital:

Lower decile	*Median*	*Mean*	*Upper decile*
$0.18	$1.22	$1.87	$4.32

However, the mean ratio is almost $2, and more than 10% of the deposits have ratios greater than $4. These results indicate that economic gold deposits typically represent very profitable exploration targets and mine development opportunities.

Rate of return. The histogram of rate of return results is shown in Figure 13.18. Most of the economic deposits have rates of return in excess of 20%:

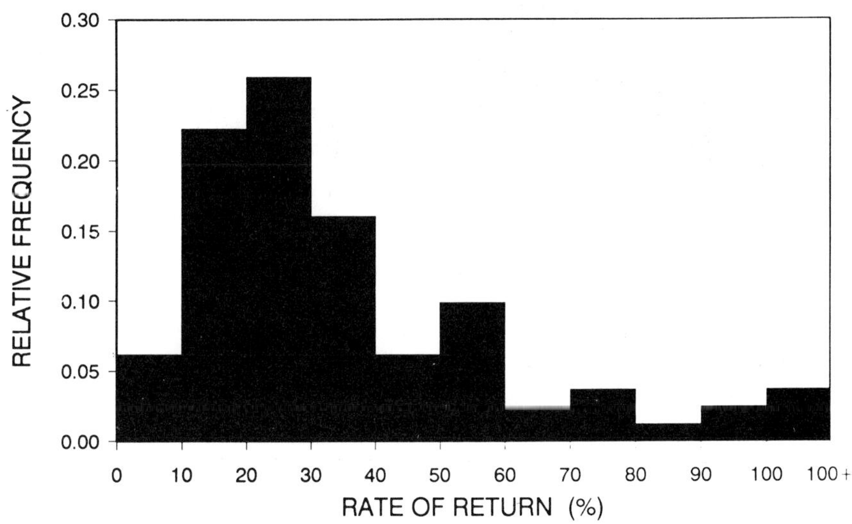

Figure 13.18 Rate of return.

Lower decile	*Median*	*Mean*	*Upper decile*
12%	29%	38%	70%

A high degree of variability is associated with the 38% mean value. Based on this evidence, all we can say in advance of discovering an economic gold deposit in Canada is that we are 80% sure that its before-tax rate of return will be between 12% and 70% !

Net present value – rate of return relationship. The correlation between net present value and rate of return for the 81 economic gold deposits is examined in Figure 13.19. There appears to be one large but fairly marginal deposit towards the bottom right of the scatter diagram. On the other hand, there are several in the top-left area which are very profitable but quite small. The five deposits located towards the top right of the figure have the ideal combination – large and very profitable.

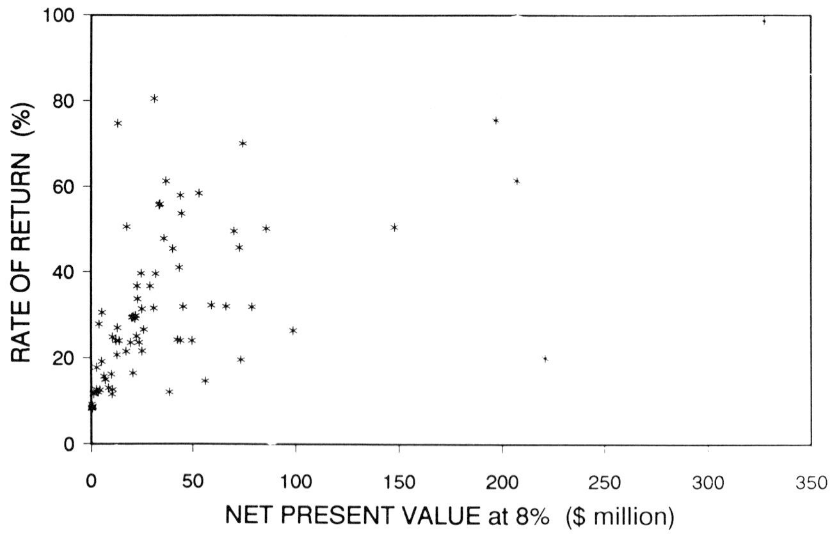

Figure 13.19 Relationship between net present value and rate of return.

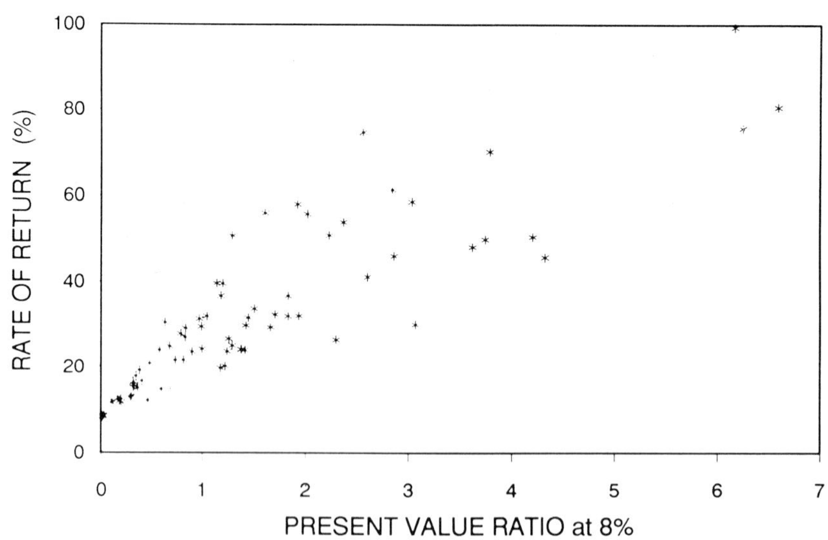

Figure 13.20 Relationship between present value ratio and rate of return.

Present value ratio – rate of return relationship. The relationship between these two main measures of profitability is presented in Figure 13.20. Not surprisingly, a direct relationship between present value ratio and rate of return can be discerned. However, there are also specific instances of contradictory results which occur whenever there is a negative slope between two deposit points. In other words, one deposit has a higher present value ratio and the other a higher rate of return. These contradictions are most pronounced for some of the most profitable deposits. In these circumstances, the discount rates used to calculate the respective DCF criteria are very different. When this type of contradiction occurs, determination of the most economic deposit requires looking beyond the projects themselves to make a judgement about the rate at which the respective cash flows would be reinvested. Since reinvestment will typically be at rates closer to the cost of capital than to the rate of return, the project offering the highest present value ratio will normally be the most economic.

13.6 Conclusion

The chapter examines the economics of gold deposits in terms of historical production trends, evaluation procedures, and illustrative deposit characteristics. As in the past, future growth potential and the pattern of world supply will be shaped by changing the relationships between overall market forces and the geological environment in which gold deposits occur.

As gold projects proceed through exploration towards development, an economic overlay is essential to translate the scientific-technical basis for exploration success into economic criteria that can be related to organizational objectives. Since exploration is such a high-risk activity, we must have a clear understanding of economic target conditions (or, in other words, what we are looking for) so that the search can be narrowed as quickly and as judiciously as possible to those few opportunities which have wealth-creating potential. When major investment in mine capacity is being considered at the development decision point, a comprehensive economic evaluation including probabilistic risk analysis is essential to minimize the chance of error. Otherwise, in gold exploration and development it is all too easy to let economic deposits slip through our fingers and to be misled into backing uneconomic projects. The 1980s is littered with the debris of many such mistakes. Although economic gold deposits are elusive, they are, as illustrated in this chapter, a prize worth seeking.

References

Market setting

Anders, G. (1987) Gold price forecasting: commodity and monetary perspectives. In *Long-Term Prospects for Gold Mining*, Proceedings no. 20, eds. Mackenzie, B. and Gesing, R. Centre for Resource Studies, Kingston, 61-83.

Enzig, P. (1972) *The Destiny of Gold*. Macmillan, London.

Jastram, R.W. (1977) *The Golden Constant*. Wiley Interscience, New York.

Gold mine production

Antrobus, E.S.A. (ed.), *Witwatersrand Gold - 100 Years*. Geol. Soc. of South Africa, Johannesburg, 1986.

Araujo, Homero de (1989) *Potencial Aurifero Brasileiro*. Departmento Nacional da Producao Mineral, Brasilia.

Blainey, G. (1978) *The Rush That Never Ended: a History of Australian Mining*. Melbourne Univ. Press.

Cartwright, A.P. (1962) *The Gold Miners*. Purnell, Johannesburg.

Doggett, M. and Mackenzie, B. (1987) *Economic Trends in Canadian Gold Supply.* Working paper no. 41. Centre for Resource Studies, Kingston.

Green, T. (1987) *The Prospect for Gold.* Walker, New York.

Hoffman, A. (1947) *Free Gold: The Story of Canadian Mining*, Associated Book Service, New York.

Middleton, C.N. (1986) *Brazilian Gold Exploration.* Unpublished report, Western Mining Corp., Rio de Janeiro.

Milling-Stanley, G. (1989) *Gold 1989.* Consolidated Gold Fields, London.

Stone, I. (1956) *Men to Match My Mountains: The Opening of the Far West, 1840-1900.* Doubleday, New York.

Thomas, P.R. and Boyle, E.H. (1986) *Gold Availability – World*, Information Circular 9070, US Bur. of Mines, Washington.

Woodall, R. (1975) *Gold Resources of Australia.* Unpublished report, Western Mining Corp., Kalgoorlie.

Economic evaluation of gold deposits

Mackenzie, B.W. (1989) *Economic Guidelines for Mineral Exploration.* Unpublished Course Notes, Queen's Univ., Kington, 1989.

Characteristics of economic gold deposits

Mackenzie, B. and Doggett, M. (1989) *Economic Aspects of Gold Exploration: How Much is Too Much?* Working paper no. 44, Centre for Resources Studies, Kingston.

Mackenzie, B., Bilodeau, M. and Doggett, M. (1989) *Mineral Exploration and Mine Development Potential in Ontario.* Technical Paper no. 9, Centre for Resource Studies, Kingston.

Index